Textbook of Plant Genetic Resources

Kuldeep Tripathi • Veena Gupta
Gopalareddy Krishnappa
Jyoti Kumari • Gyanendra Pratap Singh
Editors

Textbook of Plant Genetic Resources

Springer

Editors
Kuldeep Tripathi
ICAR-National Bureau of Plant Genetic Resources
New Delhi, Delhi, India

Veena Gupta
ICAR-National Bureau of Plant Genetic Resources
New Delhi, Delhi, India

Gopalareddy Krishnappa
ICAR-Sugarcane Breeding Institute
Coimbatore, Tamil Nadu, India

Jyoti Kumari
ICAR-National Bureau of Plant Genetic Resources
New Delhi, Delhi, India

Gyanendra Pratap Singh
ICAR-National Bureau of Plant Genetic Resources
New Delhi, Delhi, India

ISBN 978-981-96-5023-1 ISBN 978-981-96-5024-8 (eBook)
https://doi.org/10.1007/978-981-96-5024-8

This Springer imprint is published by the registered company Springer Nature Singapore Pte Ltd.
The registered company address is: 152 Beach Road, #21-01/04 Gateway East, Singapore 189721, Singapore

Dedicated to Nature and Farmers

Preface

The conservation and utilization of plant genetic resources (PGR) are essential to addressing critical global challenges such as food security, environmental sustainability, and climate change. This textbook, "Plant Genetic Resources," aims to provide a comprehensive understanding of PGR-related issues like exploration, conservation, tissue culture, cryopreservation, quarantine and bio-security, characterization, evaluation, pre-breeding, utilization, crop wild relatives, genomics, Geographic Information System (GIS), and statistical tools for PGR management. It highlighted the basic facts, scientific principles, methodologies, and policies that are essential for PGR management. It has catalogued comprehensive information on the strategies and tools for efficient management and utilization of the PGRs to cater to the needs of various stakeholders including students, researchers, academicians, and policymakers. It will be beneficial for agriculturists, plant breeders, seed technologists, plant pathologists, biotechnologists, biochemists, pharmacologists, agronomists, botanists, entomologists, social scientists, policy analysts, and any other persons interested in getting information about PGR at a single place. The need for this textbook arises from the growing importance of plant genetic resources in sustaining agriculture and ecosystems. With a rising global population and increasingly unpredictable climatic conditions, preserving and utilizing plant genetic diversity are warranted.

This textbook is a collaborative effort by leading experts, contributing specialized knowledge and experience. It provides a balanced mix of theoretical foundations and practical applications, making it a valuable resource for students and professionals in agriculture, botany, genetics, environmental science, and related fields.

We hope this book enhances the understanding of PGR and inspires the readers to actively participate in their conservation and sustainable use. The future of our food systems, ecosystems, and environmental security depends on preserving genetic diversity.

We are assured that the present book will be a landmark, as this kind of compilation on PGR is the first of its kind and the chapters contain updated information as well as the views conveyed by the researchers based on their vast experience in PGR science. We thank all the contributors and institutions that supported the development of this textbook. Undoubtedly, textbook on "Plant Genetic Resources" is a timely and much-warranted book considering the significance of PGR for global food and nutritional security. We also thank the readers whose interest and commitment to plant genetic resources will drive future innovations in PGR research. We hope that the book will serve as a useful guide for scientists, teachers, policymakers, and students.

New Delhi, India — Kuldeep Tripathi
New Delhi, India — Veena Gupta
Coimbatore, Tamil Nadu, India — Gopalareddy Krishnappa
New Delhi, India — Jyoti Kumari
New Delhi, India — Gyanendra Pratap Singh

The original version of the back matter of this book has been updated: minor typographical and grammatical errors have been corrected, and redundant definitions have been removed from the Glossary section.

Acknowledgements

The textbook, "Plant Genetic Resources," has been a collaborative effort involving various individuals with great expertise in their research area. We are profoundly grateful to everyone who has contributed to this project. First, we extend our deepest gratitude to the authors who have shared their extensive knowledge and expertise. Their contributions have provided the comprehensive content that forms the core of this textbook. Their commitment to excellence and the advancement of plant genetic resources (PGR) is truly admirable. We acknowledge the invaluable assistance of our editorial and Springer's production team. Their efforts in coordinating contributions, managing the review process, and ensuring high standards for the final publication have been essential to the success of the upbringing of the book.

Finally, we thank our PGR families and friends for their understanding and encouragement throughout the development of this textbook. Their support has been a source of inspiration and motivation. This textbook represents the collective efforts of all the contributors to promote the conservation and sustainable use of plant genetic resources. We hope this book will serve as a valuable resource and inspiration for future generations in this vital field.

Contents

About the Editors and Contributors

About the Editor

Kuldeep Tripathi is a Senior Scientist at the Division of Germplasm Evaluation, ICAR-National Bureau of Plant Genetic Resources, New Delhi. He has made significant contributions to plant genetic resources management, particularly in pre-breeding and genetic enhancement of pulses, a field in which his work is considered unparalleled. Since joining ICAR-NBPGR, his notable research on legumes has led to the development of three Indian pulse core sets in lentil, pea, and cowpea from the entire Indian national genebank collections, 26 registered legume genetic stocks. Tripathi has published over 80 research papers in peer-reviewed national and international journals and has extensive experience in managing externally funded projects funded by ICARDA, DBT, the Alliance of Bioversity International and CIAT, and ICAR. He presented his research at international conferences held in France, Australia, Belgium and China. Besides his research, he has been a faculty member at the Graduate School of ICAR-IARI, where he has taught and guided MSc and PhD students. Tripathi was conferred with the NAAS Associate (2025), NAAS Young Scientist Award (2024), ISPRD Young Scientist Award 2023, ISPGR Young Scientist Award 2020, Dr. KL Mehra Best Student Award, etc.

Veena Gupta, PhD, Principal Scientist, ICAR-NBPGR, has done her Bachelor's Degree (1980), Master Degree (1982), and Doctorate Degree in Botany from Delhi University. She is actively engaged in the area of Plant Genetic Resources Conservation for the last 35 years. Her scientific contributions in the area of Plant Genetic Resources are well appreciated. She has been bestowed with Dr SABUJ SAHOO Life Time Achievement Award for her contributions to the Plant Genetic Resources Conservation and Management, and fellow of several scientific societies such as Indian Society of Seed Technology (2005), Indian Society for Plant Genetic Resources (2007), Hind Agri-horticultural Society (2010), Society for Plant Research (2012), and International Society for Noni Science (2015). She has visited various laboratories in United States of America engaged in the conservation and utilization of plant genetic resources. She had been trained about the IPR governing regulations during advance training on Genetic Resources and Intellectual Property at Swedish International Developmental Co-operation Agency, Uppsala, Sweden; Agricultural Research Council of South Africa, Pretoria and NIAB, Cambridge,

UK. She has been nominated as Honorary Member of Indian Industrial Hemp Association and Medicinal Plants Expert for CCRAS, CCRUM, Ministry of AYUSH and DD Kisan channel of Delhi Doordarshan. As Professor of the Post Graduate School (PGR Discipline), ICAR-IARI, she had guided two MSc (PGR) students and six PhD students for their degrees in Plant Genetic Resources discipline. She had published more than 80 research papers and two books in her area of expertise.

Gopalareddy Krishnappa, PhD, Senior Scientist, ICAR-SBI, Coimbatore, completed his Bachelor's Degree during 2006 and Master Degree during 2008 from UAS, Bangalore, and Doctorate Degree in Genetics from ICAR-IARI, New Delhi, during 2016. Reddy developed 17 wheat varieties and 27 wheat genetic stocks to benefit multitude stakeholders including farmers, consumers, researchers, and industry. His research is largely focused in the development of high yielding wheat varieties with improved quality attributes. Reddy utilized modern breeding tools like genomic selection, MARS, MAS, and GWAS to complement conventional breeding techniques to dissect complex traits in wheat. Reddy published more than 40 research and review articles of national and international repute with high impact journals, 03 books, 25 book chapters, and 30 technical or popular articles. Reddy was conferred with Nanaji Deshmukh Award, AIASA Young Scientist Award 2020, Best Worker 2020, NEEDEF Young Scientist Award 2021, and Dr. VS Mathur Memorial Award 2022.

Jyoti Kumari, PhD, Principal Scientist, ICAR-NBPGR, New Delhi, obtained her Master's and Doctoral Degrees from Indian Agricultural Research Institute, New Delhi, in the discipline of Genetics in 2002 and 2005, respectively. Her major scientific contribution includes mega characterization and evaluation of ~22,000 wheat germplasm, development of core (2226 acc.) and minicore (224 acc.) set from the entire wheat germplasm conserved in the Indian national genebank. She has identified several accessions with unique traits including high popping accession in pearl millet, waxy lines in maize, high grain protein, high zinc content, rust, powdery mildew, and spot blotch tolerant accessions in wheat. She has also identified QTLs for terminal heat tolerance, salt tolerance, and seedling root architecture traits using genome wide association strategy. She has contributed in augmentation of rice and wheat germplasm from the drought and heat prone areas of Madhya Pradesh, Rajasthan, and Gujarat. She has published more than 140 publications, 20 book chapters, three books, and one crop catalogue. She has developed four varieties in lentil and mung bean and registered 40 genetic stocks with Plant Germplasm Registration Committee. She is fellow of Indian Society of Genetics and Plant Breeding and recipient of prestigious award, Dr. Harbhajan Singh Memorial Award from ISGPB. As a faculty of Division of Genetics, she has guided twelve MSc and six PhD students of Graduate School, ICAR-IARI, New Delhi.

Gyanendra Pratap Singh, PhD, Director, ICAR-NBPGR and former Director, ICAR-IIWBR, Karnal, has more than three decades of agriculture research including 10 years of teaching and more than eight years of administration experience. Singh is instrumental in the development of 60 wheat and 03 barley varieties and 01 potato variety benefitting multitude farmers, consumers, and industries. Singh largely focused in intrinsic and strategic research on abiotic stress that led to the publication of 300 peer-reviewed research articles of high impact, 18 books, 65 book chapters, 57 technical bulletins, 55 popular articles, and 05 Policy/Strategy papers. His strenuous efforts on advancing the R and D led to several accolades at national and international level by honoring with 17 individual awards. He is an elected fellow of three prestigious academies, that is, INSA, NASI, and NAAS and also a fellow of three registered professional societies. His acumen on leading research programs and visionary perspective garnered recognitions like Gene Stewardship Award from the Borlaug Global Rust Initiative, USA.

Contributors

Sneha Adhikari Division of Genetics, ICAR-Indian Agricultural Research Institute, New Delhi, India

S. P. Ahlawat Division of Plant Exploration and Germplasm Collection, ICAR-National Bureau of Plant Genetic Resources, New Delhi, India

Jameel Akhtar Division of Plant Quarantine, ICAR-National Bureau of Plant Genetic Resources, New Delhi, India

J. Aravind Division of Germplasm Conservation, ICAR-National Bureau of Plant Genetic Resources, New Delhi, India

Sunil Archak Germplasm Exchange and Policy Unit, ICAR-National Bureau of Plant Genetic Resources, New Delhi, India

C. Bhardwaj ICAR-Indian Agricultural Research Institute, New Delhi, India

Rakesh Bhardwaj Division of Germplasm Evaluation, ICAR-National Bureau of Plant Genetic Resources, New Delhi, India

K. C. Bhatt Division of Plant Exploration and Germplasm Collection, ICAR-National Bureau of Plant Genetic Resources, New Delhi, India

Pratibha Brahmi Germplasm Exchange and Policy Unit, ICAR-National Bureau of Plant Genetic Resources, New Delhi, India

Richard S. Cahoon Department of Global Development, Cornell College of Agriculture and Life Sciences, Ithaca, NY, USA

Debashis Chakraborty CIMMYT, New Delhi, India

Debasish Chakraborty ICAR-Research Complex for NEH Region, Umiam, Meghalaya, India

V. Celia Chalam Division of Plant Quarantine, ICAR-National Bureau of Plant Genetic Resources, New Delhi, India

Subhash Chander Division of Germplasm Conservation, ICAR-National Bureau of Plant Genetic Resources, New Delhi, India

Puran Chandra Division of Germplasm Exchange and Policy Unit, ICAR-National Bureau of Plant Genetic Resources, New Delhi, India

Debjani Roy Choudhury Division of Genomic Resources, ICAR-National Bureau of Plant Genetic Resources, New Delhi, India

Dalamu ICAR-Central Potato Research Institute, Shimla, Himachal Pradesh, India

Bappa Das ICAR-Central Coastal Agricultural Research Institute, Old Goa, Goa, India

O.P. Dhariwal Division of Plant Exploration and Germplasm Collection, ICAR-National Bureau of Plant Genetic Resources, New Delhi, India

S. Elumalai Tamil Nadu Ambedkar Law University, Chennai, Tamil Nadu, India

Raj Kumar Gautam Division of Germplasm Evaluation, ICAR-National Bureau of Plant Genetic Resources, New Delhi, India

Padmavati G. Gore Division of Germplasm Conservation, ICAR-National Bureau of Plant Genetic Resources, New Delhi, India

R. Gowthami Division of Germplasm Conservation, ICAR-National Bureau of Plant Genetic Resources, New Delhi, India

Arun Gupta ICAR-Indian Institute of Wheat and Barley Research, Karnal, Haryana, India

Kavita Gupta Division of Plant Quarantine, ICAR-National Bureau of Plant Genetic Resources, New Delhi, India

N.C. Gupta ICAR-National Institute for Plant Biotechnology, New Delhi, India

Sandhya Gupta Division of Germplasm Conservation, ICAR-National Bureau of Plant Genetic Resources, New Delhi, India

Veena Gupta Division of Germplasm Conservation, ICAR-National Bureau of Plant Genetic Resources, New Delhi, India

G.D. Harish ICAR-NBPGR RS, Shillong, India

K.S. Hooda Division of Germplasm Evaluation, ICAR-National Bureau of Plant Genetic Resources, New Delhi, India

Sherry Rachel Jacob Division of Germplasm Conservation, ICAR-National Bureau of Plant Genetic Resources, New Delhi, India

Antil Jain Division of Germplasm Evaluation, ICAR-National Bureau of Plant Genetic Resources, New Delhi, India

Roomesh Kumar Jena ICAR-Indian Institute of Water Management, Bhubaneswar, Odisha, India

Girish Kumar Jha Division of Agricultural Bioinformatics, ICAR-Indian Agricultural Statistics Research Institute, New Delhi, India

Racheal John Division of Germplasm Evaluation, ICAR-National Bureau of Plant Genetic Resources, New Delhi, India

Simardeep Kaur Division of Crop Science I ICAR-Research Complex for North Eastern Hill Region (NEH), Umiam, Meghalaya, India

Z. Khan Division of Plant Quarantine, ICAR-National Bureau of Plant Genetic Resources, New Delhi, India

Mahesh Kumar Division of Plant Physiology, ICAR-Indian Agricultural Research Institute, New Delhi, India

Pradeep Kumar ICAR-Indian Institute of Wheat and Barley Research, Karnal, Haryana, India

Vishnu Kumar ICAR-National Bureau of Plant Genetic Resource, New Delhi, India

Jyoti Kumari Division of Germplasm Evaluation, ICAR-National Bureau of Plant Genetic Resources, New Delhi, IndiaICAR-National Bureau of Plant Genetic Resource, New Delhi, India

Sapna Langyan Division of Germplasm Evaluation, ICAR-National Bureau of Plant Genetic Resources, New Delhi, India

Ulrike Lohwasser Department of Genebank, Research Group Resources Genetics and Reproduction, Leibniz Institute of Plant Genetics and Crop Plant Research (IPK), Gatersleben, Germany

Anju Mahendru Singh Division of Germplasm Conservation, ICAR-National Bureau of Plant Genetic Resources, New Delhi, India

Pavan Kumar Malav Division of Plant Exploration and Germplasm Collection, ICAR-National Bureau of Plant Genetic Resources, New Delhi, India

Era Vaidya Malhotra Division of Germplasm Conservation, ICAR-National Bureau of Plant Genetic Resources, New Delhi, India

Nand Lal Meena Division of Germplasm Evaluation, ICAR-National Bureau of Plant Genetic Resources, New Delhi, India

Dwijesh Chandra Mishra Division of Agricultural Bioinformatics, ICAR-Indian Agricultural Statistics Research Institute, New Delhi, India

Sneha Murmu Division of Agricultural Bioinformatics, ICAR-Indian Agricultural Statistics Research Institute, New Delhi, India

S. Nagaraju Ministry of Environment Forest & Climate Change, Indira Paryavaran Bhavan, New Delhi, India

E. Roshini Nayar Division of Plant Exploration and Germplasm Collection, ICAR-National Bureau of Plant Genetic Resources, New Delhi, India

R. K. Pamarthi Division of Plant Exploration and Germplasm Collection, ICAR-National Bureau of Plant Genetic Resources, New Delhi, India

Rinky Resma Panda The Graduate School, ICAR-Indian Agricultural Research Institute, New Delhi, India

Division of Plant Genetic Resources, Graduate School, ICAR-Indian Agricultural Research Institute, New Delhi, India

Graduate School, Division of Plant Genetic Resources, ICAR-Indian Agricultural Research Institute, New Delhi, India

ICAR-National Bureau of Plant Genetic Resource, New Delhi, India

Anjula Pandey Division of Plant Exploration and Germplasm Collection, ICAR-National Bureau of Plant Genetic Resources, New Delhi, India

Priya Pardeshi Division of Germplasm Evaluation, ICAR-National Bureau of Plant Genetic Resources, New Delhi, India

ICAR-National Bureau of Plant Genetic Resource, New Delhi, India

Bhaskar C. Patra ICAR-National Rice Research Institute, Cuttack, Odisha, India

K. Pradheep ICAR-National Bureau of Plant Genetic Resources, Thrissur, Kerala, India

Pragya Division of Germplasm Exchange and Policy Unit, ICAR-National Bureau of Plant Genetic Resources, New Delhi, India

K. Prasad Department of Botany, Sri Venkateswara College, University of Delhi, New DelhiIndia

Pooja Rathore ICAR-IARI, New Delhi, India

Sebastian Raubach Department of Information and Computational Sciences, The James Hutton Institute, Dundee, UK

Prasenjit Ray Division of Soil Science and Agricultural Chemistry, ICAR-IARI, New Delhi, India

Suman Roy Division of Germplasm Conservation, ICAR-National Bureau of Plant Genetic Resources, New Delhi, India

Vaidurya Pratap Sahi Department of Genetics and Plant Breeding, Sam Higginbottom University of Agriculture, Technology and Sciences, Prayagraj, Uttar Pradesh, India

Lekshmy Sathee Division of Plant Physiology, ICAR-Indian Agricultural Research Institute, New Delhi, India

Dinesh Prasad Semwal ICAR- National Bureau of Plant Genetic Resources, New Delhi, India

Laxmi Sharma Division of Genomic Resources, ICAR-National Bureau of Plant Genetic Resources, New Delhi, India

Paul D. Shaw Department of Information and Computational Sciences, The James Hutton Institute, Dundee, UK

Anju Mahendru Singh Division of Germplasm Conservation, ICAR-National Bureau of Plant Genetic Resources, New Delhi, India

Gyanendra Pratap Singh ICAR-National Bureau of Plant Genetic Resources, New Delhi, India

Gyanendra Singh ICAR-Indian Institute of Wheat and Barley Research, Karnal, Haryana, India

M. C. Singh Division of Plant Quarantine, ICAR-National Bureau of Plant Genetic Resources, New Delhi, India

Praveen K. Singh Division of Plant Exploration and Germplasm Collection, ICAR-National Bureau of Plant Genetic Resources, New Delhi, India

Rakesh Singh Division of Genomic Resources, ICAR-National Bureau of Plant Genetic Resources, New Delhi, India

Sudhir Srivastava Division of Agricultural Bioinformatics, ICAR-Indian Agricultural Statistics Research Institute, New Delhi, India

Vartika Srivastava Division of Germplasm Conservation, ICAR-National Bureau of Plant Genetic Resources, New Delhi, India

S. B. Suby ICAR-Indian Institute of Maize Research, Ludhiana, India

Kanchit Thammasiri Department of Gardening and Horticulture, Xishuangbanna Tropical Botanical Garden, Chinese Academy of Sciences, Menglun, Mengla, Yunnan, China

Maharishi Tomar ICAR-Indian Grassland and Fodder Research Institute, Jhansi, Uttar Pradesh, India

Kuldeep Tripathi Division of Germplasm Evaluation, ICAR-National Bureau of Plant Genetic Resources, New Delhi, India

B. S. Tyagi ICAR-Indian Institute of Wheat and Barley Research, Karnal, Haryana, India

Vandana Tyagi Division of Germplasm Exchange and Policy Unit, ICAR-National Bureau of Plant Genetic Resources, New Delhi, India

M. Vetriverithan ICAR-International Crops Research Institute for Semi-arid tropics, Hyderabad, India

Swayam Vid ICAR-IARI, New Delhi, India

Acronyms

A	Adenine
AAD	Arbitrarily Amplified DNA markers
ABD	Augmented Block Design
ABS	Access and Benefit Sharing
AFLP	Amplified Fragment Length Polymorphism
AGG	Australian Grains Genebank
AGRDB	Amaranth Genomic Resource Database
AICRP	All India Coordinated Research Project
AINP	All India Network Project
ALA	Alanine
ANN	Artificial Neural Networks
ANOVA	Analysis of Variance
APIs	Application Programming Interfaces
ApTransDB	Amaranth Genomic Resource Database
AVRDC	Asian Vegetable Research and Development Center (Now, World Vegetable Centre)
AWCC	Australian Winter Cereals Collection
BBSRC	Biotechnology and Biological Sciences Research Council
BDA	Biological Diversity Act
BMCs	Biodiversity Management Committees
BP	Breeding Population
C	Cytosine
CAC	Codex Alimentarius Commission
CBD	Convention on Biological Diversity
CDDM	Conserved DNA-Derived Polymorphism
CDM	Conserved DNA-Based Markers
CGIAR	Consultative Group on International Agricultural Research
CGN	Center for Genetic Resources, the Netherlands
CGRFA	Commission on Genetic Resources for Food and Agriculture
CIAT	International Center for Tropical Agriculture
CIFOR	The Center for International Forestry Research
CIMMYT	International Maize and Wheat Improvement Center
CIP	International Potato Center
CIRAD	French Agricultural Research Centre for International Development

CITES	Convention on International Trade in Endangered Species of Wild Fauna and Flora
CNN	Convolutional Neural Networks
CNPs	Copy Number Polymorphisms
CNVs	Copy Number Variations
CoRAP	Conserved Region Amplification Polymorphism
CPB	Cartagena Protocol on Biosafety
CRISPR	Clustered Regularly Interspaced Short Palindromic Repeats
DAFW	Department of Agriculture and Farmer's Welfare
DALP	Direct Amplification of Length Polymorphisms
DArT	Diversity Array Technology
DBMS	Database Management System
DBT	Department of Biotechnology
DIP	Destructive Insects and Pests
DL	Deep Learning
DOI	Digital Object Identifier
DPPQ&S	Directorate of Plant Protection, Quarantine & Storage
DSI	Digital Sequence Information
DSIR	Department of Scientific and Industrial Research
DSS	Decision Support Systems
EC	Exotic Collection
ECPGR	European Cooperative Programme for Plant Genetic Resources
EDXRF	Energy Dispersive X-Ray Flourescene
ELISA	Enzyme-Linked Immunosorbent Assay
ETAAS	Electrothermal Atomic Absorption Spectroscopy
EURISCO	European Search Catalogue for Plant Genetic Resources
EXIM	Export and Import
FAO	Food and Agriculture Organization
FGB	Field Genebank
FIGS	Focused Identification of Germplasm Strategy
FLIP	Food Legume Improvement Program
FRI	Forest Research Institute
G	Guanine
GBIF	Global Biodiversity Information Facility
GBIS	Genebank Information system
GBS	Genotyping by Sequencing
GC	Gas Chromatography
GCDT	Global Crop Diversity Trust
GEAC	Genetic Engineering Appraisal Committee
GEBVs	Genomic Estimated Breeding Values
GEPU	Germplasm Exchange and Policy Unit
GEXQIS	Germplasm Exchange and Quarantine Information System
GFAAS	Graphite Furnace Atomic Absorption Spectrometry
GIS	Geographic Information System
GMO	Genetically Modified Organism

GNDVI	Green Normalized Difference Vegetation Index
GPS	Global Positioning System
GRDC	Grains Research and Development Corporation
GRIN	Germplasm Resources Information Network
GRU	Genetic Resources Unit
GWAS	Genome Wide Association Studies
He	Expected Heterozygosity
Ho	Observed Heterozygosity
HPLC	High-Performance Liquid Chromatography
IARCs	International Agriculture Research Centres
IAS	Invasive Alien Species
IBSC	Institutional Biosafety Committee
IC	Indigenous Collection
ICC	ICRISAT Chickpea
ICAR	Indian Council of Agricultural Research
ICARDA	International Center for Agricultural Research in the Dry Areas
ICRAF	International Centre for Research in Agroforestry
ICRISAT	International Crops Research Institute for the Semi-Arid Tropics
IEDCs	Induced Embryonic Determined Cells
IFPRI	International Food Policy Research Institute
IFLA	International Food Legume Accession
IITA	International Institute of Tropical Agriculture
ILL	International Legume Lentil
ILRAD	International Laboratory for Research on Animal Diseases
ILRI	International Livestock Research Institute
InDels	Insertions-Deletions
INPGRS	Indian National Plant Genetic Resources System
INRA	Institut National de la Recherche Agronomique
INGR	Indian National Germplasm Registration
IOE	International Office of Epizootics
IP	Import Permit
IPBS	Inter-Primer Binding Site Amplification
IPGRI	International Plant Genetic Resources Institute
IPK	Leibniz Institute of Plant Genetics and Crop Plant Research
IPPC	International Plant Protection Convention
IQC	Internal Quality Control
IRAP	Inter-Retrotransposon Amplified Polymorphism
IR	Infrared (IR)
IRRI	International Rice Research Institute
ISPM	International Standards for Phytosanitary Measures
ISSRs	Inter Simple Sequence Repeats
ISTA	International Seed Testing Association
ISVs	Intermediate-Sized Variants
ITPGRFA	International Treaty on Plant Genetic Resources for Food and Agriculture

IUPGR	International Undertaking on Plant Genetic Resources
KNN	k-Nearest Neighbours
LCV	Large-Scale Copy Number Variants
LFIA	Lateral Flow Immune Assay
LiDAR	Light Detection and Ranging
LMO	Living Modified Organism
LOB	Limit of Blank
LOD	Limit of Detection
LOQ	Limit of Quantification
LSTM	Long Short-Term Memory
LWIR	Long Wave Infrared
MABC	Marker Assisted Backcrossing
MAS	Marker Assisted Selection
MCPD	Multi-Crop Passport Descriptor
MIR	Mid Wave Infrared
ML	Machine Learning
MLS	Multilateral System (of Access and Benefit-sharing)
MoAFW	Ministry of Agriculture and Farmers Welfare
MOS	Most Original Sample
MP	Microwave Plasma
mRNA	Messenger RNA
MSC	Multiplicative Scatter Correction
MSI	Membrane Stability Index
MTA	Material Transfer Agreement
MTS	Medium Term Storage
NAGS	National Active Germplasm Sites
NARO	National Agriculture and Food Research Organization
NARS	National Agricultural Research System
NBA	National Biodiversity Authority
NBPGR	National Bureau of Plant Genetic Resources
NCGRP	National Center for Genetic Resources Preservation
NDRE	Normalized Difference Red Edge
NDVI	Normalized Difference Vegetation Index
Ne	Effective Population Size
NERPIS	Northeast Regional Plant Introduction Station
NGB	National Genebank
NGS	Next-Generation Sequencing
NIR	Near-infrared (NIR)
NIRS	Near-Infrared Spectroscopy
NLGRP	National Laboratory for Genetic Resources Preservation
NPCI	Normalized Pigment Chlorophyll Index
NPGS	National Plant Germplasm System
NPPO	National Plant Protection Organization
OECD	Organisation for Economic Co-operation and Development
OP	Open Pollinated

PAAP	Promoter Anchored Amplified Polymorphism
PAMP	Pathogen-Associated Molecular Patterns
PBA	P450-Based Analogue markers
PCA	Principal Component Analysis
PCR	Polymerase Chain Reaction
PEG	Polyethylene Glycol
PGC	Pyrolysis Gas Chromatography
PGR	Plant Genetic Resources
PGRFA	Plant Genetic Resources for Food and Agriculture
PRA	Pest Risk Analysis
PRI	Photochemical Reflectance Index
PRR	Pattern Recognition Receptors
QCM	Quartz Crystal Microbalance
QTL	Quantitative Trait Locus
RAPD	Random Amplified Polymorphic DNA
RBD	Randomized Block Design
RCGM	The Review Committee on Genetic Manipulation
RDA	Rural Development Administration
RDBMS	Relational Database Management System
REMAP	Retrotransposon-Microsatellite Amplified Polymorphism
RFLP	Restriction Fragment Length Polymorphism
RGAP	Resistance Gene Analog Polymorphism
RGM	Resistance Gene Based Markers
RMSE	Root Mean Square Error
RNA	Ribonucleic Acid
RNN	Recurrent Neural Networks
ROS	Reactive Oxygen Species
RSQ	R-Squared
RT-PCR	Reverse Transcription Polymerase Chain Reaction
RWC	Relative Water Content
SAVI	Soil-Adjusted Vegetation Index
SBB	State Biodiversity Board
SCoT	Start Codon Targeted Polymorphism
SDG	Sustainable Development Goals
SDM	Species Distribution Modelling
SD	Standard Deviation
SEC	Size Exclusion Chromatography
SECV	Standard Error of Cross-Validation
SES	Standard Evaluation System
SINGER	System Wide Information Network for Genetic Resources
SMTA	Standard Material Transfer Agreement
SMV	Support Vector Machines
SNPs	Single Nucleotide Polymorphisms
SNV	Standard Normal Variate
SOP	Standard Operating Procedures

SRMs	Standard Reference Materials
SSAP	Sequence-Specific Amplification Polymorphism
SSRs	Simple Sequence Repeats
STMS	Sequence Tagged Microsatellite Analysis
STRs	Short Tandem Repeats
SVM	Support Vector Machines
SWIR	Short Wave Infrared
T	Thymine
TBIA	Tissue Blot Immunoassay
TBP	Tubulin-Based Polymorphism
T-DNA	Transfer Deoxyribonucleic Acid
TEM	Transposable Element-Based Markers
TFM	Targeted Fingerprinting Markers
TILLING	Targeting-Induced Local Lesions in Genomes
TLC	Thin-Layer Chromatography
TP	Training Population
TRAP	Target Region Amplification Polymorphism
tRNA	Transfer Ribonucleic Acid
TWAS	Transcriptome Wide Association Studies
UPOV	International Union for the Protection of New Varieties of Plants
USAID	United States Agency for International Development
USDA	United States Department of Agriculture
VIR	NI Vavilov All-Russian Institute of Plant Genetic Resources
VNIS	Visible and Near-Infrared Spectroscopy
WGR	Whole Genome Resequencing
WGS	Whole Genome Sequencing
WHO	World Health Organization
WIEWS	World Information and Early Warning System
XRF	X-Ray Fluorescence

An Introduction to Plant Genetic Resources

1

Gyanendra Pratap Singh

Learning objectives

- Importance of PGR management
- Status of PGR conservation and use
- Related topics

Keywords

Plant genetic resources · Germplasm · Digitization

1.1 Introduction

To ensure the food and nutritional security of the increasing population, today we are again looking towards biological diversity to unveil new sources of food. To meet the challenges imposed due to climate change, land degradation, and increasing human population, plant breeders need to develop new varieties/cultivars for which plant genetic resources (PGRs) are the building blocks. Plant genetic resources (PGRs) encompass the genetic material found in plants, valued for its utility to both current and future generations. According to the International Undertaking on Plant Genetic Resources (FAO 1983), PGR comprises various types of reproductive or vegetative propagating material, such as cultivated varieties (cultivars), obsolete cultivars, primitive cultivars (landraces), wild and weedy species, genetic stocks, elite lines and mutants. The Convention for Biological Diversity (CBD 1992) broadens this

G. P. Singh (✉)
ICAR-National Bureau of Plant Genetic Resources, New Delhi, India
e-mail: GP.Singh@icar.org.in

K. Tripathi et al. (eds.), *Textbook of Plant Genetic Resources*,
https://doi.org/10.1007/978-981-96-5024-8_1

definition, incorporating any living material with present or potential value to humans. This broad characterization includes crops and some of their wild relatives due to their possession of valuable traits. At times, genetic resources are expanded to cover elements such as genes, DNA fragments, and RNA. These are fundamental to the foundation of agriculture and the assurance of global food and nutritional security. PGRs, including traditional varieties and landrace populations, often exhibit specific traits such as early or late maturity, adaptability to particular soil types, and diverse uses, typically identified by local names. These traits have enabled their survival amidst various biotic and abiotic stresses within centers of diversity, alongside wild progenitors of crop plants and their wild and weedy relatives. Different terms have been employed to describe these geobotanical diversity patterns in crop plants, such as primary and secondary centers of diversity, gene centers, cradles/subcradles of agriculture, megacenters (Vavilov 1926; Zhukovsky 1968), non-centers, microcenters (Harlan 1975), and cradles of angiosperm diversity (Takhtajan 1969). These centers of diversity are the outcomes of ongoing processes of evolution and domestication, thus subject to persistent change (Singh et al. 2020). Throughout crop evolution, plants originating in their centers of origin have undergone an evolution from their wild progenitors, exhibiting physiological, morphological, and agronomic alterations to adapt to human preferences and cultivation practices.

PGRs play an indispensable role in the continuous search for specific economically important traits to enhance crop performance, encompassing both yield and nutritional value. Their dynamic exchange and exploration form the backbone of international germplasm collection and sharing, reflecting each nation's drive to secure diverse and superior genetic resources for conservation and utilization. PGRs provide the essential genetic diversity for breeding crops intending to increase yield, resistance to pests and diseases, and resilience to climatic shifts to fulfil the United Nations Sustainable Development Goal of food and nutritional security. Moreover, PGRs facilitate the adoption of sustainable agricultural methodologies by diminishing reliance on chemical inputs and fostering the preservation of biodiversity. In addition to their pivotal role in ecological sustainability, they are instrumental in combating malnutrition and enhancing human health by elevating the nutritional profile of staple crops. In today's context, the genetic richness inherent in crop diversity faces a significant threat known as "genetic erosion." This phenomenon involves the gradual loss of specific alleles and combinations of genes that are available in locally adapted landraces. The Food and Agriculture Organization (FAO) identifies the widespread adoption of modern varieties, leading to the displacement of indigenous varieties, as the primary driver of genetic erosion. This process not only diminishes the absolute number of cultivated varieties but also erodes the intricate genetic diversity essential for crop resilience. The situation is compounded by the emergence of new pests, weeds, and diseases, coupled with environmental degradation, urban expansion, and land conversion. The cumulative effect is a decline in agricultural biodiversity, putting at risk the stability of food systems.

Therefore, conserving PGRs emerges as a critical strategy to counteract this problem, ensuring the continued availability of diverse and adaptable crop varieties. By safeguarding these genetic resources, we fortify food security not only for the current generation but also for posterity. Apart from this, the management of PGRs encompasses a multifaceted approach, involving exploration, quarantine, conservation, characterization and evaluation, genomic research and database management.

1.2 PGR Collection and Conservation

PGR exploration involves the systematic search for valuable genetic material in diverse ecosystems, ensuring the inclusion of rare and unique traits. Conservation efforts focus on conserving this genetic wealth through methods such as seed banks, field genebanks and on-farm conservation to maintain the integrity and viability of diverse plant populations. Priority collection trips are vital to identify unexplored areas, with repeated visits necessary to document evolving diversity. Integrating modern tools like Geographical Information Systems (GIS) and remote sensing complements traditional ground data collection, particularly in challenging terrain. Digitization of herbarium plays a pivotal role in the taxonomic identification of collections conserved in the genebank and newly collected plants.

Over the years, 11 Consultative Group on International Agricultural Research (CGIAR) genebanks have accumulated 736,210 accessions of cereals, grain legumes, forages, tree species, root and tuber crops, and bananas. Many of these accessions are crop wild relatives. India has developed a very strong scientific management strategy for *ex situ* conservation and access to its genetic resources (Dhillon and Saxena 2003). ICAR-NBPGR (Indian Council of Agricultural Research–National Bureau of Plant Genetic Resources), New Delhi, is the nodal agency for the coordination of all PGR-related activities in India and provides access to various crop improvement programs. At present, the National Genebank (NGB) at NBPGR holds more than 4.6 lakhs accessions of 1762 species including 5034 Released Varieties and 4316 Genetic Stocks. The Cryobank Facility in the National Genebank has accessions of orthodox, intermediate and recalcitrant seed species and also of pollen samples. The *in vitro* genebank conserves various crops including tuberous and bulbous crops, tropical fruit species, spices and industrial crops, and medicinal and aromatic plant species under short- to medium-term storage periods. Key parameters assessed for genetic resource conservation include ensuring the uniqueness of accessions to avoid redundancy, maintaining seed quality based on global standards, ensuring seed viability of at least 85% (relaxed for vegetable, forage and wild species genetic resources), prioritizing pest-free conservation, and ensuring the availability of comprehensive passport information. Global genebank standards prescribe a minimum of 2000 seeds for self-pollinated crops and 4000 seeds for cross-pollinated crops. However, for wild germplasm accessions, this requirement is reduced to 500 seeds.

1.3 Plant Quarantine in PGR Management

For pest-free conservation, plant quarantine plays a critical role in safeguarding PGRs by preventing the introduction and spread of pests and diseases that could devastate agricultural ecosystems. As genetic resources are often sourced globally for research, breeding, and conservation purposes, the risk of inadvertently introducing harmful pathogens or pests is significant. Plant quarantine measures involve rigorous inspection, testing and regulation of imported plant materials to ensure they are free from harmful organisms. By effectively screening incoming plant material, quarantine protocols mitigate the risk of introducing exotic pests and diseases that could threaten both cultivated crops and natural ecosystems. Adherence to quarantine and biosafety regulations ensures the safe exchange of germplasm, mitigating the risk of introducing pests and diseases.

1.4 PGR Evaluation, Informatics and Use

Characterization and evaluation entail comprehensive study and assessment of genetic resources to generate characterization data profiles and identify traits of interest for breeding programs and sustainable agriculture to develop new varieties as per the requirements of the farmers. The collected or introduced germplasm is characterized and evaluated to assess its potential, by recording data on different qualitative and quantitative parameters based on descriptors developed by global organizations. The germplasm is also evaluated for novel traits using molecular tools to identify the genes linked to target traits. Genetic diversity in large collections has been determined using morphological and DNA fingerprinting markers. Thus, genomic tools allow for in-depth analysis of genetic diversity and facilitate the selection of traits crucial for crop improvement and adaptation to changing environmental conditions.

Finally, efficient database management ensures accessibility and dissemination of valuable genetic information, facilitating collaborative research and informed decision-making to safeguard global food security. By organizing vast repositories of genetic information, databases facilitate the systematic cataloguing of diverse germplasm collections, streamlining access to invaluable genetic material for researchers, breeders, and policymakers. These databases not only store data on germplasm characterization, evaluation, and trait profiles but also enable the integration of diverse datasets, including molecular markers and genomic sequences, thereby enhancing our understanding of genetic diversity and facilitating targeted breeding efforts.

For efficient utilization of germplasm, core sets, mini core sets and trait-specific reference sets are being developed to identify unique and promising donor germplasm for use in breeding programs. Evaluation of Focused Identification of Germplasm Strategy (FIGS set), multi-environmental trial, screening under artificially controlled conditions, high-throughput phenotyping, and phenomics- and genomics-assisted trait discovery are several approaches for mining of genebank genetic resources.

1.5 PGR Policy

Policy-making organizations play indispensable roles in governing the management, conservation, and utilization of PGRs on a global scale. The International Treaty on Plant Genetic Resources for Food and Agriculture (ITPGRFA) establishes a framework for the fair and equitable sharing of benefits derived from the use of PGRs, promoting access to genetic resources while ensuring the rights of farmers and indigenous communities. It facilitates international cooperation in the conservation and sustainable use of PGRs, fostering the exchange of germplasm for food and agriculture. Similarly, the Convention for Biological Diversity (CBD) emphasizes the conservation of biological diversity, including PGRs, and the fair and equitable sharing of benefits arising from their utilization. It provides a broader framework for biodiversity conservation, encompassing various ecosystems and species, with provisions for the sustainable use of PGRs to support agriculture, food security, and livelihoods. The Cartagena Protocol on Biosafety oversees the safe handling and transfer of living-modified organisms (LMOs) from modern biotechnology, focusing on mitigating risks to biodiversity during transboundary movements. In contrast, the Nagoya Protocol concentrates on access to genetic resources, ensuring fair and equitable benefit-sharing from their utilization through prior informed consent and mutually agreed terms. Both protocols regulate plant genetic resources, managing their access, transfer, and use while promoting conservation and equitable benefit-sharing, fostering international cooperation for biodiversity sustainability. In conclusion, the global landscape of PGRs necessitates a collaborative approach across borders and regions. No single country can independently satisfy its PGR requirements, highlighting the importance of cooperation at local, regional, national and international levels for acquisition and conservation efforts.

1.6 Summary

Rich PGR diversity fosters the development of novel crop varieties capable of thriving across diverse environmental conditions, thereby enhancing agricultural productivity and resilience. Complementary conservation strategies encompassing both *in situ* and *ex situ* approaches are crucial, with a focus on genetically rich hotspots and engaging stakeholders, including local communities. Characterization and evaluation are essential for promoting germplasm utilization, demanding decentralized networks and streamlined database searches. Generating awareness among policymakers, scientists, administrators, and farmers about the significance of PGR conservation is paramount, fostering innovative management approaches through stakeholder collaboration.

The National Bureau of Plant Genetic Resources manages PGR activities including collection, characterization, evaluation and maintenance of various crops with a national network comprising 10 Regional Stations located in different agro-ecological regions along with 49 national active germplasm sites. At headquarter, the Bureau has five Divisions (Plant Exploration and Germplasm Collection, Plant Quarantine, Germplasm Evaluation, Germplasm Conservation, Genomic Resources) and two units (Germplasm Exchange and Policy Unit, Agriculture Knowledge and Management Unit).

Suggested Reading

Dhillon BS, Saxena S (2003) India chapter. In: Plant genetic resources in SAARC countries, SAIC publication. SAARC Agricultural Information Center, Bangladesh, pp 241–296

Food and Agriculture Organization of the United Nations. (1983). International Undertaking on Plant Genetic Resources. https://www.fao.org/3/aq535e/aq535e.pdf

Harlan JR (1975) Geographic patterns of variation in some cultivated plants. J Heredity 66:182–191

Singh K, Gupta K, Tyagi V, Rajkumar S (2020) Plant genetic resources in India: management and utilization. Vavilov Journal of Genetics and Breeding 24(3):306

Takhtajan A (1969) Flowering plants: origin and dispersal (Engl. trans: Jeffrey C). Oliver & Boyd, Edinburgh

United Nations. (1992). Convention on Biological Diversity. https://www.cbd.int/convention/text/

Vavilov NI (1926) Centres of origin of cultivated plants (Trudy po Prikladnoy Botanike, Genetike i Selektsii). Proc Appl Bot Genet Breed 16(2):139–248. (in Russian)

Zhukovsky PM (1968) New centres of origin and new gene centres of cultivated plants including specifically endemic microcentres of species closely allied to cultivated species. Bot J (Russ Bot Z) 53:430–460

Plant Taxonomy and its Role in Studies of Plant Genetic Resources

2

R. K. Pamarthi, S. Nagaraju, K. Prasad, Pavan Kumar Malav, E. Roshini Nayar, Rinky Resma Panda, and K. C. Bhatt

Learning objectives

- Introduction and objectives of plant taxonomy
- Systems of classification
- Botanical nomenclature and types of nomenclature
- Herbarium methodology and NHCP (National Herbarium of Cultivated Plants)
- Taxonomic literature
- Methods of plant identification and preparation of descriptions, illustrations, line diagrams, and photoplates
- The role of taxonomy in PGR (Plant genetic resource) studies

Keywords

Plant genetic resources · Germplasm · PGR · Plant taxonomy · Herbarium · Digitization

R. K. Pamarthi · P. K. Malav · E. R. Nayar · K. C. Bhatt (✉)
Division of Plant Exploration and Germplasm Collection, ICAR-National Bureau of Plant Genetic Resources, New Delhi, India
e-mail: ravi.pamarthi@icar.org.in; pavan.malav@icar.org.in; kailash.bhatt@icar.org.in

S. Nagaraju
Ministry of Environment Forest & Climate Change, Indira Paryavaran Bhavan, New Delhi, India

K. Prasad
Department of Botany, Sri Venkateswara College, University of Delhi, New Delhi, India

R. R. Panda
The Graduate School, ICAR-Indian Agricultural Research Institute, New Delhi, India

K. Tripathi et al. (eds.), *Textbook of Plant Genetic Resources*,
https://doi.org/10.1007/978-981-96-5024-8_2

2.1 Introduction

Plant taxonomy primarily focuses on identifying, nomenclature and classifying plants. It follows a set of rules for naming a species and using a system (binomial nomenclature) for accessible and accurate communication; taxa are classified expressing the natural relationship of organisms (systems of classification, on a global scale), and used to detect evolutionary processes at work (biosystematics study and phylogeny). Taxonomy is a system that has developed and evolved over 250 years, a testament to the enduring human curiosity and dedication to understanding the natural world. Pioneering efforts were made by the Greeks (384 BC–200 AD), followed by the Romans (100–1500 AD) and the Elizabethan period (eighteenth and nineteenth centuries), each contributing to the rich tapestry of plant taxonomy. Taxonomy consists of three integral parts of the taxonomic study: (1) the herbarium (holding specimens of collections of plants over time), (2) the garden where plants are raised for comparative analysis of all features of the plant, and (3) the library bearing taxonomic literature (floras, monographs, etc.) and referring with the herbarium holdings. Modern classification systems are a testament to the collaborative nature of scientific progress. They aim to use evidence from various fields to study plants, with each contribution adding a new layer of understanding. Our existing knowledge molds taxonomic concepts—taxonomy has no data of its own, hence all developments and data generated are used to improve it. Explorations and colonization during the Elizabethan period have added to the known number of species, facilitated by the development and use of the microscope in the seventeenth and eighteenth centuries and chromatography and computers in the early twentieth century. Anatomical, embryological, palynological, cytological, cytogenetical, and molecular studies have all contributed to modern classification systems. In the process, plant taxonomy has evolved to accommodate the present state of knowledge in all aspects of botany.

Taxonomy depicts that all organisms are grouped into different structural categories based on their physical characteristics, and it is determined to be the oldest discipline of science in the present world. The term taxonomy combines two Greek words: *Taxis* (orderly arrangement) and *Nomos* (discourse or study or law, rule). Augustin Pyramus de Candolle used the term "*Taxonomy*" for the first time in 1813 in his book entitled "*Theorie Elementaire de la Botanique*". Later on, several taxonomists defined the term in their way (Box 2.1). Plant taxonomy emphasizes describing, naming, and classifying plants distributed worldwide. In a broader sense, its primary goal is to reconstruct organisms' phylogeny based on evolutionary life history. Nowadays, to study the modern trends in taxonomy, one should thoroughly know other fields like chemotaxonomy, numerical taxonomy, computer science, and modern scientific tools and techniques. Some researchers/scientists assume taxonomy and systematics are interchangeable words. Plant taxonomy aims to keep the grouping and arrangement of plants systematically concerning a well-recognized classification system. Generally, the primary goal of systematics is to understand the history of phylogeny and the evolutionary pattern of a group of organisms.

Box 2.1

- **Aristotle (384–322 BC):** Greek philosopher, known as the Father of Biology and Zoology, and the key person who first classified plants into herbs, shrubs and trees.
- **Theophrastus (371–287 BC):** Greek native of Eresos in Lesbos; was the successor to Aristotle. He is known as the "Father of Botany" for his works on plants. Theophrastus's multiple talents and interests ranged widely, from biology and physics to ethics and metaphysics. He classified mainly herbs, under shrubs, shrubs and trees and developed the names annual, biennials and perennials for plants.
- **Carl Linnaeus (1707–1778:)** A Swedish biologist and physician, known as the "Father of Modern Taxonomy," who organized binomial nomenclature, the modern organism naming system. He has written "*Systema Naturae*".
- **William Roxburgh (1751–1815):** A Scottish surgeon and botanist who worked in India, described and documented the plant wealth of India in his book "*Flora Indica*," and was well-known as the "Father of Indian Botany." He was the first superintendent of the Indian Botanic Garden in Kolkata.

The major trends of change include observational to experimental and qualitative to quantitative:

- Describing a species
- Naming a species, and the use of the tools of nomenclature
- Comparative study to work out diagnostic characters
- Study of populations: biosystematics study to recognize and distinguish infraspecific variation, introgression, polyploidy, etc. and genealogical tools (uniform conditions for growth) for recognizing ecotypes, physiological clines of variation, etc.

It is fascinating to observe that the taxonomy of organisms has evolved through a series of stages known as alpha (analytical phase), beta (synthetic phase), and gamma (biological phase) taxonomy. "Alpha," "orthodox," or "classical taxonomy" is the level at which the taxa/species are meticulously described and characterized (morphologically), "Beta taxonomy" deals with the arrangement of the species within the natural system of lower and higher categories and "Gamma taxonomy" is the study of intraspecific variations and evolutionary studies, that is, the study of speciation. "Omega taxonomy" is the modern or neo-taxonomy, enriched with data from ecology, phytochemistry, phytogeography, cytogenetics, and physiology. This represents the biosystematic or experimental phase in the development of taxonomy involving detailed cytological and genetic studies. Through the extensive approaches or uses of molecular tools and computer applications, systematics has become a dynamic science for studying organisms. Recently, plant taxonomy has seen tremendous changes, with advancements in microscopy, cytology, and molecular tools. It creates a new platform for characterizing organisms with the modern applications of different allied sciences of biology. The terms "New systematics" (Huxley 1940) and "Biosystematics" (Camp and Gilly 1943) entered a new dynamic in taxonomy. Biosystematics, a comprehensive scientific discipline, comprises the discovery, identification, description, naming, classification, and cataloguing of life forms. It also records their diversity, life histories, living habitats, roles in an ecosystem, and spatial and geographical distributions, providing a holistic understanding of the natural world. The objectives of the plant taxonomy are given in Box 2.2.

According to Mayer and Ashlock (1991), taxonomy can be further classified into two levels: (a) micro-taxonomy—the level that deals only with the troubles related to species; and (b) macro-taxonomy—the level that mentions errors and principles of taxa having higher group only. They have also recognized three schools of macro-taxonomy:

1. **Phenetics (Numerical Taxonomy):** Classify organisms according to overall characteristics; alternatively, evolution from the same ancestor. Phenetics finds that knowing phyletic lineage (relationship) depends on similarities and dissimilarities.
2. **Cladistics (Phylogenetic Systematics):** In this, the organisms are ranked and classified according to "recency of common descent." Taxa are based on entirely shared, derived (synapomorphic) characters evolving from the same ancestor. This technique is used to determine clades or monophyletic lineages. Julian Huxley (1958) introduced the term "clade" and the term "cladistic" was introduced by Cain and Harrison (1960).
3. **Evolutionary Taxonomy (Evolutionary Systematics):** It is proposed on the principles of Darwinism. According to this concept, each valid taxon evolves from the same ancestor. In this context, the characteristic features of an organism are not compulsory to include a common ancestry. The organisms are classified as coalescence of both phylogenetic relationships and overall similarity. It determines taxa as being more complex than a single species.

In the present chapter, four integrated facts of the plant taxonomy as well as its role in plant genetic resource (PGR) studies have been explained:

(a) Principles of Botanical Nomenclature
(b) Description and Identification
(c) Herbarium Preparation Methodology
(d) Taxonomic Literature
(e) Role of Plant Taxonomy in PGR Studies

Box 2.2 Objectives of the Plant Taxonomy
The taxonomists are involved with:

1. Determining what is a species (or its subdivisions)
2. Distinguishing these species from others using keys, descriptions, and geographic boundaries, and mapping their distributions
3. Investigating their interrelationships
4. Assigning proper names to species and higher-order ranks (such as genera or families) in accordance with international rules of nomenclature
5. Additionally, modern taxonomists investigate evolutionary processes to understand species' existing pattern and their interrelationships

2.2 Systems of Classification

It is defined as "the systematic arrangement of similar organisms into categories based on their structural or evolutionary relationships." Mainly, it is categorized into three types of classifications.

Artificial Classification This classification is based on convenient or necessary diagnostic characters without involving characters indicating a relationship. It was developed primarily based on a single arbitrarily selected characters, such as habitat, habit, and flower color, rather than evaluating the totality of characters. The pre-Darwinian systems of classification were predominantly artificial. The most significant systems are those proposed by Andraea Caesalpino (1519–1603), Gaspard Bauhin (1560–1624), John Ray (1628–1705), J.P. de Tournefort (1656–1708), Carolus Linnaeus (1707–1778), etc.

Natural Classification It relies on overall resemblances in external morphology and incorporates as many characters as possible, in contrast to artificial systems. It is assumed that the larger the number of characters shared by different plants, the more closely they are related. Later pre-Darwinian systems were natural; they were based on overall resemblances in gross morphology. The critical systems are of Michel Adanson (1727–1806), Antoine Laurent de Jussieu (1748–1836), A.P. de Candolle (1778–1841), Robert Brown (1773–1858), John Lindley (1799–1865), George Bentham (1800–1884), Joseph Dalton Hooker (1817–1911), etc.

Phylogenetic Classification This system depends on hypothesized evolutionary relationship. With the introduction of Darwin's *Origin of Species* (1859), the theory of evolution gradually restored the theory of the unique creation of species. It determines species are not fixed or unchanging but have evolved from pre-existing species over geological time. In general, this classification was based on structural similarities, which became evidence of evolutionary relationships. Thus, modern phylogenetic classification systems have arisen based on relationships by descent. The critical systems include those proposed by A.W. Eichler (1839–1887), H.G. Adolf Engler (1844–1930), K. Prantl (1849–1893), C.E. Bessey (1845–1915), John Hutchinson (1884–1972), Armen L. Takhtajan (1910–1997), Arthur Cronquist (1919–1992), etc.

Considering the facts of classification, in-depth knowledge of the impact of current technology, reflection of past events, and use of modern classification systems are clearly understood thoroughly. Experimental gardens, laboratories, libraries, herbariums, field visits, and computer applications have changed the taxonomical approach to studying taxonomy, shifting from purely traditional to modern ways.

2.2.1 Angiosperm Phylogeny Group Classification (APG)

This classification concept developed a unified view of the taxonomy of flowering plants and their relationships depending on phylogenetic studies. They set the most recent classification of flowering plants based on phylogenetic data—the Angiosperm Phylogeny Group Classification (APG). The principles of the APG's approach to angiosperm classification were first published in 1998 (APG I), and it has changed in the subsequent revisions, APG II, APG III and APG IV, which were published in 2003, 2009, and 2016, respectively. Each version supplants the previous version. Molecular taxonomy is the key concept in developing the APG III system of plant classification. Members of the Linnaean Society put forth a formal phylogenetic classification of all land plants compatible with the APG III, which recognizes 415 families. APG IV (2016), the most recent update, identifies 64 orders and 416 families, 259 of which are represented in India.

Merits of APG system

- The system demonstrates the monophyletic origin of groups.
- The system integrates multiple data sources, including morphology, anatomy, embryology, phytochemistry, and molecular data.
- Group names up to orders have been assigned.
- There is no priority for traditional divisions of angiosperms into monocotyledons and dicotyledons. Most of the monocots are positioned between primitive angiosperms and eudicots. It can solve the problem of paraphyly among monocots and dicots in the early stages. Several cladograms in the classification show general affinities between the various groups of organisms.
- Primitive families are positioned at the beginning of the angiosperms.
- The concept of bracketed families used in earlier APG was removed in APG III.

Demerits of APG system

- This classification is only applicable to families. This system is not suitable for the proper identification of flora.

- The status of some unplaced families and a few replaced genera remains uncertain.
- The botanical nomenclature Code has not been applied to the new group.

2.3 Botanical Nomenclature

Botanical nomenclature is the established system for giving accurate and recognized names to plants, guided by the *International Code of Botanical Nomenclature* (ICBN). The ICBN focuses on two main functions: naming new plant taxa and confirming or revising the names of previously identified taxa.

The primary purpose of botanical nomenclature is to assign a unique scientific name to each taxon, providing a universal reference for all organisms. This system enhances clarity in communication about plants and is essential for scientific exchanges. According to Rao (2004), naming aids in delivering unambiguous taxonomic information. The international community of botanists established guidelines through International Botanical Congresses (IBCs), which resulted in various "Codes" that outline the rules for naming plants. Scientific names are preferred over common names to prevent confusion.

- Common names for plants vary across languages, places, and countries.
- Scientific names are universal and easily recognizable.
- Common names do not indicate the relationships between different plant species.
- A single plant species may have multiple common names.
- Some plants have similar common names, which can be confusing.
- Many rare or obscure plant species lack common names altogether.

To address the issues related to common names in plant identification, scientists recommend adopting globally accepted names. Botanists have established specific rules and guidelines for this naming system, emphasizing that Latin serves as the fundamental language for nomenclature:

1. The language is not a national language of any country at present.
2. European languages are derived from Latin only.
3. Past European scholars studied their subjects in the Latin language.

The *International Code of Botanical Nomenclature* (ICBN) established the formal rules for naming plants, beginning with Linnaeus's "*Species Plantarum*" in 1753. This system ensures scientific names (Box 2.3) are precise and universally recognized, enabling people worldwide to identify and learn about various plant species. The primary goal of botanical nomenclature is to create a consistent and straightforward method for naming taxonomic groups, reflecting both their characteristics and taxonomic rank. While grammatical correctness and aesthetic considerations are acknowledged, the main focus remains on standardizing plant naming practices across different countries.

(a) **Polynomial Nomenclature**: It was common among botanists in the eighteenth century, including Linnaeus. This naming system consisted of multiple descriptive words for plants (e.g., *Salix pumila angustifolia altera*), making them difficult to remember and challenging for creating identification keys. Linnaeus eventually replaced this complex system with the simpler binomial nomenclature.

(b) **Binomial Nomenclature:** A scientific name consists of a generic name and a specific epithet, forming a binomial known as the species name. While "species name" is often mistakenly used to refer to just the specific epithet, it actually includes both components. To complete the binomial, it is also necessary to include the name of the person or individuals who originally described the plant.

Example The complete scientific name of mango is *Mangifera indica* L.; here *Mangifera* is genus, *indica* is specific epithet, and L. (Linnaeus) is the author or authority. Thus, species name of mango is *Mangifera indica* L.

Box 2.3

Generic Name

1. It is a singular Latinized noun or a word treated as a noun.
2. It is always written with an initial capital letter.
3. If a generic name is spelled more than once in a paper, it may be abbreviated by using an initial capital letter. For example: *Brassica alba, B. nigra, B. juncea.*
4. The generic name can be taken from any source; it can be an aboriginal name. Many ancient common Greek names like *Asparagus* and *Narcissus* have been adopted as generic names.
5. It may be given in honor of a person. Example: *Linnaea*, *Bauhinia*, *Caesalpinia*.
6. Generic names may depict a certain characteristic of a plant. Example: *Rhododendron* (the color of the wood is red), *Zanthoxylum* (yellow wood).

Specific Epithet (or Descriptor)

1. A specific epithet is often an adjective and usually depicted with a small initial letter (except for a few exceptions like some names commemorating a person).
2. It usually depicts distinguishing features of the species like color, shape, size, habit, and habitat. Example: *Morus nigra* (black mulberry), *Calotropis gigantea* (big size), *Mentha arvensis* (field mint).
3. It may be derived from any source and may be given in honor of a person, or may be derived from an old common name, a geographic locality or some characteristic of plant, or they may be even composed arbitrarily.
4. Specific epithets consisting of two words are hyphenated. Example: *Hibiscus rosa-sinensis* L.
5. The specific epithet usually agrees with the gender of the generic name if the specific epithet is an adjective and masculine, feminine or neuter. The names ending with -us are masculine, -a are feminine and -um are neuter.

2.3.1 History of Botanical Nomenclature and Formation of *Code*

The early literature on botanicals, primarily "herbals," originated in ancient Egypt, China, India, and Europe. Herbalists, apothecaries, and physicians of the time contributed to the medical knowledge. For centuries, plant names were long and descriptive, making them hard to remember. Dutch Governor Van Rheede (1636–1691) published "*Hortus Malabaricus*" between 1678 and 1693, featuring descriptions, illustrations, and practical information about various plants under their Malayalam names. In 1620, Swiss botanist Gaspard Bauhin introduced "Binomial nomenclature," which uses two parts for each species name—*genus* and *specific epithet*—but he applied it only to a few species. Linnaeus later simplified this naming system in "*Species Plantarum*" (1753) and established early nomenclature rules in "*Critica Botanica*" (1737), further developing them in "*Philosophia Botanica*" (1751).

Augustin Pyramus de Candolle introduced the classification system and the term "taxonomy" in his 1813 work "*Théorie Élémentaire de la Botanique*." He addressed nomenclature laws and emphasized good practices, advocating for priority in naming, with some exceptions. In 1821, Steudel published "*Nomenclator Botanicus*," listing Latin names and synonyms for flowering plants. Schisms in classification arose after the 1843 British Association endorsed Zoological rules, culminating in the 1867 Paris Congress, which approved Alphonse de Candolle's botanical "*laws.*" A.P. de Candolle presented the "*International Code of Botanical Nomenclature* (ICBN)" on August 1, 1867, during the International Botanical Congress (IBC) held in Paris. This work laid the foundation for botanical nomenclature. This 60-page document included an introduction, laws in 68 articles, and commentary, making the rule of priority fundamental in plant nomenclature. The latest edition, titled "*International Code of Nomenclature for Algae, Fungi, and Plants* (ICN)," was published in 2012 at the XVIII International Botanical Congress in Melbourne in 2011.

Nicolson (1991) identified three overlapping periods in botanical nomenclature: the schismatic period (1840–1930), the Dark Ages (1915–1950), and the IAPT (International Association of Plant Taxonomy) renaissance (1950s). McNeill (2012) highlighted three primary phases since 1867 in the history of nomenclature rules. The "*International Code of Nomenclature for Algae, Fungi, and Plants* (ICN)," formerly known as the "*International Code of Botanical Nomenclature*," consolidates the rules and recommendations for naming plants and other organisms. The term "*International Botanical Congress*" was established in Melbourne in 2011 with the adoption of the Melbourne Code, which replaced the Vienna Code (2005). The Code is revised periodically at 6-year intervals during *International Botanical Congresses*. Changes to the ICN occur through these Congresses, supported by the International Association of Plant Taxonomy (IAPT), with each new edition superseding the previous ones.

The development of agro-horticultural varieties led to the establishment of a specific naming code for cultivated plants, known as the *International Code of Nomenclature for Cultivated Plants* (ICNCP). This code is updated every 5–6 years during the *International Botanical Congress*, which is named after its host location.

Over the past century, the code has undergone significant changes, and the nomenclature rules are now largely stabilized. A comprehensive list of the congresses and associated rules is provided in Fig. 2.1.

In addition to the change in the title of the *Code*, the fallowing few other major changes are included in the rules of nomenclature in the 18th IBC held in Melbourne.

1. New taxa names through electronic publication are permitted (from January 1, 2012). As per this rule, it is compulsory to effectively publish new names of fungi, algae, and plants (and designated types) in printed matter. All such new names are published online in PDF (Portable Document Format) along with the International Standard Book Number (ISBN) or International Standard Serial Number (ISSN).
2. At present (Starting in January 2012), a new taxon description or diagnosis must be published in English or Latin to confirm a validly published name.
3. **Fungal names registration**: All new fungal names (January 2013), including new taxa, new combinations and replacement names, must have identifier issued by an authenticated repository at new ranks. The new fungal names could not be validated without a proper identifier issued by the repository. The MycoBank, Fungal Names, and Index Fungorum are the designated official repositories of fungal names.
4. **One fossil, one name**: Usually, separate names are denoted to differentiate the parts of fossils (morphotaxa) by the code. According to the Melbourne Code, one scientific name is used for fossils if they have two or more morphotypes.

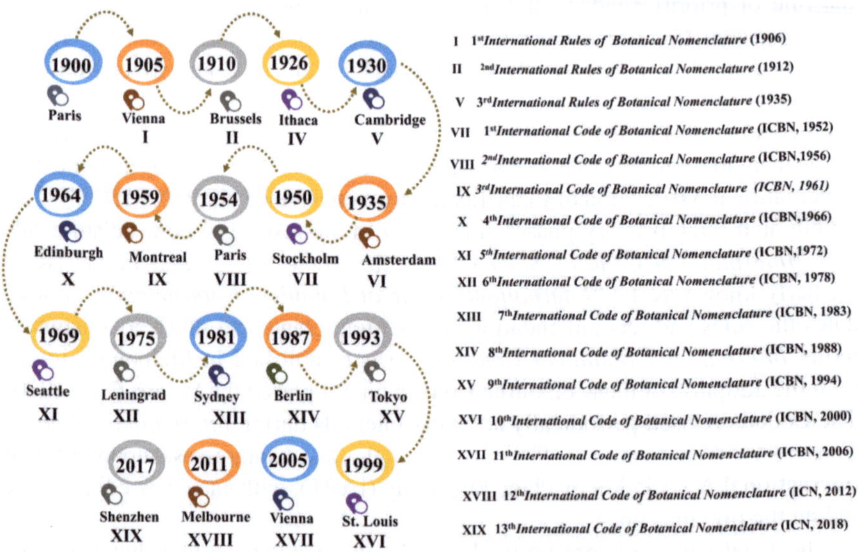

Fig. 2.1 Pictorial depiction representing International Botanical Congress and International Rules/Codes

The code is categorized into three divisions and five appendices (Box 2.4). The divisions are divided into principles, rules and regulations and provisions for the governing code. The principle leads to the initiation of the basis for botanical nomenclature. Further, the code is categorized into rules, and framed with the articles and recommendations. The rules are to put nomenclature in order and designed for the future. The rules are mandatory to follow. Names contrary to rules are considered illegitimate and cannot be maintained. The recommendations deal with subsidiary points, and are laid down to bring uniformity and clarity, particularly in future nomenclature. Names contrary to recommendations cannot be rejected, but are not examples to be followed.

Box 2.4

Divisions

I. Principles
II. Rules and Recommendations; further divided into 7 Chapters and 62 Articles
III. Provisions for the governance of the Code

Appendices

I. Names of hybrids
IIA. Conserved and rejected family name of fungi, pteridophytes and fossils
IIB. Conserved bryophyte and spermatophyte family names
IIIA. Conserved and rejected generic names
IIIB. Conserved and rejected specific names
IV. Rejected names and all combinations based on these names
V. List of publications and the category of taxa that are not validly published

Principles of ICN

There are six principles on which ICN is based:

I. Botanical nomenclature is independent of zoological and bacteriological nomenclature.
II. The application of the taxonomic groups is determined by means of nomenclatural types.
III. The nomenclature of a taxonomic group is based on priority of publication.
IV. Each taxonomic group with particular circumscription, position and rank can bear only one correct name, the earliest that is in accordance with the rules, except in specified cases.
V. Scientific names of taxonomic groups are treated as Latin regardless of their derivation.
VI. Rules of nomenclature are retroactive unless expressly limited.

2.4 The Taxonomic Hierarchy

This concept was introduced by Linnaeus, in which each organism is positioned into a layered hierarchy of groups or taxonomic categories. Different groups of plants classified for taxonomic purposes are called taxa. Species is the basic unit of the taxonomic unit. The principle ranks of taxa in descending order are Kingdom-Division-Class-Order-Family-Genus-Species. Thus, each species is assignable to a genus, each genus to a family, and so on.

2.4.1 Features of Taxonomic Hierarchy

- It shows that each taxon name above the family rank is depicted with initial capital letters and marked as plural nouns. Such names are generally based upon the name of an included genus, called the type genus. Each rank has a distinctive ending that is attached to the stem of the type genus. Suffixes used to form these names are:

  - *aceae* for families (e.g., Magnoliaceae, ending in the genus Magnolia)
  - *ales* for order (Magnoliales)
  - *opsida* for class (Magnoliopsida)
  - *phyta* for division (Magnoliophyta)

- Names of genera are treated as nouns in the nominative singular, underlined (or italicized), and the first letter is capitalized. They may be taken from any source whatsoever, and may even be composed arbitrarily.
- Usually, a species' scientific name is mentioned in binary combination, which bears the genus name followed by a specific epithet.
- The specific epithet is usually considered to be an adjective; it is also italicized or underlined and written in all lower case. However, species named after people may be capitalized. The specific epithet may be derived from any source, or may even be composed arbitrarily.
- To be complete, the scientific names include authority (name of the person who described the species). The author's name is never italicized or underlined. To make standard format, the author's names are generally abbreviated (e.g., L. for Linnaeus; Roxb. for Roxburgh; Hooker son as Hook.f.), and if the author's name is short, they are cited as such (Vahl, Nees). When two authors publish a single plant name jointly, both are cited with "*et* or &" between the two. If more than two authors publish a single plant, the first author's name is followed by "et al." (and others). Sometimes, a plant name is mentioned by one author and the same published by another author, then "ex" is cited between both authors.
- Alternative family names: There are some family names, which were not based on any included genus, and their ending was also not according to rules. The Code has suggested alternative names for such families. Use of both is allowed by the Code.

Old names	Alternative names	Old names	Alternative names
Cruciferae	Brassicaceae	*Compositae*	Asteraceae
Guttiferae	Clusiaceae	*Labiatae*	Lamiaceae
Leguminosae	Fabaceae	*Palmae*	Arecaceae
Umbelliferae	Apiaceae	*Gramineae*	Poaceae

2.4.2 Rule of Priority

Priority of publication is a crucial and essential part of the rules of nomenclature and makes a lead in one of the principles of the Code. In general, each taxon or family of lower rank must have one correct name. The earliest legitimately published name is the correct name. Combining the earliest validly published generic name with the earliest validly published specific epithet delivers the proper name of a species, except under the case priority limitation by conservation. Conserved names are legitimate; initially, they are illegitimate. The conserved names may be at the family, genus, or species level. Different workers may have given the same taxon different names; the latter names are called "Synonyms".

Example *Malus pumila* Mill., 1768; *Pyrus malus* L., 1753; *Malus domestica*, Bork., 1863; *Malus communis* Poir., 1884. Here, the name *Pyrus malus* L. has priority over other names, and all other names are synonyms. The priority principle is limited to the rank of family and other groups by ICN.

2.5 Type Method or Typification

The "Type" concept, also known as typification, is a method of describing "type." It plays a crucial role in the nomenclature part of taxonomy, being one of the six principles of the Nomenclatural Code (Vasudeva Rao 2017; Turland et al. 2018). Typification, a complex step in the nomenclature of taxa, involves systematically and permanently associating, particularly with the name of taxon, to the specimen of the specific species. The majority of these "types" can be traced using the relevant protologues and "*Taxonomic Literature, Vol. 1–7*" of Stafleu and Cowan (1976–1988) and its eight supplements (Suppl. 1–6 by Stafleu and Mennega, 1992–2000; Vols. 7 and 8 by Dorr and Nicolson, 2008, 2009).

Nomenclature types were supported strongly by mentioning the principles of names of taxonomic groups in both VI IBC held at Amsterdam (The Netherlands) and VII IBC held at Stockholm, Sweden, 1950. The application of the name of any taxon, up to the level of the family, is based on nomenclature type. Nomenclature type is a specimen or element conserved on a herbarium sheet on which the original description of the taxon is based. When not having a herbarium sheet, one can have an illustration of the same. According to the Code, the nomenclatural type of the family is the genus, and that of a genus is the species.

2.5.1 Different Nomenclatural Types

"Holotype" is defined as "one specimen or illustration of a taxon name either (a) indicated by the author(s) as nomenclatural type or (b) used by the author(s) when no type was indicated' (Article 9.1)." The holotype always is a singular and single specimen under any conditions. "Isotype" is defined as the duplicate specimen of the holotype that is from a single gathering (made at one time) of the "Holotype" (Article 9.5). "Paratype" is a type in which any specimen(s) other than holotype or isotype is included in the protologue (Article 9.7), whereas "Syntype" is any specimen(s) cited when no "Holotype" was designated (Article 9.6). If the author did not indicate a holotype in the publication process or it is missed, the term "Lectotype" is defined for the specimen on the support of protologue (Article 9.3). "Neotype" is designated as a nomenclatural type (Article 9.8) if the original material is missing or extant. The neotype loses its importance and identity in the typification methodology, if original material is found. Generally, the "Epitype" will be selected as interpretative type when type material (holo-, lecto- or neo-) is ambiguous or insufficient or inadequate for precise application or identification of the name of designated specimen or illustration (Article 9.9; Brickell et al. 2016). The management of handling type specimen is mentioned in Box 2.5.

Box 2.5

1. Type specimen preservation with utmost care, mostly in different places in a herbarium.
2. Type specimen requires special attention and regular inspection in the herbarium compacters/storage space.
3. Mostly allowed for virtual images for usage. Not allowed for routine use, and only made available to researchers after special permission, especially to taxonomists.

2.6 Herbarium and Herbarium Methodology

Herbarium is a repository of plant specimens (whole or part), which are explored from different places and subsequently identified, dried, pressed and mounted on paper sheets with label data as to its provenance and other information, but depending upon the material, may also be kept in alcohol or another preservative. The same term is often used in mycology to describe an equivalent collection of preserved fungi. Herbarium (plural: Herbaria) is an institution or division consisting of the exploration of plant specimens (dried) either arranged alphabetically or according to an accepted system of classification. It is an important source of data used by researchers in many plant-related sciences and is most often used for taxonomic research. Herbaria are centers of research and learning pertaining to identification, nomenclature, classification, distribution and use. The Latin term "*herba*," which originally meant for grass/pasture, was derived from Greek (*forbh*), which in turn appears to have been derived from Sanskrit: "bhar" (pasture/forage, Latin Dictionary) (Lewis and Short 1879). The Latin substantival suffix "-arium" refers to a place wherein something is housed. The word herbarium in its original sense referred to a book about medicinal plants. Tournefort first used the term "herbarium" to describe a collection of dried plants in *c*. 1700. Herbarium specimens are very valuable for botanists in general because they provide a long-term scientific record of a plant collection intending to identify the botanical species observed in the field. The herbarium has revolutionized the taxonomy and become its most important tool (Box 2.6) for material recorded. The type specimens in herbarium are the backbones of nomenclature. Notes on label associated with specimens are useful for identification, vernacular names, ethno-botany, phenology, ecology, distribution, phytogeography, geobotany, etc.

Box 2.6 Significance of Herbarium

- A herbarium is a comprehensive data bank representing the plant diversity of a region in one place and a useful tool to students and researchers for studying a wide range of collections.
- It helps in the identification of a plant specimen.
- Important data recorded on herbarium labels aid in planning field trips.
- The specimen data are used for monitoring climate change studies and producing checklists for floras.
- The field data of Herbaria specimens and ancillary collections (e.g., photographs) are the authenticated source of information regarding plants and the phytogeography.
- They provide the comparative material that is crucial for studying other integrated subjects such as ecology, anatomy, morphology, conservation biology, biodiversity, ethnobotany, and palaeo-biology used for both research and academic purposes.

The Oldest Herbarium

- Professor Luca Ghini (University of Bologna, Italy) is thought to have been the first person to dry plants under pressure and mount them on paper to serve as a lasting record. The oldest herbarium specimen is preserved in Florence, Italy. It was collected by Gherard Cibo (Student of Luca Ghini) in 1532. First herbarium of the world was established in 1545 and first botanical garden at the University of Padua, Italy.

Oldest Herbarium Specimen in India

- The oldest herbarium specimen is *Cyperus procerus* Rottb. collected by Samuel Brown (Assistant Surgeon) in 1696 in the Madras region of Southern India (Prasad 2014).

2.6.1 Herbaria in the World

At present, approximately 3400 recognized herbaria are located around the world. Collectively, the world's herbaria contain an estimated 400,000,000 specimens. For details, the "*Index Herbariorum: A Global Directory of Public Herbaria and Associated Staff*" can be referred to at http://sweetgum.nybg.org/ih/. Each herbarium is assigned an official acronym (Code) that is used as a standard for referring to the institution and its specimens. *Index Herbariorum* gives a particular herbarium's brief details (web address, physical location, specimen number and type of specimens, history, names, and contact information) (Table 2.1).

2.6.2 Herbaria in India

In India, there are approximately 100 herbaria, with the Central National Herbarium (CNH) in Kolkata being the largest and oldest. The CNH holds about two million herbarium sheets representing 350 families of higher plants. It specializes in taxonomy through various studies, including floristic, monographic, and revisionary research. Additionally, the CNH offers digital herbarium facilities and serves as a center for taxonomic work and consultation of herbarium specimens (Table 2.2).

2.6.2.1 National Herbarium of Cultivated Plants (NHCP)

The National Herbarium of Cultivated Plants (NHCP) is conserving the herbarium of cultivated taxa. In general, all herbaria usually consist of all types of specimens, whereas NHCP (Box 2.7) represents variability in crop plants, mainly primitive types/landraces, cultivars, wild/semi-domesticated forms and crop wild relatives (CWRs), weedy types, and also the minor economic species explored from different agro-ecological regions of India. Mainly, NHCP preserves herbarium specimens that have a PGR value. Additionally, herbarium vouchers for exotic germplasm (introduced material imported under different research programs), local flora, weed

Table 2.1 Five important herbaria of the world

Sl. no.	Herbarium	Acronym	Country	No. of specimen
1	Herbarium, Royal Botanic Gardens, Kew	K	United Kingdom	8,125,000
2	Herbarium, V.L. Komarov Botanical Institute, Saint Petersburg	LE	Russia	7,160,000
3	Herbarium, Missouri Botanical Garden, Saint Louis, Missouri	MO	USA	6,500,000
4	Herbier, Laboratorie de Phanerogamie & Cryptogamie, Museum National d' Histoire Naturelle, Paris	P	France	6,000,000
5	William and Lynda Steere Herbarium, New York Botanical Garden, New York	NY	USA	7,921,000

Source: Index herbariorum (2023)

Table 2.2 Important herbaria in India

Sl. no.	Herbarium	Acronym	Place	Number of specimens	Source
1	Central National Herbarium (CNH)	CAL	Howrah	c. 2,086,650	Thiers (2021)
2	Forest Research Institute (FRI)	DD	Dehradun	c. 340,000	Thiers (2021)
3	National Botanical Research Institute (NBRI)	LWG	Lucknow	c. 276,786	Thiers (2021)
4	Botanical Survey of India, Eastern Regional Centre	ASSAM	Shillong	c. 260,000	Thiers (2021)
5	Botanical Survey of India, Southern Regional Centre	MH	Coimbatore	c. 259,073	Thiers (2021)
6	National Herbarium of Cultivated Plants	NHCP	New Delhi	c. 26,627	Annual Report (2024)

flora, and vouchers for experimental material deposited in the repository by the researchers are preserved in the NHCP. The herbarium materials are a reference collection for identification, taxonomic studies, and teaching purposes. Currently, the NHCP has 26,627 herbarium specimens representative of 267 families, 1551 genera and 4394 taxa (as of 31 March, 2025) of essential plant genetic resource (PGR) relevance, including over 500 crop taxa and 550 species of crop wild relatives (CWR)/weedy relatives (Pandey et al. 2015, 2021). In addition, the NHCP also preserves economic products like seeds and carpological items (dried plant fruits and other products) by utilizing plant genetic resources.

Box 2.7 National Herbarium of Cultivated Plants (NHCP)

- National Herbarium of Cultivated Plants is conserving the herbarium of cultivated taxa.
- The mandate of NHCP preserves a wide variability in crop plants (cultivars, primitive types/landraces, wild/semi-domesticated forms, and crop wild relatives (CWR)/weedy types) and other minor economic species collected from different agro-ecological regions of India. NHCP preserves herbarium specimens having PGR value.
- Currently, the NHCP has 26,627 herbarium specimens representative of 267 families, 1551 genera, and 4394 taxa (as of 31 March 2025).

2.6.3 Steps in Herbarium Methodology

The herbarium preparation involves a series of steps to make a good herbarium specimen, such as collection, drying, mounting, identification, and storage. A detailed methodology along with a digital herbarium is given in Fig. 2.2.

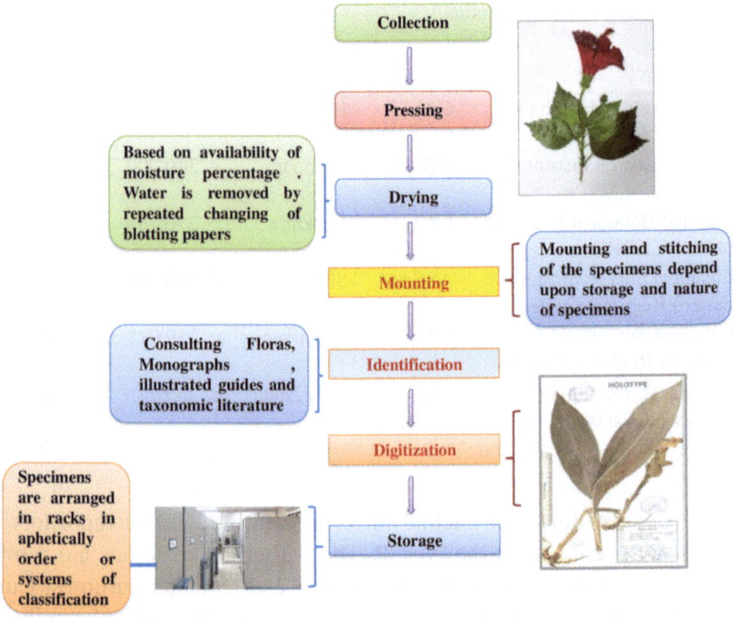

Fig. 2.2 Pictorial representation of Herbarium methodology

2.6.3.1 What and How Much to Collect?

- Collection of certain specific groups of plants (Box 2.8).
- Collections should contain at least flowers or fruits or preferably both. Sterile twigs or plants are of little value. In the case of small terrestrial species, the whole plant, including the underground parts, should be collected (Box 2.9).
- If the taxon is common and represented by a good number in a population, collect five specimens against a single field number.
- If the taxon is found in only one locality and represented by less than five specimens, then the collection must be avoided. In that case, try to take good photographs and record the characters in the field. Without collecting a twig or whole plant, collect one to two flowers for further field and laboratory studies. Avoid over-collection.

Box 2.8 Collection of Plants: With Special Reference to Crop Taxa

Objectives:

- Building new herbaria for enriching old herbaria
- Collecting plant material for writing a flora
- Collecting live material for introduction in gardens/hybridization
- Collecting members of particular genus/species for revisionary studies
- Ethnobotanical studies
- Biosystematics studies

Equipment for collection When planning for specimen collection, it is essential to consider various factors that influence the necessary equipment. These include the weather conditions, the duration of the trip, the work area, the mode of transportation, the locations where you will be staying, and the specific purpose of the collection. By evaluating these elements, you can effectively determine what equipment is needed and in what quantity to ensure a successful collection process.

Essential articles Camera, vasculum, cutter/hand pruner or secateurs, knife, khurpi, tree pruner, polythene bags, field book, hand lens, old newspaper, drying sheets/blotting sheets, field press, straps, ropes, global positioning system (GPS), altimeter, tents, first aid box, etc.

Size of the plant specimens In India, the standard size for herbarium sheets is approximately 42 × 28 cm. For small herb specimens, the entire plant, including roots or underground parts, can typically fit on a single mounting sheet or in one dryer. However, for larger herbs, it is essential to collect the whole plant, but it often needs to be cut into two or three sections to fit properly on the mounting sheet.

Field notes The field notes recorded on a herbarium sheet are crucial and should be treated with the same importance as the plant specimen itself. According to Jain and Rao (1977), an essential part of plant collection involves meticulously documenting information in field books at the time of collection. This documentation should include key details such as the date, vernacular names, plant habit, locality, type of vegetation, associated flora, abundance distribution, uses, significant observations, and the name of the collector. Properly recording this information enhances the value and utility of the specimen for future research and reference.

2.6.3.2 Processing of Specimens

The entire processing is divided into six phases:

- Pressing specimens
- Drying of the specimens
- Poisoning of specimens
- Mounting
- Stitching of specimens
- Labelling of specimens

To preserve non-succulent plant species, follow a standard method that includes pressing, drying, and treating with preservatives. Shortly after collection, immerse specimens in boiling water for 1–2 min to prevent succulent leaves from dropping. For pseudo-bulbs, excess tissue can be scooped away, while thicker organs may be preserved using an alcohol/formalin solution or boiling water. Delicate floral parts, such as those of Orchids and Legumes, should be treated with an 80:20 alcohol–formaldehyde solution to maintain their three-dimensional features essential for identification. Remember to document key field characteristics, including the overall growth habit, flower parts, and fruits, in a field notebook, and take color photographs of each plant in its natural habitat to capture both their appearance and details.

Box 2.9 Best Tips Followed During the Collection and Processing of Specimens

- Select a good twig or plant, which is free from disease.
- Place a single plant within a single folder of drying paper; if the plants are minute, more than one may be kept in one folder.
- Plants longer than 30 cm should be folded in the form of V, N, M or W, and then pressed.
- If necessary, some pruning is done to prevent overlapping of parts, leaving their basal part to indicate their position.
- Some of the leaves should be pressed with their lower side up.
- Tubers, bulbs, and other bulky parts can be pressed after removing extra tissues.

Identification of Specimens

Identification is the process of recognizing an unknown plant specimen by comparing its characteristics to those of known plants. This typically involves directly comparing the unknown specimen with already classified and named plants. Additionally, identification may utilize various tools, such as identification keys (including those that are computer-based), to assist in the process.

Filing of Specimens

After identification, the specimens are incorporated into the herbarium. All sheets of the same species are placed in a "species cover" followed by "genus cover." The specimens are arranged according to a recognized system of classification (Bentham and Hooker 1888) and kept in almirah (herbarium case) with pigeonhole.

Digital Herbarium and Virtual Herbarium

The digital herbarium is an innovative concept where digital images or photographs of herbarium specimens are stored in databases typically developed using PHP software tools. Each image is labelled with a unique identity number corresponding to the herbarium specimen and linked to various databases across different institutes. This virtual herbarium enables users to access these digital images at any time, which is particularly beneficial for remote users, allowing access from anywhere in the world. The images are scanned at resolutions of 300 ppi and 600 ppi using high-resolution scanners and are stored in either JPEG or TIFF formats. The primary goal of the digital herbarium is to minimize damage to original herbarium vouchers caused by frequent manual handling.

2.7 Taxonomic Literature

Various forms of taxonomic literature, including descriptions, illustrations, identification keys, and photographs, are essential for accurately identifying unknown plants. Some of these works are among the oldest and most complex in the field of science. The three major types of taxonomic literature that form the foundation for studies and research in plant taxonomy are flora, revisions, and monographs, which are crucial not only for accurate plant identification but also for understanding and conserving biodiversity. A flora serves as an inventory of the plants within a specific region or area, while revisions and monographs are more comprehensive and typically focus on specific genera or supra-generic taxa. Additionally, indexes, abstracts, manuals, journals, conspectuses, synopses, and bibliographies also play important roles in conducting taxonomic research.

2.7.1 Flora

An inventory of plants of a defined geographical region.

Example: "*The Flora of British India*, Vol. I–VII," by J.D. Hooker (1888, 1890, 1895); "*Flora of Gangotri National Park Western Himalaya, India*" by P.K. Pusalkar and D.K. Singh.

Objectives

- Survey and collection of the plants from the study area.
- Identify the species collected from the study area.
- Describe the identified species.
- Preparation of a detailed account of the plants found in the study area based on own collections, consultation of herbaria and study of relevant literature.
- Preparation of a comprehensive list of economically as well as medicinally important species/taxa for their conservation and sustainable utilization.
- Evaluation of threatened/near-threatened plant species of the area.
- Study of biotic interferences on plant diversity of the study area.

2.7.2 Revision

Exhaustive taxonomic account of a taxon/taxa; usually, the scope is not worldwide like that of a monograph.

Example: "Orchidaceae: Genus *Coelogyne*" in *Fascicles of Flora of India* (Fascicle 5) by Sandhyajyoti Das and S.K. Jain (1980).

Objectives

- To prepare an up-to-date checklist of the different members of the taxon/taxa.
- To draw detailed taxonomic descriptions of all taxa based on the critical study of the protologues, type material, herbarium specimens, fresh collections, and authentic literature.

- To provide up-to-date nomenclature, plates/illustrations, photographs (wherever possible), uses, and distribution in India and the world.
- To prepare taxonomic keys for any taxa and help in identification.
- To assess the present population status of the species in the field based on field survey, and literature, and herbaria consultation to propose effective conservation measures.
- To deliver other information regarding taxonomy (cytology, biochemistry, palynology, etc.) available for different taxa of the taxon/taxa.

2.7.3 Monograph

Comprehensive taxonomic account (including data from related fields) of a taxon with a worldwide geographical scope. It contains all the information of a taxon in a single publication.

Example: "Turmeric, The genus *Curcuma*" by Ravindran et al. (2007).

Monographs are similar to revision but more exhaustive in the below features.

- Number of species/taxa (all members of the world belonging to that taxon/taxa)
- Review literature (world)
- Synonyms (complete synonymy)
- Specimens examined/herbaria consulted
- Information from allied fields

2.7.4 Manual

A manual is a more exhaustive treatment than the flora, having keys for identification and brief description, but generally covers a specialized group of plants.

Example: "*Manual of Cultivated Plants*" by L.H. Bailey.

2.7.5 Conspectus

A conspectus is an effective outline of a revision listing all the taxa, with all or major synonyms, with or without a short diagnosis, and briefly mentioning its geographical range.

Example: "*Species Plantarum*" by Carl Linnaeus.

2.7.6 Journal

A taxonomic journal provides information on the results of ongoing research and is published at regular intervals.

Examples: *Taxon, Phytotaxa, Nelumbo, Rheedea.*

2.7.7 Abstract and Bibliography

The *Abstract* provides a summary of various articles published in various journals throughout the world. The bibliography is a list of names of publications of books and scientific literature on specific topics during a specific period.

2.7.8 Checklist

List of plants (often with information on distribution but lacking description/diagnosis and other details) of an area/region, sometimes dealing with a particular group of plants. Example: Flowering Plants of India—Dicotyledons, Vol. 1 (Acanthaceae—Avicenniaceae) by Karthikeyan et al. (1989). Flowering Plants of India: An Annotated Checklist, Dicotyledons (Vol. 1 & 2) and Monocotyledons (Mao and Dash 2020).

2.7.9 Botanical Dictionary

A botanical dictionary may include a list of all known genera/taxa (sometimes with brief information on distribution) of certain plant groups.

Example: *A Dictionary of the Flowering Plants and Ferns* by J.C. Willis.

2.7.10 Botanical Glossary

A glossary is an alphabetical list of different terms with their explanations.

Example: "*A Glossary of Botanical Terms*" by B.D. Jackson (1893, 1894).

2.8 Taxonomic Databases

The taxonomic databases are the comprehensive information on taxonomic identification in the diverse sciences of the scientific species literature. These databases give clarity to the different taxonomic groups of the species. Currently, different models represent the taxonomic tools in the diverse taxonomic groups. The taxonomic databases contain taxa information, accepted species, synonyms, misapplied names, basionyms, nomenclature data, protologue links, etc. (POWO 2023). Some databases are designed for a particular group of general or crop taxa, such as the Global Compositae Database and ILDIS (ILDIS 2022), and these databases are linked to the specific ranking groups in particular flora. Important taxonomic databases are listed in Table 2.3.

Table 2.3 Important taxonomic databases with respect to crop taxa

Sl. no.	Database	Link
1	Biodiversity Heritage Library (BHL)	https://www.biodiversitylibrary.org/
2	Genetic Resources Information Network (GRIN)	https://www.ars-grin.gov/
3	Global Compositae Database (GCD)	https://www.compositae.org/
4	Kew Herbarium Catalogue	http://apps.kew.org/herbcat/navigator.do
5	Indian Virtual Herbarium	https://ivh.bsi.gov.in/
6	International Plant Names Index (IPNI)	https://www.ipni.org/
7	International Legume Database and Information Service (ILDIS)	http://www.ildis.org/
8	National Centre for Biotechnology Information (NCBI)	https://www.ncbi.nlm.nih.gov/taxonomy
9	Plants of the World Online (POWO)	http://www.plantsoftheworldonline.org/
10	United States Department of Agriculture (USDA)	https://plants.usda.gov/about_adv_search.html

Source: Respective website/URL

2.9 Basic Techniques of Plant Identification and Preparation of Descriptions/Illustrations/Line Diagrams/Photoplates

The most basic concept of taxonomy includes description, identification, nomenclature, and classification. The general subjects of study are taxa (singular, taxon), which are a taxonomic group or category of any rank, such as a species, family, order, and class.

2.9.1 Description

Description refers to the comprehensive information about the features or attributes of a particular taxon. These features are known as characters, and when there are two or more variations of a character, they are called character states. For example, petal color is a character that can have character states such as blue and yellow. Similarly, leaf shape is another important character, with possible character states including elliptic, lanceolate, and ovate. In plant systematics, a variety of terms related to characters and character states are used to describe general plant morphology and specialized data types.

2.9.2 Identification

Identification refers to the practical approach of confirming a specific organism as belonging to a recognized taxon, or the method used to determine an unknown taxon based on available literature or resources. This confirmation of the unknown organism represents a significant advancement in the history of science, supported by formal naming and description. To understand an unknown organism, it is essential to first observe and describe its characteristics. Identification provides initial information necessary for describing the plant and contributes to developing a comprehensive morphological description throughout its various stages of growth. Regional floras serve as valuable tools for confirming the identity of native taxa. However, identifying cultivated plants can be challenging due to the multitude of cultivars, hybrids, landraces, and other variable forms. The following methods are commonly used in plant identification.

2.9.3 Taxonomic Keys

The taxonomic key is one of the most helpful tools in the identification of taxa. The key is an artificial or analytical tool provided to the specific taxon. Sometimes, couplet statements also follow in key development. Leads are one of the best differentiating features or characters and are known as secondary key characters. There are two types of keys: (a) dichotomous, and (b) polyclave (also called *multiple access* or *synoptic*) (Box 2.10). The diagnostic features/key characters of some of the families are given in Table 2.4.

Box 2.10 Identification Keys

Types of identification keys

1. **Punched Cards Keys**

These consist of cards having names of all the taxa (i.e., all species, genera or families for which the key is meant) printed on all of them. On one of the corners of each card is printed any one character and a definite number.

2. **Dichotomous Keys**

These keys consist of pairs of contrasting characters or couplets, each statement of which is a lead.

Both the leads are numbered, and begin with the same word as far as possible. Dichotomous keys are of two general types: (a) indented key, and (b) bracket key.

Indented keys (also called yoked): These indent the choices (leads) of the couplet an equal distance from the left margin. The two choices of the couplet are usually labelled; for example, 1 and 1′ or 1a and 1b. The choices do not need to be numbered, but it helps. The user goes to the next indented couplet following the lead that was selected.

(continued)

Example: Selected genera of Saxifragaceae are listed below:

- Shrub or woody vine
  - Woody vine; petals 7 or more .... *Decumaria*
  - Shrub; petals 4 or 5

    Leaves alternate or on short spur branches

    Leaves pinnately veined; ovary superior; fruit a capsule .... *Itea*

    Leaves palmately veined; ovary inferior; fruit a berry ....*Ribes*

    Leaves opposite

    Petals usually 4; stamens 20–40; fruit longitudinally dehiscent, not ribbed .... 4. *Philadelphus*

    Petals usually 5; stamens 8–10; fruit poricidally dehiscent, 10–15 ribbed................................ *Hydrangea*

Bracketed keys: These designate both choices side by side. The choices of the couplet must be numbered (or lettered). It is very helpful if the previous couplet is given. Note: in some bracketed keys, alternate couplets are indented; in others, all couplets begin at the left margin. The user proceeds to the couplet that is indicated by the lead selected. For example,

- 1. Shrub or woody vine 2
- 1. Herbs 6
  - 2. Woody vine; petals 7 or more .... *Decumaria*
  - 2. Shrub; petals 4 or 5 3
- 3. Leaves alternate or on short spur branches 4
- 3. Leaves opposite 5
  - 4. Leaves pinnately veined; ovary superior; fruit a capsule *Itea*
  - 4. Leaves palmately veined; ovary inferior; fruit a berry *Ribes*
- 5. Petals usually 4; stamens 20–40; fruit longitudinally dehiscent, not ribbed .... *Philadelphus*
- 5. Petals usually 5; stamens 8–10; fruit poricidally dehiscent, 10–15 ribbed .... *Hydrangea.*
  - 6. Staminodia present; petals more than 10 mm long........................... *Parnassia.*

3. **Polyclave Key**

A polyclave key bears a gist of numerous character states, whereby the user selects all of the states that match the specimen. Correct taxon (or closest match) usually conformed or narrowed down to a smaller subset of the possibilities depending on many similar matched characters. At present, all polyclave keys in use are programmed by computer algorithms. Ultimate application over the dichotomous ones permits the use of a limited subset of information to at least narrow down the possibilities.

Example: If a dichotomous key entrusts only floral characters, its usage is limited if the plant specimen is not having flowers.

Table 2.4 Diagnostic features or key characters of some important economic families

Sl. no.	Family	Key features
1	Apiaceae	Herbs; leaves alternate with sheathing, inflated leaf-stalk bases; flowers pentamerous and fragrant; umbels type of inflorescence
2	Brassicaceae	Herbs; leaves alternate, no stipules; flowers have four petals and petals in a cross
3	Caryophyllaceae	Herbs; leaves in opposite pairs, unlobed and untoothed; flowers usually have five petals and five sepals; cymes (group of flowers, terminal flower opens first); single capsule fruit
4	Lamiaceae/ Labiatae	Herbs; square stems; leaves opposite and often toothed, no stipules; tubular flowers; flowers usually have a hood and prominent lower lip
5	Asteraceae/ Compositae	Herbs; leaves without stipules; composite flowers small in dense heads; petals always joined into a corolla-tube; pappus found
6	Cucurbitaceae	Herbaceous vines; climbers with tendrils at 90° to petiole; leaves alternate with palmately compound leaflets; plants usually monecious; flowers 5-merous, synandrous condition; ovary inferior; pepo fruit
7	Euphorbiaceae	Latex often produced; leaves simple, alternate stipulate; flowers unisexual; ovary superior, syncarpous, tricarpellary; fruit dehiscent with central columella
8	Poaceae	Mainly herbs (annuals or perennials); adventitious fibrous roots, branched or stilt; stem is underground rhizome, cylindrical, prominent nodes and internodes, herbaceous or woody; leaves alternate, simple, exstipulate, sessile, tubular sheath forming by leaf base, sheath open, surrounding the internodes completely, rough or hairy, linear, parallel venation; inflorescence compound spikelets in open or contracted panicles, racemes or spikes; caryopsis fruit
9	Malvaceae	Presence of epicalyx; petals with twisted aestivation; stamens indefinite and monoadelphous; anthers reniform and monothecous; ovary two, many carpels with axile placentation
10	Leguminosae	Leaves are pinnate (trifoliate); zygomorphic flowers; papilionaceous corolla

Source: Utteridge and Bramley (2015)

2.9.4 Written Description

A major challenge in identifying unknown taxa is comparing them with written descriptions of potentially known taxa. A reliable method for confirming an unknown plant is to match it with the descriptions of known plants, which is crucial for accurate identification. However, the vast number of literature sources containing written descriptions of flora can make this process less efficient and hinder the chances of confirming a taxon. Therefore, effectively utilizing these written descriptions can create more opportunities for further identification and classification.

2.9.5 Specimen Comparison

A third method for identifying plants is to compare the specimen in question with a live or preserved plant collection, typically an identified herbarium specimen. This is an excellent identification method, as many features of a plant taxon may not be captured in written descriptions, images, drawings, or illustrations. Comparing with a herbarium specimen is particularly useful for confirming the identity of a plant in order to verify other herbarium data. In such studies, synoptic collections play a vital role because they contain one specimen of each taxon for a specific region, such as a county. It is crucial to ensure that you are looking at correctly identified herbarium specimens, as this can help to verify the written descriptions of those specimens.

2.9.6 Image Comparison

This method involves identifying or confirming an unknown plant by visually comparing it to photographs or illustrations of previously published taxa. Resources such as literature, flora books, and online images are particularly useful for this task. However, a significant limitation of this approach is that photographs and illustrations are typically available only for a limited number of taxa. Additionally, finding the correct image can be time-consuming, requiring a thorough examination of all available options. Despite these challenges, visual comparison remains the most effective way to identify a plant by assessing all the characteristics of the specimen.

2.9.7 Illustrations

Illustrations/line diagrams/photoplates are important visual tools to describe the characters of the plants. These not only allow to record an image of the taxon but also help to remember important features of the plant for a long time. An accurate scientific representation of pictorial depiction of plants and their morphological features for the Botanical scientific illustration is the accurate pictorial depiction of plants and plant traits for a scientific purpose (Rix 2012), as opposed to flower painting, which has no further purpose than to be admired. "The flower painter fails if the work lacks beauty; the botanical artist fails if the work lacks accuracy" (Stearn 1990).

2.9.7.1 Line Diagrams

Line diagrams or line illustrations are the drawings of the original botanical live specimen. In earlier days, they were more prominent and usual method applicable for drawings or pictures. Generally, lines are illustrated by the isograph technical ink pens of different sizes. The shades and dots are drawn by using different sizes (nibs) of isograph pens (Figs. 2.3 and 2.4).

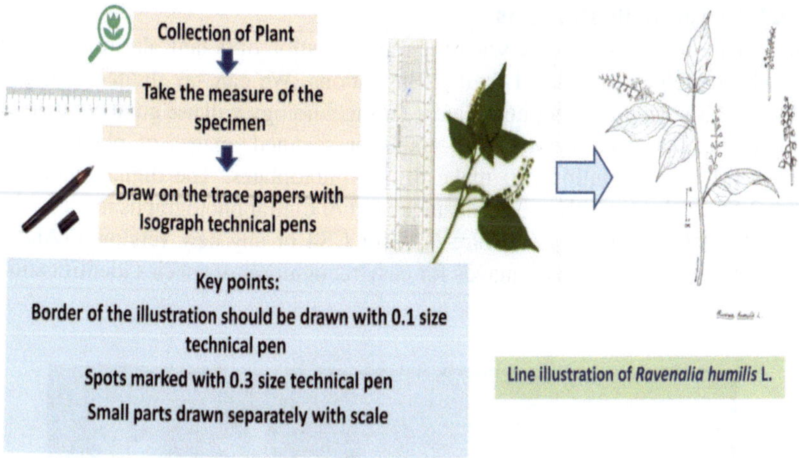

Fig. 2.3 Representation of step-wise manner for drawing line illustration

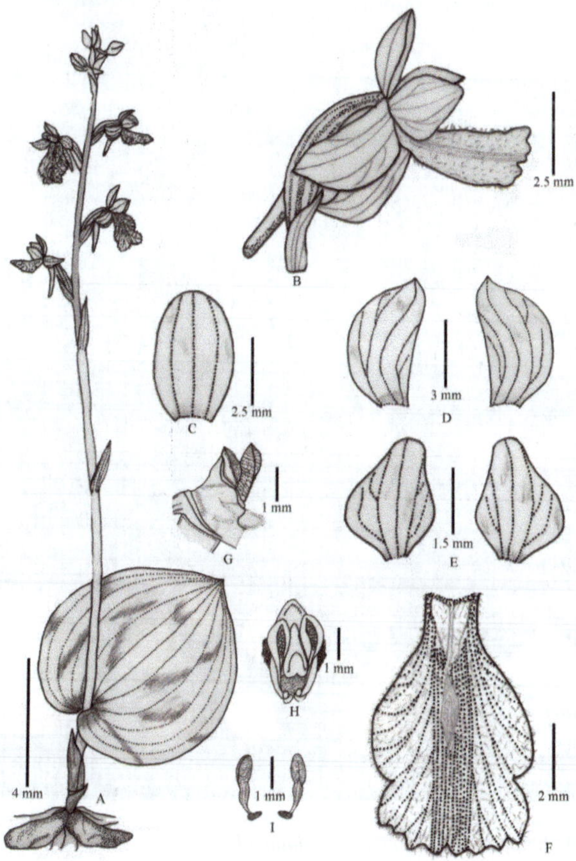

Fig. 2.4 Line diagram of *Hemipilia cordifolia* Lindl. (**a**) Habit; (**b**) Flower; (**c**) Dorsal sepal; (**d**) Lateral sepals; (**e**) Petals; (**f**) Labellum; (**g–h**) Column; (**h**) Column and (**i**) Pollinia

2.9.7.2 Digital Illustrations

The illustrations in which the scientific representation of botanicals are made using digital technology are called digital illustrations. We can say digital photoplates instead of line diagrams or photoplates. This technology will use good contrast digital photos, generally having 5–25 pixels or scanned images of plants (usually 150–1200 ppi, resolution), to make digital photoplates. The digital photoplates (Fig. 2.5) are drawn with a serial arrangement of digital photographs made with the help of Adobe Photoshop software (CS3 or CS4 or any new versions) (Andrews 2007). These are the instant images for easy recognition of species identification.

Fig. 2.5 Digital photoplate of *Solanum virginianum* L.

2.9.8 Practical Identification

Identifying plant taxa effectively relies on a person's expertise and the application of specific techniques, as detailed in most floras that describe plant families. Understanding the general characteristics of these families aids in recognizing key genera. Essential tools for identification include knowledge of the taxa within each family, along with access to keys, illustrations, and herbarium specimens. Data verification is critical for validating descriptions and confirming that a plant does not match any known taxa; this thorough evaluation may reveal the plant as a new record for a particular geographic range or potentially new to science, necessitating the publication of a new taxon.

2.10 Role of Plant Taxonomy in PGR Studies

Plant taxonomy plays a crucial role in the study of plant genetic resources (PGRs) by aiding in the morphological screening and characterization of germplasm, which is essential for identifying different species. The delineation of species relies on various taxonomic tools, with two significant contributions being the characterization of species and the determination of their evolutionary relationships (Hanelt 1988). These applied botanical studies offer valuable information to resolve taxonomic issues. Furthermore, plant taxonomy is instrumental in analyzing the relationships between cultivated plants and their closely related wild species, as well as the interrelationships among these groups. The primary functions of characterization and evaluation are vital in the systematics of both cultivated and wild crop species, ultimately enhancing our understanding of their evolutionary connections and the broader relationships between cultivated and wild species.

2.10.1 Field Notes and Digital Field Notes

Field notes on herbarium sheets play a crucial role in preserving vital information about plant specimens. The initial step in exploration work involves collecting detailed raw data, which must be recorded in a field book at the time of collection. Essential details to include are the date of collection, common names, plant growth habit, precise location, type of vegetation, nearby plants, abundance, potential uses, noteworthy observations, and the collector's name. With advancements in digital technology, tools like tablets and smartphones have revolutionized this process, enabling electronic data recording. Mobile applications now simplify the documentation of various types of data, making the process more efficient and organized, often referred to as digital field notes.

2.10.2 Characteristics Often Considered Before Plant Identification

- Herbaceous or woody, annual or perennial
- Leaf type, phyllotaxy, and venation
- Flower and flowering parts
- Separate or fused sepals and petals
- Presence of nectar-secreting disc in flower
- Actinomorphic or zygomorphic flower
- Number and attachment of stamens
- A number of pistils, styles, and stigma
- A number of carpels

2.10.3 Cytotaxonomic Studies

Cytotaxonomy plays a vital role in understanding the genetic status of various species through the study of chromosomes, while karyomorphology examines the number, size, shape, and internal structure of chromosomes in plants. This information is essential for identifying and classifying species across different plant groups, as karyomorphological studies help determine ploidy levels, indicating the number of chromosome sets in a cell. Molecular evidence further supports species classifications, which is particularly valuable in conservation efforts. Additionally, the examination of seed morphology provides critical insights for classifying certain taxa, making these approaches integral to plant taxonomy and biodiversity conservation.

2.10.4 Morphological Key Characters

In the field of PGR studies, taxonomical keys have traditionally been used for plant characterization and evaluation, but recent advancements in molecular and biochemical techniques have improved our understanding of intraspecific taxonomy (Lohwasser et al. 2010). The significance of taxonomy in cultivated plants is emphasized as a vital resource for PGR studies, although it is important to recognize that molecular and biochemical data may not always align with morphological observations (Lohwasser et al. 2010). Morphological keys effectively identify plant taxa; for instance, Pandey et al. (2019) noted that an additional morphological taxonomic key aids in distinguishing tepal characteristics in *Allium ramosum* and *Allium tuberosum*. Moreover, morphological characteristics contribute to the development of unique landraces or wild species, as observed in *Lens culinaris*, where a distinct seed morphotype was identified (Tripathi et al. 2019). The importance of morphological features is further illustrated in their role in determining the taxonomic identity of *Vigna trilobata* (L.) Verdc. and *Vigna stipulacea* (L.) Kuntze within the section Aconitifoliae (Gore et al. 2019). Overall, morphological keys play a significant role in clarifying the taxonomic relationships among closely related species.

2.10.5 Numerical Taxonomy

Numerical taxonomy quantifies values to assess the similarities and differences among taxa in taxonomic and systematic studies. It provides an alternative to traditional taxonomic methods by using statistical techniques for a more objective analysis of relationships among species.

- All characters used, both discrete and continuously variable, are assigned numerical values (examples given below).
- All data (vegetative/reproductive; observed/derived through experimental study; cytological, phytochemical, molecular, etc.) are integrated into a single table/database.
- Ambiguity in defining and scoring is avoided by using "unit" characters (characters that cannot be further divided) for taxa: "operational taxonomic unit" (OTUs), that is, the units of study.
- For making statistically significant conclusions, large number of characters are to be used.
- A 2 × 2 similarity matrix is used to determine similarities versus differences.
- The individual units/subunits defined from the study are "phenons" and the phenetic study is presented in the form of a cluster analysis.

2.10.6 Biosystematics

"Biosystematics" determines the taxonomy hierarchy by considering not only morphology but also other disciplines such as ecology, phytogeography, cytology, and molecular studies. It offers a broader perspective by utilizing modern biological tools. The methodology of biosystematics is illustrated in Fig. 2.6. A crucial step in biosystematics is selecting crop taxa, which helps in identifying taxonomic issues. These studies emphasize the importance of interdisciplinary applications and tools, including molecular, statistical, and computational methods, along with morphological data, to address taxonomic challenges effectively.

Objectives

- Species delineation
- To study the morphological relationship among the closest taxa
- Development of field diagnostic keys in all vegetative and reproductive stages

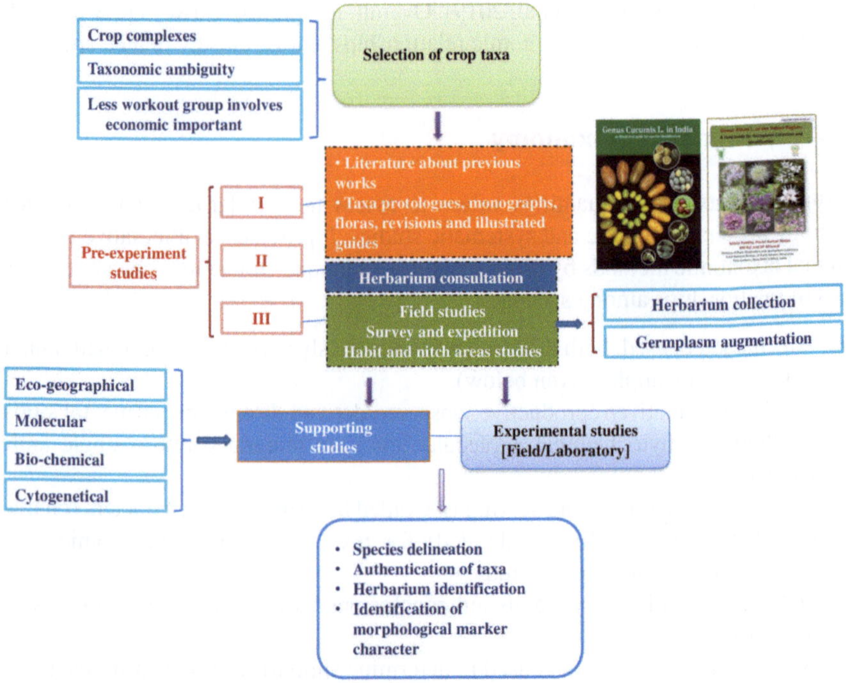

Fig. 2.6 Flowchart in biosystematics studies

2.11 Summary

Taxonomic research in India is facing a decline despite the country's rich biodiversity. To address this, it is essential to focus on neglected taxonomic groups and strengthen collaborations with research and conservation organizations. Inventorying biological components is crucial for human benefit, especially given the increasing pressures on biodiversity.

Recent publications highlight the need to enhance research capacity in taxonomy and the systematics of biodiversity. Establishing centers of excellence in systematic biology at universities can help train systematic biologists. Additionally, incorporating systematic biology courses into undergraduate programs and integrating it with other sciences will improve the understanding of species diversity and its evolutionary implications. Strengthening research and development (R&D) facilities and promoting integrated approaches can significantly advance plant taxonomy and biosystematics in India.

Multiple-Choice Questions

1. The system of classification widely accepted and followed in India is the one proposed by
   (a) Carolus Linnaeus
   (b) Adolf Engler and Karl Prantl
   (c) Lamarck
   (d) Bentham and Hooker
2. Bentham and Hooker classification is a
   (a) Phylogenetic system of classification
   (b) Artificial system of classification
   (c) Natural system of classification
   (d) Sexual system of classification
3. The specimen that is used by the author in the original publication as the nomenclatural type is called
   (a) Isotype
   (b) Holotype
   (c) Syntype
   (d) Neotype
4. Herbarium is
   (a) A preserved plant part used for identification
   (b) A collection of dried plant specimen
   (c) Twig of a plant collected for identification
   (d) All of the above
5. The Taxonomic key with each pair of contrasting choices (leads) that are equally indented and followed by the appropriate couplets, and thus often with very widely separated leads, is called
   (a) Bracketed key
   (b) Indented key
   (c) Determinate key
   (d) Indeterminate key
6. Biosystematics includes the study of
   (a) Morphology
   (b) Cytology and genetics
   (c) Ecology
   (d) All of the above
7. Carolus Linnaeus classified plant kingdom on the basis of
   (a) Floral morphology
   (b) Overall morphology of plant
   (c) Type of roots
   (d) Anatomical characters
8. National herbarium of Cultivated Plants is located at
   (a) BSI, Kolkata
   (b) FRI, Dehradun
   (c) NBPGR, Delhi
   (d) NBRI, Lucknow

9. Floral diagram is
   (a) The figure of a flower
   (b) The three-dimensional appearance of a flower
   (c) L.S. of a flower
   (d) T.S. of a floral bud
10. Taxonomic databases include
   (a) IPNI
   (b) POWO
   (c) BHL
   (d) All of the above

Answers

1. (d)
2. (c)
3. (b)
4. (d)
5. (b)
6. (d)
7. (a)
8. (c)
9. (b)
10. (d)

Suggested Reading

Andrews P (2007) Adobe Photoshop CS3 A-Z, tools and features illustrated ready reference. Elsevier, Jordanhill, Oxford, Burlington, MA, p 320

Bentham G, Hooker JD (1888) Genera plantarum, 3 vols. Various Publishers, London

Brickell CD, Alexander C, Cubey JJ, David JC, Hoffman MHA, Leslie AC, Malécot V, Jin X (2016) International code of nomenclature for cultivated plants. Scripta Hortic 18, 171 p. International Society for Horticultural Science (ISHS)

Camp WH, Gilly CL (1943) The structure and origin of species with a discussion of intraspecific variability related nomenclatural problems. Brittonia 4:323–385

Cain AJ, Harrison GA (1960) 'Phyletic weighting', Proceedings of the Zoological Society of London 135:1–31

Das S, Jain SK (1980) Orchidaceae: Genus *Coelogyne*. In: Fascicle 5 of the "Fascicles of Flora of India". Botanical Survey of India, Howrah

Gore PG, Tripathi K, Pratap A, Bhat KV, Umdale SD, Gupta V, Pandey A (2019) Delineating taxonomic identity of two closely related Vigna species of section Aconitifoliae: *V. trilobata* (L.) Verdc. and *V. stipulacea* (Lam.) Kuntz in India. Genet Resour Crop Evol 66:1155–1165

Hanelt P (1988) Taxonomy as a tool for studying plant genetic resources. Die Kulturpflanze 36:169–187. https://doi.org/10.1007/BF01986959

Hooker JD (1888) Flora of British India 5. L. Reeve & Co., London

Hooker JD (1890) Flora of British India 6. L. Reeve & Co., London

Hooker JD (1895) A century of Indian orchids. Ann R Bot Gard Calcutta 5(1), 1–68, tt. 1–101

Huxley JS (ed.) (1940) The new systematics. Oxford: Clarendon Press, viii + 583 pp

Jackson BD (1893) Index Kewensis Plantarum Phanerogamarum 1(1–2). Oxford University Press, London
Jackson BD (1894) Index Kewensis Plantarum Phanerogamarum 2(3–4). Oxford University Press, London
Jain SK, Rao RR (1977) A handbook of field and Herbarium methods. Today & Tomorrow's Printers and Publishers, New Delhi, 157 p
Karthikeyan S, Sanjappa M, Moorthy S (1989) Flowering plants of India–Dicotyledons (Acanthaceae–Avicenniaceae), vol 1. Botanical Survey of India, Calcutta
Lewis CT, Short C (1879) A Latin dictionary. LL.D. Oxford. Clarendon Press
Linnaeus C (1753) Species plantarum, vols 1 & 2. Impensis Laurentii Salvii, Holmiae
Lohwasser U, Dittbrenner A, Budahn H, Marthe F, Börner A (2010) Taxonomy of plant genetic resources—use of morphological, molecular and phytochemical data in order to verify existing classifications. Agrice Conspec Sci 75(4):175–178
Mao AA, Dash SS (2020) Flowering Plants of India: An Annotated Checklist (Monocotyledons). 3:467
Mayer E, Ashlock PK (1991) Principles of Systematic Zoology. McGraw-Hill, Inc., New York, illus 475 pp
McNeill J (2012) The type method: an introduction. http://www.tela-botanica.org. Accessed 12 Nov 2020
Nicolson DH (1991) A history of botanical nomenclature. Ann Missouri Bot Gard 78(1):33–56
Pandey A, Pradheep K, Gupta R (2015) Manual on National Herbarium of Cultivated Plants. NBPGR, New Delhi, 50 p
Pandey A, Pradheep K, Gaikwad AB, Gupta R, Malav PK, Rai M (2019) Systematics study on a morphotype of *Allium tuberosum* Rottler ex Spreng. (Alliaceae) from Ladakh, India. Indian J Plant Genet Resour 32(2):223–231
Pandey A, Pamarthi RK, Pradheep K, Gupta R, Ahlawat SP (2021) Catalogue of the type specimens in the National Herbarium of Cultivated Plants. ICAR-National Bureau of Plant Genetic Resources, New Delhi, 59 p + i–iii
POWO (2023) Plants of the World Online. Facilitated by the Royal Botanic Gardens, Kew. Published on the Internet; http://www.plantsoftheworldonline.org
Prasad VP (2014) On a 17th Century herbarium specimen and an archive letter in CAL. Nelumbo 56:255–258. https://nelumbo-bsi.org/index.php/nlmbo/article/view/86689
Rao MKV (2004) The importance of Botanical Nomenclature and synonymy in Taxonomy and Biodiversity. Curr Sci 87(5):602–606
Ravindran PN, Nirmal BK, Sivaraman K (2007) Turmeric: The genus curcuma. Boca Raton, FL: CRC
Rix M (2012) The golden age of botanical art. Andre Deutsch, Carlton Publishing Group, London
Stafleu FA, Cowan RS (1976–1988) Taxonomic literature: a selective guide to botanical publications and collections with dates, commentaries and types, 2nd edn, 7 vols. Utrecht, Antwerp, The Hague and Boston
Stearn WT (1990) Flower artists of Kew: botanical paintings by contemporary artists. Herbert Press, in association with the Royal Botanic Gardens, Kew, London
Thiers B (2021) Index Herbariorum: a global directory of public herbaria and associated staff. http://sweetgum.nybg.org/ih
Tripathi K, Gore PG, Pandey A, Bhardwaj R, Singh N, Chawla G, Kumar A (2019) Seed morphology, quality traits and imbibition behavior study of atypical lentil (*Lens culinaris* Medik.) from Rajasthan, India. Genet Resour Crop Evol 66:697–706
Turland NJ, Wiersema JH, Barrie FR, Greuter W, Hawksworth DL, Herendeen PS, Knapp S, Kusber WH, Li DZ, Marhold K, May TW, McNeill J, Monro AM, Prado J, Price MJ, Smith GF (2018) International Code of Nomenclature for algae, fungi, and plants (Shenzhen Code). Regnum Vegetabile 159. Koeltz Botanical Books, Glashütten
Utteridge T, Bramley G (2015) The Kew tropical plant families identification handbook. Royal Botanic Gardens, Kew, 240 p
Vasudeva Rao MK (2017) Type concept and its importance in Plant Nomenclature. J Econ Taxon Bot 41(5–6):91–94. Scientific Publishers

Plant Domestication, Centres of Origin and Diversity of Cultivated Plants

3

Anjula Pandey, Pavan Kumar Malav, Rinky Resma Panda, and Vaidurya Pratap Sahi

Learning objectives

- Understanding of fundamental concepts of plant domestication in the light of existing crop species.
- Distinguishing between wild relatives of crops and cultivated plants with references to various stages of domestication.
- Visualizing plant genetic resources activities in relation to crop domestication and identification of novel crops.
- Identifying sites of crop domestication and evaluating the degree of domestication syndrome traits recorded in each case.
- Comparing the centres of origin with centres of diversity of crops.

Keywords

Plant genetic resources · Germplasm · PGR · Domestication · Centres of origin · Diversity

A. Pandey (✉) · P. K. Malav
Division of Plant Exploration and Germplasm Collection, ICAR-National Bureau of Plant Genetic Resources, New Delhi, India

R. R. Panda
Division of Plant Genetic Resources, Graduate School, ICAR-Indian Agricultural Research Institute, New Delhi, India

V. P. Sahi
Department of Genetics and Plant Breeding, Sam Higginbottom University of Agriculture, Technology and Sciences, Prayagraj, Uttar Pradesh, India
e-mail: vpsahi@shiats.edu.in

K. Tripathi et al. (eds.), *Textbook of Plant Genetic Resources*,
https://doi.org/10.1007/978-981-96-5024-8_3

3.1 Introduction

Plant domestication is the process by which wild plants have evolved into crop plants through artificial selection. This usually involves an early hybridization event followed by selective breeding. The domesticated plants are phenotypically and genetically different from their wild progenitors. Domestication being an evolutionary process brings distinct changes or modifications in characteristics of plant species for which the selections were done. Understanding the domestication process in plants has been possible with deep insight into different theories proposed in the pioneering works of Vavilov, A de Candolle, Harlan and many more scientists, leading to the concepts of centre of origin and diversity. During domestication from wild species, the crop species have evolved with many adaptive traits to meet the challenges arising due to climate change and environmental conditions allowing wider expansion of species in cultivation areas and its survival for longer periods. This chapter focuses on various aspects of plant domestication to enable readers to have a better understanding of the concepts of centre of origin, role of crop wild relatives (CWRs), neo-domestication, de domestication, de novo domestication and re-domestication processes in the context of crop diversification for global food security and sustainability.

3.2 Plant Domestication

The evolution of plants and animals has been interdependent, a result of which is what we see today in the form of agriculture. The relationship between human beings and plants has gradually evolved with the beginning and spread of agriculture about 12,000 years ago. Less than 200 species are commercially important globally, where cereals (wheat, rice, maize, sorghum, millets etc.), root and tubers (potato, sweet potato and cassava), pulses/legumes (beans, pea, peanuts, soybean etc.), sugar yielding crops (sugarcane, sugar beet) and others (coconut and banana) provide food to supply energy. Three crops, rice, wheat and maize, which account for the major part of the plant-derived nutrients for human consumption, were among the earliest domesticates. Besides, there are several of the species domesticated for minor value as nuts, fruits and vegetables.

Understanding the domestication process in plants was viewed with the origin of agriculture after early humans selected and gathered plant species as a source of high-value food reserve which were identified and colonized the barren areas around their dwellings. With an increase in population, the human families would have migrated from their original inhabitation and reached areas where the wild ancestors of crops were not seen. This would have started with a tradition of developing new human habitations, field preparation for cultivation, weeding to decrease competition of unwanted plants, seed sowing in the right season, protecting crops from predators and finally collecting seeds from fresh harvest for self-consumption as well as for conservation.

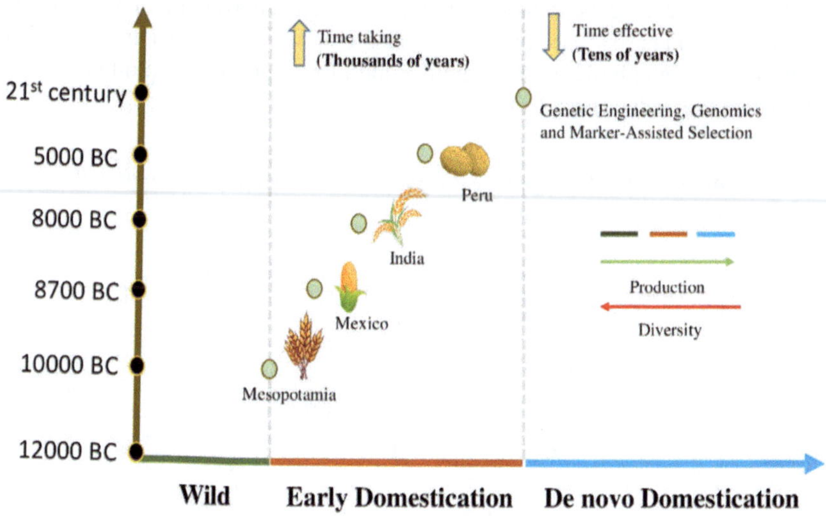

Fig. 3.1 Insights into the domestication process

Crop domestication and agricultural evolution are the backbone of modern civilization witnessed in traditional and modern cultivation sites globally. Archaeological excavations have revealed evidence of human settlements in Mesopotamia as early as 10,000 years ago. The plant domestication traces its roots to Mesopotamia, the area between the Tigris and Euphrates rivers in present-day Iraq. The development of technologies to support irrigation systems allowed the rivers to sustain agricultural activities. Many of the globally important and popular crops, namely wheat, almond, beetroot, chickpea, fig, barley, walnut and apricot have originated there.

The domestication process (Fig. 3.1) can be understood through the study of plant–animal co-evolution, dynamics of cultivation practices, operative selective pressures, genes associated with domestication, archaeological sciences and genetics. Pieces of evidence through archaeo-botanical and plant sciences, study of carbonized seeds and plant parts, radio-dating, mass spectro-photometry and modern techniques such as DNA markers have unfolded the process of domestication.

3.2.1 Concept of Plant Domestication

Plant domestication is defined as the process of hereditary reorganization of wild plants into cultivated forms, which are of interest to human beings. Artificial selection has led to the development of domesticated crops from a wild ancestor having phylogenetically distinct genotypic traits that are suited for different agro-climatic regions. It is a dynamic and ongoing process greatly influenced by ecological, evolutionary, cultural and technological factors. The process of domestication has resulted in a narrow genetic base of crop plants, hence making them prone to various diseases and pests. Since the domesticated crops have crossed a threshold, there is an increased emphasis on the identification of rare or under-exploited plants of future potential value (Box 3.1).

Box 3.1 Plant domestication, genetic diversity, and human population growth (Wilkes 1984)

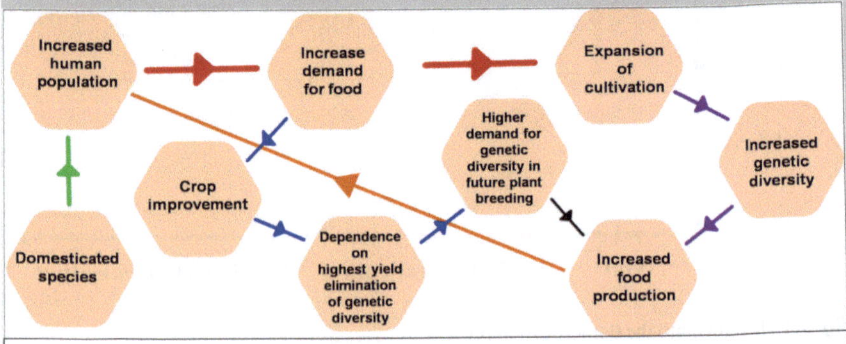

Box 3.1: The Potential Wild Species

Our fundamental knowledge of the processes of domestication has helped us to understand the underlying mechanism associated with phenotypic modifications and domestication syndrome. Such studies have highlighted our knowledge of breeding traits and other basic studies such as self-compatibility and polyploidization. Domestication-based studies are available on crops of major economic importance and well-studied crops largely from the global perspective. The studies of geographical centres of origin with an understanding of ecology, physiology and genetics are essential to fully understand the conceptual utilization in crop improvement. Wild species of little importance today may attain great use tomorrow. Worldwide, about 100,000 plant species have been used by humans for multiple purposes; cultivated plants amount to 35,000 species of which only about 100 species of higher plants have contributed as valuable domesticates. Total 2489 species of 173 families show maximum representation of domesticated plant species (Fig. 3.2).

3.2.2 Selection Process in Plant Domestication

During the process of domestication, the artificial selection (human-caused) process works in reverse direction to the natural selection. In contrast, in the unconscious (unintentional) selection process, human-made interference results from shifting plants from the natural to the human domain which causes selection pressures

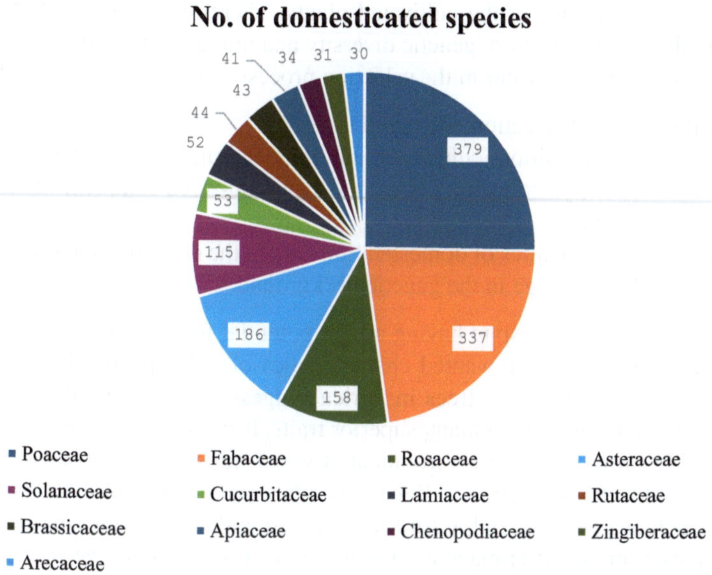

Fig. 3.2 Major plant families with largest number of domesticated species (Zeven and de Wet 1982)

resulting from plant care during growing and harvesting. This entire process results in the adaptation of the phenotypes to the human-made environment. In the case of conscious (intentional) selection, phenotypes are identified, and selected, while undesirable ones are rejected or less preferred which automatically decreases in the population. Selected traits preferred by the consumers such as ease of harvest, good palatability (lack of bitter taste), good nutritional qualities and absence of defensive organs lead to reduced success in natural environments. Therefore, the domesticated crops generally find difficulty in surviving under natural conditions.

3.2.3 Plant Domestication, Genetic Diversity and Human Population Growth

The domestication process is an ongoing process which continues in future to meet human demands. The response of wild species to the domestication process is primarily dependent on the species' behaviour in the selection process. Domestication is likely to be enhanced if the species is adaptable with higher market demand than its production in natural population, and higher profitability from cultivation. In the global scenario, the rising population has resulted in more reliance on the high-yield characteristics and thus of the cultivated plants. Therefore, a close linkage exists between plant domestication, genetic diversity and human population growth.

The following steps occur in the selection process of domesticated species:

1. Identification and selection of desired types
2. Change in soil condition (cultivation-environment) and cultivation methods
3. Differentiation–hybridization cycles between crop-weed complex, selections of wanted genotype.
4. Therefore, the dynamics of domestication lead to the desired phenotypic modifications without change in the genetic background.

Many plant species despite having a high potential for domestication are undomesticated due to many unwanted characteristics of wild species. Findings from domestication of indigenous trees in Africa suggested that identification of trees combining a few, rather than many superior traits, is preferable to create new cultivars suitable for a particular environment, a set of traits and a single product. For example, for the domestication of *Choerospondias axillaris* (lapsi), the combination of traits such as sweet and pulpy fruit recombined with larger fruit types may fetch a high market in Nepal Himalaya. During domestication of *Brassica*, progenitor, *Brassica oleracea* genotypes were selected for alternative purposes: cabbage (*Brassica oleracea* var. *capitata* L.) and kale (*Brassica oleracea* var. *alboglabra* (L.H.Bailey) Musil for leaves; kohlrabi (*Brassica oleracea* var. *gongylodes* L.) for stem; broccoli (*Brassica oleracea* var. *italica* Plenk) and cauliflower (*Brassica oleracea* var. *botrytis* L.) for shoots and brussels sprouts (*Brassica oleracea* var. *gemmifera* Zenker) for buds.

A domesticated crop species may be of single or multiple origins. It may have a single origin, if the crop is domesticated once from its wild ancestor(s), which is further dispersed. The crop is of multiple origin if domestication has occurred independently from the same ancestor(s), at different geographical regions and at different times. Crop can be of doubtful or undefined status of origin when there is insufficient data to determine the type of origin (single or multiple) (Box 3.2).

Box 3.2 Species Response to Selection Pressure

Oak tree species, the most important wild food plant in many parts of Eurasia and North America, were never domesticated despite its wide use for edible purpose; while wild almonds, with rarely non-poisonous mutant types, were selected for use by humans. In oak, it is a **polygenic trait,** whereas in almond, this trait is controlled by a **single gene**. In almond, the offsprings are occasionally non-poisonous, whereas in oak, they are inedible thus preventing the selection of varieties for human consumption till date.

3.3 Changes Under Domestication

The domestication process leads to changes or modifications in plant species with distinct morphological and genetic characteristics that make them distinct from their wild ancestors (Box 3.3), for example, seed shattering at maturity in wild wheat, rice and barley. A genetic mutation responsible for preventing shattering in wild type is generally lethal. Therefore, a non-shattering type was selected as a result of unconscious selection during the preliminary stage. The domesticated types can be recognized based on modification in morphological, physiological and phenological traits and their distribution beyond the wild progenitors. Some patterns of crop domestication of food crops have been highlighted in this section.

Box 3.3 Changes Under Domestication: Some Trends

1. Reduction in genetic diversity due to artificial selection and population bottlenecks
2. Appearance of frequent mutations related to domestication syndrome in populations that pre-existed at low frequency
3. Accumulation of deleterious mutations
4. Gene flow between wild and cultivated populations
5. Phenotypic convergence

3.3.1 Domestication Syndrome

Domesticated species result in the development of more productive and adaptive crops in contrast to their wild ancestors in the centre of origin, which generally have low genetic variability and tolerance to abiotic stresses. The domestication syndrome is defined as a set of trait(s) that develop during plant domestication and distinguish a crop from its wild progenitor(s). Domestication traits are selected over an agricultural environment, after several generations and get fixed within the crop genome to develop domestication syndrome. They may be combinations of several traits, like retention of seed (non-shattering), gigantism in fruit or seed or other parts, change in plant habit/stature, reproductive strategy and changes in secondary metabolites (Box 3.4).

Box 3.4 Some of the Apparently Distinct Changes or Modifications in Crop Divergence (Crop/Crop Groups in Parenthesis)

(a) Changes in habitat—perennial to annual habit.
(b) Change in plant form—decrease in plant height (cereals, millets); increase in height/width of culm—*Corchorus* species (jute), sugarcane, forages; tillering, branching, vegetative/reproductive parts of economic importance (cereals, leafy vegetables); change or increase in seed/useful part as size of grain or fruit, leaves; reduced shattering of seeds, spikes, pods etc.
(c) Decrease in undesirable components, improved palatability—alkaloids, toxins, bitters (solanum, cucurbits), acridity (yams and taros)
(d) Loss of seed dormancy—Cereals, pulses etc.
(e) Uniformity in economic parts—leaf, fruit, pod and seed
(f) Increase in economic part—all crops
(g) Change in mode of reproduction—breeding system (compatibility), seedless/parthenocarpic fruits; asexual reproduction—sugarcane, potato and sweet potato
(h) Change in ploidy level—higher polyploidy preferred (potato, wheat, tobacco etc.)
(i) Synchronization in maturity and uniform seed germination—seeded crops.
(j) Modifications in quality of economic product—thinner skin or peel, low seed number, pulpier, sweeter and non-fibrous fruit, loss of acridity (roots/tubers), protein/oil rich, non-shattering of fruits, pods.
(k) Increased susceptibility for diseases and pests—loss of protective coverings, sturdiness, adaptative features (epidermal wax and coatings, hairs, spines, thorns etc.).
(l) Greater adaptability—to wider environments and geographical conditions.

The domestication process may occur at a slower rate in vegetatively propagated root/field crops, perennial or tree species having fewer sexual generations and fewer domestication syndromes in comparison to the sexually reproducing crops: annual seed crops (cereals and pulse crops/legumes). The process of polyploidization facilitates domesticated crops to sustain under ever-changing (disturbed) habitat conditions unsuitable for the sustenance of the wild ancestor. This process leads to the development of fixed traits, thereby limiting the gene flow between wild and cultivated types. The change in ploidy levels leads to modification in self-fertilizing or out-crossing plant behaviour to vegetative propagation. A number of traits associated with polyploidization in domesticated crops are gigantism (larger plant parts, fruit, seed size), increased resistance to disease and pests and decreased rate of multiplication. The fruit and seed crops generally show domestication syndrome by shifting from out-crossing to self-fertilization traits. This allows the crop to produce fruit with few or no seeds. Generally, a number of domestication syndrome traits in a crop depends largely on the stage of domestication and selection pressure; the earlier the stage of domestication, the lesser the syndrome trait(s). For example, a semi-domesticated type in a crop may have a smaller number of domestication syndrome traits than its fully domesticated counterpart.

3.3.2 Stages of Domestication

Domestication in general is viewed as the end-point of an ongoing process that begins with the identification and selection of wild potential plant diversity and continues as protected or semi-domesticated types which are not yet genetically distinct from the wild plants. This can also be called as initial stage of cultivation, which after modifications results in fully domesticated plant species. The whole process operates through the following stages exhibiting various degrees of management (Box 3.5).

Box 3.5 Stages of Plant Domestication

1. Un-domesticated populations thriving with no care or without human interference
2. Semi-domesticated populations showing protection of plants with least human interference and care
3. Domesticated populations thriving under full protection with highest degree of human care and management

Undomesticated populations are the uncultivated plants that are routinely wild-harvested with no conscious artificial selection pressures operative and no evident morphological and/or genetic differentiations to distinguish them from their domesticated counterpart (e.g. *Ziziphus nummularia*). Wild species of jujube, *Z. nummularia*, a small bush, inhabitates in arid and dry areas of India, Pakistan and Iran and its fruits are used as food and traditional folk medicine. The semi-domesticated population grows under human protection but has not diversified for morphological or genetic traits as available in their wild counterparts. The selected population is subjected to regular monitoring, and maintenance in open areas around the human domain (protected populations). The fruits are regularly collected from the plants; the plants are protected from animals/other vagaries of nature. Sometimes, the fruits are identified with desired traits- palatability, taste, size, yield and other traits for future use or planting. The domesticated plant populations are subjected to agricultural management practices such as sowing seeds in well-prepared soil at the right time and carefully protecting them at all stages, harvesting with great care and conserving them for use and future cultivation (as seed). Any species may be seen as wild-harvested, or protected/semi-domesticated populations (*Buchanania lanzan*) and semi-domesticated as well as cultivated type (*Allium tuberosum*) in parts of India (for better understanding of the concept refer to Fig. 3.3).

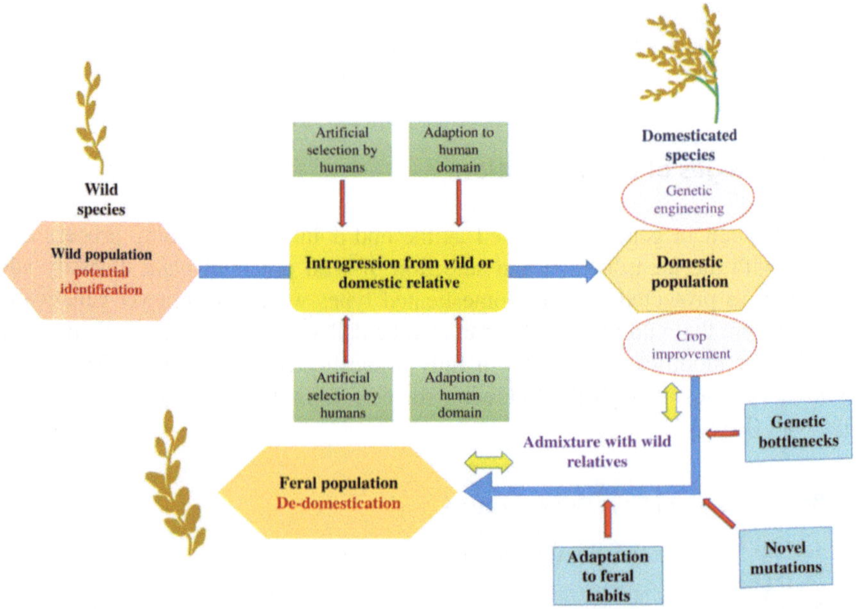

Fig. 3.3 Schematic representation of various stages of plant domestication

The domestication process is hastened when the selected population freely crosses among themselves, which facilitates the maintenance of and survival of the selected genotype with a higher frequency of selected alleles. Through the isolation of potential wild populations, there are no chances of crossing between the selected population and their wild counterparts, thus leading to the generation of domesticated species. Wild plant populations identified for economic use, still growing in their natural habitats or not isolated from their natural habitat remain at the first stage of the domestication process, that is, as the undomesticated (wild, potentially useful type). For example, selected plants grown along margins or on clear lands are protected or semi-domesticated populations. age crops and many medicinal, avenue trees and ornamental plants. The plant species is considered to be fully domesticated when there is the highest level of genetic modification in plants and human care required for its survival.

The crop evolution is significantly influenced by changes in cultivation practices that play a role in providing different environments to the wild plant. Greater potential variability is generated as a result of altered field conditions, operative selection pressure, generation of weedy races (they retain some traits of the wild plants and compete with the cultivated populations) and occasional hybridization takes place between them. The limited exchange of genes between the two populations results in variability, though gene flow barriers maintain a clear identity of the two. The generated variability is subjected to selection practices under human management thus providing a source for novel traits. Deliberate selection results in different cultivars chosen for specific use, for specific ecological niches of the agri-system and for different characters. Some of the examples of crops are, glutinous and non-glutinous types, long and short grained types, aromatic types in rice; types with sweet tender cob, popping, boiling, flour quality in maize etc.

3.3.3 Domestication and Speciation

Speciation and domestication have great analogies as processes of natural and artificial selection. In the speciation process, a species develops into another one, if the latter is reproductively isolated from the former species, while in the domestication process, a transformation occurs of a former species into an anthropogenic-driven (man-selected) entity (generally called form). Domesticated plants are usually phenotypically distinctive, and therefore often considered to be distinct species from their wild counterparts. In literature, the definition of a species is a subject of much debate. Historical views were presented by Darwin and Wallace (1858), and subsequently by the later workers, where there were no underlying agreements. Despite some physical differences observed between wild relatives and domesticated species at a basic level, these are crossable to produce fertile F_1 offspring. Domesticated crops and their wild relatives are to be classified as subspecies of one another if a reproductive isolation model is followed to define a species and the apparent ecological and phenotypic differences between these two taxa. Wild sub-species ancestors are defined based on the absence of domesticated traits.

The beginning and end-point of the process of domestication are very difficult to define and so is the path between the wild and domesticated species. In the absence of archaeological records, clear identification is possible generally at the genus level rather than species, and subspecies. For example, domesticated soybean (*Glycine max*) and its wild relatives (*G. soja*) are instead named as *G. max* subsp. *max* and *G. max* subsp. *soja*, respectively. Species-based exclusively on reproductive isolation fail to consider ecology or phylogeny or seed dispersal mechanisms (for example) as evidence, but using these alternative systems as bases for a definition also fails to take other aspects such as phenotypic expression and morphology. The study of cytogenetic and molecular affinities recognized polyploidy, leading to speciation and genetic incompatibility, as one of the main tools in the domestication of wheat. Different data sources pose strong debates across archaeology regarding the timing of genetic manipulation and morphological changes indicating the time when a species becomes domesticated. It is, therefore, suggested to avoid the taxonomic debate on the status of species and wild/domesticated types.

3.3.4 Neo-domestication and Re-domestication

Domestication is and will be playing a role in neo-domestication with conventional crossing with domesticated crops and wild species, including crop wild relatives (CWRs). Neo-domestication of plant species is the process of developing, potentially important, wild plant species and crop wild relatives into a crop. The wild plant species are at the pioneer stage of domestication where they have been identified for their economic importance. In other words, the potential domesticated plant species are undomesticated and routinely collected for use wild-harvested without involving any kind of selection. The process of neo-domestication (also called contemporary domestication) has resulted in the development of a novel tetraploid from crosses between the oilseed crop, *Helianthus annuus,* an annual diploid and a hexaploid tuber crop, *H. tuberosus* which is a perennial type. De-novo domestication is a type of non-conventional neo-domestication, in which domesticated genes have been incorporated into the non-domesticated species resulting in a new crop species. Genome editing or introgression techniques have proved to be a quicker method of de-novo domestication to accelerate the development of desirable traits in novel crop species. However, re-domestication is the process of selecting a domesticated crop initially selected for some specific use but later repurposed (other use identified) for another objective.

Neo-domesticated macadamia nut tree, *Macadamia ternifolia,* was identified for its potential use only 100 years ago during an expedition to Queensland rain forests of Australia. Kiwi fruit, *Actinidia chinensis*, is a recent domesticated fruit tree, still collected in large quantities from natural habitats in China. Recently, a species used as spice and condiment jamboo (*Allium negianum*) was identified as a neo-domesticate from the western Himalaya, India. Some recently domesticated crops like kiwi, cranberry have evolved within a short time frame, probably in the past century. Species of wild *Vigna*, *V. stipulacea* and common vetch (*Vicia sativa*) are potential species identified for drought tolerance or salinity tolerance, making them the best candidate for neo-domestication. With the advancement of molecular tools like gene editing, new crops can be rapidly generated through the process of neo-domestication.

The process of discovering useful plants in diversity-rich areas by identifying neo-domesticates needs continuous surveys followed by research and development. The selection of any species, based on its domestication potential, can be assessed to ease of response to domestication. Thus, world gene centres offer great biodiversity with an enormous number of useful plants for domestication. With the availability of the existing tools, it is possible to domesticate a wild species or to genetically improve the known/established crops through a selection of traits for climatic parameters, and increase food security to bring neo-domestication for climate-ready crops.

As an example of re-domestication, during maize domestication, it was selected as a cereal crop before it was valued as an important feed and crop of industrial values emphasizing seed production. With the growing need for the production of vegetative matter (vegetative plant parts-stems, leaves), varieties with reduced seed harvest index need to be selected. Common radish, *Raphanus sativus,* was initially domesticated as a root tuber crop. Cultivar serpent radish, snake-like radish, and rat-tailed radish (*Raphanus sativus* var. *mougri* Helm. and *R. caudatus* L.) which have small fruits (pods) and glabrous leaves grown in Java to Northwestern India were selected for longer pods used as a vegetable.

3.4 Crop Wild Relatives

Understanding the process of crop domestication, origin and its relationships with wild taxa is important for the utilization of wild relatives in crop improvement (Box 3.6). Evidences through germplasm characterization and breeding programmes in the past decades have confirmed that in general, the cultivated plants have a lower tolerance to biotic and abiotic stresses as compared to their crop wild relatives (CWRs). Unlike the domesticated species, CWRs keep on evolving with changing climate conditions. It is generally believed that artificial selection during the domestication process has greatly reduced genetic variation in present-day crops, and also the loss of tolerance genes derived from CWRs. For plant breeding, crop wild relatives (CWR) are crucial sources of adaptive variety. The genes for stress tolerance have been considered to be negatively correlated with yield, and therefore breeding for multiple plant defence traits can be achieved without affecting the yield trait. With current climate change projections, more crop productivity losses are suspected to occur due to environmental stresses-extremely hot and cold weather, rainfall and drought and salinity. To meet these challenges, wild relatives have been utilized for traits, such as disease and insect resistance, heat, cold and salinity tolerance introgressed into crops. The resistant traits of CWR could be used to improve susceptible modern crop varieties through traditional breeding, transgenesis or other modern technologies. Transgenesis though is an expensive and complicated tool, it could be utilized in case of existing barriers to introgressing desirable loci from the CWRs to crops. Presently, with available modern tools such as genomics, modelling and geospatial analysis along with improved phenotyping methods, CWR can prove to serve as potentially valuable sources of stress tolerance for crop improvement.

Box 3.6 Wild Relatives of Crop Plants (WRCPs)

- **Cultivated plants**: Derived from the wild relatives over the process of selection followed by hybridization for fetching desired improvement.
- **Wild related species**: Contributed towards improvement in crop plants, for example, paddy, sugarcane, wheat, potato, tomato and brinjal.
- **Wild relatives**: Economically important for edible, medicinal and other uses: *Allium, Citrus, Cicer, Coix, Dioscorea, Crotalaria, Prunus, Rubus* etc.

3.4.1 Wild Relatives: Source of Desirable Traits

The crop plants and their wild progenitors may spontaneously hybridize with each other when grown side by side and may result in infiltration of wild genes to the cultivated germplasm and vice-versa. This process is called introgression or introgressive hybridization. The accomplishment of inter-specific crosses between crops and their related types has introduced new opportunities for utilizing wild relatives in various crop improvement programmes. The importance of wild relatives became more evident with the discovery of *Zea diploperennis*, a new teosinte from Mexico and *Oryza nivara* from India. Additionally, crosses between *Allium cepa* and *A. sativum* with *A. roylei*, a species of Indian origin, have opened new avenues for utilizing Indian species. Similarly, the muskmelon crop, threatened by a global downy mildew outbreak, was rescued by a gene for resistance to downy mildew derived from a wild species of Indian muskmelon (*Cucumis*).

Wild relatives, semi-domesticated crops and distantly related taxa have significantly contributed to the improvement in major crop species during the process of domestication. For example, the identification and utilization of a single important gene have led to character advancements in crops, such as the high sugar content in modern tomatoes derived from a wild and weedy tomato accession from the Peruvian Andes.

Some of the notable examples of wild relatives, with potential traits, from India include:

Oryza nivara: A wild relative of cultivated *Oryza* (paddy) with resistance to grassy stunt virus, found in Basti and Gonda, Uttar Pradesh, India.

O. coarctata: A weedy relative with hardiness traits for saline/marshy habitats from the Sunderban delta region in West Bengal.

Vigna mungo var sylvestris **and** ***V. radiata var sublobata:*** Wild relatives of cultivated *V. mungo* (urd) and *V. radiata* (mung), respectively, from the Western Ghats and adjacent areas, showing tolerance to yellow mosaic virus.

Sesamum laciniatum: From coastal Andhra and Tamil Nadu, a wild relative of cultivated *Sesamum indicum* (sesame) with resistance to leaf phyllody disease.

Linum perenne**:** A wild relative of cultivated *Linum usitatissimum* (linseed) with cold hardy traits from the Lahaul and Spiti region and other parts of the Western Himalayas.

Abelmoschus tuberculatus **and** ***A. manihot*****:** Wild relatives of *Abelmoschus esculentus* (lady's finger), tolerant/resistant to yellow vein mosaic virus and fruit borer, found in northern India.

The deliberate introduction of genes from wild progenitors is now a common practice in major plant improvement programmes. Genetic affinities between wild relatives and many crop plants have enabled breeders to utilize these relatives as potential sources of genetic variation. Although successful introgression of useful genes from distantly related species to wild or cultivated species can be challenging, it depends on factors such as cross-compatibility, hybrid seed production, normal development of F_1 hybrids, partial seed production (through natural or back-crossing methods) and the absence of hybrid breakdown in segregating generations. The traits of economic value that have been successfully transferred mainly include those for biotic and abiotic stress resistance from distantly related species to cultivated types.

Role in adaptive evolution Artificial hybridization in recent times between crop plants and their wild progenitors, as well as wild relatives, has led to variation in many crops. The wild relatives of crop plants have been the donors of many useful traits such as resistance/tolerance to diseases, insect pests and other stress, some of which include wild lady's finger (*Abelmoschus tuberculatus*) for yellow vein mosaic virus and wild urd (*Vigna mungo* var. *sylvestris*) for resistance to yellow mosaic virus.

Wild-related species constitute a part of the gene pool of that crop. The crop wild relatives possess a big reservoir of untapped genes that have the potential to be utilized in the improvement of crops. Based on their classification, the species have been grouped into primary, secondary and tertiary gene pools. The wild species in primary gene pool produce fertile hybrids with cultivated types. The wild progenitors of crops and wild and weedy forms were called the primary gene pool of crops. The species in the secondary gene pool can only result in some degree of fertility, in these, gene transfer being difficult. This includes the distantly but cross-compatible wild-related species, leading to partly fertile cross-progenies. In tertiary gene pool, the crossing is rather difficult. In terms of utilization of wild relatives, primary and secondary gene pools are easy to exploit because they are wild progenitors/closer/relatively closer to their respective crop groups. The tertiary gene pool is difficult to exploit and represent distantly related species or unrelated taxa of different genera/species. Though this gene pool has little direct value, modern tools can be used to trap valuable genes and such genes could be applied for crop improvement using gene transfer methodologies. For example, wild apricot (*Prunus armeniaca*, locally called chulli) growing in the high altitudes of the Western Himalayas is a source of genes for cold hardiness to the cultivated varieties of apricot (*Prunus armeniaca*). Thus, wild apricot constitutes the primary gene pool for cultivated apricot. Other distantly related species, *P. cerasus* and *P. nepaulensis* belong to the tertiary gene pool. Similarly, for cultivated species *Sesamum indicum* (til), *S. mulayanum* and *S. malabaricum* constitute the primary gene pool and *S. laciniatum* and *S. prostratum*, the tertiary gene pool. The wild relatives of crop plants with some important traits are given in Box 3.7.

Box 3.7 Important Traits of Crop Wild Relatives

Crop(s)	Wild relative (s)	Important trait (s)
Allium cepa, A. sativum	*Allium roylei, A. vavilovii*	Resistance to powdery mildew and leaf blight
Abelmoschus esculentus	*Abelmoschus manihot, A. tuberculatus*	Resistance to yellow vein mosaic virus and fruit borers
Arachis hypogaea	*Arachis chacoense, A. correntina, A. villosa, A. cardenastii*	Immunity to rust diseases, insect and mites, resistance to leaf spot disease, peanut bud necrosis and peanut clump virus
Avena sativa	*Avena sterilis, A. strigosa, A. barbata*	Resistance to yellow dwarf virus, rust and mildew, crown rust disease
Cajanus cajan	*Cajanus scarabaeoides, C. sericeus, C. acutifolium* and *Rhynchosia bracteata*	Resistance to pod fly damage, general resistance to diseases
Cicer arietinum	*Cicer reticulatum, C. echinospermum, C. bijugum*	Resistance to disease and pest, cold and drought, Ascochyta blight, fusarium wilt
Helianthus annuus	*Helianthus debilis, H. argophyllus*	Powdery mildew and drought resistance

(continued)

Box 3.7 (continued)

Crop(s)	Wild relative (s)	Important trait (s)
Hordeum vulgare	*Hordeum bulbosum*	Resistance to various diseases
Lycopersicon esculentum	*Lycopersicon pimpinellifolium, L. hirsutum* f. *glabratum; L. Peruvianum*	Resistance to verticillium, fusarium and bacterial wilt, bacterial canker, grey leaf spot, leaf mould, septoria leaf spot, curly top virus, mosaic virus and spotted wilt virus, root-knot nematode, tobacco mosaic virus (TMV)
Nicotiana tabacum	*Nicotiana sylvestris, N. knightiana, N. debneyi, N. repanda*	Cytoplasmic streptomycin resistance, blue mould resistance
Oryza sativa	*Oryza nivara, O. officinalis, O. rhizomatis, O. granulata*	Resistance to insects, grassy stunt virus and brown plant hopper, green leaf hopper
Pennisetum americanum	*Pennisetum squamulatum*	Resistance to floral disease
Sesamum indicum	*Sesamum laciniatum, S. prostratum*	Resistance to phyllody and powdery mildew
Solanum melongena	Wild relatives of brinjal, *Solanum incanum*	Resistance to diseases and borer
Solanum tuberosum	*Solanum tarijense, S. berthaultii, S. demissum,* wild potatoes	Resistance to aphids, mites, thrips, cyst nematode, mosaic virus and late-blight diseases
Zea mays	*Zea diploperennis*	Immunity to major diseases of maize
Camellia sinensis	*Camellia* spp.	High flavour content
Gossypium spp. (upland cotton)	*Gossypium raimondi, G. thurberi, G. sturtianum*	Fibre strength, glanded plants and glandless seeds
Lycopersicon esculentum	Wild tomato, *L. cheesmanii, L. peruvianum, L. pennellii*	High sugar contents, salt tolerance
Triticum aestivum	*Triticum turgidum, T. dicoccoides* (wild tetraploid wheat)	High protein content
Vicia faba	*Vicia narbonensis*	Early pod and seedling growth
Abelmoschus esculentus	*Abelmoschus tuberculatus*	Heavy bearing
Avena sativa	*Avena sterilis*	Grain yield
Cajanus cajan	*Cajanus cajanifolia*	Higher pod and seed yield
Pennisetum americanum	Wild relatives of *Pennisetum*	Increased grain yield and growth rate
Triticum aestivum	*Triticum turgidum, T. dicoccoides* (wild tetraploid wheat)	High kernel weight
Vigna radiata and *V. mungo*	*Vigna sublobata* var. *radiata*	Higher yield

(continued)

Box 3.7 (continued)

Crop(s)	Wild relative (s)	Important trait (s)
Cereals	Wild and weedy relatives of cereals	Yield character
Beta vulgaris	*Beta corolliflora*	Incorporation of apomixes
Brassica juncea; B. campestris	*Brassica oxyrrhina, Diplotaxis siifolia, D. catholica, Sinapis alba*	Cytoplasmic male sterile lines
Helianthus annuus	*Helianthus petiolaris*	Cytoplasmic male sterile based traits
Secale cerale	*Secale vavilovii*	Cytoplasmic male sterile and restorer system, cleistogamous and self-fertile
Zea mays	*Tripsacum dactyloides*	Development of apomixes by distant hybridization
Glycine max	*Glycine soja*	Adaptability to colder regions and shorter-season
Lycopersicon esculentum	Wild tomato, *Lycopersicon hirsutum*	Salt tolerance, higher carotenoid contents
Triticum spp. (cultivated)	*Agropyron* spp.	Winter hardiness

3.4.2 De-domestication

De-domestication or feralization in a way is a reversal process of domestication (Box 3.8). It is an evolutionary process in which the domesticated crop species revert and reacquire wild-like traits (feral crops) after escaping from the human domain and develop an ability to perpetuate outside cultivation, for example, weedy rice, Tibetan semi-wild wheat, weedy/feral populations of olive (*Olea europaea*) and perilla [*Perilla frutescens* (L.) Britton var. *frutescens*]. This process of a reversal of crop origin originates from endo feral source as a result of spontaneous mutations in genes responsible for the traits; selections made of trait(s) with mutations in crop population; and through the process of hybridization between crops and their wild relatives, resulting in de-domesticated weed. A cross between wild and cultivated sunflower (*Helianthus annuus* L.) may facilitate the evolution of populations having weedy traits (competitive ability, seed dormancy, herbicide resistance etc.).

Box 3.8 Feral Crops Serve as

- Vectors for unintended transgene spread into cultivated and wild relatives
- Source of some significant desirable traits as genetic resource for crop improvement
- Tool for understanding general evolutionary processes, particularly role of reversal genes

De-domestication process has resulted in the reversal of seed shattering trait that was lost during rice crop domestication resulting in weedy rice from domesticated rice. Feral rice populations have been used to identify as a source of stress tolerance represented among genetic variation. Despite their high potential value in crop improvement as a donor source, the feral crops are poorly represented in global germplasm repositories since the material is collected and identified as a wild species and not as a feral form.

3.5 Centres of Origin and Diversity in Cultivated Plants

A centre of origin is a geographical area where a group of organisms, either domesticated or wild, first developed their distinctive properties. Centres of origin were first identified in 1924 by Nikolai Vavilov. The study on the geographical distribution of centres of plant domestication goes hand in hand with the study on the origin of agriculture. The archaeologists discovered the early farming sites (called the cradle of agriculture) in Thailand, the Near East and Mexico. Later, agriculture could have spread to other regions resulting in crop diversification. Largely, the regions of ancient civilizations throughout the world are those where the variation in crops has built up. However, with new archaeological studies of civilizations and the genomics tolls, we have deepened our studies on centres of origin.

Charles Darwin (1868) was known as the architect of the theory of evolution by natural selection and he dealt with diversity in all living organisms known to science; while Alphonse de Candolle and Vavilov did the pioneering work on domesticated plants. The concept of the study of origin and history of cultivated plants perhaps for the first time was proposed by A de Candolle (1890) who suggested the probable regions of initial plant domestication. Nevertheless, our understanding of cultivated plant diversity had hardly begun until the time of the world-famous Russian geneticist and plant breeder, NI Vaviov (Box 3.9). He was called "Darwin

Box 3.9 Vavilov's Detailed Steps for Studying the Origin of Cultivated Plants

(a) Differentiate the plant into specific and intraspecific taxa by morphological and genetical criteria
(b) Establish distribution of genetic diversity for endemic forms located in geographical centres
(c) Determine the centres of accumulation of diversity of genetically allied species
(d) Check presence of wild relatives in locating a centre of origin and areas of concentration of the nearest wild relatives
(e) Locate presence of specialized parasites.
(f) Support from evidences using archaeological, historical and linguistic data on existing knowledge.

Box 3.10 Contributions of Vavilov to Plant Breeding

Pioneered the genetic resources explorations: He pioneered the concept of genetic resources exploration to serve as a model for genetic diversity of crops plants and related wild species.

Germplasm evaluation and breeding: Collected germplasm was evaluated at the Institute of Plant Industry in Leningrad in diverse environmental conditions in parts of the Russia (erstwhile USSR).

Established the concept of a Linnean species: Species of cultivated plants during their spread from actual areas of origin had become differentiated into geographical and ecological types and varieties.

Developed the concept of centre of origin

Distinguished primary and secondary crops: He was the first to establish this hypothesis of primary and secondary crops (though credit should go to Engelbrecht, a plant geographer, who pointed out that most crops in earlier times had attendant weeds mimicking very closely such features as seed size and maturity time).

Established law of homologous series: If a character was known for an area in the crop but not in another, a similar variant might be predicted for those second crops. Examples include, Ligule less rye and wheat, naked grained forms of barley, oats and millet. Well-marked differences between Mediterranean races of wheat, barley, flax with large fruit, seed and flower, southern Asiatic races of peas, lentils, broad beans, chickpea and castor oil plants possessing smaller fruits, seeds and flowers.

Importance of weedy races of crops: Wild and weedy forms are taxonomically close to the cultigens and form one ecological group with them and therefore prove to be valuable resource for genetic variation for breeders.

Nikolai Ivanovich Vavilov introduced the 'Law of Homologous Series in Variation' in 1922, which revealed that similar morphological traits recur across different species, genera and families due to common genetic and developmental pathways. Vavilov's observations suggested that variation in nature follows specific, predictable patterns, rather than being entirely random.

He meticulously cataloged variations in numerous plant species, noting consistent traits like seed shapes and leaf textures within related groups. This pattern extended to fungi and animals, indicating a universal application of his law across the biological kingdom. Vavilov's law challenged the neo-Darwinian view that variation is purely random, proposing instead that intrinsic biological constraints limit the range of possible forms organisms can take, introducing predictability into evolution.

This idea implies that evolutionary pathways are guided by intrinsic rules, resulting in similar traits emerging independently in different species under similar environmental pressures. This predictability has significant implications for evolutionary biology and practical applications in agriculture. Understanding genetic variation patterns can enhance breeding programs, allowing breeders to predict and select desirable traits more efficiently, such as focusing on recurring disease-resistant traits to develop resilient crops.

of the 20th century" because of his enormous understanding and innovative vision, ability to recognize basic similarities to several quite distinct phenomenon as Darwin had done in the nineteenth century. His major contributions to plant breeding were summarized in Box 3.10. Similarly, Dr. Harbhajan Singh led numerous systematic exploration and evaluation programmes both in India and internationally, during which he introduced several new species of crop plants to India. He also undertook various crop improvement programmes using the crossability of crops with wild relatives (Box 3.11).

Box 3.11 Dr. Harbhajan Singh (1914–1974)

- Dr. Harbhajan Singh (Born at Pusa, Bihar in 1916) was the most distinguished plant explorer, renowned as the **'Indian Vavilov'** in the field of Plant Genetic Resources in India.
- He organized a number of systematic explorations and evaluation programmes in India and abroad.
- Dr. Singh introduced several new species of crop plants such as soybean, sunflower, low chilling and temperate fruits (peaches, apples, West Indian cherry, Chinese gooseberry) and a wide range of ornamentals in India.
- The President of India conferred on Dr. Singh the 'Padma Shri' award in 1971. Punjab Agricultural University decorated him with the degree of Doctor of Science (*Honoris causa*).

Vavilov (1951) conceived the concept of cradles of agriculture in an exhaustive way (Box 3.14). He emphasized the fact that time is the only factor that controls the spread of a plant species and its enhanced character variation. The areas or centres of greatest diversity are also called 'centres of origin' for the crops; he postulated that the dominant genes are found in the centres of origin of a crop plant and the recessive genes in the periphery. In 1926, five centres were initially identified based on differential phyto-geographical approach. These include - the Mediterranean, Central and South America, the Far East and South-western Asia. He recombined some and added others from time to time, until he reached eight, with three subcentres where agriculture developed independently (Fig. 3.5).

In 1935, Vavilov employed the 'differential phytogeographical methods' to identify the centres of origin or diversity of cultivated plants. The method involves the following steps:

- Conducting a rigorous differentiation of the plants into Linnaean species and intraspecific groups using all available disciplines such as morphology, agrobotany, phyto-pathology, cytology, and more recently, molecular methods.
- Mapping the areas of distribution of these plants, including historical distribution areas when communication and seed exchange were more restricted.
- Analysing the composition of varieties and races of each species to develop a general system of genetic variability within different species.
- Determining the distribution of genetic variability across regions and identifying geographical centres where these varieties are concentrated. Regions with maxi-

mum diversity, often containing numerous endemic types and characteristics, can also serve as centres of type-formation.
- For a more precise definition of centres of origin and type-formation, identifying geographical centres of closely related botanical species is essential.

Finally, establishing the diversity areas of wild subspecies and species closely related to the cultivated species can refine and enhance the defined area of origin, using the differential method to study races.

Moreover, Vavilov explained his ideas by comparing from other scientific fields. Just as the periodic table shows predictable properties of elements, biological traits also follow predictable patterns because of genetic and developmental rules. This comparison emphasizes Vavilov's belief that biological variation is not random but follows certain laws that we are still discovering.

Vavilov's 'Law of Homologous Series in Variation' support how important his ideas are for understanding evolution. His work provides a way to study the predictable patterns in how traits evolve. This challenges us to think about how both genetic constraints and natural selection shape life. Vavilov's insights reveal that while natural selection is key to evolution, the paths available for evolution are also guided by inherent rules. This more detailed view of evolution continues to inspire and guide modern biological research, showing the complex relationship between genetics and evolution.

Harlan (1951) used the term 'micro-centres' in a different sense than Zhukovsky, to distinguish very small areas or pockets of racial diversity concentrated in a Vavilovian centre. Small areas in Turkey and Africa contain such rich varietal diversity of many crops, with primitive or advanced crop husbandry. Based on pieces of evidence from archaeological and plant genetics, Harlan recognized centres and non-centres of agricultural origin, unlike relying only on phytogeography as done by Vavilov. There were three main independent centres of agriculture origin and spread; each with a corresponding non-centre, spread and diffusion of domesticated crops (Box 3.12).

Box 3.12 Centres and Non-centres (Harlan 1971)

Centre	Non-centre
North Chinese Centre—BI	Southeast Asian and South Pacific non-centre—B2
Near East Centre—Al	African non-centre—A2
Meso-American Centre—CI	South American non-centre—C2

Hawkes (1969) has postulated that the main crops, when moved to the cooler region at higher altitudes, or drier region, became less adapted to changed habitats whereas the weeds performed better; in this process, the weeds replaced the crops. Tomato (*Lycopersicon esculentum*) is a secondary crop as it spread from its centre of origin as a species in Peru, northwards and eastwards throughout the New World tropics as a weed of phaseolus beans and maize. After reaching Mexico, it developed tremendous fruit variability. Hawkes's concept visualised that the centres of agricultural origin led to the spread or diffusion points for farming into other regions, called nuclear centres and regions of diversity (Box 3.13). Based on archaeological evidence, he pointed out that agriculture began multiple times simultaneously across diverse regions of the world. He further suggested that the regions of paucity or absence of wild progenitors and lack of archaeological evidences were the areas of no crop origin.

Box 3.13 Nuclear Centres and Region of Diversity

(a) Northern China (Loess Plateau)
(b) Near East (Fertile Crescent)
(c) Southern Mexico to Guatemala (Tehuacan to southwards)
(d) Peru (Andes and coastal belt, as well as the eastern Andean slopes)
(e) West Africa (nuclear centre tentative due to lack of ancient archaeological data)

Zhukovsky (1965), a close colleague of Vavilov, brought forward the concept of mega gene centres. Based on Vavilov's centres of origin theory, he proposed 12 mega-gene centres of crop-plant diversity which contained domesticated plant species and showed areas of high diversity in crop and wild relatives (Figs. 3.4 and 3.5). Vavilov's eight centres were enriched by the addition of Australia, whole Africa and Siberia which was followed by revision of the boundaries. Zhukovsky defined the primary micro-gene centres as the areas where the cultigens first originated and secondary mega-gene centres where the crop later spread and diversified. He also described micro-gene centres within the mega-gene centres with the unique richness of regional diversity in crops and wild species. While distinguishing the distribution pattern between the wild and cultivated species, he emphasized the former having very limited distribution on the whole, while the latter had widespread through human migration and had developed tremendous variability. He attempted to meet the criticism of those who complain that the modern gene centres of wheat, for instance, could now be thought of as Mexico, the United States, Australia and

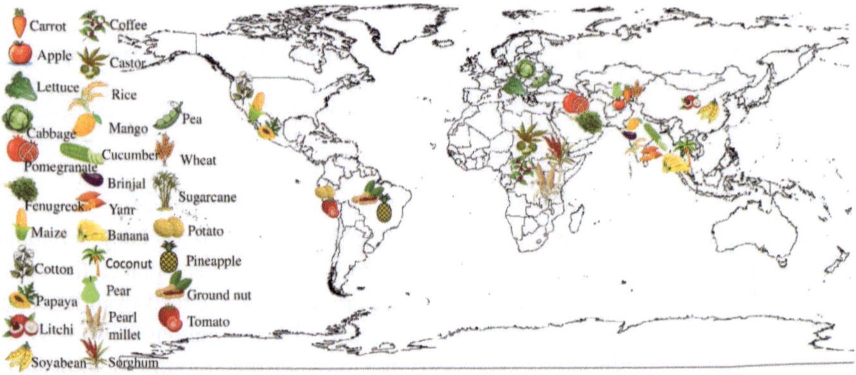

Fig. 3.4 World centres of diversity of cultivated plants

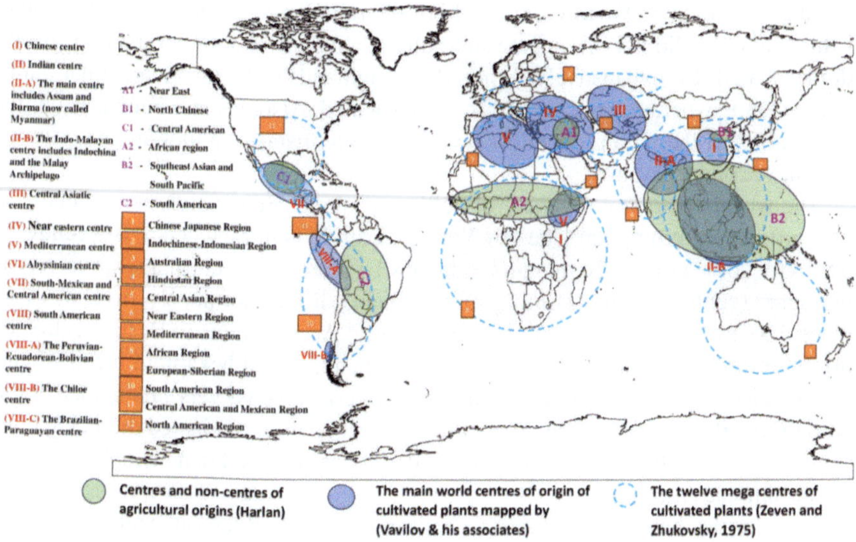

Fig. 3.5 Schematic representation of centres and non-centres, centre of origin and mega centres of cultivated plants

Argentina, or that of tomatoes as Europe, due to the breeding, selection and cultivation of varieties that have taken place in those areas in the last century.

Zeven and Zhukovsky (1975) in their publication entitled 'Dictionary of cultivated plants and their centres of diversity' have elaborately listed crop species for different mega-gene centres, and the range and extent of the distribution of crop diversity (genetic/variety/specific diversity). Zeven and de Wet (1982) justified to call the centres as the regions of diversity possibly because of the extensiveness of the referred areas as centres. They also recognized all 12 broader diversity regions as given by Zeven and Zhukovsky with larger coverage and more acceptability (Fig. 3.5).

Hymowitz (1972) gave the concept of trans-domestication also called tertiary domestication which is generally accepted for cluster bean/*guar* (*Cyamopsis tetragonoloba* (L.) Taub.) which has an unclear centre of origin. This crop is grown in India, Pakistan and the United States. Recent studies support that *C. senegalensis,* the crop wild relative of African origin migrated along with Arab traders between the ninth and the thirteen centuries and got domesticated in the Indian region (Indo-Pakistan), selected as a fodder crop that later developed and improved into the actual form for dual purpose.

3.6 Modifications and Criticisms of Gene Centre Theories

Despite the pioneering concept, Vavilov's gene centre theory on the study of origin and probable regions of initial plant domestication history of cultivated plants, the gene centre theories and domestication concepts invited criticism on a global level.

During exploration, Vavilov explored some areas/centre of origin in Africa and Ethiopia but he did not investigate sub-Sahara, where several important crops were domesticated. He exhaustively explored the highlands of South America but omitted lowlands where many crops were domesticated. Also, his less emphasis on the distribution of the wild progenitors invited several mistakes in judging the primary and the secondary centres.

The world centres of diversity of cultivated plants are certainly not a centre of origin for every crop plant listed by Vavilov, even though each centre is a centre of origin for some crops. The mutation rate may be more or less constant, but selection pressures (ecological and farming management) differ enormously from place to place. Vavilov claimed the Ethiopian and central Asian centres to be the centres of origin of durum wheat and barley, on the one hand, and of hexaploid bread wheat, on the other. Even though diversity for the crops concerned is very great, none of the wild progenitors were present. Such areas though no crop originated there, but concentration of genetic variation over the millennia has resulted in developing them as the accumulation centres.

Some critics have pointed out that centres—the Indian region, central Asia, the Near East and the Mediterranean—should be considered as a continuous band. Perhaps the most serious criticism of the gene centre theory has been advanced by Harlan who questioned whether there are more than three true centres connected as very large regions and diffused non-centre—Brazil and South East Asia and the Mediterranean. As Harlan pointed out for potatoes, the dominant alleles in the centre and recessives in the periphery of distribution (e.g. in wheat and sorghum) could not be made evident. Hawkes believed Harlan's theory of centres and non-centres was also somewhat rigid in terms of the dimension of a centre. There is also a problem of distinguishing between places where agriculture originated (Harlan's Center Al, B1 and C1) and places where cultivated plants were domesticated. Hawkes purposely avoided naming minor centres, where one or two crops/species (in parenthesis) seem to have originated, for examples, Solomon Isles to Fiji (Fe'o bananas), Europe (*Avena strigosa*, *Secale cereale*, *Ribes* spp., *Rubus* spp.) and Brazil (*Manilhot annuus*, *H. tuberosus*). Such crops may have arisen as a result of diffusion from adjacent region. He obviously needed more archaeological data for Africa, India and South-East Asia before he could have ascertained about nuclear centres for these areas. However, he would have related the nuclear centres about which he was reasonably certain to the region (or centres) of diversity, which equate more or less with Vavilov's centres.

The centres of origin were defined by Vavilov on the basis of the co-occurrence of wild ancestors with their domesticated crops, rich varietal diversity and a long history of use of the crop. However, all centres did not show a rich diversity of crop varieties. Further evidence from archaeology led to re-define the existing centres of origin which include independent centres of origin. For example, India was initially considered as one centre of origin, but based on fresh evidences it was re-defined into five independent centres.

3.6.1 Region of Diversity

Regions of diversity are places where the crop diversity got developed, with migration from the nuclear centres; here both unconscious and conscious selection processes took place in developing the regional diversity in crops. They equate with Vavilov's centres fairly well but it was necessary to modify some names slightly. Twelve centres of diversity are mentioned in Box 3.14.

Box 3.14 Cradles of Agriculture and Centres of Diversity

1. Japanese Centre
2. Indo-Chinese-Indonesian Centre
3. Australian Centre
4. Hindustani Centre
5. Central Asian Centre
6. Near Eastern Centre
7. Mediterranean Centre
8. African Centre
9. European-Siberian Centre
10. South American Centre
11. Central American and Mexican Centre
12. North American Centre

3.6.2 Centre of Domestication and Diversification: Indian Gene Centre

In the context of crop diversity the terms "Indian gene centre" and "Hindustani gene centre" are used interchangeably, referring to the same geographic region encompassing India and surrounding geography. The Indian gene centre, a cradle of crop origin of many economically important agri-horticultural plants, is a domestication and diversification centre for many economically important plant species used as wild edible types, ornamentals, medicinal types, forestry plants, forages, grasses etc. which constitute the genetic wealth of the region. The Indian gene centre is considered to be one of the 12 regions of diversity representing several crop plants, wild relatives and useful taxa. About 160 species of cultivated plants represent the primary and secondary centre of diversity along with 320 wild relatives. In recent works, the number of wild relatives has been updated to 861 taxa (769 species). Among over 1000 species of higher plants in the category of wild edible plants, the edible types and the semi-domesticated/domesticated diversity are represented by over 150 species. The wild relatives of cultivated plants and wild valuable species along with the weedy relatives constitute an integral part of native wealth and constitute a useful gene pool.

3.6.3 Useful Diversity in Tribal Pockets

The Indian sub-continent holds a wide spectrum of life forms, and rich cultural and ethnic diversity. About 53 million tribal people are belonging to over 550 tribal communities of 227 ethnic groups spread over 5000 forest villages. Their lifestyle differs from forest hunting, primitive hill cultivation (shifting), plains agriculture, artisan, pastoral and cattle breeder and urban industry worker types. Areas of high ethnic diversity are located in the distinct pockets of the North-eastern region, part of western Himalayas, central India and the western and eastern peninsular tracts and in the island regions. These are still the storehouse of rich diversity of crops, wild useful potential plant species and indigenous knowledge on plant-based uses. This wealth is protected, selected for useful species and conserved in their domain and surrounding habitats. Over 500 important potential species awaiting domestication offer a broad approach for the Indian region. Tribal/rural areas in a real sense are one of the most appropriate centres for demonstrating domestication activity by way of developing, shaping and utilizing useful plant diversity and also conserving it in situ or on farm. The advent of modern civilization in tribal pockets is leading to erosion of all kinds of diversity—cultural, plant as well as ethnic. Wild potential plant species (also called wild economic plant or under-utilized plant; WEP, UUP) are selected and gradually brought under folk domestication (Fig. 3.6). The primitive communities hold an array of potential plant diversity ranging from wild, semi-wild and domesticated forms and bring intermediate forms of potentially useful material near to their dwellings. Besides, the protected diversity is conserved in their backyards, home gardens, field margins, fallow land and marginal land. From pre-adapted or partially adapted types, domesticated forms are evolved through the local folk selection that operates through necessity-based and/or ritual-based or tribal customs/culture and socio-economy-based requirements. They are morphologically similar to their wild types, with several promising attributes. An explorer may witness the index of domestication while identifying areas of rich wild useful plant diversity (Box 3.15), where species at various stages of transition are evident.

Among food plants, conscious folk domestication is prevalent in tuberous rooted legume, sohphlang, *Moghania vestita* grown by the Khasi and Jaintia tribals. A tall grass *raisin* (*Digitaria cruciata* var. *esculenta*) is used as a minor cereal and as fodder, selected and grown as cultigen by Khasi tribes in the high hills of Shillong. In coix (*Coix lacryma-jobi*), the folk selection has resulted in the identification of soft,

Box 3.15 Wild Useful Diversity

About 3000 species are of edible value; 4000 species are known reputed medicinal plants; 700 are of traditional and socially significance; 500 used as fibre yielding species; 400 are of fodder value; 40 for insectivorous uses; 300 as gum and dye yielding plants; 100 as aromatic and essential oil yielding species

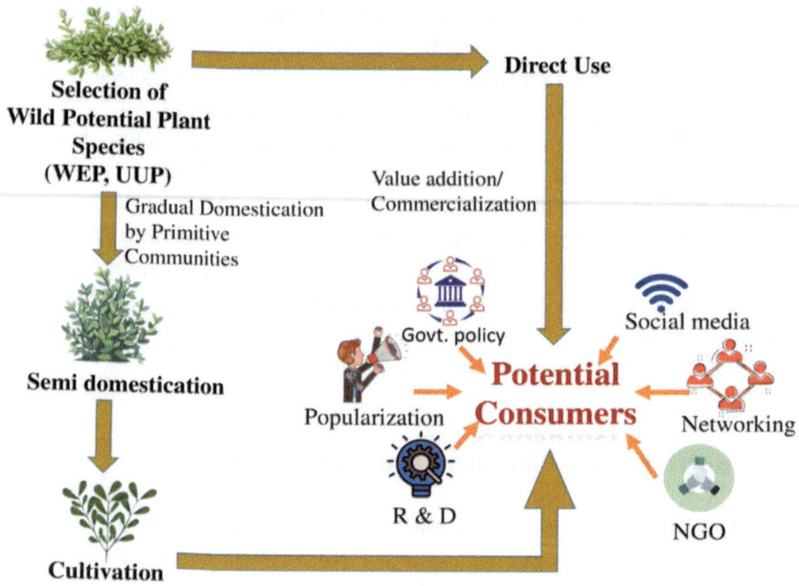

Fig. 3.6 Schematic chart showing flow of potential domesticates to consumer

thin-shelled and easy hulling types, with bold seeds and high yield by the tribals of Meghalaya, Nagaland and Manipur. Similar trends are evident in the rice bean, *Vigna umbellata,* a tribal pulse; leafy vegetable, *Phytolacca acinosa* and perilla, *Perilla frutescens* used condiment. In the western Himalaya, the Lahaul tribals have domesticated the aromatic oil yielding plant, *Inula racemosa;* forage/energy plant/ multipurpose tree *Parkia roxburghii*, in Manipur, the fodder tree and *Prosopis cineraria* in Rajasthan. Many high trade value medicinal plant species include *Aconitum heterophyllum, Dactylorhihza hatagirea, Hedychium spicatum, Nardostachys jatamansi* and *Rheum webbianum, Saussurea lappa* (costus), *Taxus baccata, Valeriana jatamansi, Berberis asiatica, Bergenia ligulata, Picrorhiza kurroa, Podophyllum hexandrum* and *Zanthoxylum armatum* that are still collected from wild sources.

3.6.4 Orchard and Home Gardens

Species diversity can generally be seen in the small plots of individual farmers, growing populations in home gardens (kitchen gardens, dooryard gardens) or orchard gardens. Anderson in his well-known book "plants, man and life" described in great detail the diversity found in these orchard gardens with lack of uniformity. The tremendous diversity of food and condiment plants in Indian gardens in Guatemala followed the pattern of the natural ecosystems. Hawkes investigated (in Indonesia, central Java), the total flora of cultivated species in an orchard garden, 45 species of cultivated plants with about 25 wild species used for medicinal purposes. The ecosystem of such orchard gardens closely follows that of the forests themselves, with tall trees (coconuts) forming the highest level, interspersed with shorter trees (mostly fruits), and shrubs, climbers and herbs filling up the spaces in between, right down to ground level, ecologically forming complex system in which cultivation proceeds by the substitution of selected domesticates.

Home gardening is being considered as a model for understanding the agroecosystem functioning in tropical as well as temperate conditions. Home gardens essentially hold plants of high nutrition and food value. As compared to the cultivated agro-ecosystem, rich diversity is conserved in home gardening systems which occasionally contain about 50–80% species representation. In India, most of the home gardens are located in marginal and tribal lands that usually hold wild potential and semi-domesticated plant species being conserved for day-to-day use to meet food, vegetables, medicine and other economic uses.

3.6.5 Bottleneck in Plant Domestication Study

The evidence on the process of domestication derived from ancient crops over the past decade challenges some of our most basic assumptions. The information resources on the historical account of the use of the plant species, process of protection, selection and domestication of a crop are meagre and difficult to access, particularly in many minor food crops, less-known crops and regionally neglected taxa. Studies in the distant past have largely focused on regional crops or specific groups (Asteraceae, Poaceae) and perennials, vegetatively propagated and underutilized plants (UUP). Domestication studies have a major gap due to lack of literature and quality data on the place of origin, time of domestication and presence of the wild ancestors, crops with disputed origins, non-availability of wild ancestors etc. Various aspects relevant to the theory of domestication need large datasets and critical scrutiny of historical trends choosing a range of plant species from well-worked-out crop species to understudied/less-known, neglected and regionally important crops at the global level. The advantages and disadvantages of plant domestication have been shown in Fig. 3.7.

3.7 Way Ahead

Domestication is a co-evolutionary interaction between plants and people which may involve diverse institutional networking using a multi-disciplinary approach. The overall changes happening during the process of domestication are less crop diversity, more input requirement and environmental impact, etc. The food plant species have reached their threshold and therefore new traits can be upgraded using advanced technological tools. De novo domestication involves the interaction of genetic, agronomic and cultural factors, so as to benefit at ecological and social level.

Modern tools have allowed a new insight into the genomic evolution during plant domestication. With advancements in a better understanding of domestication as a process and, the availability of modern tools, it is technically possible to

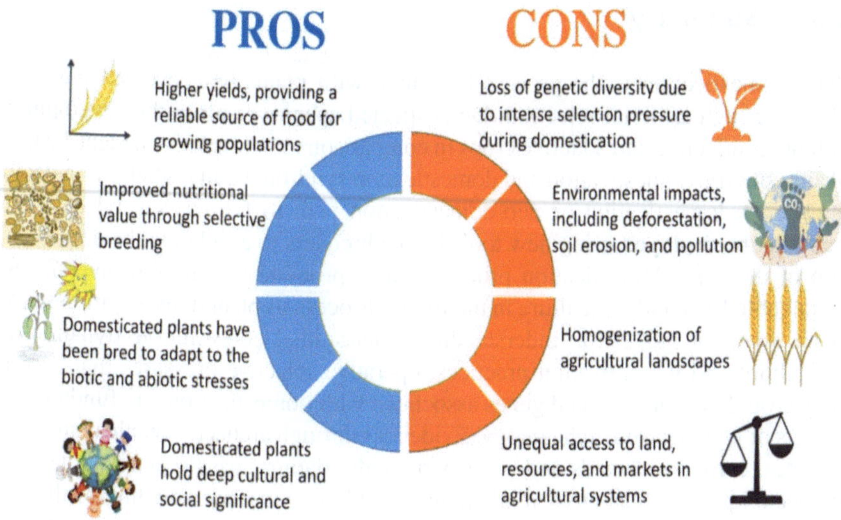

Fig. 3.7 Pros and cons of domestication

domesticate the undomesticated species. These tools facilitate de novo domestication by modifying genomes of wild plant species under controlled manipulation through mutation and recombination. Such methods also help in accelerating the improvement of traditional, semi-domesticated, less-known crops that have been neglected by modern farming systems. However, the desired impetus to generate new crops or modify the existing ones needs a thorough deep knowledge of the basic aspects of the domestication process. The gene editing method for de novo domestication does not bypass the necessary steps involved in classical breeding, agri-horticultural practices, crop protection research and extension, farmer participation and socio-cultural considerations. Gene editing approaches used to domesticate a species are typically a one-time event in wild plants, targeting a single gene or generation. Genotypes with desired traits were produced through de novo domestication using gene editing tools in wild relatives of the tomato (*Solanum lycopersicum* or *Lycopersicon esculentum*), and in the polyploid population of wild relatives of rice (*Oryza* species).

During the last decades, the development of molecular markers-based studies has thrown light on a better understanding of the domestication and breeding programmes, the domestication process, allowing to access information on history and identifying genes affecting the whole process. Currently, data on various aspects of centres of crop domestication are being upgraded significantly, seeking evidences from eco-geographic, and archaeology, and using modern methods and genomic scientific tools. Several desirable traits can be offered through new tools by the breeders to accelerate the domestication of the semi-domestication process and promote agricultural diversity for sustainable food and agriculture in the future.

3.8 Summary

Plant domestication is the process by which wild plants have evolved into crop plants through artificial selection. Domesticated species result in the development of more productive and adaptive crops in comparison to wild relatives. Identification of sites of crop domestication and domestication syndrome traits is helpful in deciding the de-domestication or neo-domestication path to follow. Several desirable traits can be deployed using new tools by the breeders to accelerate the domestication of the semi-domestication process and to promote agricultural diversity for sustainable food and agriculture in the future. Process of plant domestication serves as an excellent model for understanding plant-animal co-evolution, dynamics of agriculture and its cultivation practices, operative selective pressures that accompany crop domestication and genes associated with domestication and fundamental aspects of genetics and archaeology. Evidences through archaeo-botanical and plant sciences, carbonised seeds and plant parts, radio dating, mass spectro-photometry and techniques such as DNA markers have facilitated in understanding of the process of domestication. Domesticated species result in the development of more productive and adaptive crops in a set of environments opposite to those in wild relatives that possess low productivity and tolerance to biotic/abiotic stress under natural environment in their centre of origin. The domesticated species exhibit a reduction in genetic diversity due to selection pressure and population bottlenecks; wild species in their natural habitats exhibit more frequent mutations and low frequency of domestication syndrome traits. The changes under domestication result in developing a distinct type or variety greatly unidentifiable from their parental wild types (progenitors). These include several traits, like seed retention (non-shattering), increased fruit or seed size, changes in branching and stature, change in reproductive biology and changes in secondary metabolites (called domestication syndrome). Stages of domestication exhibit different degrees of human management: undomesticated populations are the ones that thrive with no care; semi-domesticated populations show protection of plants with little care or human interference; domesticated populations under the protection of plants with high human care and management. Neo-domestication of plant species works on wild species to develop traits for specific need based agronomic or nutritional requirements. On the other hand, re-domestication develops a new cultivar based on repurposed, thus allowing wider diversification from the initially domesticated type. The process of de-novo domestication allows genome editing or introgression to accelerate the process of domestication, thereby leading to the fast development of novel crop species having desirable traits. Vavilov defined eight centres of origin based on the co-occurrence of wild ancestors with their domesticates, high varietal diversity and a long history of crop use. Small independent centres were identified within pre-established centres of origin based on archaeological evidence. Despite the pioneering concept, the gene centre theories and domestication concepts invited criticism on a global level. During exploration, he explored some areas/centre of origin in Africa and Ethiopia but he did not investigate sub-Sahara where several important crops were domesticated. He exhaustively explored the highlands of South America but omitted the lowlands where many crops were domesticated. Also, his less emphasis on the distribution of the wild progenitors invited several mistakes in judging the primary and secondary centres.

Multiple-Choice Questions

1. The origin of a cultivated plant is a process, not an event (Anderson 1960) was stated because
   (a) It, being an evolutionary process, is largely long, under which the characteristics of plants change as per the selection process
   (b) It constitutes the agricultural evolution, the signs of which are still visible in traditional cultivation sites in the developing world.
   (c) It is a phylogenetically diverse assemblage of wild ancestors through artificial selection for different traits in different agro-climatic regions.
   (d) Distinct changes/modifications in characteristics of plant species that are morphologically and genetically distinct from their wild ancestors
2. In determining the place of origin of cultivated plants, Alphonse de Candolle attached great importance to the presence of
   (a) Crop wild relatives
   (b) Rich species diversity
   (c) High frequency of dominant alleles
   (d) Frequent gene flow between wild and cultivated populations
3. Which of the following statements is not related to the contribution of Vavilov
   (a) Distinguished primary and secondary crops
   (b) Established law of homologous series:
   (c) Emphasised on importance of weed races of crops
   (d) Established concept of centres and non-centres of agricultural origin
4. Vavilov hypothesized that in a crop plant's area of origin, we should expect
   (a) Greater diversity for trans-domesticated species, but lower diversity for species domesticated in their area of origin.
   (b) Greater genetic diversity than in areas outside of the plant's area of origin.
   (c) Lower genetic diversity than in areas outside of the area of plant origin.
   (d) Lower diversity for trans-domesticated species, but greater diversity for species domesticated in their area of origin.
5. The term 'microcentres' was used to distinguish very small areas of varietal richness within a Vavilovian Centre
   (a) Harlan
   (b) A de Candolle
   (c) Zhukovsky
   (d) Hawkes
6. A domestication syndrome may evolve over thousands of generations, as desirable traits are selected in a set of agricultural environments. Which of the following traits is odd
   (a) Seed shattering
   (b) Reproductive biology
   (c) Secondary metabolites
   (d) Population diversity

7. Vavilov suggested that crop plants are found away from their primary area of origin or diversity are generally not have the following main features
   (a) Lesser genetic diversity than primary centre
   (b) Large number of recessive genes - dominant gene
   (c) Mostly desirable characters
   (d) Both natural and artificial selection operate
8. The origin of cultivated plants is not the same as the origin of agriculture. Which of the following statement is true
   (a) Centres of origin are the points of beginning of agriculture
   (b) Centres of agriculture origin preceded the centre of origin
   (c) Two always go together as far as crop domestication goes
   (d) Two events are not related to each other
9. Which of the following processes is involved with the reversal of traits in the domestication of crops?
   (a) Neo-domestication
   (b) Re-domestication
   (c) De-novo domestication
   (d) De-domestication
10. Which of the following new area was not added to Vavilov's eight centres
    (a) Australia
    (b) Whole Africa
    (c) Siberia
    (d) Abyssinia

Answers

1. (a)
2. (a)
3. (d)
4. (b)
5. (a)
6. (d)
7. (b)
8. (a)
9. (d)
10. (d)

Suggested Reading

Anderson E (1952) Plants, man and life. University of California Press, p 251
Anderson E (1960) The evolution of domestication. In: Tax. pp 67–84
Darwin C (1868) The variation of animals and plants under domestication. John Murray, London
Darwin C, Wallace A (1858) On the tendency of species to form varieties; and on the perpetuation of varieties and species by natural means of selection. J Proc Linn Soc Lond Zool 3:45–62

de Candolle A (1890) Origin of cultivated plants. American Association for the Advancement of Science, Appleton, Wisconsin

Grimm A, Sahi VP, Amann M, Vidotto F, Fogliatto S, Devos KM et al (2020) Italian weedy rice—a case of de-domestication? Ecol Evol 10(15):8449–8464

Harlan JR (1951) Anatomy of gene centers. Am Nat 85:97–103

Harlan JR (1971) Agricultural origins centers and noncenters. Science 174:468–474

Harlan JR (1975) Crops and man. American Society of Agronomy, Crop Science Society of America, Madison, Wisconsin

Harlan JR (1992) Crops and man. American Society of Agronomy and Crop Science Society of America, Madison

Hawkes JG (1969) The ecological background of plant domestication. In: Ucko PJ, Dimbleby GW. pp 17–29

Hawkes JG (1983) The diversity of crop plants. Harvard University Press, Cambridge, MA

Hymowitz T (1972) The trans-domestication concept as applied to guar. Econ Bot 26:49–60

Pradheep K, Ahlawat SP, Nivedhitha S, Gupta V (2022) Crop Wild Relatives in India: Inventorization, prioritization and conservation. Indian J Plant Genet Resour 35(3):124–130

Vavilov NI (1926) Studies on the origin of cultivated plants. Bull Appl Bot 16(2):248

Vavilov NI (1935) The phyto-geographical basis for plant breeding. In: Theoretical basis for plant breeding, pp 117–175

Vavilov NI (1951) The origin, variation, immunity and breeding of cultivated plants (translated by K. Starr Chester). Chronica Bot 13:1–366

Vavilov NI (1992) Origin and geography of cultivated plants (transl: Love D). Cambridge University Press, Cambridge

Wilkes G (1984) Germplasm conservation towards the year 2000. Potential for new crops of enhancement of present crops. In: Yeatman C, Kefton D, Wilkes G (eds) Plant genetic resources: a conservation imperative. American Association for the Advancement of Sciences, Washington, DC, USA

Zeven AC, de Wet JMJ (1982) Dictionary of cultivated plants and their regions of diversity, Wageningen

Zeven AC, Zhukovsky PM (1975) Dictionary of cultivated plants and their centres of diversity excluding ornamentals, forest trees, and lower plants. Centre for Agricultural Publishing and Documentation, Wageningen

Zhukovsky PM (1965) Main gene centers of cultivated plants and their wild relatives within the territory of the USSR. Euphytica. 14:177 88.

Plant Germplasm Introduction and Exchange Activities

4

Vandana Tyagi, Pragya, Puran Chandra, Richard S. Cahoon, and Ulrike Lohwasser

Learning objectives

- The plant introduction from the distant past to the present.
- Guidelines and procedures for plant germplasm import, export and domestic supply for research purposes.
- Legislative framework for facilitating access to plant germplasm from other countries and exporting the germplasm to other countries.
- The role of international organizations and genebanks working for conservation and sustainable utilization of plant genetic resources (PGR) for food and agriculture.

Keywords

Plant genetic resources · Germplasm · PGR · Introduction · Exchange · SMTA · Import permit

V. Tyagi (✉) · Pragya · P. Chandra
Division of Germplasm Exchange and Policy Unit, ICAR-National Bureau of Plant Genetic Resources, New Delhi, India
e-mail: Vandana.Tyagi@icar.gov.in; Pragya@icar.org.in; puran.chandra@icar.org.in

R. S. Cahoon
Department of Global Development, Cornell College of Agriculture and Life Sciences, Ithaca, NY, USA
e-mail: rsc5@cornell.edu

U. Lohwasser
Department of Genebank, Research Group Resources Genetics and Reproduction, Leibniz Institute of Plant Genetics and Crop Plant Research (IPK), Gatersleben, Germany
e-mail: lohwasse@ipk-gatersleben.de

K. Tripathi et al. (eds.), *Textbook of Plant Genetic Resources*,
https://doi.org/10.1007/978-981-96-5024-8_4

The entirety of all genetic material, including both cultivated and wild species, crop plant relatives, landraces, present and obsolete cultivars, breeding stocks and wild varieties of cultivated crops, can be referred to as plant germplasm. Germplasm exchange activities include the acquisition of superior germplasm from other countries, distribution of the germplasm to researchers for crop improvement and facilitating access to germplasm from other countries under provisions of different memorandum of understandings/work plans/collaborative research programmes, or any other ongoing research while protecting our rich diversity of plant genetic resources. PGR may be conserved by storing seeds in genebanks or complete plant/tree in the field genebanks such as arboretum/botanical garden/seed orchards and research farms. The exchange of germplasm is important for ensuring food security and achieving the targets of the sustainable development goal of zero hunger by the year 2030. The factors that affect the germplasm introduction and exchange activities include ownership issues (public/private/protected/not protected), extant import/export regulations (national obligations) and bilateral or multilateral exchange.

4.1 History of Plant Germplasm Introduction

An introduction is the earliest as well as the easiest method followed in plant breeding. In India, with the dedicated and consistent efforts of the Plant Introduction Division at IARI, New Delhi and later as the National Bureau of Plant Genetic Resources (NBPGR), the significant introduction of new crops, new varieties, breeding material and wild species of cultivated crops has been made since 1940. The introduction of different germplasm offers plant breeder desirable genes for use in crop improvement. It results in greater increases in agricultural yields and quality in shorter period of time. Introduction has played a pivotal role in the diversification of the food basket of the country. Introduction has resulted in the adoption of large number of new crops and improvement of existing crops. Introduction and exchange of crops have given the way for the crops to spread out from their origin to newer places. For example, the United States of America (USA), a major exporter of wheat and soybeans now, did not have old-world wheat and soybeans approximately 400 years ago. Rice is also an old-world crop that was not known earlier to new world countries. Similarly, the old-world countries also now depend on various crops introduced from old world for food and nutritional security. Prominent of these introduced crops are maize, potatoes, sweet potatoes, tomatoes, groundnuts etc. Travelers, settlers, traders, invaders, explorers, pilgrims and naturalists, either knowingly or unknowingly, have been the primary agent of introduction. The last five centuries have witnessed the exchange of useful crops in different regions of the world. Present plant wealth of various nations is the result of plant introductions from other part of the world during the several centuries after domestication of crop in the original place. Plant introductions of historical importance are mentioned in Box 4.1.

Box 4.1 Important Introductions

Botanical gardens have always been repositories providing precious plants for introduction. Kew Botanic Garden has played important role in introducing quinine and rubber trees to India from South Africa in the nineteenth century. The Royal Horticultural Society of Great Britain sponsored collecting expeditions to different parts of the world such as Himalayas. Some of the most beautiful garden plants have been brought to cultivation due to the efforts of Royal Horticultural Society. The first formal introduction recorded in India is *Cynodon plectostachyus* (Giant star grass) from Royal Botanic Gardens, Kew England on 1 August 1940. Documented records of all these introductions are maintained at the Germplasm Exchange Division, which is now functioning as the Germplasm Exchange and Policy Unit (GEPU) at NBPGR, New Delhi. The ushering in of the Green Revolution in India is the outcome of the introduction of wheat and rice germplasm, breeding lines and strains for the development of high-yielding, short-statured varieties responsive to fertilizers and intensive cropping.

The major new crops introduced and facilitated by NBPGR are kiwi fruit, sunflower, soybean, hops, sugar beet, pepper mint, sea buckthorn, prickly pear and very recently heeng and monk fruit. Promising varieties of temperate fruits such as apple, pear, plum, peach, grapes and vegetables like cabbage, carrot, tomato, radish, peas, and watermelon have been introduced and directly released as varieties. Several medicinal plants, industrial crops and potential crops have also been introduced for popularization among farmers like jojoba, guayule, pyrethrum and paradise tree. Recently emphasis is on selected wild relatives and wild species of different crops which are used as wild gene pool for breeding new lines.

4.2 Plant Introduction

Plant introduction involves taking a genotype or a group of genotypes of plants into new environments where they have not been grown before. It includes new crops, new crop species, new varieties of crops already grown in that area or wild relatives of the crop species already grown in the area. A few definitions of plant introduction are provided in Box 4.2. The materials are frequently imported from other continents or nations. But moving crop kinds across the nation from one habitat to another is also an introduction. The spread of rice in Punjab, wheat in West Bengal and grape production in Haryana are a few instances of introductions inside the nation.

Box 4.2 Definitions of Plant Introduction

Transposition of genetic entity from an environment to which it is attuned to one in which it is untried as defined by *Frankel (1957)*.

Plant introduction is the acquisition of superior varieties by importing them from other areas as defined by *Allard (1960)*.

Plant introduction in broadest-sense may be taken to mean the introduction of wild plants into cultivation as defined by *Whyte 1958, 1959*.

Introduction of wild plants into cultivation and the successful transfer of cultivars, with their genotypes unaltered, to new environments would together be designated as primary plant introduction and the rest as secondary, as defined by *Erna Benett, 1965*.

Plant introduction also encompasses the modern techniques of plant breeding, namely transfer, addition or substitution of genes, near or distant hybridization. The ultimate objective is to get diverse material/gene from other part of the world. When a species is found outside of its natural range and has been brought there by intentional or unintentional human activity, it is referred to as an imported, alien, exotic, non-indigenous or non-native introduction. Plant introduction may be classified into two categories.

4.2.1 Primary Plant Introduction

Sometimes the introduced variety is well suited to the new environment, and if the variety is released directly, without any alteration in the original genotypes, this is referred to as primary introduction. Primary introduction is less common, particularly in countries like India, where suitable infrastructure and organized crop improvement programmes are available. In India, the introduction of input responsive semi-dwarf wheat Sonora 64, Lerma Rojo and rice Taichung Native 1, IR 28, IR 36 are the examples of primary introduction (Table 4.1).

Table 4.1 Some examples of primary introductions

Crop	Variety (introduction from)
Wheat	PV 18 (Mexico), Ridely (Australia)
Barley	Clipper (Australia), Dolma (USA)
Rice	IR 8 and IR 50 (Philippines)
Oats	Rapida (USA), Kent (Australia)
Sunflower	Aramvirikij and Peredovik (USSR)
Groundnut	Asiriya Mwitunde (Tanganyika), Rehovot 33-1 (Israel)
Soybean	Bragg and Lee (USA)
Tomato	Sioux and Labonita (USA)
French bean	Contender and Kentucky wonder (USA)
Cabbage	Golden Acre and Drum Head (Denmark)
Water melon	Ashahi Yamato and Sugar baby (Japan)
Banana	Grand Nain M. S. (France), Valery (West Indies)
Papaya	Sunrise (USA), Cariflora (EC 300205, USA)
Pomegranate	Shirin Anar (USSR), Wonderful (USA)
Pear	Flemish Beauty (USA), Max Red Bartlet (Italy)
Peach	Stark Early Glo (USA), Flordasun (USA)
Plum	Methley (Kenya), Kanto-5 (USA)
Apricot	Nugget (USA), Coninos (Italy)
Walnut	Lake English (USA), Hansen
Pecan nut	Mahan (USA)
Kiwi fruit	Allison (New Zealand)

4.2.2 Secondary Plant Introduction

Secondary introduction refers to a variety/hybrid developed by hybridization and subsequent selection from segregating material, where one of the parents is an exotic introduction. In secondary introduction, one or a few traits from an exotic variety may be transferred to the existing variety for the improvement in the particular trait. Secondary introduction is more common as compared to the primary introduction. Some examples to cite secondary introduction are wheat varieties namely Kalyan Sona and Sonalika which were selected from material introduced by CIMMYT, Mexico; tomato varieties Pusa Ruby (Sioux (USA) × Improved Meeruti); cowpea variety Aseem (Pusa Phalguni × Exotic line from Philippines); okra variety G2 (Pusa Sawani × Ghana Red); Carrot variety Pusa Kesar (Local Red × Nantes); Water melon variety Arka Manik (Local selection × Crimson Sweet).

4.2.3 Merits of Plant Introduction

Introduction of new germplasm is the easiest way to enhance the overall germplasm wealth of a country. It is quick, less expensive and protects the variability. The major objectives of plant introduction are listed below:

1. To obtain a new plant/crop:
   Examples include maize, potato, tomato, tobacco, soybean, sunflower, rapeseed, kiwi fruit, tree tomato, hazel nut, guayle, jojoba, geranium are all introduced crops.
2. To obtain elite and promising varieties of crops already established:
   Examples include Fuji apple, Santa rosa plum, Lee and Bragg soybean, Clipper barley, and Nantes carrot.
3. To obtain natural variability useful for broadening germplasm pools for crop improvement, new trait specific germplasm may be introduced and incorporated genetically to develop a usable hybrid or synthetic variety.
4. To enable scientific/research studies/studies on biosystematics, evolution and origin and that of homologous series in variation from the study of a vast collection of plant types.
5. To satisfy social and recreational needs, ornamental shrubs and lawn grasses have been introduced for decoration and are of great social value.

4.2.4 Demerits of Plant Introduction

The local ecosystem may be affected due to the introduction of new species. The newly introduced species that harm the local ecosystem are known as invasive species or invasive alien species (IAS). Thus, the demerits associated with the introduction activities may result in the introduction of weeds, disease and pests. Some examples include *Eichhornia crassipes, Phalaris minor* and *Lantana camara* were introduced as ornamentals; however, they turned into noxious weeds. Diseases like late blight of potato came from Europe in 1883, coffee rust from Ceylon in 1876 and bunchy top of banana from Ceylon in 1940 along with the crop introduced. Several new insect pests also got introduced from different countries, for example, potato tuber moth came from Italy in 1900, woolly apple aphis came from England in 1909 and fluted scale of citrus from Sri Lanka in 1928. Sometimes the introduced species may disturb the ecological balance in their new home and alter the microclimate of the introduced area. *Eucalyptus* sp. introduced from Australia is the classical example of such species which caused a rapid depletion of the subsoil water reserves. However, most of these cases occurred before quarantine regulations. Strict quarantine and phytosanitary measures ensure safe introduction to ensure that the material which is to be introduced should not carry diseases or pests and eliminate the associated risks.

4.2.5 Plant Introduction Services

To obviate the risks and lost opportunities while introducing plant germplasm, establishment of plant introduction services at the national level is needed to facilitate their efficient operation. These services in different countries have common features that account for the successful utilization of the introduced material. Some important features of these services are listed here

1. Plant introduction services usually form part of the department of agriculture or a national level research organization.
2. Plant introduction services always work in collaboration with the quarantine authority. Either they have a good quarantine facility for conducting quarantine or they may send the material to quarantine authority for conducting quarantine examination.
3. Plant introduction services generally have testing facilities in all the major climatic zones of the country concerned.
4. Plant introduction services have an efficient record-keeping system to maintain records and documents and they ensure that national requirements are met without unnecessary duplication.
5. They serve as a contact point for international collaborations.

Plant introduction services maintain and conserve extensive collections of germplasm, which is used for the need-based supply to the researcher, and have a close collaboration with its counter parts abroad. Services for plant introduction have been crucial to the transfer of beneficial plant material between nations. They make it possible to explore gene centres for collection conservation and utilization of primitive cultivars and their wild relatives of crop plants.

4.2.5.1 Plant Introduction Agency in India

In 1946, at the Indian Agricultural Research Institute (IARI), in the Division of Botany, a Centralized Plant Introduction Agency was established with funding from the Indian Council of Agricultural Research (ICAR). In 1956, the agency was converted into the Plant Introduction and Exploration Division, and later, in 1961, into the Division of Plant Introduction of IARI, New Delhi. In the year 1976, the division was upgraded to an independent institute and was named as the National Bureau of Plant Introduction (NBPI). In 1977, the institute was renamed the National Bureau of Plant Genetic Resources, New Delhi. The bureau is now serving as the nodal agency for introducing and maintaining agricultural and horticultural crop germplasm in India. Other agencies involved in germplasm introduction are Forest Research Institute (FRI), Dehradun, which acts as plant introduction organization for the introduction, maintenance and testing of germplasm of forest trees and Central Research Institute for crops, namely tea, coffee, sugarcane, potato, tobacco, rice which introduce test and maintain plant material of interest. However, the activities of FRI and Central Research Institutes are coordinated by ICAR-NBPGR, the nodal agency responsible for germplasm exchange activities. Plant materials may also be introduced by scientist, universities and other government and private research organizations; however, all the introductions must be routed through the ICAR-NBPGR, New Delhi.

4.2.5.2 Agro-Climatic Concepts

The introduction of plant genetic resources from regions having similarity in climate (solar radiation, rain fall, temperature etc.), edaphic (soil physical and chemical properties) and topographic (attitude, latitude, slope and aspect etc.) conditions have better possibilities of success in the establishment of new crop. India is located between 8° 4' N to 37° 6' N latitude and 68° 7' E to 97° 25' E longitude and exhibits extreme variation in edapho-climatic conditions/agroclimatic regions and floristic diversity. Although India is mainly a tropical country, the great Himalayas and other mountain ranges provide subtropical, temperate and alpine conditions. The extreme climatic conditions result in extreme variation in vegetation which ranges from arid desert in the west, tropical rain forest in the southwest and humid and dry inland regions to glaciers in the north. Therefore, every agricultural or horticultural plant has a chance to survive in the varied Indian conditions. India holds today plants from all parts of the world adapted to micro- or macro-climatic. Phytogeographical similarity, adaptability behaviour of crop variation, inadaptability behaviour of new crops or new introduction of different crop plants to abiotic and biotic stresses can be viewed at three different stages: (1) Domestication of wild species, (2) stabilizing semi-domestic/semi-wild crop species and (3) breeding for improved varieties. The introduction of PGR is required at all three stages.

4.3 Plant Germplasm Exchange Activities

Increased food demand due to exploding human population requires increase in production and productivity of food and fodder crops and the exchange of germplasm plays an important role in the crop improvement programmes. To attain this objective exchange of plant genetic resources at the global level and its efficient and sustainable utilization is very critical. Germplasm exchange before 1993 was free and biological resources were considered as the common heritage of mankind. This unrestricted exchange resulted in expanding the food basket worldwide. Various Indian crops, such as sugarcane and mango, are now distributed and utilized globally. Indian germplasm has also provided important genes for stress breeding in various crops. For instance, a gene from wild rice (*Oryza nivara*) in Odisha, India, protected the crop from a growth-stunting virus that threatened rice in Indonesia a few decades ago. The finest salt tolerance in the world has been provided by the rice varieties Pokkali and Kharchia. Powdery mildew resistance from Indian cucumber germplasm was used for resistance breeding of cucumber crop in USA. The dwarf wheat and rice varieties from Mexico and the Philippines provided the initial impetus for increasing the productivity of rice and wheat In India which ultimately resulted in the 'Green Revolution'. This diversity in biological resources helped the world to fight the challenges of pandemics such as coronavirus disease 2019 (COVID-19), as food and nutritional security are the key for good health and immunity. The Convention on Biological Diversity (CBD), which came into effect in 1993, limited this unrestricted and ongoing movement of germplasm. With the enactment of CBD, the biological resources including plant germplasm became the sovereign rights of the nation hence implementation of Access and Benefit Sharing (ABS) Systems, Nagoya Protocol (NP), regulated this plant germplasm exchange. The genebank curators and researchers/users of the germplasm had to comply with extant national and international legislations in exchanging plant germplasm. Germplasm exchange activities in India are regulated by the Plant Quarantine (Regulation of Import into India) Order, 2003 (PQ order 2003), CBD, 1993 and the International Treaty on Plant Genetic Resources for Food and Agriculture (ITPGRFA) also known as international seed treaty/plant treaty.

The germplasm need of various researchers and scientists are met by the Germplasm Exchange and Policy Unit (GEPU) at ICAR-NBPGR, New Delhi. The main responsibilities of GEPU include introducing (importing), promoting export and providing the plant germplasm for domestic research purposes. The Director of ICAR-NBPGR is authorized by PQ Order 2003 to provide import permits for the import of germplasm for research purposes and to accept imported materials from customs officials for quarantine examination, clearance and subsequent distribution to the nation's researchers. However, germplasm sharing with foreign collaborators for research use is facilitated under the provisions of Biological Diversity Act (BDA), 2002 of India which was enacted in compliance with CBD. GEPU is also responsible for the supply of plant germplasm within the country for research purposes. The commercial import of plant genetic resources in bulk quantity including the seeds/stock material of fruit plant species are permitted only on the recommendations of Export-Import (EXIM) Committee of the Department of Agriculture, Cooperation and Farmers Welfare (DAFW), Ministry of Agriculture and Farmers Welfare. The bulk import of PGR for commercial use is handled by Department of Plant Protection, Quarantine and Storage (DPPQS), NH-IV, Faridabad.

4.3.1 Procedure for Import of Germplasm

Germplasm import is facilitated by ICAR-NBPGR, New Delhi, by issuing Import Permit (IP), which is an official document authorizing the import of the consignment in accordance with specified phytosanitary measures. Only the Director of NBPGR has the authority to grant import permit for import of seed for research use. This applies to transgenic seed material too. According to the Plant Quarantine Order 2003, import permits from ICAR-NBPGR, New Delhi, and phytosanitary certificates from the place of origin are required for the import germplasm in India for research use. All consignments of seed/planting material imported for research use must accompany these two documents (see flowchart in Annexure 1).

4.3.2 Application and Issuance of Import Permit

The Director of ICAR-NBPGR is authorized by Clause 6(2) of the PQ Order 2003 to grant import permits for the import of germplasm, transgenic organisms or genetically modified organisms for research purposes. ICAR-NBPGR receives imported materials from customs authorities for quarantine inspection, clearance and subsequent distribution to researchers within country. For issuing the import permit, the applicant is required to fill in the application form (Plant Quarantine Form 08) and also need to pay the processing fee. The fee is revised from time to time and is non-refundable.

The procedure for issuing import permit is also available online at ICAR-NBPGR (www.nbpgr.ernet.in.geq). The import permit, which is issued in triplicate using Plant Quarantine Form 09, is valid for subsequent shipments for a period of 6 months as long as the importer and exporter, the bill of entry, the country of origin and the phytosanitary certificate remain the same for the entire shipment. Permits for imports are non-transferable. Once the import permit has been obtained, the applicant must forward it to the relevant source that has consented to provide the necessary germplasm, requesting that the import permit in duplicate should be shipped with the planting material or seed. Private seed companies are required to furnish R&D certification of their lab by the Department of Scientific and Industrial Research (DSIR).

However commercial import consignment of coarse cereals, pulses, oilseeds, fodder and seeds/stock material of fruit species can only be permitted on the recommendation of EXIM Committee, Department of Agriculture, Cooperation and Farmers Welfare's (MoAFW) (as per Clause 3(3)(4)). However, before other crops as stated in the above line, the trial/testing seed material may be imported through ICAR-NBPGR as specified in Schedule-XII of PQ Order, 2003. For the import of plants and plant products for consumption and processing, a permit application must be submitted to the Plant Protection Advisor or the issuing Authority as specified in Schedule X of the PQ Order, 2003 in Plant Quarantine Form 01; for the import of seeds and plants for propagation covered by Schedule V and VI, a permit application must be submitted in Plant Quarantine Form 02.

4.3.3 Issuance of Phytosanitary Certificate

A phytosanitary certificate is a document that a government official from the country of origin issues about the health status of a shipment in the format specified by the Food and Agriculture Organization (FAO). A phytosanitary certificate (original copy) issued by an authorized officer in the supplier or country of origin, along with additional declarations attesting to the shipment's absence from particular pests and diseases, should be provided with every shipment. This must be supported by documentary evidence that the pests specified do not occur in the country or state of origin as specified in the import permit. These declarations are crop and country-specific and the phytosanitary certificate shall be compliant with these additional declarations. ICAR-NBPGR is also authorized to issue a phytosanitary certificate for the supply of the germplasm to foreign countries and is issued in compliance with the respective import permit issued by the importing country.

4.3.4 Procedure for Import of Transgenic Seed Material/ Genetically Modified Organisms

The guidelines for the production, use, import, export and storage of hazardous microorganisms or genetically engineered organisms or cells created under sections 6, 8 and 25 of the Environment Protection Act, EPA 1986 (29 of 1986) have been defined by Authority No. 621, New Delhi, and published in the Gazette of India extraordinary Part II Section 3 subsection (1). Every transgenic plant is a regulated item under this Act. The EPA's regulations are crucial for protecting the environment, reducing the risk of pollutants and toxins harming wildlife and plants, and protecting human and animal health. The import of transgenic seeds is also subject to the PQ Order, 2003's requirements. The Department of Biotechnology (DBT) and the Ministry of Environment, Forests, and Climate Change (MoEFCC) have provided a set of guidelines for granting authorization for the importation of transgenic. The Institutional Biosafety Committee (IBSC) receives proposals for the import of transgenics, which are then sent to the Review Committee on Genetic Manipulation (RCGM), a DBT-operating authorized government entity. Representatives from DBT, the Council of Scientific and Industrial Research (CSIR), the Indian Council of Medical Sciences (ICMR), the Indian Council of Agricultural Research (ICAR) and other specialists in their respective fields are part of RCGM. From a biosafety perspective, RCGM investigates transgenic lines import considering biosafety issue The application for the issuing of an import permit is sent to the Director, ICAR-NBPGR, using the required Plant Quarantine Form 08, following technical clearance for the Seed Transfer Clearance Letter given by DBT. However, in accordance with the updated, simplified procedure/guidelines on import, export and exchange of genetically engineered organisms and their products for research and development purposes issued by DBT in January 2020, the import permit may be granted for laboratory use only for model organisms like plants (Arabidopsis), common laboratory models and other model organisms that are routinely used in laboratories worldwide, subject to IBSC certification. The IBSC must attest that the plants and model organisms do not contain alien gene insertions from non-model species and that they contain only regular and standard experimental insertions.

4.3.5 Quantity of Seed Permitted for Import of Plant Germplasm

Minimum quantity of seed or planting material needed for establishment is considered and permitted for import. Details regarding permitted seed quantity for research purposes are available at the ICAR-NBPGR website. The DBT permission letter specifies the amount of seed allowed for transgenic plants. From the imported germplasm 5–10 seeds /planting material are separated as voucher specimen and stored in the National Genebank at ICAR-NBPGR for future use.

4.3.6 Germplasm Exchange and Quarantine Information System

An online information system named as the Germplasm Exchange and Quarantine Information System (GEXQIS) was created to keep track of all the data related to germplasm exchange. Everything from receiving a consignment to assigning accession numbers, fee details, dispatching the item to the indentor and getting feedback from the indentor can be entered into the system.

Through GEXQIS one can apply for the issuance of an import permit by registering on the system. A unique identity number, that is, exotic collection (EC) number, is assigned to all the germplasm which is introduced or imported into the country and is registered and documented in this system. Indigenous collection (IC) number is also a unique ID which is assigned to all collections which are made indigenously and collected through exploration trips or other programmes within Indian territory. GEXQIS maintains the records of all imports; however, gradually it will include all the information related to the export of germplasm that has been to other countries under various options available.

4.4 National and International Regulations Governing the Germplasm Export from India

4.4.1 Convention on Biological Diversity (CBD) and Nagoya Protocol (NP)

The first legally binding agreement to ensure the preservation and sustainable use of all biological diversity and to establish a fair process for sharing the benefits that result from using biodiversity was the Convention on Biological Diversity (CBD), which came into effect in 1993. It provided guidelines for bilateral access and exchange between nations on mutually agreed terms (MAT) and prior informed consent (PIC). India has also ratified the Nagoya Protocol (NP) which aims at sharing the benefits arising from the utilization of genetic resources fairly and equitably, including by appropriate access to genetic resources and by appropriate transfer of relevant technologies, taking into account all rights over those resources and to technologies, and by appropriate funding. NP is an international agreement which came into force on 12 October 2014. Nagoya Protocol provides a strong base for legal certainty and transparency for both providers and users of genetic resources. It recognizes the importance of promoting equity and fairness in negotiation between providers and users. India is a party to CBD and Nagoya Protocol. In compliance with CBD and NP, Indian Parliament has enacted the Biological Diversity Act, 2002, which was amended in 2023.

4.4.2 The Biological Diversity Act

India enacted the Biological Diversity Act, 2002, and notified the Biological Diversity Rules, 2004. The objectives of the Biodiversity Act, 2002 are (1) conservation of biodiversity, (2) sustainable use of its components and (3) fair and equitable sharing of benefits arising out of the use of biological resources, which is commonly known as access and benefit sharing (ABS). The Act, 2002, Rules, 2004, and the Guidelines, 2014, are aimed at implementing the three objectives of the CBD and the obligations under the NP. The Act was revised in 2023 for certain sections. The major sections of the Act that are directly related to germplasm exchange activities are described below.

4.4.2.1 Section 3: Regulation of Access to Biological Diversity

This provides for the definition of a non-Indian entity. Accordingly, as per the definition provided, a non-Indian under the Act can access biological resources occurring in India or knowledge associated for research, commercial utilization, bio-prospecting or bio-utilization only with the approval of the National Biodiversity Authority (NBA). The definition is provided under Box 4.3.

Box 4.3 Section 3 (2) of BDA, 2002

Defines the requirement of NBA approval for accessing any germplasm occurring in India by

1. A person who is not a citizen of India;
2. A citizen of India who is a non-resident as defined in Clause (30) of Section 2 of the Income tax Act, 1961; (43 of 1961);
3. A body corporate, association or organization
   (a) Not incorporated or registered in India,
   (b) Incorporated or registered in India under any law for the time being in force, which is controlled by a foreigner within the meaning of Clause (27) of Section 2 of the Companies Act, 2013

Clause (27) of Section 2 of the Companies Act, 2013 (describes the phrase 'Control' & control shall include the right to appoint majority of the directors or to control the management or policy decisions exercisable by a person or persons acting individually or in concert, directly or indirectly, including by virtue of their shareholding or management rights or shareholders agreements or voting agreements or in any other manner).

4.4.2.2 Section 4: Transfer of Results of Research

The research results which are related to any biological resource occurring in, or obtained from India, for monetary consideration or otherwise can be transferred to a non-Indian entity only after receiving the approval of NBA.

4.4.2.3 Section 5: Exemption for Collaborative Research Projects (CRPs)

Collaborative research projects involving the exchange of biological resources or information between two countries are exempted from Sections 3 and 4. However, such CRPs must confirm to the policy guidelines issued and approved by the central government.

4.4.2.4 Section 6: Application for Intellectual Property Rights (IPRs)

Without the NBA's approval, no one may file for any intellectual property right (IPR) in India or abroad for an innovation based on research or data on a biological resource acquired from India.

4.4.2.5 Section 7: Prior Intimation to SBB for Obtaining Biological Resources for Certain Purposes

Any biological resource can only be accessed for commercial use, bio-survey or bio-utilization by an Indian citizen or a body corporate, association or organization that is registered in India after obtaining permission/providing information to the relevant State Biodiversity Board (SBB).

4.4.2.6 Section 21: Determination of Equitable Benefit Sharing

The prescribed terms and conditions are in place to ensure that the benefits of using biological resources, their by-products and the innovations and practices related to their utilization are shared fairly. Benefit sharing may include granting joint ownership of IPR to the NBA, technology transfer, establishing research and development facilities to improve the quality of life for benefit claimants, creating a venture capital fund to support them, providing monetary compensation, and offering other non-monetary benefits.

4.4.2.7 Section 40: Power to Exempt Certain Biological Resources

In consultation with NBA, the central government may notify from time to time that the provisions of the Act shall not apply to any items; such exemption is already provided for biological resources normally traded as commodities. Other exemptions provided under the Act are detailed in Box 4.4.

Box 4.4 What Is Excluded Under the BDA, 2002?

- Indian citizens/entities for research use
- Publication of research papers/presentation in seminar/workshops
- Conventional breeding or traditional practices in agriculture, horticulture, poultry, dairy farming, animal husbandry, bee keeping in India
- Local people and communities of the area, growers and cultivators of biodiversity, and *vaids* and *hakims* and registered AYUSH practitioners only who have been practicing indigenous medicines, including Indian systems of medicine as profession for sustenance and livelihood except for obtaining intellectual property rights;
- NTCs normally traded commodities (NTACs) so far material is used only as commodity)
- Collaborative research projects conforming to policy guidelines issued and approved by the central government
- Crops covered under the multilateral system of ITPGRFA (Annex 1 crops)

(List attached as Annexures 3 and 4)

4.4.2.8 Section 55: Penalties

The penalty for contravention or attempt to contravene Section 3, 4 or 6 is imprisonment for a term which may be extended up to 5 years or with a fine of up to ten lakh rupees. Contravention or attempt to contravene Section 7 is imprisonment for a term which may extend up to 3 years or a fine of up to five lakh rupees.

To implement the Biodiversity Act, a centralized authority named the National Biodiversity Authority (NBA) was established at Chennai, Tamil Nadu in 2003. The NBA comprises of Secretariat, State Biodiversity Boards (SBBs), Biodiversity Management Committees (BMCs) and Expert Committees. The SBBs advise the state governments on matters related to conservation, sustainable use and equitable sharing of the benefits arising out of the utilization of biological resources. The SBBs also regulate approvals for access to Indians for commercial utilization or bio-survey and bio-utilization of any biological resource. The BMCs are working for the documentation and conservation of biological diversity and preserving habitats, land races, folk varieties, cultivars, domesticated stocks and breeds of animals and microorganisms. The NBA is implementing the provisions of the Biodiversity Act, 2002 and it has supported for the creation of SBBs in 28 states and union territories. In the country, total 2,66,499 BMCs have been created.

4.4.3 International Treaty on Plant Genetic Resources for Food and Agriculture (ITPGRFA)

The FAO established the International Undertaking on Plant Genetic Resources (IUPGR) in 1983. Its main objective was to investigate, conserve and assess PGR of economic use and made it accessible for plant breeding and research. The FAO Commission on Genetic Resources for Food & Agriculture (CGRFA) monitored the IUPGR's execution. Many of the developed nations with major private sector breeding companies were not signatories to IUPGR as this was non-legally binding which led to the revision of IUPGR under the guidance of CGRFA. The revised text of IUPGR was submitted in 31st session of FAO Conference and International Treaty on Plant Genetic Resources for Food and Agriculture (ITPGRFA) held on 3 November 2001. Thus, the legally binding ITPGRFA was thus negotiated in 2001, and came into force in 2004. It facilitates the access to PGRFA through an efficient mutually agreed system of access and benefit sharing. Access to PGR under the ITPGRFA is only for research, breeding and training and not for chemical, pharmaceutical or non-food/feed industrial use and it is always considered in harmony with CBD. No intellectual property rights can be claimed on the material received through the multilateral system. 64 plant genera—35 food and 29 forage crops—that were selected based on interdependence and food security are covered by the multilateral system and are known as Annex 1 crops. Adopted by the ITPGRFA governing body, the standard material transfer agreement (SMTA) lays out the terms for access and benefit sharing. As per the Biological Diversity Act and obligations of the treaty procedure, the export of germplasm from India is based on the following principles.

4.5 Procedure for Export of Plant Germplasm

The Biological Diversity Act (BDA), 2002, enacted in compliance with CBD, defines the non-Indian entity. As per the definition provided, a non-Indian entity can access biological resources or associated knowledge only with prior approval of NBA. Collaborative projects are exempted under Section 5 of the Act. These exemptions are as per the policy guidelines issued/approved by the central government. The request for research use of the germplasm may be classified into four categories

(Refer flowchart at Annexure 2). The Phytosanitary Certificate is, however, issued by the Division of Plant Quarantine at ICAR-NBPGR after the indenter gets approval in any category as follows.

4.5.1 Export of Plant Germplasm for Research Use Under Collaborative Projects

As mentioned above, Section 5 of the BDA, 2002, provides for exemption of export under collaborative research projects which are compliant with MoEFCC Guidelines issued in year 2008 (http://nbaindia.org/uploaded/pdf/notification/7%20%20collaborative%20guidelines.pdf). The requests for export under the provisions of collaborative research projects are facilitated by ICAR-NBPGR; however, for seeking approval, competent authority is the Department of Agricultural Research and Education (DARE), MA&FW, GOI.

4.5.2 Export of Annex 1 Crops Under ITPGRFA and FAO Designated Accessions of CG Centres Present in India

The ITPGRFA, 2001, provides for facilitated access to the crops mentioned in Annex 1 of ITPGRFA to all its member countries. The purpose of access is solely for utilization and conservation for research, breeding and training. India has signed and ratified the ITPGRFA in 2003. Under Section 40 of the BDA, 2002, Annex 1 crops are exempted from its purview. Any request for accessing Annex 1 crops (MLS) of the Treaty is submitted to Department of Agricultural Cooperation and Farmers Welfare (DAFW) as per the guidelines for the implementation of the ITPGRFA. These requests are processed by ICAR-NBPGR and based on the recommendations, the approval is granted by the competent authority DACFW, MA&FW, GOI. The Department of Agriculture and Cooperation (DAC), MA & FW, GOI is the national focal point (NFP) for the implementation of the ITPGRFA in India and has developed guidelines for access to PGRFA from India. According to these rules, NFP will have the authority to determine whether requests for access to PGRFA from contracting parties of the treaty are in compliance with BDA, 2002, and the terms of ITPGRFA. The supply of PGRFA of Annex 1 crops for the treaty is coordinated by NBPGR, the nodal body for the management and exchange of PGR for research purposes in India, in coordination with other pertinent national research institutions/organizations. Following agreements between IARCs and FAO, PGRFA gathered prior to 1993 and stored in 'trust' by International Agricultural Research Centres (IARCs) will also be made available in line with the terms of SMTA. On February 16, 2014, the Department of Agriculture & Cooperation, Ministry of Agriculture, on behalf of the Government of India, issued a notification on the exemption of Annex 1 crops, together with the approved Guidelines for the implementation of the ITPGRFA (Office Memorandum No. 13-5/2013 SD-V). A notification regarding the exemption of Annex crops under the treaty has also been released by the MoEFCC. Notice of exemption of crops listed in Annexe I of the ITPGRFA.

http://nbaindia.org/uploaded/pdf/Gazette_Notificaiton_on_exemption_of_crops_listed_in%20the_Annex-I_of_the_ITPGRFA_1.pdf. Therefore, according to the notification, access under the multilateral system is exempt from BDA, 2002 (18 of 2003) Sections 3 and 4, and it is comparable to Section 5, which exempts collaborative research projects from Sections 3 and 4.

4.5.3 Indian Researcher/Government Institution to Carry/Send Germplasm for Non-Commercial Research or Research for Emergency Purposes Other Than Collaborative Research

An Indian researcher or government organization wishes to transport or send germplasm for non-commercial study or research for urgent needs other than collaborative research; they must complete Form B and Undertaking and get permission from NBA, Chennai. Such applications are required to be submitted online as per the ABS guidelines notified in the year 2014. Approval for such exports is granted within 45 days of submission. The applicant is also required to submit a voucher sample to the designated repository and obtain an acknowledgement certificate for deposition within 30 days after receiving approval from NBA.

(http://nbaindia.org/uploaded/pdf/Gazette_Notification_of_ABS_Guidlines.pdf).

The designated repository for plant germplasm is ICAR-NBPGR (http://nbaindia.org/uploaded/pdf/notification/1%20designated%20repositories.pdf).

4.5.4 Access to the Biological Resource Other Than Annex 1 Crops by Non-Indian Entity

Any non-Indian entity (as per the definition provided under section 3 (2) of the Biological Diversity Act, 2002) for accessing any biological resource occurring in India is required to apply in Form 1 and submit the requisite fees to NBA, Chennai (https://absefiling.nic.in/NBA/login/auth). NBA grants approval within 6 months for such exports. The services for the same are available online.

4.6 Procedure for Indenting Germplasm for Research Use from ICAR-NBPGR

All the requests for the supply of PGR conserved/maintained by ICAR-NBPGR/NAGS are to be submitted online (http://pgrinformatics.nbpgr.ernet.in/ds/). The germplasm is supplied as per the terms of the material transfer agreement (MTA). Indenter has to pay the germplasm handling charges as per the fee approved by

ICAR NBPGR. The details are available on NBPGR website. (http://pgrinformatics.nbpgr.ernet.in/ds/Text/Domestic_Fee_Stru_2023.pdf). The private seed companies have to provide an undertaking that the company is wholly Indian and it does not fall under Section 3 (2) of BDA, 2002. Additionally, they need to submit the certificate of R&D recognition from the Department of Scientific & Industrial Research (DSIR), Ministry of Science & Technology, GOI. The private entities falling under Section 3 (2) of BDA, 2002, will have to take additional approval from NBA, Chennai.

4.7 Material Transfer Agreement (MTA)/Standard Material Transfer Agreement (SMTA)

A contract known as a material transfer agreement (MTA) regulates the exchange of physical research items between two organizations when the receiver plans to utilize them for research. Regarding the material and derivatives, the MTA outlines the recipient's rights and obligations as well as the provider's rights. Both the terms MTA and standard material transfer agreement (SMTA) can be used interchangeably. The SMTA adopted by the governing body of ITPGRFA is generally used for sharing or exchange of germplasm under the multilateral system of the treaty. SMTA is available in six official languages.

4.7.1 Material Transfer Agreement (MTA)

Issues of ownership, access, use, fair benefit sharing and intellectual property rights are all relevant when germplasm is shared or exchanged. The terms under which material is transferred for a specific purpose are outlined in an MTA, which is a contract. MTA is legally binding, covers a number of topics, including ownership of the transferred material and its derivatives, liability, confidentiality, and the rights of the provider and the recipient with regard to material transfer.

Each accession of germplasm is exchanged after the signing of MTA. An ideal format of MTA shall include a brief background, purpose for access, the definitions, scope, IPR on derivatives, benefit sharing to be determined on potential value, joint registration of varieties (IPR sharing), royalties sharing on commercialization, confidentiality, indemnity and liability section, validity, arbitration and signatories. These can be negotiated between the parties when being signed.

4.7.2 Standard Material Transfer Agreement (SMTA)

A standardized contract known as the standard material transfer agreement (SMTA) lays out the terms and conditions for transferring plant genetic material of crops that are part of the Annex 1 multilateral system (MLS) while guaranteeing adherence to the pertinent provisions of the international treaty. The treaty makes it easier for researchers, breeders and agricultural and food training professionals to access the genetic materials of the 64 crops in the MLS. Access to the documents is restricted to ratifying countries of the treaty and requires full use for agricultural and food research, breeding and training. The treaty guarantees that access to genetic resources already covered by international property rights are compliant with both national and international laws, and it prohibits recipients of genetic resources from asserting intellectual property rights over those resources in the form in which they were received. Through benefit-sharing procedures set out by the treaty, those who get genetic materials through the multilateral system commit to sharing any benefits arising from the use of accessed material.

SMTA ensures that the provisions of the treaty are followed by providers and recipients of plant genetic resources for food and agriculture. SMTA developed by the secretariat in the six official languages of the Treaty. SMTAs are available in four forms:

4.7.2.1 Click Wrap SMTA

To click wrap SMTA, the indenter needs to log in, click the 'SMTAs ready for acceptance' link from the menu 'As Recipient', A new page will appear listing any SMTAs waiting for review and acceptance.

4.7.2.2 Shrink Wrap MTA

Shrink wrap MTA is printed and attached to the parcel being shipped, and by accepting the parcel, the recipient is accepting the terms of the SMTA attached.

4.7.2.3 Signed SMTA

SMTA is presented to the recipient for physical signature and the material is shipped when the document has been signed. This type of SMTA is the traditional way of doing things.

4.7.2.4 EASY-SMTA

User-friendly information system called EASY-SMTA (http://mls.planttreaty.org) allows users to generate SMTA online, and in six official languages, for Annex 1 crops, without the need to negotiate a new agreement each time a transfer is made.

4.8 International Agricultural Research Centres and Germplasm Exchange

The international agricultural research centres (IARCs) play a very important role in global agricultural research with a focus on research, training and information exchange. The IARCs are non-profit organizations, independent, autonomous and have a special mandate, international in character and non-political in nature. To accomplish their goals, the IARCs draw on a wide range of fields, including economic policy and management, molecular genetics and agronomy. They work on a variety of crops and play a significant role in the global movement of germplasm. The Consultative Group on International Agricultural Research (CGIAR) is a consortium of donors that funds the majority of these research centres. Established in 1971 by international organizations and donor nations, the CGIAR offers a platform for proponents of global agricultural research to collaboratively evaluate and finance studies on food production, reducing hunger and conserving natural resources. The organization is a coalition of governments, private foundations, and regional and international organizations. A number of developing nations, including Brazil, India, the Philippines, and Mexico are among the donor countries.

The main objective behind CGIAR support is safe maintenance and free availability of diverse germplasm for their sustainable use in global research/crop improvement programmes. The CGIAR seeks to achieve the objective both through the institutions of support and through integrating national capitals. CGIAR, thus, supports diverse activities that include exploration, collection, characterization, evaluation, conservation, data management, information sciences and supply of germplasm to research workers. The foundation of the One CGIAR transition is the idea that when CGIAR employees and partners are united under fewer institutional barriers and supported by empowered management, transparent governance and an integrated operational structure; they can make a greater impact. Aiming for better integration and effect in the face of the interconnected challenges facing the modern world, the one CGIAR transition is a dynamic reformulation of CGIAR's partnerships, knowledge, assets and global presence. All CGIAR centres' boards accepted the CGIAR Integration Framework Agreement (IFA) on 22 February 2023, and their board chairs signed it. The contribution made by the various CGIAR centres for achieving food security are as follows.

4.8.1 International Rice Research Institute (IRRI), Los Banos Philippines

IRRI, which was founded in 1960 with funding from the Ford and Rockefeller Foundations, develops rice cultivars and crop management strategies to help rice farmers increase rice productivity and quality in a sustainable manner.

4.8.2 International Center for Maize and Wheat Improvement (Centro Internacional de Majormento de Maize y Trigo) CIMMYT, Apartado Postal, Lisboa, Mexico

CIMMYT was established in the year 1964. The mission of CIMMYT is to reduce poverty, improve food and nutrition security, and improve natural resources. It is a worldwide research partnership for a food-secure future.

4.8.3 International Institute of Tropical Agriculture (IITA), Ibadan, Nigeria

Since its inception in 1967, IITA has collaborated with both domestic and foreign partners to raise living standards, strengthen food and nutrition security, boost employment and protect the integrity of natural resources. The mandate crops include food legumes (cowpea, lima beans, soybean), root and tubers crops (sweet potato, cassava, yams).

4.8.4 International Center for Tropical Agriculture (Center Internacional de Agricultura Tropical), CIAT, Apartado, Cali, Columbia

The organization was established in 1968, and now it is known as Alliance of Biodiversity International and CIAT, Rome, with a mandate to advance the conservation and use of genetic diversity for the well-being of present and future generations. In order to enhance people's lives in the face of a climate crisis, the Alliance of Biodiversity International and CIAT provide research-based solutions that capture agricultural biodiversity and restructure food systems in a sustainable manner.

4.8.5 AfricaRice, Earlier West-African Rice Development Association

It was founded in 1971 by 11 African countries. AfricaRice headquarter is based in Côte d'Ivoire, and it undertakes exploration/collection of rice genetic resources.

4.8.6 International Potato Center (Centro Internacional de la Pappa) CIP, Apartado, Lima, Peru

It was established in 1971 and it mandate includes crops such as potato, sweet potato and Andean roots and tubers.

4.8.7 International Crops Research Institute for the Semi-Arid Tropics (ICRISAT)

ICRISAT was founded in 1972, and its headquarters is located in Hyderabad, India. It also has country offices in Niger, Nigeria, Zimbabwe, Malawi, Ethiopia and Mozambique, as well as two regional hubs: Nairobi, Kenya and Bamako, Mali. ICRISAT is a World repository of groundnut germplasm and the mandate crops include chickpea, pigeon pea, pearl millet, sorghum and groundnut.

4.8.8 International Livestock Centre for Africa (ILRA), Adidas Ababa, Ethiopia 1974 & International Laboratory for Research on Animal Disease (ILRAD), Nairobi, Kenya, 1974

These two organizations were merged to form International Livestock Research Institute (ILRI), Nairobi, Kenya. ILRI scientists lead the CGIAR Initiative on livestock and climate.

4.8.9 International Food Policy Research Institute (IFPRI), Washington DC

It was founded in 1975 and it offers evidence-based policy recommendations to eradicate hunger and malnutrition and reduce poverty in a sustainable manner. In order to meet the demand for research on food policy and provide comprehensive assistance for development led by the nation, the institute's regional and national projects are essential.

4.8.10 International Centre for Agricultural Research in Dry Areas (ICARDA), Syria

ICARDA has carried out research-for-development initiatives in 50 countries throughout the arid regions of the world since its inception in 1977. With a global mandate to produce barley, lentils, and faba beans, ICARDA also works to increase water-use efficiency in agriculture, rangeland concerns, and small-ruminant production in non-tropical dry regions. ICARDA helps to develop bread and durum wheat, kabuli chickpea, pasture and fodder legumes, and related farming systems across the Middle East, North Africa and Central Asia.

4.8.11 World Agro-Forestry (International Centre for Research in Agro-Forestry (ICRAF), Nairobi, Kenya)

It was founded in 1991 with the goal of using better agroforestry methods to reduce rural poverty, land depletion, and tropical deforestation. The only organization conducting internationally relevant agro-forestry research for all emerging tropical regions is ICRAF.

4.8.12 The Center for International Forestry Research (CIFOR), Indonesia

CIFOR investigates the world's most important forest and landscape management issues. The CGIAR Research Program on Forests, Trees, and Agroforestry (FTA) is headed by CIFOR. With offices in Nairobi, Kenya; Yaounde, Cameroon; Lima, Peru; and Bonn, Germany, it has its headquarters in Bogor, Indonesia.

4.8.13 International Management Institute, Colombo (Sri Lanka) Now International Water Management Institute (IWMI)

It was established in established in 1984 for providing scalable agricultural water management solutions that significantly improve food security, ecosystem health and poverty reduction, IWMI collaborates with governments, civic society and the commercial sector.

4.8.14 International Center for Living Aquatic Resources Management, ICLARM Now WorldFish

World fish leads the CGIAR Fish Research Programme. With regional offices in Bangladesh, Cambodia, Egypt, Myanmar, Solomon Islands and Zambia, WorldFish operates from its headquarters in Penang, Malaysia.

4.9 Centres Involved in Plant Introduction and Germplasm Activities Other Than CGIAR

Apart from CGIAR network and the FAO, important centres involved in plant genetic resource introduction and exchange activities are as follows:

4.9.1 World Vegetable Center Genebank, Taiwan

The World Vegetable Centre (formerly Asian Vegetable Research and Development Center - AVRDC) preserves the greatest public vegetable germplasm in the world. Distribution and protection of vegetable germplasm are important undertakings. The centre is a member of Genesys, a global network for exchanging germplasm and information. All of the centre's accessions are listed in the Vegetable Genetic Resources Information System.

4.9.2 International Development Research Centre, Ottawa, Canada

The organization aims to conserve and use oilseeds, bananas, rattans, bamboo and small millets. Its goal is to safeguard, maintain and improve the genetic variety of both their wild relatives and cultivated plants that are significant to Canadian agriculture. Saskatoon Research and Development Centre (seed germplasm), Fredericton Research and Development Centre (potato germplasm) and Harrow Research and Development Centre (fruit germplasm) are the three genebanks in the Canadian national plant germplasm system that focuses on distinct crop groupings.

4.9.3 National Institute for Agrobiological Resources (NIAR), Tsukuba Japan

NIAR facilitates the exchange of fruit plant and grain crop germplasm. Currently, Japan's largest agricultural research institute for basic life sciences is the National Institute of Agrobiological Sciences (NIAS). The main organization in Japan that coordinates the conservation of agriculturally related plants, microbes and animals is the National Agriculture and Food Research Organization (NARO). Scientists interested in germplasm from the NARO Genebank system are referred to databases that offer comprehensive instructions on submitting requests for materials in each genetic resource category.

4.9.4 Australian Grains Genebank (AGG)

With one of the world's largest collections of grain crop species, including cultivated, landrace and wild relative species of temperate and tropical crops, the Australian Grain Genebank is the country's bio-digital resource centre that facilitates and informs access to grain crop genetic resources. Stringent guidelines with the provisions of phytosanitary and quarantine legislation, as well as rules pertaining to access and benefit-sharing, are followed while sourcing germplasm. Researchers and breeders can easily access all germplasm information using the GRIN-Global database.

4.10 Plant Germplasm Introduction and Exchange Activities at USDA, USA and IPK, Germany

USDA-ARS, Beltsville, Maryland, USA, was founded in 1862 and to formalize plant introduction, a section called seed and plant introduction was established in 1898. USDA have a large collection of animal, microbial and plant genetic resources (germplasm) crucial to food and agricultural production and the information about these collection can be accessed online using Germplasm Resources Information

Network (GRIN) database. More than 600,000 active accessions of plant genetic resources for food and agriculture are maintained by the U.S. National Plant Germplasm System. All life forms' germplasm is preserved by the National Seed Storage Laboratory, which is currently known as the National Centre for Genetic Resources Preservation (NCGRP). The Agricultural Research Service (ARS), the research arm of the US Department of Agriculture (USDA), oversees the National Plant Germplasm System (NPGS), a cooperative endeavour to preserve the genetic diversity of crops of agricultural importance. Numerous NPGS genebanks are housed in state land-grant university locations, which provide staff for technical and support services as well as lab, office, greenhouse and field space for operations. One such Plant Genetic Resources Unit (PGRU), which is a part of the National Plant Germplasm System (NPGS), is situated on Cornell University's Geneva, New York Campus. The National Clonal Germplasm Repository for Apple, Tart Cherry, and Grape (NGR) and the Northeast Regional Plant Introduction Station (NERPIS) merged to become the PGRU in 1986. One of the biggest collections in the world is the ex-situ genebank at IPK, Germany, which is housed on the main campus in Gatersleben and two branches in the country's north. Since IPK has designated all of its genebank collections under the treaty, seed and plant material are dispersed in accordance with the conditions of the SMTA. To gain an idea of the activities carried out at the global level, a brief description of the activities of PGRU, Geneva, and IPK, Germany, is provided below. The Genesys-PGR, USDA-GRIN, and National Genebank databases are online resources that provide access to all of the collections that are kept.

4.10.1 Plant Germplasm Introduction and Exchange Activities at the USDA-ARS, Plant Genetic Resources Unit (PGRU), and the Grape Genetics Research Unit (GGRU) Located on the Geneva, NY Campus of Cornell University

The Plant Genetic Resources Unit (PGRU) of USDA-ARS is a part of the National Plant Germplasm System (NPGS), which is managed by the USDA. The National Clonal Germplasm Repository for Apple, Tart Cherry and Grape (NGR) and the Northeast Regional Plant Introduction Station (NERPIS) merged to become the PGRU in 1986. PGRU and the Grape Genetics Research Unit (GGRU) are part of the USDA-ARS National Plant Germplasm System (NPGS). PGRU's goals include acquiring, documenting, maintaining, characterizing, breeding, improving and distributing improving clonal and seed crops and the breeding and improvement of apples. Over 300 species are represented by more than 20,000 accessions that PGRU maintains. Additionally, PGRU cultivates a variety of seed-propagated crops, such as tomato, radish, buckwheat, cole crops, squash, celery, onion and others. Numerous crops experience seed regeneration each year; the frequency of regeneration depends on variables like quantity, viability and genetic integrity.

Genetics Information Network (GRIN) database of USDA-ARS refers to the PGRU as site NE9 for vegetable crops and site GEN for clonally propagated crops.

These codes are to be used when placing an order for germplasm. Preservation of genetic diversity is the primary factor for the existence of germplasm repositories. PGRU utilizes both orchard-grown and cryogenic preservation. The *Malus* collection, for instance, is kept on a farm close to Geneva. For every accession, two trees are usually kept up to date. Additionally, certain accessions are kept in cryogenically maintained buds. About 20,000 grape, apple, tart cherry and vegetable accessions are kept at PGRU. The Grape Genetics Research Unit (GGRU) was founded by USDA-ARS in Geneva, New York, in 2005 with the goal of promoting grape production, multidisciplinary research, breeding and technology transfer. The programme aims to improve grape and grape product quality and utilization while lowering crop yield and quality losses due to abiotic stress, diseases and pests. Through breeding, genetics, genomics, proteomics, digital technologies, and bioinformatics, better varieties, germplasm, tools and information are created. The findings are shared with other researchers, packers, processors and grape producers. Currently, GGRU concentrates on three plant protection and production challenges that are significant to the grape sector nationally: enhanced fruit quality, tolerance to stress from unfavourable cold and drought conditions and resistance to powdery mildew disease. Present focus of GGRU are the increased fruit yield and quality of wine, juice, and table grapes; tolerance to stress from unfavourable cold and drought conditions and resistance to powdery mildew disease. The USDA-ARS National Grape Improvement Centre, located on the Cornell campus in Geneva, is home to GGRU's cutting-edge research centre, which offers genetic, genomic and biotechnological solutions.

4.10.2 Leibniz Institute of Plant Genetics and Crop Plant Research (IPK), Gatersleben, Germany

The largest plant germplasm collection in the European Union, both in terms of the number of accessions maintained and the botanical diversity represented, is the Federal ex-situ Genebank of the Leibniz Institute of Plant Genetic Resources and Crop Plant Research (IPK), which was established in 1943. Its current collection includes 2912 species, 92 families, 758 genera and 152,359 accessions. The Genebank's primary duties include gathering, preserving and characterizing its plant genetic resources as well as carrying out research and development related to the preservation of plant genetic resources. Three locations make up the collection: the main one is on the Gatersleben campus, while the other two are at Malchow (oil and fodder crops) and GroBlusewitz (potato) in northern Germany. The genebank collaborates closely with Bioversity International, the European Cooperative Programme on Plant Genetic Resources (ECPGR) and the German Federal Ministry of Food and Agriculture to support coordinated conservation of genetic resources at the national and international level. In this regard, the genebank is in charge of running and expanding the European Search Catalogue for Plant Genetic Resources (EURISCO), which also supplies data for the Genesys-PGR database. The Genebank Information System (GBIS), an internal database created by IPK, houses all pertinent data on the genebank collection. It serves as the main database and software for managing accessions in the genebank and offers passport data, as well as details on the material's identity, history, place of origin and botanical classification, to people all over the world through its website. Characterization and evaluation data (C&E data) of specific plant species are being published through GBIS as well. Additionally, it has an integrated genebank material ordering system. International PGR databases receive genebank data in MCPD format. Additionally, the system is linked to both external and IPK-internal systems through various interfaces. Using free or scientific search criteria, the IPK Genebank Information System (GBIS) provides the opportunity to learn more about the Federal ex-situ collection's holdings. The integrated ordering system is accessible after a user account has been created. The Standard Material Transfer Agreement's (SMTA) terms must be accepted before any orders may be placed. An extra agreement stating that users accept the SMTA for non-Annex 1 crops is required for non-Annex 1 crops. Additionally, shipping of the genebank material is free for developing nations, although it is subject to a handling fee. Approximately 15,000 to 20,000 accessions are distributed globally each year. Breeders and researchers order the majority of the material. Over 1.1 million samples have been provided to users in Germany and other countries since 1948.

4.11 Summary

Plant introduction is the process of procuring plant germplasm from different sources and bringing into cultivation the germplasm in an area which is new for them or they have not been grown earlier. Based on their acclimatization, they can be defined as primary introduction and secondary introduction. Germplasm exchange includes activities of importing plant germplasm from other countries and also exporting to other countries based on the requests from the applicants. The exchange of plant germplasm was unrestricted before 1993, however, during the Rio Earth Summit, 1992, a consensus was made to implement the Convention on Biological Diversity, and since then the biological resources including plant germplasm became the sovereign rights of the nation. CBD regulated the access to biological resources. The International Treaty on Plant Genetic Resources for Food and Agriculture (ITPGRFA) was negotiated and put into effect in 2004 to enable the exchange of plant germplasm, which is essential for food and nutritional security. The Nagoya Protocol then went into effect in 2014 with the objective fair and equitable distribution of advantages resulting from the utilization of biological resources. A list of 64 crops that are essential for food and nutritional security was identified and agreed upon by the contractual parties under ITPGRFA. These are known as Annex 1 crops, and the member nations of the multilateral system are required to facilitate their interchange under the standard material transfer agreement. The international agricultural research centres play an important role in global agricultural research through the conservation and free exchange of diverse germplasm for research use at a global level. In compliance with Convention on Biological

Diversity, India enacted the Biological Diversity Act in the year 2002 and came up with Biological Diversity Rules in 2004. The regulations outlined in the Plant Quarantine (Regulation of Import into India) Order, 2003, govern the import of plant germplasm into India, while the Biological Diversity Act, Biological Diversity Rules and the International Treaty govern the export of plant germplasm within India.

Multiple-Choice Questions

1. Grape, an introduced crop in India, is documented to be introduced from
   (a) Pakistan
   (b) Afghanistan
   (c) Uzbekistan
   (d) Japan
2. The National Centre for Genetic Resources Preservation (NCGRP) is located in which state of USA
   (a) Colorado
   (b) Washington
   (c) California
   (d) Maryland
3. *Kalyan Sona* is an example of
   (a) Primary introduction
   (b) Secondary introduction
   (c) Indigenous variety
   (d) None of the above
4. Which of the following introduced crops cause depletion of subsoil water reserve
   (a) *Eichhornia crassipes*
   (b) Water hyacinth
   (c) *Phalaris minor*
   (d) Eucalyptus
5. One of the following IARC is not supported by CGIAR
   (a) World forestry (CIFOR)
   (b) World fish
   (c) World vegetable centre
   (d) Africa rice
6. Which of the following CGIAR centre is the world repository for groundnut?
   (a) ICRISAT
   (b) CIAT
   (c) IITA
   (d) ILRI
7. Annato (*Bixa orellana*) is an introduction from
   (a) Tropical America
   (b) China
   (c) South East Asia
   (d) West Indies

8. To import a transgenic seed sample for research purposes, permission of the following committee is mandatory
   (a) RCGM
   (b) GEAC
   (c) EXIM
   (d) None of the above
9. The non-Indian entity in BDA, 2002, is defined under
   (a) Section 2
   (b) Section 4
   (c) Section 5
   (d) Section 3
10. The office of the National Biodiversity Authority is located in
   (a) Kolkata
   (b) Chennai
   (c) Delhi
   (d) Bangalore

Answers

1. (b)
2. (a)
3. (b)
4. (d)
5. (c)
6. (a)
7. (a)
8. (a)
9. (d)
10. (b)

Fill in the Blanks

1. Access to any biological resource occurring in India is governed by --------------
2. IRRI, Philippines, is a --------------------------------
3. I am a non-Indian entity as per the Biological Diversity Act 2002 and require some seed material for research work which is occurring in India. I need to apply to -------------------------
4. Plant Quarantine Order 2003 regulates --------------------------------- of germplasm.
5. ----------------- and --------------- are the two mandatory requirements for import of any seed/planting material into India.

6. Forest Research Institute, Dehradun, has a plant introduction organization which looks after the introduction, maintenance and testing of germplasm of ---------------------------------
7. To import seed/planting material in bulk for commercial purpose, the application should be sent to ----------------------------
8. The full form of NCGRP is ------------
9. For national supply/domestic supply of germplasm, the indentor is requested to submit --------------- and --------------------------
10. Tea and loquat were introduced from ------------------------------------

True or False

1. Director, NBPGR, is not empowered to issue import permit for import of seed/plant material for research/experimental purposes into the country. (T/F)
2. Import permit for commercial use issued by NBPGR. (T/F)
3. The full form of BDA is Biodiversity Act (T/F).
4. Sonora 64, Lerma Rojo wheat varieties are examples of secondary introduction (T/F).
5. The applicant desirous of importing germplasm into India is required to fill in PQ 08 form. (T/F)
6. Import of germplasm into India is governed by Biological Diversity Act (T/F).
7. Prefix IC is assigned to germplasm imported into the country. (T/F)
8. Water hyacinth and *Lantana camara* both introduced as ornamentals turned noxious weeds (T/F).
9. World vegetable centre is a CG institution. (T/F)
10. For importing transgenics, the application is submitted to Institute Biosafety Committee (IBSC) which has a member from Department of Biotechnology (DBT) (T/F).

Short Answers

1. Name any four national/international agreements, conventions, policies regulating access to PGR.
2. Describe the objectives of International Treaty on Plant Genetic Resources for Food and Agriculture (ITPGRFA).
3. Name exemptions under Section 3 of the BDA 2002.
4. Name salient features of Convention on Biological Diversity.
5. Describe role of CGIAR centres in germplasm exchange.
6. Describe plant introduction services.
7. Describe agro-climatic concepts in relation to germplasm introduction.
8. Describe role of review committee in exchange of transgenic germplasm.
9. Name secondary introductions and crop improvement.
10. Describe role of National Biodiversity Authority, Chennai, in regulating access to germplasm

Annexures

Annexure 1

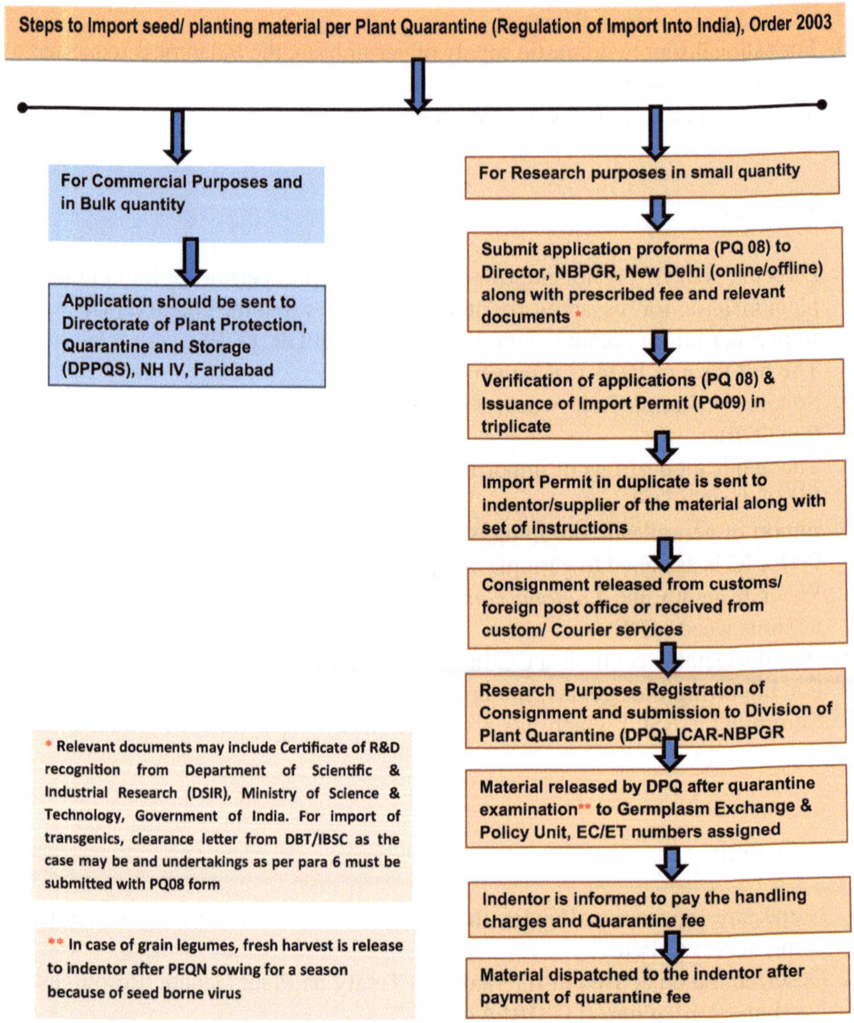

Annexure 2

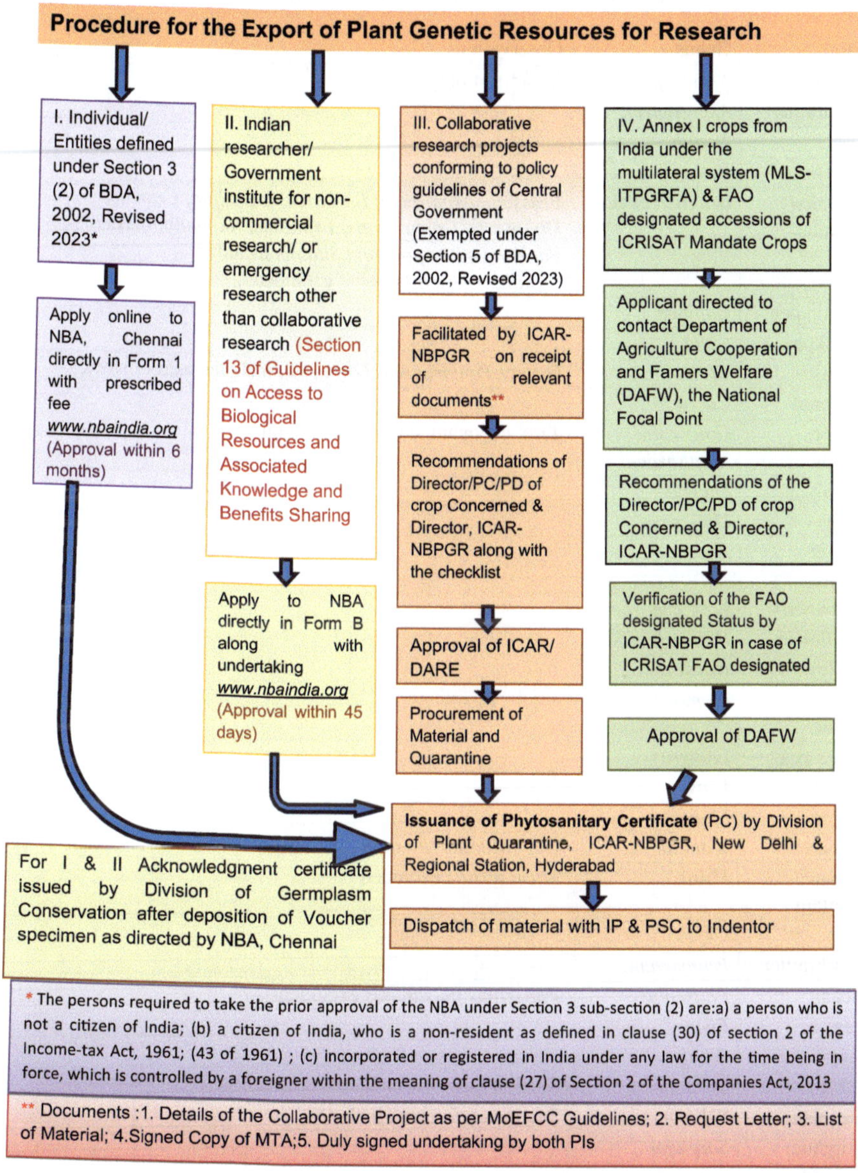

* The persons required to take the prior approval of the NBA under Section 3 sub-section (2) are:a) a person who is not a citizen of India; (b) a citizen of India, who is a non-resident as defined in clause (30) of section 2 of the Income-tax Act, 1961; (43 of 1961) ; (c) incorporated or registered in India under any law for the time being in force, which is controlled by a foreigner within the meaning of clause (27) of Section 2 of the Companies Act, 2013

** Documents :1. Details of the Collaborative Project as per MoEFCC Guidelines; 2. Request Letter; 3. List of Material; 4.Signed Copy of MTA;5. Duly signed undertaking by both PIs

Annexure 3

List of crops covered under the multilateral system of ITPGRFA (Annex 1 crops) food crops

Crop	Genus	Observations
Breadfruit	*Artocarpus*	Breadfruit only
Asparagus	*Asparagus*	
Oat	*Avena*	
Beet	*Beta*	
Brassica complex	*Brassica* et al.	*Brassica, Armoracia, Barbarea, Camelina, Crambe, Diplotaxis, Eruca, Isatis, Lepidium, Raphanobrassica, Raphanus, Rorippa, and Sinapis* included *Lepidium meyenii* (maca) excluded
Pigeon pea	*Cajanus*	
Chickpea	*Cicer*	
Citrus	*Citrus*	Genera *Poncirus* and *Fortunella* included as root stock
Coconut	*Cocos*	
Major aroids	*Colocasia, Xanthosoma*	Taro, cocoyam, dasheen and tannia
Carrot	*Daucus*	
Yams	*Dioscorea*	
Finger millet	*Eleusine*	
Strawberry	*Fragaria*	
Sunflower	*Helianthus*	
Barley	*Hordeum*	
Sweet potato	*Ipomoea*	
Grass pea	*Lathyrus*	
Lentil	*Lens*	
Apple	*Malus*	
Cassava	*Manihot*	Only *Manihot esculenta*
Banana/ plantain	*Musa*	*Musa textilis* excluded
Rice	*Oryza*	
Pearl millet	*Pennisetum*	
Beans	*Phaseolus*	*Phaseolus polyanthus* excluded
Pea	*Pisum*	
Rye	*Secale*	
Potato	*Solanum*	Section *tuberosa* included, *Solanum phureja* excluded
Eggplant	*Solanum*	Section *melongena* included
Sorghum	*Sorghum*	
Triticale	*Triticosecale*	
Wheat	*Triticum* et al.	*Agropyron*, *Elymus*, and *Secale* included
Faba bean/ vetch	*Vicia*	
Cowpea et al.	*Vigna*	
Maize	*Zea*	*Zea perennis*, *Zea diploperennis* and *Zea luxurians* excluded

Annexure 4

Forages

Genera	Species
Legume forages	
Astragalus	*Chinensis, cicer, arenarius*
Canavalia	*Ensiformis*
Coronilla	*Varia*
Hedysarum	*Coronarium*
Lathyrus	*Cicera, ciliolatus, hirsutus, ochrus, odoratus, sativus*
Lespedeza	*Cuneata, striata, stipulacea*
Lotus	*Corniculatus, subbiflorus, uliginosus*
Lupinus	*Albus, angustifolius, luteus*
Medicago	*Arborea, falcata, sativa, scutellata, rigidula, truncatula*
Melilotus	*Albus, officinalis*
Onobrychis	*Viciifolia*
Ornithopus	*Sativus*
Prosopis	*Affinis, alba, chilensis, nigra, pallid*
Pueraria	*Phaseoloides*
Trifolium	*Alexandrinum, alpestre, ambiguum, angustifolium, arvense, agrocicerum, hybridum, incarnatum, pratense, repens, resupinatum, rueppellianum, semipilosum, subterraneum, vesiculosum*
Grass forages	
Andropogon	*Gayanus*
Agropyron	*Cristatum, desertorum*
Agrostis	*Stolonifera, tenuis*
Alopecurus	*Pratensis*
Arrhenatherum	*Elatius*
Dactylis	*Glomerata*
Festuca	*Arundinacea, gigantea, heterophylla, ovina, pratensis, rubra*
Lolium	*Hybridum, multiflorum, perenne, rigidum, temulentum*
Phalaris	*Aquatica, arundinacea*
Phleum	*Pratense*
Poa	*Alpina, annua, pratensis*
Tripsacum	*Laxum*
Other forages	
Atriplex	*Halimus, nummularia*
Salsola	*Vermiculata*

Plant Exploration and Germplasm Collection

5

Praveen K. Singh, K. Pradheep, K. C. Bhatt, S. P. Ahlawat, R. K. Pamarthi, and O. P. Dhariwal

Learning objectives

- Understanding the terminologies and fundamentals of plant exploration and germplasm collection
- Practical tips or guidelines to be followed while conducting explorations
- Step-by-step procedure involved in the collection process and afterwards

Keywords

Plant genetic resources · Germplasm · Exploration · Collection · Biodiversity hotspots · Collection process

5.1 Introduction

Plant genetic resources (PGR) are defined as genetic material of plant origin of actual or potential value in the form of seed, vegetative propagule, tissue, cell, pollen, DNA molecule etc. containing the functional unit of heredity that can be utilized in crop improvement. In short, they comprise plant genetic materials of value for present and future generations. They form part of agrobiodiversity within the broader realm of biodiversity, which includes landraces, genetic

P. K. Singh (✉) · K. C. Bhatt · S. P. Ahlawat · R. K. Pamarthi · O. P. Dhariwal
Division of Plant Exploration and Germplasm Collection, ICAR-National Bureau of Plant Genetic Resources, New Delhi, India
e-mail: praveen.singh1@icar.org.in; pksingh128@gmail.com

K. Pradheep
ICAR-National Bureau of Plant Genetic Resources, Regional Station, Thrissur, Kerala, India

K. Tripathi et al. (eds.), *Textbook of Plant Genetic Resources*,
https://doi.org/10.1007/978-981-96-5024-8_5

stocks, farmers' varieties, primitive cultivars, released cultivars, parental lines and experimental hybrids, wild and weedy relatives, semi-domesticated species (Box 5.1). There has been a fourfold increase in food grain production since the 1950s and six times in horticultural crop production, thanks to the significant contribution of genetic resources in achieving the same. It is worth mentioning that dwarfing gene *Norin 10* from 'Daruma', a native Japanese short-straw wheat, helped in the development of dwarf, but high-yielding and photo-insensitive Indian wheat which led to the green revolution. Similarly, *Dee-Gee-Woo-Gen* from TN-1, Taiwan rice, helped in the development of high yielding, dwarf and photo-insensitive varieties of rice.

Box 5.1 Common Terminologies in the PGR Spectrum used in Exploration

Cultivar: A variety defined by a set of common features of a plant that has been created or selected intentionally and maintained through cultivation.

Ecotype: A population of a species that survives as a distinct group through environmental selection and isolation, maintains its genetic identity and usually is linked to one or more environmental factors.

Farmers' variety: A variety which has been traditionally cultivated and evolved by the farmers in their fields. It was developed, grown and exchanged among the farmers.

Genetic stock: Plants or populations generated and/or selected for genetic studies and represent a unique class of extremely valuable germplasm.

Plant population: The total number of plants present in a geographical area or region.

Sample: A part of a population used to estimate the characteristics of the whole (Source: Guidelines for management of PGR-2016).

Landrace: A crop cultivar that evolved with and has been genetically improved by traditional agriculturalists, but has not been influenced by modern breeding practices. It comprises a population of individuals that have become adapted to specific local environmental conditions in which it evolved and is usually genetically heterogeneous.

Primitive cultivar: An early cultivated form of a crop species evolved from a wild population.

Semi-domesticated species: Species that are not clearly morphologically or genetically distinct from their wild counterparts, but are under cultivation and subjected to conscious artificial selection pressures.

Wild relative: Plant species that are taxonomically related to crop species and serve as potential sources for genes in the breeding of new varieties of those crops.

Wild species: Species occurring in relatively undisturbed or less disturbed habitats and areas that have not been subjected to breeding to alter them from their native (wild) state

Main source: NBPGR (2014)

Agricultural biodiversity has a significant role in sustaining and strengthening the food and nutritional security and health of the growing human population. Of the 2,50,000 known species of higher plants, only about 3000 species are grown worldwide for human use and only 30 species account for most of the food and other basic needs of human beings; therefore, the present food base is very narrow. In most of the economic crops, a yield plateau has already been attained. The rapid change in climatic conditions in recent decades has caused a significant impact on crop production, which in turn necessitates climate-resilient crop breeding approaches and alternatively, looking for new crops. The alarming depletion of local landrace diversity in crops due to changes in land use patterns, changes in food habits and population pressure, all necessitate the collection of PGR before getting lost. Therefore, the main purpose of exploration and germplasm collecting is to enhance the genetic variability (within the crop/species) and species diversity in genebank collections and thus provide useful materials for breeding, research and restoration. Augmentation of germplasm is the first and foremost activity in the PGR management system; therefore, meticulous planning and exploration following scientific principles governing the diversity distribution are crucial. Ideally, appropriate collecting strategies need to be designed specifically for each crop and its wild relatives.

The objective of plant exploration and germplasm collecting is to assemble the desired genetic material in viable and sufficient quantity along with adequate documentation, following relevant legislative protocols. International Code of Conduct for Plant Germplasm Collecting and Transfer may also be followed (https://www.fao.org/4/aj634e/aj634e.pdf) in case of explorations aimed at foreign countries.

Exploration: Definition

- Exploration refers to the activity of surveying and searching for new or useful things in a particular locality while germplasm collecting encompasses the activities of identifying, locating, acquiring, organizing, cataloguing and post-collection handling.

5.2 History of Germplasm Exploration in India

There were reports of sporadic collections of indigenous crop germplasm during the earlier part of the twentieth century, for instance, wheat (from north-western plains), jute and few pulses. Dr. BP Pal in his classic paper 'Search for new genes' emphasized the importance of germplasm augmentation in crop improvement programmes. However, systematic plant exploration and germplasm collecting activities was initiated in India with the establishment of a nucleus Plant Exploration and Collection Unit in 1946 in the then Division of Botany, Imperial Agricultural Research Institute, New Delhi, under the leadership of Late Dr. Harbhajan Singh; and later as a separate Division of Plant Introduction in

IARI. Dr. Harbhajan Singh, also known as 'Indian Vavilov', significantly contributed to the field of PGR collection; he initiated many international collaborative explorations for the collection of wild rice, minor millets, legumes, sesame, wild crucifers, forage grasses, *Musa*, cucumber and melons, tea clones and ornamental orchids from diverse regions of the country. The period between 1946 and 1976, hence, has witnessed several national and international collaborative explorations in the country (Fig. 5.1).

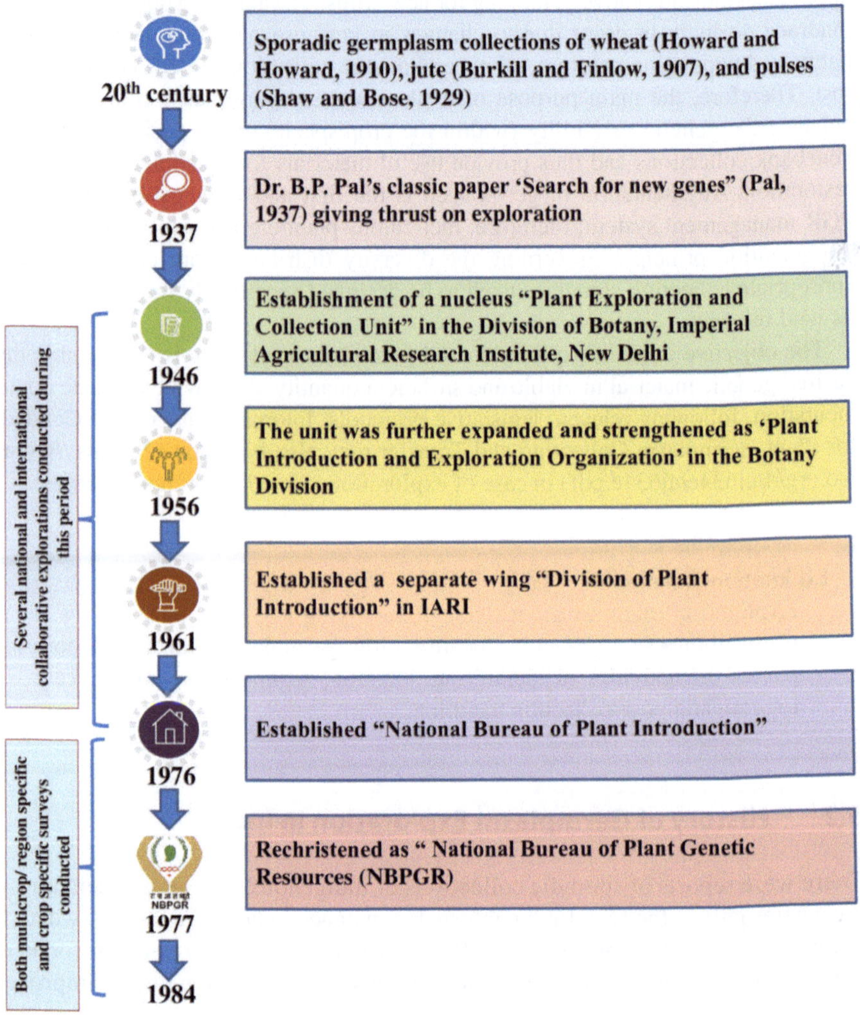

Fig. 5.1 History of germplasm collections and establishment of ICAR-NBPGR, New Delhi

5.3 Distribution and Extent of Prevalent Genetic Diversity

A biodiversity hotspot is a designated region with significant levels of biodiversity but at the same time threatened by human inhabitation. According to Norman Myers, who conceptualized biodiversity hotspot, a region must comply two important criteria, (1) it must contain at least 1500 species of vascular plants (more than 0.5% of the world's total) as endemics; (2) it has to have lost at least 70% of its primary vegetation. Globally, 36 biodiversity hotspots have been identified so far (Table 5.1). These sites with just 15.7% area support nearly 60% of the world's species, with a high share of endemic species. India is a gene-rich country. It constitutes a part of four out of the 36 biodiversity hotspots—Western Ghats, Himalayas, Indo-Burma (NE India and Andaman) and Sundaland (Nicobar).

According to the EnviStats India 2022, 21,984 species of angiosperms occur in India, of which 4,556 species are endemic. India is acclaimed as one of the 12 mega-diversity gene centres (Table 5.2). In 1968, Zhukovsky put forth his idea about 'Megagene Centres'. Zhukovsky also identified over 100 micro-gene centres within the megagene centres. These micro-gene centres demonstrated exceptional local diversity and a rich variety of wild species related to cultivated crops (Zhukovsky 1975). Since many crops originated outside the Vavilov's centres of origin, it was necessary to expand the areas in which species were domesticated.

About 168 crop taxa have originated and/or developed diversity in India including rice, sugarcane, pigeon pea, yam, taro, eggplant, banana, cucumber, cardamom, black pepper, mango and citrus. About 1000 wild edible plant species and 861 taxa of wild relatives of crops occur here. Various phytogeographic regions/ecological zones and associated richness in crop diversity of this country are given in Table 5.3.

Table 5.1 List of biodiversity hotspots in the world

Sl. no.	Biodiversity hotspots
1	Tropical Andes (South America)
2	Tumbes-Choco-Magdalena (South America)
3	Madrean Pine–Oak Woodlands (North and Central America)
4	Cerrado (South America)
5	Chilean Winter Rainfall and Valdivian Forests (South America)
6	Atlantic Forest (South America)
7	Mesoamerica (North and Central America)
8	Caribbean Islands (North and Central America)
9	California Floristic Province (North and Central America)
10	Guinean Forests of West Africa (Africa)
11	Cape Floristic Region (Africa)
12	Succulent Karoo (Africa)
13	Maputaland–Pondoland–Albany (Africa)
14	Coastal Forests of Eastern Africa (Africa)

(continued)

Table 5.1 (continued)

Sl. no.	Biodiversity hotspots
15	Eastern Afromontane (Africa)
16	Horn of Africa (Africa)
17	Madagascar and the Indian Ocean Islands (Africa)
18	Mediterranean Basin (Europe and Central Asia)
19	Caucasus (Europe and Central Asia)
20	Irano-Anatolian (Europe and Central Asia)
21	Mountains of Central Asia (Europe and Central Asia)
22	Western Ghats and Sri Lanka (South Asia)
23	Himalaya (South Asia)
24	Mountains of Southwest China (East Asia)
25	Indo-Burma (South Asia)
26	Sundaland (Southeast Asia and Asia-Pacific)
27	Wallacea (Southeast Asia and Asia-Pacific)
28	Philippines (Southeast Asia and Asia-Pacific)
29	Japan (East Asia)
30	Southwest Australia (Southeast Asia and Asia-Pacific)
31	East Melanesian Islands (Southeast Asia and Asia-Pacific)
32	New Zealand (Southeast Asia and Asia-Pacific)
33	New Caledonia (Southeast Asia and Asia-Pacific)
34	Polynesia–Micronesia (Southeast Asia and Asia-Pacific)
35	Forests of East Australia (Southeast Asia and Asia-Pacific)
36	North American Coastal Plain (North and Central America)

Table 5.2 List of mega-diversity gene centres (Zeven and Zhukovsky 1975)

Sl. no.	Mega-diversity gene centre	Sl. no.	Mega-diversity gene centre
1	China	7	Mediterranean coastal and adjacent regions
2	Indochina-Indonesia	8	Africa (8a Ethiopia)
3	Australia-New Zealand	9	Europe-Siberia
4	Indian subcontinent	10	Central America
5	Central Asia	11	Bolivia-Peru-Chile
6	West Asia	12	North America

Table 5.3 List of phyto-geographical regions of India

Sl. no	Region	Crops
1	Western Himalayas	Barley, wheat, maize, rice, buckwheat, amaranth, chenopod, proso millet, finger millet, French bean, lentil, black gram, peas, pumpkin, cucumber, *Allium* spp., ginger, brassica, pome, stone, soft and nut fruits, medicinal and aromatic plants
2	Eastern Himalayas	Rice, barley, maize, buckwheat, amaranth, chenopod, finger millet, foxtail millet, French bean, cowpea, black gram, peas, scarlet bean, pumpkin, cucumber, *Allium* spp., ginger, brassicas, pome, stone fruits

(continued)

Table 5.3 (continued)

Sl. no	Region	Crops
3	North-Eastern Region	Rice, maize, sorghum, finger millet, foxtail millet, Job's tears, French bean, pigeon pea (perennial), black gram, rice bean, lablab, winged bean, pumpkin, cucumber, okra, eggplant, chilli, pointed gourd, ash gourd, taro, yams, citrus, banana, tea, tree cotton, jute, kenaf, roselle, large cardamom
4	Gangetic Plains	Rice, sorghum, barnyard millet, little millet, proso millet, chickpea, cowpea, mung bean, okra, eggplant, bitter gourd, melons, *Cucumis* spp., *Luffa* spp., jackfruit, mango, citrus fruits, jujube, Indian gooseberry, jamun, linseed, niger, sesame, *brassicas*, sugarcane, mulberry
5	Indus Plains	Durum wheat, pearl millet, moth bean, cluster bean, chickpea, black gram, okra, *Cucumis* spp., jujube, *khirni*, *phalsa*, sesame, taramira, cotton
6	Peninsular India (except Western Ghats)	Rice, sorghum, finger millet, pearl millet, foxtail millet, little millet, proso millet, kodomillet, black gram, green gram, cowpea, horse gram, mucuna, pigeon pea, lablab, rice bean, taro, yam, elephant-foot yam, banana, mango, citrus, jackfruit, niger, Brassicas, sesame, ginger, turmeric, chilli, kenaf, sugarcane, coconut, cotton
7	Western Ghats	Rice, sorghum, finger millet, little millet, black gram, green gram, cowpea, Dolichos bean, horse gram, sword bean, okra, eggplant, cucumber, chilli, taro, yam, elephant-foot yam, jackfruit, mango, breadfruit, banana, citrus fruits, jamun, Malabar tamarind, kokum, sugarcane, black pepper, turmeric, ginger, cinnamon, coconut, arecanut, cotton
8	Island regions	Coconut, arecanut, breadfruit, chilli, yams, taro

5.4 Purpose of Collection Missions

In general, the very purpose of germplasm collection is to have maximum genetic representation of the population without any damage to it (except in rescue-collecting). Germplasm collecting is mainly done for three purposes, namely genetic diversity study and conservation, use in breeding programmes and/or evaluation and in taxonomic, biosystematics and phylogenetic studies. Analysing the gap existing between available genetic resources (in nature) and how many and to what extent they are collected/conserved (area/topography/ecosystem-wise) would help to decide on undertaking exploration missions. Biosystematic studies would help unearth the close relatives of crops, which in turn aid in prioritizing species for collection, conservation and use in breeding.

Other reasons for collecting PGR includes:

- For immediate utilization
- Search for novel/unique genetic material
- In danger of erosion or extinction (e.g. rare landraces, wild species, rare, endangered and threatened (RET) species)
- Gap filling in existing ex-situ collections

- Rescue collecting in case of natural calamities
- Fine-grid sampling of already collected sites for trait-specific germplasm
- Maintenance of active collection for supply to users

5.5 Planning Collection Trips

Exploration and germplasm collecting programmes are planned well in advance, except in some emergent cases like undertaking rescue missions. Planning enables the team leader to chalk out the exploration mission and inform the team members well ahead as well as to arrange for the essentials like equipment and transport services. Generally, the team leader will inform the team members regarding crop/crop groups/species to be collected, the area to be surveyed/covered and the approximate time required to complete the mission and tour itinerary/collection route. Collection activity is largely determined by the gaps in past collection/conservation from the particular area that might have been inadequately covered or altogether uncovered due to various logistic and geographic constraints.

A 'classic' gap analysis for PGR collection involves a comparison of genebank collection with the sites data on crop diversity/species occurrence from other sources (such as checklists, agriculture records, biodiversity register, herbaria, monograph/revisions/conspectus/synopsis/flora). Detailed study of genebank material may give clues about patterns of geographical diversity. If required, the area should be surveyed for genetic diversity to define a subsequent collecting mission. It is to be kept in mind that germplasm collecting differs from floristic survey works in being concentrated on some selected taxa for collection (vs. wide spectrum of all species available in a particular area); collecting in the form of seed, hence field visits during seed maturity phase (vs. flowering phase) and preserved as live material (vs. herbarium); and build-up of diversity within species (vs. only representation of extreme variants). This has relevance when choosing collection partners from institutes dealing with floristic/taxonomic works.

5.5.1 Acquiring Knowledge of Species/Areas and their Prioritization

Knowledge of the biological characteristics of the target species such as breeding system, ploidy level, pollen efficiency, seed dispersal strategies, desiccation tolerance, seed viability and dormancy and response to seed multiplication is essential. Insect and bird-pollinated species have more genetic diversity among populations. Seed storage behaviour information may be obtained from the Seed Information Database managed by RBG Kew (http://www.kew.org/data/sid/). In general, large, fleshy seeds with thin seed coats from long-lived woody perennials native to tropical regions are likely to be recalcitrant. Information on the important diseases or pests that might damage the targeted crop/species is also useful. For wild species, ecogeographic data help to identify the probable sites of collection, the distribution range of target taxa and richness in species and genetic diversity.

Information on topography, climatic conditions, vegetation, locality of availability, diversity pattern and seed maturity time can be gleaned through floristic and botanical literature, visiting local herbaria and earlier exploration reports. For fruit crops, knowledge of seed maturity time is a must. The collector should also know the average seed viability/propagability of the germplasm and, accordingly, adequate sample size may be met. For cultivated species, knowledge of indigenous use, farmers' classification of landraces, history of cultivation, extent of cultivation and genetic erosion is important.

Prioritization of species/crops may be made according to their economic importance, relative closeness to crops (in case of CWR), threat level/genetic erosion status and conservation requirement (in seed genebank/in vitro/cryo/FGB). Collecting samples from crop gene pool is desirable for genebank conservation. For example, the snake gourd gene pool consists of not only cultivated germplasm, that is, *Trichosanthes cucumerina* subsp. *cucumerina* 'Anguina', but also its wild progenitor *Trichosanthes cucumerina* subsp. *cucumerina* (wild/weedy small-fruited forms), close relative, *T. c.* subsp. *sublobata*, and the related taxa fall under $GP_{2/3}$ namely *T. c.* subsp. *villosula* and *T. nervifolia*, therefore collecting germplasm of all these taxa would ensure its gene pool conservation.

There are two kind of surveys, based on the intensity of collection. A coarse grid survey aims at covering unexplored areas to capture the overall variability and here sampling is made at wide intervals over the entire region. Fine grid survey yields the build-up of more collections for a specific trait(s) known to exist in identified pockets in previously explored areas. For example, a landrace of wheat called 'Kharchia' (originated from Kharchia Tehsil in the Pali district of Rajasthan), known for its salinity tolerance, was collected earlier. Intense sampling from this site to augment more collections would amount to a fine-grid survey. Such a kind of revisit can also be planned when there is an unavailability of conserved materials and poor ecological coverage earlier.

By and large, germplasm-collecting missions are of two kinds - crop/species specific and broad-based. Crop/species-specific exploration aims to collect variability in a particular crop, or material of specific attributes and/or to collect specific wild relatives, weedy types and related taxa of agri-horticultural relevance. Here also, approaches for collecting wild relatives are different from those of crop germplasm collecting as mentioned in Table 5.4. Broad-based missions (multi-crop collecting missions) aimed to tap the maximum diversity in different crops which are maturing almost at the same time, occurring in the region to be explored. Their choice depends upon the priorities assigned to crops/regions, the extent of earlier collections, the specific needs of the breeders, the extent of threat/rarity, provision of the fund, availability of time and number of expert members in the team, etc. In remote and isolated localities with very little access, it is better to make the most of the trip through opportunistic collecting of a broad range of potential taxa as possible and spend as many working days to cover the entire area as possible. However, it is to be noted

Table 5.4 Comparison of strategies adopted for germplasm collecting of crop and crop wild relatives

Sl. no.	Particulars	Crop	Crop wild relatives
1.	No. of intended species in a collection trip	A few	Quite more
2.	Collecting locality	Farmers' field, garden, farm store, market, threshing yard	Field bunds, disturbed sites, grasslands, also in interior forest (for perennial shrub/tree species)
3.	Determination of population	Easy; usually considered restricted to fairly well-defined fields	Difficult; pragmatically taken to be all the individuals in a fairly homogenous place
4.	Problem of seed-shattering	Minimum	Maximum
5.	Phenology	Uniform flowering and maturation	Erratic flowering, hence staggered maturity
6.	Problem in taxonomic identity	Clear-cut	High, especially with species complexes; in many cases need to be confirmed later; warrants additional trip(s)
7.	Probability of obtaining sufficient seed quantity	High	Less
8.	Importance of collecting herbarium vouchers	Less	More

that germplasm collection efficiency will greatly be reduced in these explorations, besides yielding a huge number of collections which are often unmanageable.

5.5.2 Geospatial Analysis using GIS Tools

Nowadays, GIS tools based on ecogeographic data help in identifying probable sites of collection. Geographical coordinates (incl. altitude) and environmental parameters (rainfall, humidity, temperature) are used to assess the geographical and taxonomical gaps to prioritize taxa and locations for collecting missions. Maps superimposed with topographical features, climate zones, soil types and other readily available geographical data can help the collector decide upon niche areas to concentrate. Climate variables are obtained from WorldClim. There are softwares available to predict the probability of distributions (e.g. ENVIREM, FloraMap). Maps showing a high probability (>75%) of respective crop/species occurrence in a particular area, which with few or no collection sites, indicate geographical gaps in the collection.

5.5.3 Relevance of Checklists and use of Literature

The use of checklists in germplasm collecting is well-recognized. It will not only maximize the collecting efficiency but also help to assess the genetic erosion that happened over the years in a particular region. Important checklists are as follows

- Checklist of crops and their landraces in the target area (different names in use by different tribes)
- Checklist of targeted wild species and wild relatives (and their local names); their images, field identification keys
- Checklist of tribal groups and other traditional farming communities, custodian farmers and their villages
- Checklist of Tehsils/similar altitudinal areas/areas with similar topographic feature
- Prominent markets in the target area, their opening day and time

Local and regional floras, databases and monographs help to find detailed descriptions and information on how to differentiate between related taxa. A perusal of passport data of earlier collected germplasm, district agricultural contingency plans, crop calendar, the package of practices developed by central and state agencies and herbarium information from regional and national herbaria would form useful sources of information for obtaining precise localities and phenological (flowering/seed maturation time) data. However, there is a need to verify this data from several sources, especially if it is old. In general, for herbaceous species, the time between peak flowering and peak seeding would be around 2 months; accordingly trip may be planned.

5.5.4 Establishing Contacts

The team should be small and should comprise crop scientists familiar with the basics of agriculture, pathology, taxonomy, ecology, genetics etc. Ideally, the collecting team may comprise three to four members including a team leader from the collecting organization (preferably a botanist or breeder/agronomist/crop curator), collaborator, local experts from the target area (e.g. Krishi Vigyan Kendra Staff) and driver (with some knowledge of the target region). Collaborators need to be identified and contacted well in advance and there is a need to have a prior discussion with team members on collection priorities, methodologies and strategies, data/information to be gathered while on the collection process, sociological considerations, soundness of the political climate of the region to be explored, processing and conservation arrangements and financial aspects of the mission. Also, they need to be informed of germplasm rejuvenation/maintenance/characterization at their end if required. Area and route of explorations should be fine-tuned in consultation with the subject experts of local organizations. Contacts with grass-root level personnel and local institutions/partners would help in assessing the extent of the population, confirming taxonomic identity (when flowers are present), knowing about exact seed maturity, social news and arrangement of local guides. This would particularly be useful for collecting germplasm of tuber crop species (aerial portion senesce at maturity), of rare species whose plants are not easily visible in the field at the fruit maturity stage or plants that may be confused with other species at the fruiting stage.

Prior permits should be obtained from the concerned authorities for collecting in protected areas and restricted areas (e.g., areas requiring inner line permit as in some northeastern states, landing permit in Nicobar). John et al. (2020) detailed the logistics involved in undertaking exploration in the Nicobar Islands.

5.5.5 What to Collect?

The type of germplasm and its propagule to be collected depends upon the very purpose of collecting mission. For taxonomic/phylogenetic study, perhaps all species in the crop genus may be of interest. For genetic diversity conservation, crop taxa and its gene pool will be on focus, while for immediate use in breeding, trait-specific germplasm, reputed cultivars/landraces and the material with proven potential, plus/elite trees would be the targets. Sometimes, taxonomically unique/isolated or threatened/narrowly endemic species which is of local use is given high priority (e.g. *Momordica cymbalaria*). In general, cosmopolitan and polymorphic species yield more variants of our choice and often have intraspecific taxa/ecotypes, which may possess the target trait.

Germplasm is generally collected in the form of seeds/fruits/vegetative propagules/meristems/pollen, which mainly depend upon their storage behaviour. Seeds of most of the crop genepool are bankable in the seed genebank; their seed collection is preferred owing to the ease of handling. However, in the case of recalcitrant species, fruits/live plants/vegetative propagules are generally collected. In some cases, crops can be conserved through more than one means (e.g. *Moringa*, fruit trees). Here, for genetic diversity augmentation, collection in the form of seeds (sampling from the population) is preferred, while for elite/unique types, vegetative material is preferred to ensure true-to-type.

Unlike seed-bearing crops/species, basic knowledge on diversity (cult./wild) and distribution from flora, reports etc. is essential to acquire on propagation method (asexual/sexual), period of collection of cuttings, propagules etc.; sampling method; availability of rootstocks etc. However, surveys/explorations for identification and marking of desired mother plant at the time of flowering/fruiting stage (to record desired characters/traits: good size, shape, colour, flavour, taste, disease resistance and market value etc.) as well as collecting 1-year-old bud-sticks/cuttings/propagules from the middle portion of marked mother wood plant. During the survey, observations on the climatic condition and collection site data; in-situ site observations; nature of habitat; edaphic data; observations on plant type, fruit characters, associated information of ethnobotany/traditional knowledge and other relevant data need to be taken and propagules should be labelled properly with botanical name, common name, collector number, date of collection etc.

In general, before proceeding with the expedition, the explorers should have an understanding and knowledge of tree crops including the local types and their wild relatives, a good eye for observing the variation in plants and environment, an understanding of the gene pool concept and knowledge of the need for random as well as selective sampling methods. Before execution of the exploration, the curator should be informed about the itinerary/tentative programme of material to be collected. The material should be collected in the dormant stage. Sufficient care is required to avoid any damage during collection. The fruit plant has a long juvenile period, hence all data on plant and fruit characters should be recorded at sites only; while fruit should be kept for recording of quality characters in the laboratory. Collected material should be sent to the nearest site in the proper season for initial establishment and maintenance.

5.5.6 Where to Collect?

Areas of germplasm collection in a broader perspective depend upon the purpose of collection. For genetic diversity study and conservation purposes, collecting representative germplasm from diverse habitats across altitudinal and distributional ranges is recommended, while for taxonomic, systematic and phylogenetic study, collecting from areas where a large number of species occur (with less focus on genetic variability within species) is preferred. For breeding or immediate use, collecting from institutes, nurseries and innovative/progressive/custodian farmers is preferred. For abiotic stress tolerance, the habitat/environment (e.g. saline habitat, unirrigated/drought conditions, desert (cold/hot), flood-prone areas) under which species grew can be taken as a yardstick. Indicator species for particular stress would help identify different habitats. Hotspots for particular pests and diseases shall be looked into for resistant material, if any. According to Leppik (1970), the centre of origin offers scope for getting the desired trait controlled by dominant genes; therefore, landraces and wild relatives of crops of the Indian gene centre assume greater significance.

Priority should be given to collecting from fragile ecosystems like coastal areas, sholas, swampy areas and cold arid tracts. Populations in marginal areas or unusual, distinctive or isolated habitats may have a high likelihood of representing distinct taxa or at least unique traits (von Bothmer and Seberg 1995). Inaccessible valley areas, isolated hills and villages located at the edge of deserts, forests, and mountains usually hold rich and novel diversity in crop, landraces and their relatives. In the contact zones between different subspecies, a lot of morphological variations can be found. In the case of medicinal and aromatic plants, altitudinal, latitudinal and longitudinal changes do impart variation in active ingredients. High altitudinal types are usually rich in active ingredients. Polyploids (e.g. some Rosaceae, Malvaceae members, forage grasses, namely *Bromus*, *Festuca*) can be searched in regions where drought, insolation, ice and snow exist, and areas away from the centre of origin.

For crop gene pool collection, it is always suggested to survey and collect from standing crops to understand the population structure and extent of morphological variability thus expressed. Therefore, fields, backyard/kitchen/homestead gardens, orchards and community and social forests are preferred sites for sampling. Other sources/sites from which samples are often collected include farm stores and threshing yards (in case the crop has been already harvested), field boundaries, fallow areas and roadsides (in case of CWR). In remote tribal localities, if time and other constraints warrant, local markets offer enormous diversity and scope for collecting. It is advised to avoid collecting from main roads, as in general, the introduction of advanced cultivars begins in regions close to main roads.

5.5.7 When to Collect?

Choosing the appropriate time for germplasm collecting is most important. Collection trip shall be planned when most of the target crops/species come to the physiologically mature stage (say 5–7 days before actual shattering/shedding). However, it is easier said than done owing to the multiple dimensions concerning the number of target crops/species, altitudinal/ecosystem/topographic variation, vagaries of monsoon, differences in maturity time of landraces etc. In general, collection trips are planned after the peak rainy season (Oct.–Dec.) for *Kharif* season crops or March–April for *Rabi* season crops, with little deviation suiting to local needs. For vegetatively propagated species (e.g. tuber crops, fruit trees), often it is desirable for two times survey, first one for identification/marking during peak flowering/fruiting stage and later for collection of live plants/scion/tubers. Germplasm of temperate fruits can be collected as dormant buds on twigs of last year's growth during peak winter months. Pollen should be collected during peak flowering period from flowers with freshly dehisced anthers on bright sunny days between 8 and 10 AM.

Ideally, the period of crop exploration should be a minimum of 10 working days depending upon the target area and crops to be covered, financial liability and so on. Long-duration missions (2–3 weeks) and repeat visits are suggested for collecting an adequate sample size of wild species. Important collecting kits and reference materials helpful for the seamless collection process are mentioned in Table 5.5. These may vary depending on the nature of the germplasm (fruit/seed/vegetative propagule/in vitro/live plants), method used for collecting and the area(s) to be explored.

5.6 Execution of Exploration Trips

As you reach the starting point, fine-tune the target area/exploration route with subject experts at the local level, in a view to accommodate updated information related to crops, local people and events. Local market surveys would help find the current maturation stage of target species, besides helping glean information on farmer

Table 5.5 Items and equipments helpful in collecting

Survey/ collecting items	• Global positioning system (GPS), digital camera with the additional memory card, binoculars, magnifying glasses, handheld microscope, digital vernier calliper and portable balance • Haversack/kitbag, seed envelopes, cloth bags, polythene bags, aluminium & tag labels, drying sheets, old newspapers, plant press, moss, rubber bands, packing tape, *sutli* (thick and thin), secateurs, scissors, knife, digger, torchlight, measuring tape, passport data book, field notebook, pencil, ballpoint pen and permanent marker • Others: Stapler, candle, matchbox, water bottle, formaldehyde, hunter shoes, hand (rubber) gloves, waist pouch, rain suit (shirts, trousers), rucksacks, sunglasses
Reference material	Regional/national flora, field identification aids such as digital herbarium, photocopy of herbarium sheets etc., lap-top and accessories, list of local names of plants, road-map, vegetation/climate map, list of rest-houses/lodges, hotels, resting/stay places and list of local contacts (phone, fax, e-mail)
First aid box	Anti-malaria pills, anti-allergen tablets, pain killers, anti-amoebic and anti-diarrhoeal tablets, mosquito repellent, antifungal/antibacterial/antiseptic creams or lotions, cotton packs, dressing gauze, water-purifying tablets etc.

details and specific location(s) of occurrence of target germplasm. Time management through efficient utilization of sunlight hours is critical in the fruitful execution of the mission. A detailed exploration methodology is given in Fig. 5.2.

5.6.1 How to Collect?

Within the target area, the collecting mission should start from drier tracts (vs. humid), valleys (vs. hills) and unirrigated areas (vs. irrigated) as crop maturity precedes here, therefore the relatively earlier availability of germplasm. Try to follow the route in such a way that diverse topography, soils and vegetation could more likely be covered, and prefer an alternate route for the return journey. Since micro-geographic differentiation does occur, plants from obviously different environments should be collected as separate samples, even if they seem to form part of one contiguous population. In addition to population samples, a selective sample can be collected if the expression of the target traits is obvious and stable (e.g. healthy individuals in an infected field). Non-random (selective) sampling should be done in case any distinct variant is observed and that should be given a different collection number. When a population of wild species with only a few individuals is encountered, seeds should be collected from as many as plants possible to capture maximum representativeness. In the case of certain wild and semi-domesticated species occurring in a few scattered populations, the seed should be bulked; however, the survival of its population in its natural habitat is to be assured beforehand.

Look for any natural wide hybrids or any distinguishing features. The decision to collect a particular unknown species lies in its morphological proximity to cultivated species. If cryptic forms such as chromosome races are known to be

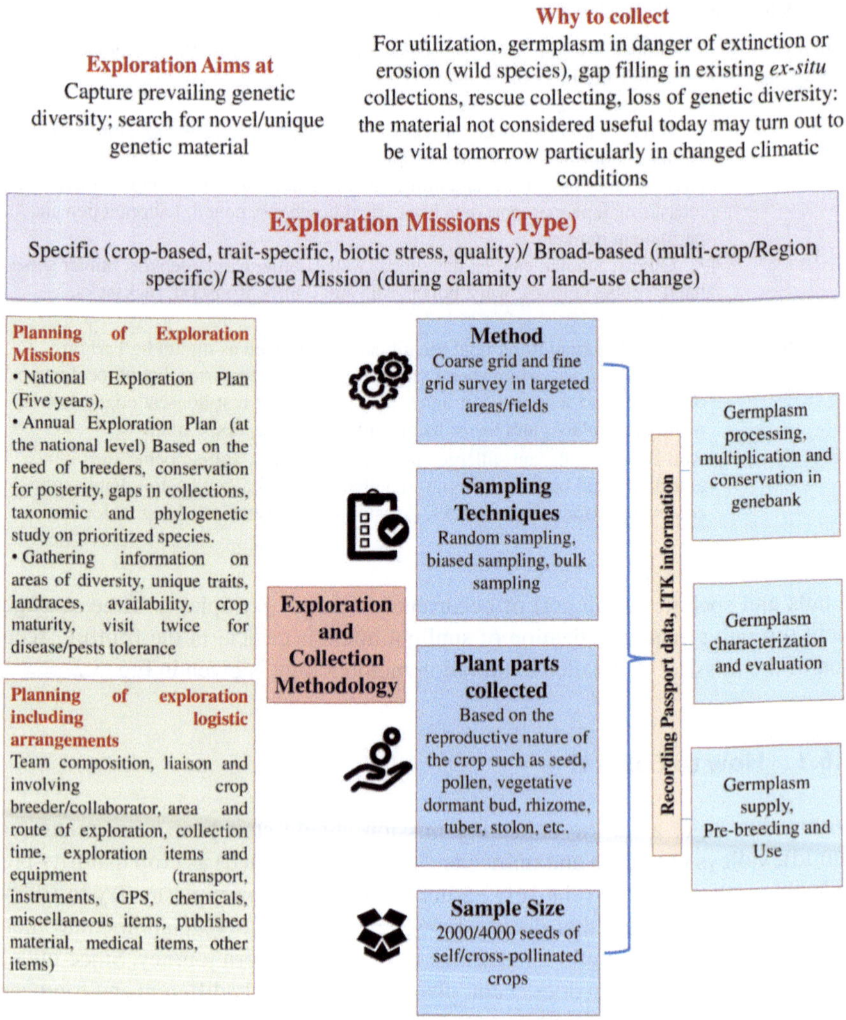

Fig. 5.2 Flowchart of exploration for germplasm collection activities

sympatric, collecting has to be done on an individual plant basis. Ensure that the population is likely to be genetically distinct (inferred by soil, climate, altitude, pollinator's range and physical barriers to genetic mixing). But the difficult question to answer is where one population starts and ends, as finding the barrier to genetic exchange, hence genetic isolation is not so easy. In the absence of any information, collections should be made from sites that are not less than 15 km apart, unless the germplasm is morphologically different, or there is a marked change in altitude or cropping systems, or the existence of a perceivable physical barrier (e.g. mountain, river) or presence of ethnically different people from the previous collection site. Most seeds disperse less than 100 m distance; however, wind-blown as well as animal-carried ones can travel a much greater distance.

5.6.2 Sampling in Self and Cross-Pollinated Species

Von Bothmer and Seberg (1995) opined that in the case of species with vast distribution, more populations should be sampled (at the expense of the number of individuals per population). In general, inbreeders show more variation among populations than outbreeders. Nevertheless, individual population of outbreeders are usually more variable than that of inbreeders; and those of widespread species more than those of narrow endemics. For species with small disjunct populations (called 'satellite' populations), it is recommended to sample a few individuals from as many populations (germplasm from marginal, isolated areas might have extreme adaptations).

The frequency of sampling (no. of samples per target area) should be decided based on on-the-spot observations on variability available. According to Marshall and Brown (1975), the optimum sample size per site should be the number of plants required to obtain, with 95% certainty, all the alleles at a random locus occurring in the target population with a frequency greater than 0.05. For use in the genetic conservation of crop plants, Brown and Marshall (1995) suggested sampling from 30 randomly chosen individuals in case of cross-pollinated species and from 59 randomly chosen individuals self-fertilizing species, which would result in capturing more than 95% of alleles composing in the target population with a frequency ≥0.05.

For general purposes, random sampling is achieved by collecting spike/panicle or fruit/berry/pod from at least 50 plants along several transects throughout the field/site (Hawkes 1980) to obtain a representative and adequate sample. The collector may walk across a site or a field twice in a cross or zigzag manner avoiding sampling from the borders. If environmental variation is clinal, sampling should be done systematically along the gradient. Ecotypes can be sampled separately (through stratified random sampling) to ensure the collection of alleles of ecological significance. If doubtful of cline/ecotype/variant, better collect separately and their bulking can be decided later as the collector gains knowledge about the species. For clonally propagating species, each distinct morphotype shall be collected as a distinct sample.

5.7 Collection Process

It is best to collect a larger sample keeping in mind that seeds may be needed for base and active collections as well as for duplicate conservation. A larger quantity is also needed in the case of genetically heterogeneous samples than for more uniform samples. In the case of species with extremely small-sized seeds, low seed set, asynchronous maturity and low seed viability, care should be taken to collect adequate sample size. A fine balance needs to be maintained between the number of samples to be collected and collecting adequate sample sizes for conservation (without the need for regeneration). For self-pollinated crops, collect a minimum of 2000 seeds, while for cross-pollinated ones, it should be 4000, keeping in view the requirements of conservation/characterization and study. Suitable harvesting techniques and equipment shall be employed keeping in view the number of seeds required, stage of maturity/ripeness, fruit size, characteristics of individual trees, stand and size. Some suggested tips while collecting are as follows:

- Before collecting, check for empty, immature seeds or infestation/infection, if any.
- Capturing seeds with shattering tendency by spreading a piece of fabric/net/tarpaulin below (esp. in case of wild legumes).
- Fleshy fruits are assembled in plastic bags (only for temporary purposes).
- Legumes in heavy-duty paper bags/cloth bags.
- Collecting fallen fruits and seeds for species with non-winged and heavy fruiting. However, it is to be done with caution (as the possibility of mixing and checking for any damage/contamination).
- Removal of fruits from the canopy using ground access by hand picking/stripping, beating and shaking or climbing.
- If the species produces heads with a large number of seeds (e.g. sorghum and millets), collect only small portion of the heads.

Herbarium collection is crucial for taxonomic, systematic and phylogenetic studies. Normally four to five individual plants/parts having representation of all parts, namely flowers, fruits or both, should be collected for preparing herbarium specimens. Locally, the date of collection and field notes should be recorded. Characters which are lost on drying or which may not be represented in the herbarium specimen (plant height), flower colour or leaflets (which may be shed on drying) should be mentioned in field notes. In the case of material with uncertain identity wherein not feasible to make herbarium specimens, efforts should be made to retain fruits, flowers and leaves along with seeds to establish their correct identity later (at work base). Healthy disease-free leaf samples may be collected (and dried over silica gel) for DNA extraction to employ in molecular systematic/diversity studies.

5.7.1 Vegetative Germplasm Collecting

Root cuttings/suckers and seedlings/saplings with minimal root injury will have a high probability of success in the establishment. However, shoot cuttings (incl. bud woods, and in vitro meristems) are the propagules commonly collected to maintain a true-to-type nature. Here, specialized wrapping is needed to maintain relative humidity till reaching the workplace. Alternately, wrapping/packing the same in moss is advised. Also, onsite grafting over portable rootstocks is often followed (e.g. pome and stone fruits). Sample size will depend upon the number of rootstocks available but not less than ten per sample so that at least eight grafts may survive. Prior arrangements should be made for the transportation of collected material to the place of its establishment/maintenance.

In the case of cuttings and rooted suckers (e.g. *Vitis*, *Musa*, *Piper*, ornamentals, passion fruit, banana, cardamom), 15–20 numbers may be sufficient. In the case of large tubers, only a portion, for example, head or proximal ends in yams, cormels or tops in taro and other aroids should be collected. If both seed and vegetative propagules are available, the supplementary sample may be collected. Materials with dubious/incomplete identity (with only genus level information) or unidentified material would quite commonly be encountered while collecting seedlings/saplings/vegetative propagules (mainly due to the absence of flower/fruit information). In such cases, vernacular names should be noted along with herbarium specimens and photographs for authentication.

5.7.2 In-Vitro Collecting

Expedition may last for weeks, and may cover remote, difficult and inaccessible areas with bad weather conditions. Sometimes there may be insufficient material because of poor growing season, general scarcity of plants in the target region and damage by pests or immaturity of the material. These problems can be faced during the collection of both seeds and vegetative propagules that have seasonal suckers, bud woods and immature seeds that often perish before they reach the genebank. Moreover, the excessive weight and bulk of large vegetative propagules and recalcitrant seeds and their fruits. To overcome the above-mentioned problems, the importance of tissue culture technology in germplasm collecting has been recognized as it can increase the efficiency of germplasm collecting by (1) increasing the sample size and reducing the bulk; (2) increasing the variability of collections by collecting different plant material, for example, vegetative propagules, ripe and unripe seeds etc.; (3) reducing the risk of loss and (4) reducing quarantine restrictions and facilitating transport of collections. Hence, to develop in vitro methods/protocols for germplasm collecting, attention should be given to (a) minimization of microbial contamination and (b) maintenance of physiological conditions and morphogenic capacity. Hence, the in vitro collecting approach is technically simple but needs preparation and planning and adequate background knowledge to integrate it into comprehensive germplasm conservation including quarantine and disease indexing, storage, distribution etc. This technique is not an alternative to current practices rather it is a supplementary technique to conventional methodology to resolve problems that may prevent the collection of valuable germplasm.

5.7.3 Handling Collections in Transit

All precautions should be taken to maintain viability of seeds and/ or survivability of vegetative propagules. Slightly under-ripe/mature seed can often be successfully conserved if entire seed heads along with short length of stem are collected, and allowed to mature in a well-ventilated environment. If transportation back to the seed bank takes several days, it is advised to partially process/extract the seeds and dry the seeds over silica gel (3 desiccant: 1 seed), or charcoal inside sealed plastic boxes (if relative humidity is >50%). Fruits of species with recalcitrant seeds may need to be stored under cool humid conditions. Frequently inspect the condition of seeds, and better shade-dry in the staying place itself, provided the halting point is the same. Necessary precautions are to be taken for pest-free germplasm samples and transportation.

5.7.4 Documentation in the Field

The collector should ensure the collection of critical data such as locality data (using GPS), sampling particulars, collection source, biological status and special visible characters of the collection while in the field. The information provided by farmers and field workers dwelling in the collecting area will be of unique value. Information on genetic erosion (at the level of crop per se as well as landrace) should be gathered from aged farmers especially on the rate of depletion of landraces over time, their distinctive morphological/agronomic traits and the reasons for their loss. Any possibility to recover the same is also to be mentioned. Indigenous Traditional Knowledge associated with plants, their use, agricultural technologies etc. is to be recorded. The meaning of the name of landraces and their characteristics should be asked from farmers and recorded. Observations on the distribution pattern and frequency status of crop wild relatives, rare, endangered and threatened species of PGR importance should be noted. Ethnobotanical observations, new uses of plants and new crops cultivated (native or exotic, to what extent), semi-domesticated species, especially those from tribal-dominated tracts, all need to be documented while in the field. (**Annexure** has details of the passport datasheet followed at NBPGR during the germplasm collection).

There is a need to mention about collector number, allotment and labelling of samples in the field. This unique number helps identify and alleviate duplicates, aid in information exchange and track down the origin of accession. It is to be noted that a separate numbering system for different members of the team should be avoided, instead 'Similar General Term Designation' be made.

5.7.4.1 Digital Documentation

The data recording is a complex task in the field notebook during the survey and collection of germplasm due to the paucity of time in gathering information and samples. Digital recording makes it very easy. Therefore, ICAR-NBPGR developed the *NBPGR-Passport Datasheet (PDS),* an android-based mobile application which is one of the most useful tools for users to manage germplasm collection information systematically (Tomar et al. 2023). The passport data sheet contains taxonomic, geographical, ecological and ethnobotanical information sections for recording various parameters. It will exercise the reduction in time during post-survey or collection programmes.

5.7.5 Post-exploration Handling of Germplasm Collections

Taking care of the collections after completion of the collecting mission and before they are processed for long-term storage is as important as collecting germplasm. Collected samples should be pre-cleaned to assess their quantity and quality. The best technique for pre-cleaning is by using a light plastic washing-up bowl where debris can be removed (by gentle blowing) and the seeds extracted into a paper bag or envelope. It is better to transport seeds and not fruits, except when they cannot be easily separated or time is limited. In the case of fleshy fruits, it might be possible to eliminate the flesh by gently macerating them in water and then air drying or drying using silica gel. Precleaned good-looking seeds should be dried in shade/controlled conditions to enable seed moisture to reach the safe level (5–8% moisture) suitable for conservation.

Seeds with short viability should be processed at the earliest and care should be taken during threshing and cleaning to avoid damage. The vegetatively propagated material should be sent for establishment in the field genebank or at a suitable site at the earliest. The material for in vitro conservation and cryopreservation should be handed over to the appropriate organizations, with prior information to the concerned curators. Observations on variability parameters on fruit/pod/seed should be recorded along with photographs for report writing, documentation and publication. Taxonomic authentication of collected samples from all possible angles is to be ensured at this stage before out for conservation/study/use. Seed packets should be labelled with the collector number, taxonomic identity and date of collection both inside (dropping a slip) and outside. They should be submitted for conservation or be shared with the collaborator or be sent for study along with passport data. All national collections augmented through exploration and collection are assigned unique 'IC' numbers (Indigenous Collection numbers). This number remains unchanged during the flow of germplasm (collection, conservation, evaluation, characterization, multiplication and supply). The information on the samples collected should be entered into a database for easy access to users. Collected herbarium specimens/economic products are to be linked with the collector number and deposited in the herbaria. There is the possibility that the collected germplasm/specimen may represent a new species to the flora or a new record to the state or country.

5.7.6 Report Writing

The mission report should be completed as soon as possible on return to base (within a month) and distributed to all concerned and also deposited in a library where it will be catalogued for future reference. This report would help in determining if follow-up collecting would be needed and provide information to users about the availability of germplasm. The collector is responsible for ensuring the correctness of passport data entered into databases. The following items would form the essence of report writing.

- Name of the organization(s)
- Name of the scientist(s)/person(s) involved
- Collaborating organization(s)
- Objectives of the collecting mission
- A description of the environment of the target area
- An account of the logistics and scientific planning
- Details of the execution of the mission (timing, itinerary, sampling strategy and collecting techniques)
- A summary of the results (areas surveyed along with route maps, germplasm and herbarium specimens collected, indigenous knowledge documented and extent and magnitude of diversity collected, along with elite germplasm, if any)
- Role of women in the conservation of diversity
- Details of sharing germplasm and information
- Photographs
- An account of the loss of germplasm, if any
- Difficulties encountered during collecting mission
- Recommendations for follow-up action(s)
- Acknowledgement(s)

5.8 Some Considerations in Exploration

In recent years, collection activities have faced challenges, which may be addressed soon for seamless augmentation of germplasm. Ahlawat et al. (2022) specified the following critical points related to exploration activities.

- Lack of trained human resources in this field calls for an emphasis on developing skilled manpower and the use of modern tools.
- Sometimes, 'fancy' names are given for landraces for raising popularity and marketability (apart from different names for the same landrace and vice versa). This necessitates thorough verification through various sources before collecting.
- Visible sensitiveness of phenology of wild species due to climate changes, therefore, make it difficult to strategize the collection activities.
- The importance given to intraspecific classification in crops (and the non-recognizing of intraspecific taxa by taxonomists) is on the decline, resulting in the loss of valuable information, which has a huge role in PGR management.
- Lack of interest among curators/collaborators in the multiplication of cross-pollinated crops following proper pollination control, as it involves huge manpower and resources.

According to Ahlawat et al. (2022), some priority and thrust areas that need sufficient attention are as below:

- Reduce collection-to-conservation gaps and strengthen regeneration facilities, enhancing the survivability of vegetative propagules/live plants during transit.
- Develop illustrative field identification keys using leaf, fruit and seed characters to enable easy identification of CWR and other economically important wild species.
- Enhance the effectiveness of collecting trait-specific germplasm—FIGS, use of portable equipment/kits (TSS, P^H meter, lysimeter, chlorophyll meter, NIR, seed driers).
- Augment representative diversity of less-attended crop groups such as forages, medicinal and aromatic plants and ornamentals.
- Develop the PGR inventory/checklist for states and the nation as well; need for a unified collection database involving all stakeholders.
- Strengthen expertise in techniques of in vitro germplasm collecting.
- Use of species distribution models for locating populations of rare CWR/rare endangered and threatened species.
- Undertaking collaborative explorations in foreign countries having primary and secondary centres of diversity, for which low diversity exists in genebank.

5.9 Indigenous Traditional Knowledge

During exploration and germplasm collecting, recording of experiences and knowledge of the farming community related to crop sowing, harvesting, storage, conservation and local uses shall help discover new plant resources and locate new germplasm for agriculturists, new ways of conservation and use of germplasm, besides many other applications. This has resulted in the reporting of some economic plants which have been selected from the wild and gradually brought under folk domestication in NE region such as *Coix lacryma-jobi*, *Moghania vestita*, *Digitaria cruciata* var. *esculenta*, *Crotalaria tetragona*, *Plukenetia corniculata* etc. Understanding farmers' classification of landraces would throw light especially in germplasm classification. The Mizo farmers used to categorize/name the maize cultivars based on grain colour: *sen*—red, *eng*—yellow, *var*—white, *dum*—black and *tial*—variegated; cob size: *te*—small, *lian*—big and *sei*—long; crop duration: *tlai*—late maturing and *mah*—early maturing. This has practical significance in cultivar classification and, therefore, in genetic resource management. This indicates that local communities put their efforts, experiences and innovation to develop, conserve and use plant diversity. Hence, public awareness and community-based management need to be encouraged at all levels alongside of urgent collection of germplasm. Format for database on local uses/ethnobotanical uses/indigenous knowledge is given in Box 5.2.

Box 5.2 Format for Database on Local uses/Ethnobotanical uses/Indigenous Knowledge

1.	Botanical name	It is a very important field where currently accepted name as per nomenclature of plant on which information is recorded
2.	Family	As per the recent classification system, the family name should be mentioned
3.	Vernacular name	It is a very essential field, which should be recorded, as the local people call in their mother tongue/dialect/ language
4.	Locality	It contains the name of village, block, tehsil, district and state from where information recorded
5.	PGR category	As per the category of use, brought from the wild and being maintained in kitchen garden/backyards
6.	Date of collection	The exact date of material/germplasm/herbarium specimens collected
7.	Part(s) used	The plant part is being used by the local people
8.	Kind of use	The plant part is being used as food, medicine, fibre, timber, fodder etc.
9.	Frequency of use	Depending on the availability of the plant in the area, how frequently it is used by the local people
10.	Method of use	Application of indigenous method/techniques/process during use of plant parts
11.	Informant/ethnic group	The person who gives the information on the use of plants and may belong to old experienced folk, local vaidya, shepherd/grazier etc.
12.	Local sale price	The local price of plant/plant products in the local market
13.	Collector's number	The number assigned in the passport data book by the collector who has collected the sample/herbarium in the field
14.	Photo/image/ sketch/herbarium	Onsite photograph/sketch of plant/plant product/ processing method of plant/plant parts and herbarium specimen collected
15.	Remarks	Information about plant characteristics/unique/specific information given by the informant etc. which is not covered in the above fields

5.10 Summary

The plant explorer should not regard himself as simply a collector of potentially useful novelties or curiosities. In fact, he/she is a researcher consciously seeking to understand the causes of variability and record the basis of the adaptation in response to climatic and environmental factors. A good explorer shall distinguish germplasm (landraces, variants, natural hybrids, wild species and so on) flawlessly in the field, effectively utilize time while in exploration, make mature decisions on the spot, besides shall have good communication ability. Plant identification in their natural habitats needs deep observation skills, taxonomic and field knowledge, and regular practice. Although collecting is intended for specific objectives or purposes, collectors should be open-minded and far-seeing to derive the best out of the exploration mission. The discovery of the amazing medicinal value of *Trichopus zeylanicus* by scientists of KSCSTE-JNTBGRI, Thiruvananthapuram through 'Kani' tribal wisdom, is an example.

In the recent scenario, the focus is on collecting a few selected materials showing desirable traits; therefore, adopting recent technologies would maximize the efficiency of collection. Detailed documentation in the field during collection itself is of paramount importance in the wake of the fullest utilization of germplasm, besides its legal protection. Collection priorities need to be reassessed at regular intervals in light of genetic erosion, threat status, breeders' requirements, ecogeographic representation and so on; accordingly, the exploration plan shall be formulated. Exploration and collection guidelines set for individual crops/species would help in the faster realization of representative crop collections from across their regions of occurrence.

Multiple-Choice Questions

1. Species diversity stands for
   (a) Number of species in a genus
   (b) Number of species in a situ or a habitat
   (c) Availability of diverse species in a given area
   (d) All of the above
2. PGR collecting primarily aims at
   (a) Tapping germplasm variability in different agri-horticultural crops
   (b) Search for new genes
   (c) Collecting germplasm for prosperity
   (d) Collecting crops along with wild and weedy relatives
3. Genepool stands for
   (a) Total number of species in a population
   (b) Total number of genes in a plant population
   (c) Total number of species in all levels of gene pool
   (d) None of the above
4. Plant genetic resources are important components used for
   (a) Crop improvement programmes
   (b) Widening the genetic base
   (c) Conserving for posterity
   (d) All of the above
5. National Bureau of Plant Genetic Resources (NBPGR) rechristened in which year
   (a) 1947
   (b) 1958
   (c) 1977
   (d) 1985

6. Which of the following are the biodiversity hotspots in India ?
   (a) Central, Western and Eastern Himalayas
   (b) Northeastern region, Western Ghats, Himalayas and Andaman & Nicobar Islands
   (c) Peninsular region, Eastern Ghats, Western Ghats
   (d) None of the above
7. The cultivars of crops developed by farmers and adapted to local environmental conditions are known as
   (a) Traditional variety
   (b) Landrace
   (c) Hybrid
   (d) Primitive cultivar
8. In plant genetic resources studies, Geographical Information System as a tool is used for
   (a) Documenting and mapping the biodiversity
   (b) Knowing the name of local landrace
   (c) Planning introduction of new species
   (d) Framing the conservation strategies
9. Number of seeds in cross-pollinated species should be collected between
   (a) 1000–2000
   (b) 2000–4000
   (c) 3500–4000
   (d) Up to 10,000
10. Herbarium is
    (a) A scientific method for sampling
    (b) A collection of dried plant specimen
    (c) Both a & b are correct
    (d) All of the above
11. The term ethnobotany was coined in 1895 by
    (a) *Vavilov*
    (b) Harshberger
    (c) Faulks
    (d) Schultes
12. The Indian gene centre holds how many number of higher plant species?
    (a) 21,984
    (b) 18,000
    (c) 17,500
    (d) 16,500
13. Who is known as the *Indian Vavilov* because of his contribution to PGR?
    (a) Dr MS Swaminathan
    (b) Dr Harbhajan Singh
    (c) Dr RS Paroda
    (d) Dr RK Arora

14. Sources of eco-geographic surveys include
    (a) Herbarium specimens and germplasm accession
    (b) Passport data, and experts
    (c) Literature, field notes and maps
    (d) All of the above
15. Indian gene centre holds diversity in which oilseed crops?
    (a) Groundnut and sunflower
    (b) Safflower and sunflower
    (c) Rapeseed-mustard and sesame
    (d) Both a & b are correct
16. The full form of FIGS is
    (a) Fungal infection in germplasm storage
    (b) Focused identification of germplasm strategy
    (c) Foreign investigation of germplasm storage
    (d) None of these
17. The Indian gene centre hosts how many hotspots of biodiversity?
    (a) Two
    (b) Four
    (c) Five
    (d) Six
18. A dictionary of economic products of India which dealt with commercial plants and their uses was written by:
    (a) JD Hooker
    (b) G Watt
    (c) NI Vavilov
    (d) HR Harlan
19. Collecting germplasm immediately after a natural calamity is termed as
    (a) Urgent collecting
    (b) Rescue collecting
    (c) Routine collecting
    (d) Special drive
20. For species delineation, which one of the following studies is required?
    (a) FIGS
    (b) Biosystematics study
    (c) Remote sensing
    (d) Ethnobotanical study
21. Binomials with identical genus names and species epithets are called
    (a) Homonym
    (b) Tautonym
    (c) Basionym
    (d) Synonym
22. Which one of the following is not observed in biodiversity hotspots?
    (a) Species richness
    (b) Endemism
    (c) Accelerated habitat loss
    (d) Lesser inter-specific competition

23. Job's tears, a minor cereal crop, is grown in which region of India?
    (a) Western Ghats
    (b) Western Himalaya
    (c) North-eastern hill region
    (d) A&N Islands

Answers

1. (d)
2. (a)
3. (b)
4. (d)
5. (c)
6. (b)
7. (a)
8. (a)
9. (b)
10. (b)
11. (b)
12. (b)
13. (b)
14. (d)
15. (c)
16. (b)
17. (b)
18. (b)
19. (b)
20. (b)
21. (b)
22. (c)
23. (c)

Annexure

दौरे का नाम /Nature/Type/ Name of Tour ....

....

....

अवधि/Duration: .... कब से/From .... कब तक/To............................

सर्वेक्षण किए गए राज्य/जिले/क्षेत्र/State(s)/District(s)/Area(s) surveyed ....

....

सर्वेक्षण किए गए कृषि पारिस्थितिकीय/वनस्पति भौगोलिक क्षेत्र/Agro-ecological/Phyto-geographical region surveyed ....

टीम लीडर का नाम व पता/Name and Address of Team Leader

....

....

सहयोगियों का नाम व पता/Name and Address of Collaborator(s)

1 ....

....

2 ....

....

3 ....

....

संग्रहकर्ता कोड़ पूरे विवरण के साथ/Collector's code with full details..

(For eg. SP Ahlawat= SPA, KC Bhatt =KCB, PK Singh= PKS: SPA/KCB/PKS/2023/.........)

....

....

....

Online passport data recording app link: https://play.google.com/store/apps/details?id=com. nbpgr.nbpgr_pds

ICAR-NATIONAL BUREAU OF PLANT GENETIC RESOURCES, NEW DELHI

PASSPORT DATA SHEET

Date.. Collector's No... Accession No....................................

Botanical Name... Common Name (English) ....

Vernacular/Local Name... Cultivar/LR/FV Name...

Meaning of LR/FV Name...Specific Characters..

Yrs. of cultivation........................ No. of villages .... Total area .... No. of farmers ....

Village.. Block/Taluk/Mandal .... District. ....

State/UTs .... Location landmarks ....

Latitude............................°N, Longitude ....°E (in dg.decimal) Altitude ....m

Farmer's/ Donor's Name..(Mob)..

Ethnic Group...

COLLECTION SITE	1. Farmer's/Jhum field, 2. Field margin, 3. Threshing yard, 4. Farm store, 5. Home Garden, 6. Orchard, 7. Natural wild, 8. Disturbed wild, 9. Grassland/ Pasture, 10. Fallow land, 11. Pond/lake, 12. Wasteland, 13. Market/Nursery, 14. Institute/KVK/Univ., 15. Others ....
HABITAT	1. Cultivated, 2. Wild, 3. Rangeland, 4. Wetland, 5. Aquatic, 6. Forest, 7. Desert (Hot/Cold), 8. Mountainous, 9. Coastal, 10.Swamp, 11. Saline, 12. Others..................
TOPOGRAPHY	1. Plain 2. Flooded, 3.Undulating, 4. Hilly, 5. Dissected, 6. Steeply dissected, 7. Valley
BIOLOGICALSTATUS	1. Landrace, 2. Cultivar, 3. Primitive cultivar, 4. CWR, 5. Wild/Weedy, 6. Domesticated, 7. Breeding line, 8. Breeding/research material, 9. Hybrid, 10. Genetic stock, 11. Inbred line, 12. Segregating population, 13. Mutant, 14. Improved cultivar, 15. Released variety, 16. Others...........
FREQUENCY	1.Abundant, 2. Frequent, 3. Occasional, 4. Endemic 5. Rare
MATERIAL COLLECTED	1. Seeds, 2. Fruits, 3. Inflorescence, 4. Panicle, 5. Roots, 6. Bulb, 7. Corm, 8. Tubers, 9. Rhizomes, 10. Suckers, 11. Live plants, 12. Budwood, 13. Cuttings, 14. Pollens, 15. Herbarium, 16. Others.............
BREEDING SYSTEM	1. Self-pollinated, 2. Cross-pollinated, 3. Often Cross-Pollinated, 4. Vegetative propagated, 5. Others................
SAMPLE TYPE	1. Population ($\geq$50 plants), 2. Individual plant, 3. Others ....
SAMPLING METHOD	1. Bulk, 2. Random, 3. Selective (non-random), 4. Individual Plant/Tree 5. No. of fields/sites sampled..................................
BIASED SAMPLE	Sub-set made: Yes/No, Sub-set no................. Reasons?.................
DISEASE/PESTS SYMPTOMS	1. Susceptible, 2. Mildly susceptible, 3. Tolerant, 4. Resistant, 5. Others.............. 2. Name of disease & pests..
PLANT HABIT	1. Annual, 2. Perennial, 3. Herb, 4. Shrub, 5. Tree, 6. Erect /Bush type, 7. Prostrate 8. Climber, 9. Lianas
CULTURAL PRACTICE	1. Irrigated, 2. Rainfed, 3. Line sowing, 4. Broadcasting, 5. Mixed cropping, 6. Others.........

Collector No.	Collector No.	Collector No.	Collector No.

SEASON	1. Kharif 2. Rabi 3. Spring-Summer/Zayad
	Sowing Month: 1 2 3 4 5 6 7 8 9 10 11 12 ($1^{st}/2^{nd}$ Fortnight)
	Flowering Month: 1 2 3 4 5 6 7 8 9 10 11 12 ($1^{st}/2^{nd}$ Fortnight)
	Harvesting Month: 1 2 3 4 5 6 7 8 9 10 11 12 ($1^{st}/2^{nd}$ Fortnight)
ASSOCIATED CROPS/ VEGETATION	1. Sole, 2. Mixed with.......... 3. Other crops grown in area:..........
SOIL COLOUR	1. Black, 2. Red, 3. Brown, 4. Yellow, 5. Forest Soil, 6. Others..........
SOIL TEXTURE	1. Sandy, 2. Sandy Loam, 3. Loam, 4.Silt Loam, 4. Clay, 5. Silt
EROSION STATUS of PGR in AREA	1. Landrace /FV Eroded.......... 2. Fruit tress/M&AP/ Others Eroded .... 3. % Changes ....
REASONS of EROSION	1. Replacement by cash crops/HYVs, 2.Climate change, 3.Change in food habit/ Demand 4. Ignorance/Reluctance, 5. Invasive species, 6. Land use change/Infrastructure-development, 7. Flood/cyclone, 8. Any other-.........
UNIQUE TRAITS OF GERMPLASM	Agronomic .... Morphological .... Value ....
PLANT CHARACTERISTICS/ IN-SITU /FIELD	Plant/ Tree height: Tall/Medium/Dwarf; Av./Range (cm/m).......... Spread/tillers: Profuse/Moderate/Solitary; Av./Range.......... Fruiting habit: Profuse/Medium/Low.......... Fruit /pod/spike/cob size: Large/Medium/Small; Av./Range .... Fruit/seed shape: .... Fruit/Seed colour: ....

ETHNOBOTANICAL/TRADITIONAL USES

PART(S) USED	1. Stem, 2. Leaf, 3. Root, 4. Fruit, 5. Flower, 6. Whole plant, 7. Seed, 8. Any other ....
KIND/PURPOSE	1. Food, 2. Fibre, 3. Fodder, 4. Fuel, 5. Timber, 6. Insecticide/ Pesticide, 7. Incense, 8. Medicinal/Therapeutic use..........9. Others..........
RECIPE NAME	(Local) .... Frequency of use: daily, weekly, occasional/festival
METHOD & INGRDIENTS	....
INDIGENOUS TRADITIONAL TECHNOLGY	....
INFORMANTS	1. Experienced person (male/female) 2. Medicineman 3. Shepherds/graziers 4. Saints 5. Others
Photograph/s	Yes/No. ....

HERBARIUM SPECIMEN LABEL

Collected by: ....

Plant description (Taxonomic) ....

....

....

....

Sl. No.	Descriptor	Definition	Letter code
1	**Wild**	All species other than cultivated of PGR Value/ potential value	WL
1a	Crop Wild Relative	All wild species that are related to cultivated species and part of the gene pools from which genetic introgression into cultivated species has occurred	WL_ CWR
2	**Weedy**	Weedy forms of cultivated species occurring in companionship with some other cultivated species	WD
3	**Landrace/ Traditional cultivar**	All cultigens in farmer's field with/without specific names frequently associated with unique traits and geographical region identified by farmers	LR_TC
3a	Landrace	All cultigens in farmer's field frequently associated with unique traits and names identified by farmers	LR
3b	Traditional cultivars/ Primitive cultivars/ folk variety	All heterogenous cultigens cultivated by farmers with specific vernacular names frequently associated either with unique traits or geographical region often with some specific traits	TC
3c	Farmers variety	Farmers variety registered under the PPV&FR Act *viz*-Varieties which have been traditionally cultivated and evolved by farmers in their fields or those that are wild relatives or land races of a variety about which farmers possess common knowledge	FV
4	**Breeding/ Research material**	Semi-finished products or segregating material generated out of breeding program to meet specific breeding objectives	BR
4a	Hybrid	It is cross breed developed by crossing of two inbred lines, varieties, species or genera	BR_H
4b	In breds	Homogenous offspring of cross -pollinated species obtained as a result of repeated forced self pollination	BR_I
4c	Segregating population	Genetically diverse progeny from crosses between parents that differ for one or more traits under genetic control	BR_SG
4d	Breeding lines	Lines developed in breeding programmes with valuable trait/ gene recombination	BR_BL
4e	Parental lines of hybrids	Parents of developed hybrids	BR_PL
5	**Genetic stock**	Trait and gene specific germplasm, experimentally developed or identified through scientific interventions (e.g. sources of resistance, cytogenetic stock, mutants, mapping population, MS lines	GS
5a	Registered germplasm	Germplasm registered for unique trait(s) with ICAR-NBPGR	GS_RG
6	**Elite line**	Crop varieties under advanced varietal trial but not yet released/ notified. Selection from population from the coordinated trial (AVT I & II line)	EL
7	**Improved cultivar**	Improved cultigens of common knowledge that are in commercial cultivation (extant varieties, and varieties released by institution/ organization/ State)	IC
7a	Released variety	Varieties released and notified by the Central Sub-Committee on Crop Standards, Notification and Release of Varieties of Agricultural and Horticultural Crops	IC_RC
7b	Released Hybrid	Hybrids released and notified by the Central Sub-Committee on crop standards, Notification and Release of Varieties of Agricultural and Horticultural Crops	IC_RH
7c	State released	Varieties released by the State Committee of Agricultural and Horticultural Crops	IC-SR
7d	Institute released	Varieties released by the Institute Committee of Agricultural and Horticultural Crops	IC-IR
7e	Exotic varieties	Exotic varieties introduced through the process of germplasm exchange	IC-EX
8	**Others**	Doubtful germplasm with unknown biological status	OT

Suggested Reading

Ahlawat SP, Bhatt KC, Semwal DP, Pradheep K, Dhariwal OP (2022) Exploration and collection of plant genetic resources in india: status and priorities. Indian J Plant Genet Resour 35(03):117–123

Arora RK (1988) The Indian Gene centre — Priorities and prospects for collection. In: Paroda RS, Arora RK, Chandel KPS (eds) Plant Genetic Resources—Indian Perspectives. NBPGR, New Delhi, 66–75 pp

Arora RK, Pandey A (1996) Wild edible plants of India: Diversity, Conservation and Use, (National Bureau of Plant Genetic Resources, New Delhi)

Brown AHD, Marshall DR (1995) A basic sampling strategy: theory and practice. In: Guarino L, Ramanatha Rao V, Reid R (eds) Collecting plant genetic diversity: technical guidelines. CAB International, Wallingford, pp 75–91

Burkill IH, Finlow RS (1907) Races jute. Agric Ledger 6:41–137

Grivetti LE, Ogle BM (2000) Value of Traditional Foods in Meeting Macro- and Micro-Nutrient Needs: The Wild Plant Connection. Nutr Res Rev, 13:31–46

Hawkes JG (1980) Crop genetic resources—a field collection manual. IBPGR/EUCARPIA, University of Birmingham, Birmingham, 37 p

Howard A, Howard GLC (1910) Investigations on wheat in India. Nature 85:249–250

ICAR-NBPGR (2016) Guidelines for management of plant genetic resources in India. ICAR-NBPGR, New Delhi, p 142

John JK, Pradheep K, Jaisankar I, Nissar VA, Jerard BA (2020) Logistics planning for plant genetic resources collecting from Nicobar Islands of India. Indian J Plant Genet Resour 33(2):132–145

Leppik EE (1970) Gene centres of plants as source of disease resistance. Annu Rev Phytopathol 8:323–344

Marshall DR, Brown AHD (1975) Optimum sampling strategies in genetic conservation. In: Frankel OH, Hawkes JG (eds) Genetic resources for today and tomorrow. Cambridge University Press, Cambridge, pp 53–80

NBPGR (2014) Guidelines for management of plant genetic resources. National Bureau of Plant Genetic Resources, New Delhi, p 170

Pal BP (1937) The search for new genes. In: Agriculture and live-stock in India, vol 7, pp 573–578

Sweeney M, McCouch S (2007) The Complex History of the Domestication of Rice. Ann Bot, 100:951–957

Tomar A, Pandey A, Bhatt KC, Panwar NS, Ahlawat SP (2023) NBPGR-PDS: a precision tool for plant exploration and germplasm collection. Indian J Plant Genet Resour 36(2):250–255. https://doi.org/10.5958/0976-1926.2023.00036.2.07

Von Bothmer R, Seberg O (1995) Strategies for the collecting of wild species. In: Guarino L, Ramanatha Rao V, Reid R (eds) Collecting plant genetic diversity: technical guidelines. CABI International, Wallingford, pp 93–111

Zeven A, Zhukovsky P (1975). Dictionary of Cultivated Plants and Their Centres of Diversity: Excluding Most Ornamentals, Forest Trees and Lower Plants. Wageningen: CAPD.

Zhukovsky PM (1975) World gene pool of plants for breeding. Mega-gene centers and endemic micro-gene centers. USSR Academy of Sciences, Leningrad

Plant Quarantine and Biosecurity for PGR

6

Kavita Gupta, V. Celia Chalam, Jameel Akhtar, Z. Khan, and M. C. Singh

Learning objectives

- Concept of biosecurity and its implications
- The significance and function of plant quarantine in ensuring the safe exchange of PGR
- Detection and identification of pests in quarantine
- Post-entry quarantine growing and inspection
- Salvaging of infested/infected/contaminated germplasm
- Biosafety regulations for transboundary movement of transgenic plants/planting material
- Various technical and policy issues related to biosecurity, biosafety and quarantine including pest risk analysis

Keywords

Plant genetic resources · Germplasm · PGR · Plant quarantine · Bio-security · Biosafety · Pest risk analysis · Post-entry quarantine

K. Gupta (✉) · V. C. Chalam · J. Akhtar · Z. Khan · M. C. Singh
Division of Plant Quarantine, ICAR-National Bureau of Plant Genetic Resources, New Delhi, India
e-mail: Kavita.gupta@icar.org.in; celia.chalam@icar.org.in; Jameel.akhtar@icar.org.in; z.khan@icar.org.in; moolchand.singh@icar.org.in

K. Tripathi et al. (eds.), *Textbook of Plant Genetic Resources*,
https://doi.org/10.1007/978-981-96-5024-8_6

6.1 Introduction

As per the Food and Agriculture Organization (FAO) of the United Nations, biosecurity means the safety of living things. Technically it is '*a strategic and integrated approach that encompasses the policy and regulatory frameworks to analyse and manage risks in the sectors of food safety, animal life and health, and plant life and health, including associated environmental risk*' (http://www.fao.org). Biosecurity covers various aspects of security related to the introduction of plant and animal pests, and pathogens and zoonoses, and also the introduction and release of genetically modified organisms (GMOs) and their products (Bhalla et al. 2018). Thus, it is an all-inclusive concept that includes a safe cross-boundary transfer of commodities, the possible threat of the pests and diseases spread in biological warfare and also from indigenous or emerging pests due to the changing cropping pattern or climate or both, and biosafety-related issues related to entry and use of transgenics. This chapter primarily focuses on the status of plant biosecurity with an emphasis on dangers from exotic plant pests, the status of national/international provisions and the necessity of a planned approach to deal with the issue.

6.2 Need for Biosecurity

The fast-paced developments since the turn of this century have created conditions which pose a new level of risk to human, plant and environmental health. Globalization, frontier biological innovations such as genetic engineering and changing climatic conditions have the possibility to impact plant, human and environment in extraordinary ways. Also, the initiation of WTO and the liberalization of global trade in agriculture since 1995 have opened new avenues for growth and diversification of agriculture, but have also brought in many new challenges. There is an ever-increasing chance of introducing exotic pests, as well as weeds, into the country with the potential threat to cause grave economic losses. Advances in genetic engineering may lead to the development and release of living-modified organisms (LMOs) or their products (GMOs) present additional risk and require proper risk management. Further, climate change has the potential to alter the habitat of known pests and even cause the introduction of new pests. Hence, there is an increasing need for biosecurity. The recent emergence and spread of transboundary plant pests such as the fall army worm, papaya mealy bug, tomato pin worm and the Ug_{99} stem rust of wheat pose novel threats to plant safety.

Owing to these factors, biosecurity is considered as one of the critical issues faced by countries, which necessitates to adopt policies and develop technological capabilities to prevent, detect early and quickly respond to such threats. Biosecurity encompasses the introduction of animal and plant diseases/pests, the introduction and release of LMOs or GMOs food safety, zoonoses and invasive alien species management. Biosecurity is a tactical and unified approach based on the recognition of critical linkages between various sectors and the need for harmonizing and integrating national biosecurity systems and controls to take benefit of synergies across sectors. This would increase the national capacity to protect human health, agricultural production and livelihood, safeguard the environment and protect against uncertain technologies. In addition, it would equip countries to meet obligations for amenability to international trade under the *Agreement for Application of Sanitary and Phytosanitary Measures* of WTO in food and agricultural products. The principal goal of biosecurity is to prevent, control and manage risks to life and health of humans, animals and plants. However, in this chapter, we will be discussing only plant biosecurity.

The introduction of pests into new regions occurs through several pathways: (a) the host could act as a carrier; (b) inert materials such as packing material could carry latent stages of the pest; (c) insect vectors and birds could transport it; (d) air currents could transport the pest over long distances or (e) there could be deliberate, illegal introductions to use them as bioweapons. The first two modes of dispersal lend themselves to restriction by quarantine measures. The next two are by and large beyond human control, and are a major constraint in the control of pests by exclusion. A different degree of alertness is required for the last category. The international trade in agricultural commodities and the transboundary movement of plant genetic resources have caused such situations which require legislative measures to regulate such trade and exchange.

Plant quarantine is an attempt imposed through statutory measures by various governments for regulating the introduction of plant products, planting materials, living organisms, soil etc. to prevent the unintentional introduction of pests damaging to the agriculture of a country/state/region, and if introduced, prevent their further establishment and spread.

The distressing effects caused due to the introduction of diseases and pests along with the transboundary movement of planting material, agricultural products and produce are well documented. Notable examples demonstrate that the introduction and establishment of quarantine pests, including weeds, in new areas can significantly harm crop production and negatively affect the economy of a country or region. Over the last two centuries, numerous such instances of damage have been recorded, and a few broadly cited examples are:

- The introduction of infected seed potatoes carrying the late blight pathogen (*Phytophthora infestans*) from Peru to Ireland led to the Irish famine of 1845, causing the total failure of the potato crop and forcing widespread migration from Europe.
- Powdery mildew (*Uncinula necator*), downy mildew (*Plasmopara viticola*) and root-eating aphid (*Viteus vitifoliae*)—all pests of grapevine were introduced in succession into France in the mid-nineteenth century from North America causing huge injury to the grapevines virtually destroying the wine industry in the country.
- Coffee rust (*Hemileia vastatrix*) in Sri Lanka in 1875 was so devastating that within a span of 15–20 years, the production reduced from 400 million pounds to just 5 million pounds from year 1870 to 1889. From Sri Lanka, the rust entered

India in 1876 and within 5–10 years, the coffee industry in Southern India was seriously affected.

- In 1898, the majority of European countries imposed a ban on the import of pome fruits from America to prevent the entry of San José scale (*Quadraspidiotus perniciosus*), a highly destructive pest. This pest of apple was introduced in India in the 1930s and caused enormous losses in Himachal Pradesh.
- Another recent example is the stored grain pest, Khapra beetle (*Trogoderma granarium*), which hampers of movement of grain across countries during international trade. Its introduction was recorded in Italy, United States and Zimbabwe in the 1940s and it continued to spread in international trade even today.

Box 6.1 Exotic Pests Introduced into India

Several pests were unintentionally introduced on various crops in India, becoming significant threats that continue to cause damage regularly.

Ascochyta Blight—Heavy losses in grain yield of *Cicer arietinum* crop in states of Haryana, Madhya Pradesh, Punjab and adjoining areas occurred during 1981–82 owing to introduction of a virulent biotype of *Ascochyta blight* from the Middle East.

Bunchy Top of Banana—Caused by the *Banana bunchy* top virus, which was introduced into India from Sri Lanka, this disease caused annual losses exceeding ₹40 million in banana production.

San Jose Scale (*Quadraspidiotus perniciosus*)—A pest of apple, which caused significant damage to apple orchards in Himachal Pradesh during the 1930s.

Wooly Aphid (*Eriosoma lanigerum*)—It is a serious pest of apple caused huge losses in apple farming of north Indian states.

Fluted Scale (*Icerya purchasi*)—It is a major pest of citrus, originally native to Australia, which was introduced into India from Sri Lanka before 1928, likely on wattles (*Acacia* sp.). It has since become a significant pest on citrus in southern India.

Golden Nematode (*Heterodera rostochiensis*)—Golden Nematode introduced from the United Kingdom in 1960s has led to severe infestation of potato tubers in the Nilgiris.

Congress Grass—The highly invasive Congress grass, *Parthenium hysterophorus*, was introduced along with imported wheat in 1956. Losses caused are to the tune of ₹16 billion per annum.

Instances of crop pests being used or threatened to be used for biological weapons include releasing of Colorado potato beetles (*Leptinotarsa decemlineata*) by Germany in United Kingdom during World War II. Cuba accused the United States of attacking its various crops using pests such as sugarcane rust, *Thrips palmi* and tobacco blue mould. Moreover, monoculture cultivation having a high degree of genetic homogeneity, the concentration of a single crop in one region and the intensive rearing of animals all aid in the disease spread. As recently revealed by the

United States, Iraq had a wheat smut bomb and stockpiles of wheat rust spores and aflatoxin and anthrax spores were also used in powered forms during the US-Afghanistan war. Further, *Clostridium tetani, C. botulinum* and *Salmonella typhimurium* organisms are being used as weapons in agro-warfare, in case of which the targets are highly vulnerable.

6.3 Present Status of Plant Biosecurity in India

6.3.1 Legislative Framework

After the World War II, the International Plant Protection Convention (IPPC) was promulgated in 1951, to which India became a party in 1956 along with Indonesia, Australia, Netherlands, Portugal, Sri Lanka, Vietnam and the United Kingdom. At present, more than 180 signatory members of the IPPC fosters international cooperation among various countries for preventing the introduction and spread of regulated pests along with international movement of planting material and plants (http://www.ippc.org). IPPC developed 47 International Standards on Phytosanitary Measures (ISPMs) dealing with various phytosanitary activities which include phytosanitary principles, pest risk analysis, phytosanitary treatments, requirements for the establishing pest-free areas and others which give specific guidance on topics related to the Sanitary and Phytosanitary (SPS) Agreement. IPPC members agreed to establish National Plant Protection Organizations (NPPO) to deal with the phytosanitary activities in the country. The Directorate of Plant Protection, Quarantine & Storage (DPPQ&S) under the Ministry of Agriculture & Farmers Welfare (MoA&FW), Government of India is the NPPO of India.

Box 6.2 The Beginning

***Anthonomus grandis* (Mexican Cotton Boll Weevil)**

The Government of India issued the first notification on quarantine in 1906 under the Sea Customs Act of 1878 to stop the introduction of the Mexican cotton boll weevil (*Anthonomus grandis*) and all imported cotton bales needed compulsory fumigation before they entered India. The **Destructive Insects and Pests (DIP) Act,** 1914 was the first quarantine law of India. A gazette notification entitled Rules for Regulating the Import of Plants etc. into India was published in 1936. Over the past century, the DIP Act was revised and amended several times to meet the changing global requirements. Presently, import is regulated under the **Plant Quarantine (Regulation of Import into India) Order** 2003. Even today, *A. grandis* is a regulated pest and the import of cotton seed or bales is required to be free of this pest. It has also been intercepted at ICAR-NBPGR on cotton seed imported from the United States

6.3.1.1 Agreement on Application of Phytosanitary Measures

The Agreement on Application of SPS Measures under the WTO has major implications for various aspects of plant health and trade. This agreement aims to overcome health-related impediments to market access by encouraging the '*Establishment, recognition and application of common sanitary and phytosanitary measures by different members*'. The primary incentive for the use of common international norms is that they are based on scientific evidences, and provide the necessary health protection while at the same time improving trade. Implementation of SPS measures at national and international levels requires a lot of scientific studies and generation of data before it could benefit the stakeholders and promote safe trade.

SPS measures are defined as any measures implemented within the territory of the Member State:

- *To safeguard human and animal health by mitigating risks posed by additives, contaminants, toxins and disease-causing organisms in food.* Restrictions on imports of oranges/grapes containing a certain level of pesticide residues are examples of typical SPS measures.
- *To protect human life from plant- or animal-carried diseases.* The measures are taken to prevent the threat of diseases carried by plant or animal, viz. the spread of rabies or the banning of imports of meat and meat products originating from foot-and-mouth disease regions.
- *To protect animal or plant life from the introduction of pest and diseases or disease-causing organisms.* Restrictions on certain fruit from areas plagued by the fruit fly would be one example of SPS measure to avoid the introduction and spread of fruit flies.
- *To protect a country from damage caused by the entry, establishment or spread of pests.* A measure taken to protect a country from damage caused by the spread of weeds would fall under this category.

6.3.1.2 International Organizations for Developing Standards

To achieve the targets, set by the SPS Agreement, international standards should be developed for which WTO has assigned the responsibility as follows:

1. *For food safety*: The Codex Alimentarius Commission (CAC) based in Rome is a subsidiary organ of the FAO and the World Health Organization (WHO). The SPS Agreement designates Codex as the authority for all matters regarding food safety evaluation and international harmonization. Codex also develops procedures relating to food additives, veterinary drugs, pesticide residues and contaminants, methods of analysis and sampling, and codes and guidelines for hygienic practices.
2. *For animal health and zoonosis:* The standards, recommendations and guidelines are developed under the patronage of the International Office of Epizootics (IOE), Paris. The IOE develops manuals on animal diseases, standards for diagnosis, vaccination, disease control, disinfection and certification.

3. *For plant health*: The IPPC, Rome is the organization recognized by the WTO under the SPS Agreement, as the body for developing ISPMs for facilitating trade. The guidelines and recommendations are developed under the auspices of the Secretariat, IPPC (http://www.ippc.in).

The above three organizations are frequently referred to as the 'Three Sisters'. They play a crucial role as both observers and key contributors to SPS committee meetings, and are frequently called upon as experts to provide guidance on dispute settlement. For matters not covered by the aforementioned organizations, appropriate standards, guidelines and recommendations are developed by other pertinent international organizations identified by the SPS Committee.

6.3.1.3 International Plant Protection Convention

The IPPC develops the ISPMs that provide guidelines on prevention, detection and eradication of pest. As of November 2023, there are 46 adopted ISPMs (ISPM 30 has been revoked), 33 Diagnostic Protocols and 45 Phytosanitary Treatments. These international standards aim to shield sustainable agriculture and improve global food security; protect the environment, forests and biodiversity and enable economic growth in trade (https://www.ippc.int/en/core-activities/standards-setting/ispms/#).

Box 6.3 International Standards for Phytosanitary Measures

- ISPM 1: Phytosanitary principles for the protection of plants and the application of phytosanitary measures in international trade
- ISPM 2: Framework for pest risk analysis
- ISPM 3: Guidelines for the export, shipment, import and release of biological control agents and other beneficial organisms
- ISPM 4: Requirements for the establishment of pest-free areas
- ISPM 5: Glossary of phytosanitary terms
- ISPM 6: Guidelines for surveillance
- ISPM 7: Phytosanitary certification system
- ISPM 8: Determination of pest status in an area
- ISPM 9: Guidelines for pest eradication programmes
- ISPM 10: Requirements for the establishment of pest-free places of production and pest-free production sites
- ISPM 11: Pest risk analysis for quarantine pests
- ISPM 12: Phytosanitary certificates
- ISPM 13: Guidelines for the notification of non-compliance and emergency action
- ISPM 14: The use of integrated measures in a systems approach for pest risk management
- ISPM 15: Regulation of wood packaging material in an international trade
- ISPM 16: Regulated non-quarantine pests: concept and application

(continued)

Box 6.3 (continued)

- ISPM 17: Pest reporting
- ISPM 18: Guidelines for the use of irradiation as a phytosanitary measure
- ISPM 19: Guidelines on lists of regulated pests
- ISPM 20: Guidelines for phytosanitary import regulatory system
- ISPM 21: Pest risk analysis for regulated non-quarantine pests
- ISPM 22: Requirements for the establishment of areas of low pest prevalence
- ISPM 23: Guidelines for inspection
- ISPM 24: Guidelines for the determination and recognition of equivalence of phytosanitary measures
- ISPM 25: Consignments in transit
- ISPM 26: Establishment of pest-free areas for fruit flies (Tephritidae)
- ISPM 27: Diagnostic protocols for regulated pests
- ISPM 28: Phytosanitary treatments for regulated pests
- ISPM 29: Recognition of pest-free areas and areas of low pest prevalence
- ISPM 31: Methodologies for sampling of consignments
- ISPM 32: Categorization of commodities according to their pest risk
- ISPM 33: Pest-free potato (*Solanum* spp.) micropropagative material and mini tubers for international trade
- ISPM 34: Design and operation of post-entry quarantine stations for plants
- ISPM 35: Systems approach for pest risk management of fruit flies (Tephritidae)
- ISPM 36: Integrated measures for plants for planting
- ISPM 37: Determination of host status of fruit to fruit flies (Tephritidae)
- ISPM 38: International movement of seeds
- ISPM 39: International movement of wood
- ISPM 40: International movement of growing media in association with plants for planting
- ISPM 41: International movement of used vehicles, machinery and equipment
- ISPM 42: Requirements for the use of temperature treatments as phytosanitary measures
- ISPM 43: Requirements for the use of fumigation as a phytosanitary measure
- ISPM 44: Requirements for the use of modified atmosphere treatments as phytosanitary measures
- ISPM 45: Requirements for national plant protection organizations if authorizing entities to perform phytosanitary actions
- ISPM 46: Commodity-specific standards for phytosanitary measures
- ISPM 47: Audit in the phytosanitary context

(Source: https://www.ippc.int/en/core-activities/standards-setting/ispms/#)

Previous to the establishment of WTO, governments could adopt international standards, guidelines, recommendations and other advisory texts voluntarily. Even though these norms remain voluntary, a new standing has been conferred upon them by the SPS Agreement. A WTO member adopting such norms is presumed to be in full compliance with the SPS Agreement.

6.3.1.4 Pest Risk Analysis

Pest Risk Analysis (PRA) is a multi-step process involving hazard identification, likelihood of occurrence and potential economic impact of pest establishment under prevailing conditions. The IPPC, through ISPM-2, ISPM-11 and the ISPM-21 supplement on PRA Guidelines (https://www.ippc.int/en/), outlines essential requirements for PRA. It can be initiated using either a pathway-based or pest-based approach. PRA relies on a mix of inductive and deductive reasoning combined with expert judgment to evaluate and characterize risks (Gupta and Irfan 2020). Databases on pest biology, ecology and expert insights play a crucial role in interpreting risks. For instance, the 'Crop Protection Compendium' by CAB International provides structured formats, questionnaires and reliable pest data to support PRA preparation (https://www.cabi.org/cpc). Commodity PRAs, a type of pathway PRA, focus on pests associated to specific pathways identified as quarantine risks. Conversely, pest PRAs consider all potential pathways and commodities associated with the pest.

Stages in PRA

PRA comprises **risk assessment**, which scientifically evaluates the likelihood and extent of pest establishment, and **impact assessment**, which estimates its potential consequences. The process follows three steps: initiation, risk assessment and risk management, all of which must be thoroughly documented. PRA records should detail information sources and the rationale behind decisions on mitigation measures. Conducting a PRA is essential to ensure that imported commodities pose minimal or no risk to agriculture and forestry.

- **Initiation of PRA** begins with identifying pests or pathways requiring assessment. Pest lists are compiled using databases, literature or expert input and prioritized based on expert judgment before proceeding. Not all pests on the list may require risk assessment, depending on the findings.
- **Pest risk assessment** determines whether each pest or pathway-associated pest qualifies as a quarantine pest by evaluating its likelihood of entry, establishment, spread and economic impact. This step involves studying pest's geographical distribution, biology and economic importance. Expert judgment is applied to estimate establishment, spread and economic potential in the PRA area, culminating in a characterization of the introduction risk.
- **Pest risk management** encompasses identifying, evaluating and selecting measures to mitigate risks. A list of options is prepared to reduce risks to acceptable levels, primarily focusing on pathway-related measures and conditions for commodity entry. Risk management actions should consider the level of risk identified in the assessment to protect endangered areas effectively.

Pest risk:
Pest Risk = (Probability of introduction) × (Magnitude of impact)
Probability of introduction:
Probability of introduction = (Probability of entry) × (Probability of establishment) × (Probability of spread)
Magnitude of impact:
Magnitude of impact = (Economic impacts) + (Environmental impacts) + (Social impacts)

The IPPC defines a 'plant pest' as any organism harmful to plants or plant products, including bacteria, fungi, viruses, insects, nematodes, mites, molluscs, other plants and animals. Pests are classified as either regulated or non-regulated, with regulated pests further categorized as quarantine pests or regulated non-quarantine pests. PRA evaluates whether a pest belongs to these categories and determines the necessary phytosanitary measures, if required.

6.3.2 Plant Quarantine in India

Quarantine measures are critical for agriculture-dependent countries like India. The government of India mandated fumigation in 1906 for imported cotton bales to check the entry of Mexican cotton boll weevil (*Anthonomus grandis*). The Destructive Insects and Pests (DIP) Act of 1914 (https://pqms.cgg.gov.in/8792aa56-5f28-400f6a202-d46697b590a6) further regulated the entry of exotic pests, pathogens and weeds, introducing domestic quarantine to control interstate movement of specific planting materials.

In 1984, the 'Plants, Fruits, and Seeds (PFS) Order' was issued under the DIP Act to streamline quarantine operations, later revised in 1989 following the New Policy on Seed Development (1988). In 2003, the PFS Order was replaced by the 'Plant Quarantine (Regulation for Import into India) Order' to address gaps and align with international agreements. The PQ Order 2003 prevents the entry and spread of invasive pests, protecting Indian agriculture and biodiversity. It regulates imports of germplasm, GMOs, transgenic materials and bio-control agents. The Order mandates declarations certifying that imported commodities, especially seeds and planting materials, are free from quarantine pests based on PRA. Key highlights of the Order include strengthened quarantine measures, biodiversity protection and risk-based procedures for agricultural imports:

- Prohibit the import of commodities contaminated with weeds or alien species as specified in Schedule VIII, and regulate the import of packaging material of plant origin unless appropriately treated.

- Provisions are included to regulate the import of soil, peat and sphagnum moss; germplasm, GMOs and transgenic materials for research purposes; live insects, microbial cultures and biocontrol agents; as well as timber and wooden logs.
- Agricultural imports have been classified as: (a) prohibited plant species listed in Schedule IV; (b) restricted species permitted for import, only by authorized institutions listed in Schedule V; (c) restricted species permitted solely with additional declarations for freedom from regulated pests and accompanied by specific treatment certifications listed in Schedule VI and (d) plant material imported for consumption/industry use allowable with regular Phytosanitary Certificate listed in Schedule VII.
- Additional declarations have been detailed in the Order for import of over 700 agricultural commodities with specific lists of over a thousand quarantine pests and 57 species of weeds.
- A total of 182 points of entry have been notified.
- Inspection charges and certification fees have been updated.

PQ Order, 2003 regulates the entire phytosanitary activities for import which includes the import of plants and plant products, germplasm, living organisms, soil, listing of quarantine pests, points of entry, inspection authorities for post-entry quarantine (PEQ), and conditions for relaxation. These are regulated through VII Chapters comprising of 15 clauses and XII Schedules:

Chapter-I: Preliminary	**Clause 1:** Short title and commencement **Clause 2:** Definitions
Chapter-II: General conditions for import	**Clause 3:** Permits for import of plants, plant products etc. **Clause 4:** Import of soil, growing media etc. **Clause 5:** Fees for inspection, fumigation etc. **Clause 6:** Permits required for import of germplasm, transgenic or genetically modified organisms **Clause 7:** Import of live insects and other arthropods/nematodes/microbial cultures including algae/bio-control agents **Clause 8:** Permit required for import of plants and plant products **Clause 9:** Requirement of import of wood and timber
Chapter-III: Special conditions of import	**Clause 10:** Special conditions for the import of plant species
Chapter-IV: Post-entry quarantine	**Clause 11:** Post-entry quarantine
Chapter-V: Appeal and revision	**Clause 12:** Appeal **Clause 13:** Revision
Chapter-VI: Power of relaxation	**Clause 14:** Relaxation conditions of import permit and phytosanitary certificate in certain cases

(continued)

Chapter-VII: Repeal and savings	**Clause 15:** Repeal and savings
Schedule I	Points of entry for the imports of plants/plant materials and other articles **Total 97 seaports, airports and LFS are notified as the point of entry (Seaport-47, Airport-26, Land Frontier Stations-24)**
Schedule II	List of Inland container depots and container freight stations for the import of plants and plant products **Total 85** Inland container depots and container freight stations **are notified as points of entry**
Schedule III	List of foreign post offices for the import of plants and plant products **Total 11** Foreign post offices **are notified as point of entry**
Schedule IV	List of plants/planting materials and countries from where import is prohibited along with justification **Entry of 15 Commodities prohibited from various countries**
Schedule V	List of plants/plant materials whose import is restricted and permitted only through authorized institutions, with additional declarations and subject to special conditions **Entry of 17 Commodities restricted in India**
Schedule VI	List of plants/plant materials import which are permitted with additional declarations and special conditions **Entry of 702 commodities is regulated under this schedule**
Schedule VII	List of plants/planting materials permitted for import upon presentation of a phytosanitary certificate issued by the exporting country, inspection by the Inspection Authority and fumigation if required, and compliance with all general conditions **Total 519 commodities are regulated under this schedule**
Schedule VIII	List of quarantine weed species **Total 57 weeds are notified as quarantine weeds**
Schedule IX	A: Inspection fees; B: Fumigation/disinfection/disinfestation/supervision charges
Schedule X	List of the permit issuing authorities for import of seeds, plants and plant products and other articles **Total 43 Authorities from DPPQ&S and ICAR-NBPGR are notified as Import permit issuing authorities for specific purposes**
Schedule XI	List of the inspection authorities for certification of post-entry quarantine (PEQ) facilities and inspection of growing plants **Total 45 Inspection Authorities for PEQ facilities certification and inspection for planting materials including Head, Plant Quarantine, ICAR-NBPGR New Delhi and Officer-in-Charge, ICAR-NBPGR Hyderabad and 3 Inspection Authorities for tissue cultured raised plants**
Schedule XII	Quantities of seeds permitted for trial purpose/accession to genebank of National Bureau of Plant Genetic Resources (**Source**: http://ppqs.gov.in/divisions/plant-quarantine/present-setup)

No plant or plant products are allowed entry without conducting a PRA. PRA is conducted by DPPQ&S on receiving requests from exporting countries NPPO or importer on submission of NPPO validated technical materials. Based on an assessment of associated phytosanitary risk, available mitigation measures and exporting countries' capabilities, commodities are regulated under different Schedules of PQ Order, 2003.

As of November 2023, a total of 116 amendments to the PQ Order, 2003, have been notified to the WTO. These amendments address revisions to definitions, clarifications on specific queries raised by quarantine authorities of various countries, and updates to the crop lists under Schedules IV, V, VI and VII. The revised lists under Schedules VI and VII now include 707 and 519 crops/commodities, respectively.

6.3.3 Infrastructure: National Plant Quarantine Set-Up

In India agriculture is a state subject and listed in the State Constitution and crop production, including plant protection is the responsibility of State Governments.

The Central Government (Department of Agriculture Cooperation and Farmers Welfare of MoA&FW) supplements States' work in pest management and surveillance through the dissemination of innovative technologies. Plant quarantine and locust control in the listed desert area, being inter-state and international matters, are managed by the Government of India. The DPPQ&S of the MoA&FW serves as the apex authority for implementing plant quarantine regulations, with the PQ Order as its operational foundation. It operates a nationwide network of 73 plant quarantine stations, distributed across 24 airports, 43 seaports and 16 land border checkpoints (http://ppqs.gov.in/divisions/plant-quarantine/present-setup).

Imported materials fall into three categories: (a) bulk consignments of grains and pulses for consumption, (b) bulk consignments of seeds and planting materials for sowing or planting and (c) small quantities of germplasm samples for research purposes. The PQ Stations under DPPQ&S handle the quarantine processing and clearance of the first two categories. However, the ICAR-NBPGR is responsible for the quarantine of all plant germplasm and transgenic materials intended for research purposes. It is equipped with advanced laboratories, a greenhouse complex and a containment facility (CL-4 level) for processing transgenics.

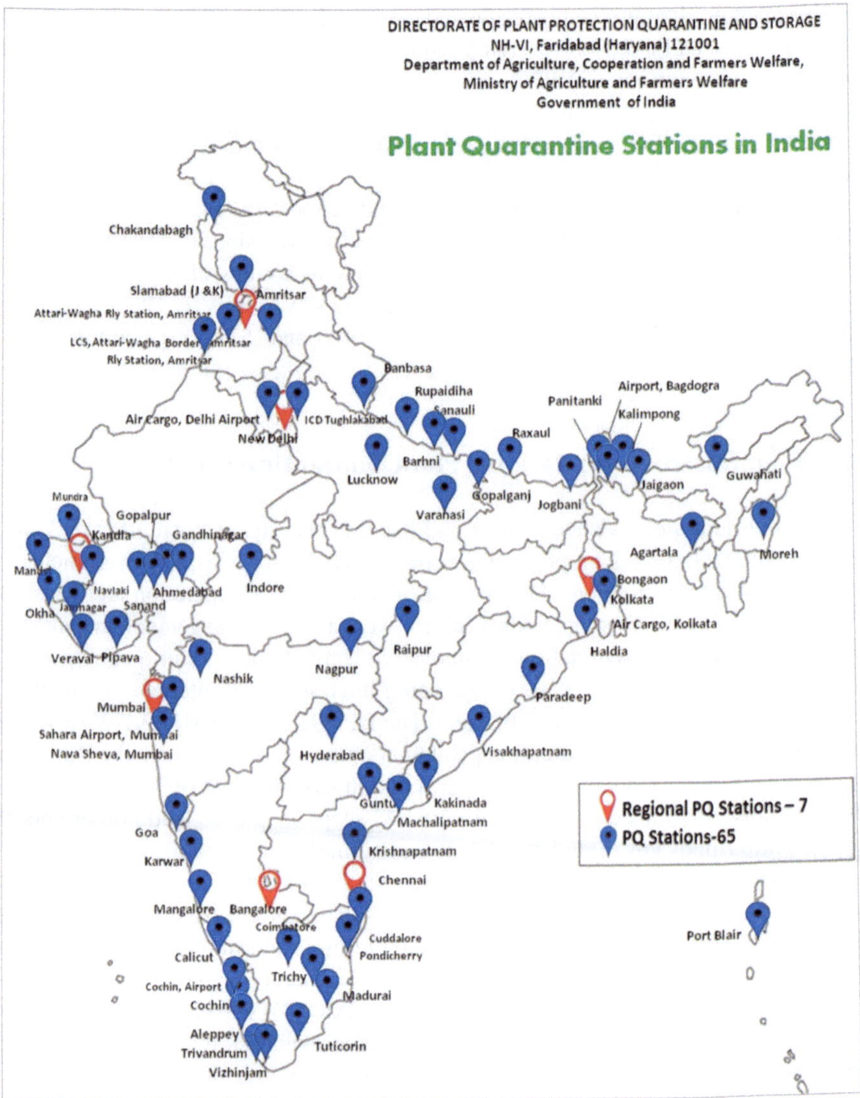

Photo Courtesy: PPA to the Government of India

6.3.3.1 Bulk Consignments for Consumption/Propagation

DPPQ&S is responsible for facilitating the issuance of import permits (IP) and ensuring quarantine clearance. As per the PQ Order 2003, the import of planting materials is subject to two types of conditions: (a) General conditions and (b) Special conditions, including post-entry quarantine requirements.

General Conditions for Imports

The import of seeds and plants for sowing or planting is permitted by the Export and Import (EXIM) Committee under the Department of Agriculture Cooperation and Farmers Welfare. Before importation, an Import Permit (IP) must be obtained, which is valid for 6 months and can be extended for an additional 6 months. Each consignment must be accompanied by a Phytosanitary Certificate (PC) issued by the quarantine authority of the exporting country. Imports are allowed only through entry points designated by the Central Government. The Plant Protection Advisor to the Government of India, or an authorized official, may grant quarantine clearance after inspection, disinfection, disinfestation or fumigation or may order the destruction or return of the consignment to the source country. Additionally, a PRA is mandatory for all materials, except those listed under Schedules V, VI and VII of the PQ Order 2003, and must be conducted in accordance with IPPC standards.

Special Conditions of Imports

The import of planting material from certain regions is prohibited due to the presence of pests listed in Schedule IV of the Plant Quarantine (PQ) Order. Materials specified under Schedules V and VI must comply with the additional declarations and special conditions outlined for them.

6.3.3.2 Germplasm

Private and public sector research institutions wishing to import plants or planting material should apply to ICAR-NBPGR for an IP, which is non-transferable. New Delhi airport is the only entry point for all germplasm consignments. On arrival, the quarantine scientists carefully inspect the material for its health (Dubey et al. 2021). If the imported material is found to be infected or infested with pests, all possible measures are taken to rescue it. Only in rare cases, when recovery is not feasible, the material is incinerated. If post-entry quarantine examination is required, the material is cultivated in the quarantine greenhouse facilities at ICAR-NBPGR, Headquarter and its Regional Stations in Hyderabad (Telangana) and Bhowali (Uttarakhand) (for temperate fruits and other horticultural crops).

Quarantine Procedures for Import

Pre-entry Plant Quarantine Requirements

(a) An Import Permit is a legal requirement of the importing country, and the exporting country must adhere to the conditions or additional declarations specified in the PQ Order.
(b) A Phytosanitary Certificate is mandatory, serving as proof that the consignment has been inspected per the importing country's requirements and is free from the quarantine pests listed in the IP.
(c) PRA is conducted by the importing country, with the exporting country providing the necessary information.
(d) Approval of the PEQ Growing Facility is essential for conducting PEQ inspections.

Post-entry Quarantine Requirements

(a) Examination at the port of entry—Screening is conducted to verify compliance with the phytosanitary conditions specified in the import permit and to ensure the material is free from exotic pests.
(b) Post-entry inspection of the material, including PEQ cultivation and treatment if necessary.

6.4 Methods for Detection and Interception of Pests in Quarantine

All imported PGR samples, including trial and nursery materials, as well as breeding materials whether true seeds or vegetative propagules are thoroughly examined using various techniques to detect any associated exotic pests. ICAR-NBPGR has developed stepwise and systematic procedures for the detection and interception of all kinds of pests, viz. the presence of insect life stages, nematodes and pathogens (fungi, bacteria, viruses, viroids, phytoplasma, weed seeds, soil, plant debris etc.) (Fig. 6.1).

Visual Inspection Seeds, vegetative propagative materials and in vitro cultures are thoroughly examined visually and with a stereo-binocular microscope to detect pests like bunt and smut balls, ergot/sclerotia, seed galls, rust pustules and spores. Seeds exhibiting shrivelling, small size, mottling, necrosis, split coats or discoloration often indicate viral infections and are immediately discarded and incinerated to prevent spread. External feeders, secondary pests and fungal infections are identified through visual examination such as malformed seeds or fungal growth, while bacterial infections are suspected from yellow discoloration around the hilum. Advanced diagnostic methods complement visual inspections. Washing tests detect downy mildews and rusts; blotter methods identify fungi, bacteria and infectivity and electron microscopy (EM), ELISA and PCR-based techniques detect pathogens, including viruses. Infected materials are either treated according to the pest type before release or incinerated if necessary.

Stringent quarantine inspections are pivotal to prevent seed-borne pests from reducing germination, introducing new pest strains or triggering epiphytotics. Specialized diagnostic techniques ensure the identification of pests of quarantine significance, with detailed methods elaborated under subsequent headings.

6.4.1 Detection of Insects and Mites

- ***X-ray Radiography*** is a non-destructive technique used to detect bruchids, phytophagous chalcidoids and other insect groups in seeds that show no external signs of infestation. Crops such as pulses, tree species, oilseeds and many grasses must undergo X-ray radiography for germplasm import. Using a real-time X-ray

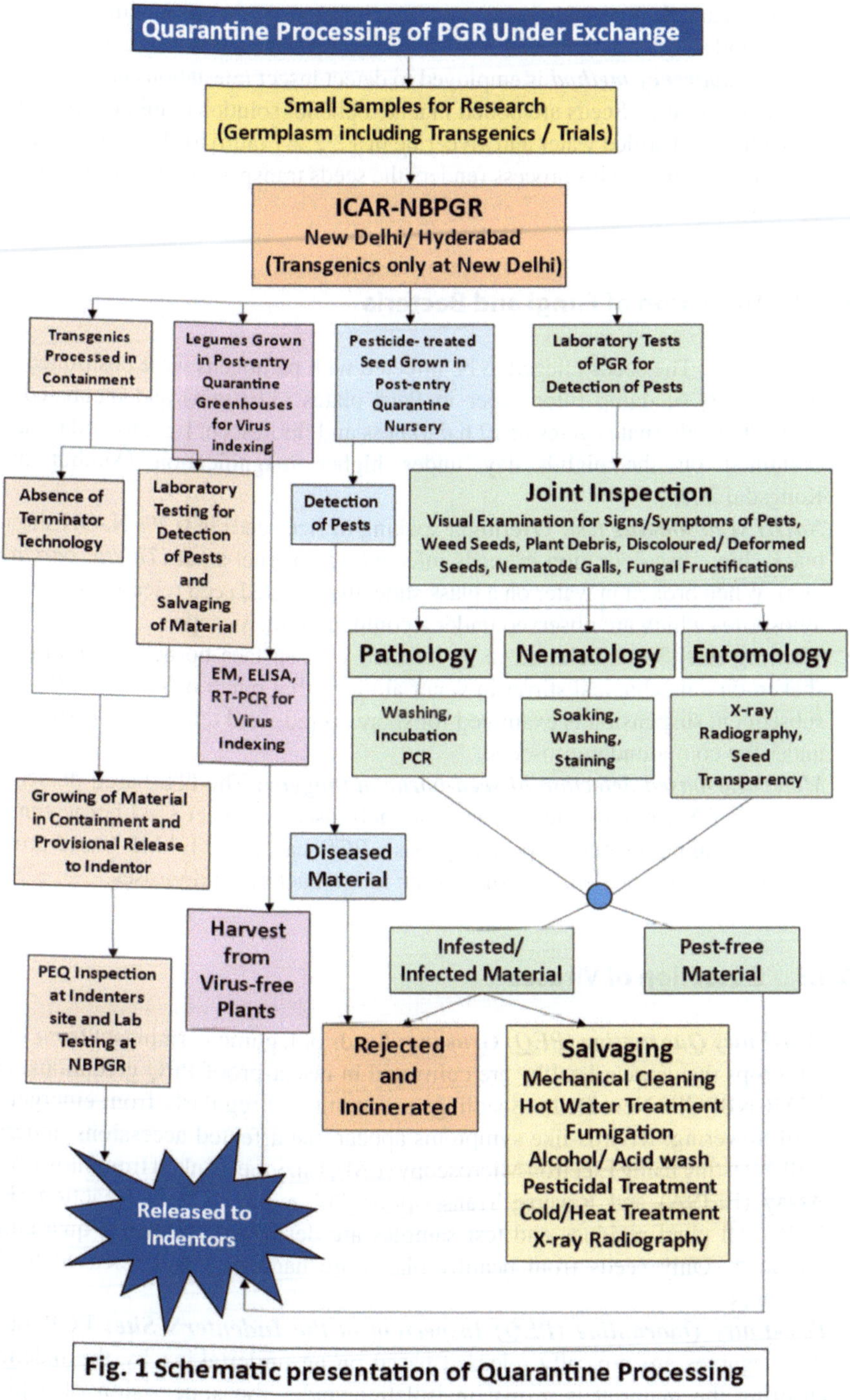

Fig. 6.1 Systematic presentation of quarantine processing

system (Kubtek Xpert 80 X-ray system), infested seeds are identified on the screen and selectively removed, while healthy seeds are released to the indentors.
- ***The transparency method*** is employed to detect insect infestations in very small seeds like sesame. Seeds are boiled in a lactophenol solution (a mixture of lactic acid, phenol, distilled water and glycerine in a 2:2:2:1 ratio) for 1–2 h, depending on seed toughness. This process renders the seeds transparent, making any internal insects visible.

6.4.2 Detection of Fungi and Bacteria

- ***Blotter test:*** The seeds alleged to be infected with pathogens were positioned on three layers of damp filter paper in Petri plates (110 mm) and incubated at 22 ± 1°C in alternate cycles of 12 h darkness and fluorescent light for 7 days and examined on the eighth day under higher magnification (Mathur and Kongsdal 2003).
- ***NaOH seed soaking test:*** Overnight soaking of rice seeds in 0.2% NaOH shows black shiny discolouration in seeds infected with kernel smut (*Tilletia barclayana*). When broken in water on a glass slide, the infected seed release a stream of teliospores which are observed under a compound microscope.
- ***Washing test***: Seeds believed to be infected with surface-borne rust spores are shaken on a mechanical stirrer in water along with a pinch of sand for 30 s and subsequent suspension is examined for spores (uredo and telio) of rust pathogen under the compound microscope.
- ***Molecular-based detection of seed-borne pathogens*:** The ITS-based detection and 16Sr RNA-based detection are routinely used to detect seed-borne fungal and bacterial pathogens, respectively. Also, PCR-based and LAMP-based protocols are used for the detection of specific fungi, bacteria and viruses.

6.4.3 Detection of Viruses

- ***Post-Entry Quarantine (PEQ) Growing*:** Seeds of legume germplasm from various crops that appear healthy are cultivated in insect-proof PEQ greenhouses at ICAR-NBPGR, New Delhi. Seedlings are monitored regularly from emergence until flowering. If virus-like symptoms appear, the affected accessions undergo further testing using Electron Microscopy (EM), Enzyme-Linked Immunosorbent Assay (ELISA) and Reverse Transcription Polymerase Chain Reaction (RT-PCR). All plant residues and test samples are destroyed following quarantine protocols. Only seeds from healthy plants are harvested and released to the indenters.
- ***Post-Entry Quarantine (PEQ) Inspection at the Indenter's Site*:** PGR other than legumes are typically released based on an undertaking by the indenter, ensuring the material is grown in isolation under the supervision of a plant pathologist. Any accessions exhibiting virus-like symptoms are collected during PEQ inspection and subjected to EM, ELISA and RT-PCR if necessary.

- ***Electron Microscopy (EM)*:** Two milligrams of tissue from symptomatic leaf samples collected during the growing-on test are processed according to standard protocols. The prepared grids are examined under a Transmission Electron Microscope (JEOL-JEM 1011).
- ***Enzyme-Linked Immunosorbent Assay (ELISA)*:** Leaf samples showing virus-like symptoms, along with healthy samples and in vitro cultures, are tested using the classical ELISA protocol with polyclonal antiserum against 52 viruses. A sample is considered infected if its optical density (OD) value exceeds twice that of the negative control (Chalam et al. 2017).
- ***Reverse Transcription Polymerase Chain Reaction (RT-PCR)*:** RNA is extracted from samples using the RNeasy Plant Mini Kit (Qiagen), following the manufacturer's protocol. cDNA is synthesized using the Thermo Fisher cDNA Kit. PCR is then performed using specific primers, and the resulting products are separated by electrophoresis. The final image is captured through a gel documentation system. RT-PCR is used when ELISA results are ambiguous or inconclusive.

6.4.4 Detection of Nematodes

- ***Nematode extraction from suspected germplasm*:** Seeds suspected or known to be infected by seed-borne nematodes are soaked in water for 24–48 h to soften them. Hard seeds are cut open/teased/crushed to enable the nematodes to come out in water. Similarly, plant debris and soil clods are also soaked in water, and strained through nematological sieves of 100, 200, 325 and 400 mesh and nematodes are taken out by modified Baermann's funnel techniques. Vegetative propagules like corms, bulbs, rhizomes, rooted cuttings, suckers, and tubers are observed for the presence of nematodes. These planting materials are rinsed properly with clean water, and small pieces of 2–3 cm were incubated for 24 h to allow mobile nematode stages to come into water suspension. Nematodes are recovered, fixed, examined and identified under a compound microscope. In addition, a few roots are stained using acid fuchsin and lactophenol solution for the detection of sedentary nematodes.

6.4.5 Detection of Weeds

- ***Naked/aided eye examination*:** Samples are inspected for weed seeds by filtering through sieves of different-sized holes. The working sample is spread in a uniform single layer on a clean sheet of white paper or in a white tray and observed using an illuminated magnifier. Weed seeds are separated and categorized as different types based on their shape, size, colour, texture and any appendages present. These seeds are kept in separate screw cap vials and labelled for future reference. Stereoscopic binoculars are used for observing extremely small seeds.

After various tests/examinations, the intercepted pests are identified. A total of 78 exotic pests including insects/mites (26), fungi (6), viruses (19), nematodes (9) and weeds (18) that are not yet reported from India were intercepted. Had they not been intercepted; they could have wreaked havoc on the Indian agriculture and caused severe losses to the Indian agriculture. Pests of quarantine significance including pests not yet reported from India intercepted from imported germplasm over the last 45 years are given in Table 6.1 (Chalam et al. 2023). The intercepted pests can be categorized as:

Table 6.1 Quarantine pests intercepted in imported PGR by ICAR-NBPGR over the years

Quarantine pests intercepted	Host	Source
Insects		
1. *Acanthoscelides desmanthi*	*Desmanthus* spp.	Columbia
2. *A. obtectus*	*Cajanus cajan*, *Phaseolus vulgaris*	Brazil, Colombia, Italy, Mexico, Nigeria, Peru, USA
3. *Anthonomus grandis*	*Gossypium* sp., *Hibiscus* sp.	USA
4. *Blapstinus histrichus*	*Carthamus tinctorius*	USA
5. *Bootanomyia* sp.	*Eucalyptus* sp.	Australia
6. *Bruchidius atrolineatus*	*Vigna unguiculata*	Nigeria
7. *B. trifolii*	*Trifolium* spp.	Egypt, Ethiopia, UK
8. *Bruchophagus gibbus*	*Medicago sativa*	USA
9. *B. roddi*	*Trifolium pratense*	USA
10. *Bruchus atomarius*	*Lens culinaris*	Lebanon
11. *Bruchus dentipes*l.	*Vicia faba*	Syria
	V. narbonensis	Afghanistan
12. *B. ervi*	*Lens culinaris*	Chile, Greece, Iran, ICARDA (Syria), Lebanon, Italy, Germany, Morocco, Turkey, Russian Federation
13. *B. nubilis*	*V. faba*	Ukraine
14. *B. rufimanus*	*V. faba*	Afghanistan
15. *B. tristis*	*Lathyrus sativus*	ICARDA (Syria), Lebanon
16. *Callosobruchus subinnotatus*	*Vigna subterranea*	Ghana
17. *C. rhodesianus*	*V. unguiculata*	Nigeria
18. *Carpophilus truncatus*	*Zea mays*	USA
19. *Ephestia elutella*	*T. aestivum*	Italy
20. *Gymnetron plantaginis*	*Plantago lanceolata*	Italy
21. *Oscinella frit*	*H. vulgare*	Sweden
22. *Pachymerus lacerdae*	*Orbignya phalerata*	Brazil

(continued)

Table 6.1 (continued)

Quarantine pests intercepted	Host	Source
23. *Quadrastichodella eucalypti*	*Eucalyptus camaldulensis*	Australia
24. *Sitophilus granarius*	*T. aestivum*	USA
25. *Trogoderma variabile*	*T. aestivum*	Australia
26. *Typhaea stercorea*	*Mangifera indica*	Israel
Viruses		
1. *Barley stripe mosaic virus*	*Hordeum vulgare*	USA
	Triticum aestivum	USA
	Zea mays	Philippines, USA
2. *Bean mild mosaic virus*	*Glycine max*	Canada, Columbia
	Phaseolus vulgaris	USA
3. *Bean pod mottle virus*	*G. max*	USA
4. *Broad bean stain virus*	*Pisum sativum*	Spain
	Vicia faba	ICARDA (Syria), Bulgaria
5. *Broad bean true mosaic virus*	*P. vulgaris*	USA
6. *Broad bean mottle virus*	*P. vulgaris*	AVRDC (Taiwan)
7. *Cherry leaf roll virus*	*G. max*	AVRDC (Taiwan), Sri Lanka, Thailand, USA
	P. vulgaris	CIAT (Colombia)
8. *Cowpea mottle virus*	*Vigna unguiculata*	Philippines
	V. subterranea	Ghana
9. *Cowpea severe mosaic virus*	*G. max*	USA
	V. radiata	Australia
	V. unguiculata	Belgium
10. *High plains virus*	*Z. mays*	USA
11. *Maize chlorotic mottle virus*	*Z. mays*	Puerto Rico, USA
12. *Pea enation mosaic virus*	*G. max*	Costa Rica
	P. vulgaris	Nepal
	P. sativum	Spain, USA
	V. faba	ICARDA (Syria)
13. *Peanut stunt virus*	*P. vulgaris*	USA
	G. max	Costa Rica
14. *Pepino mosaic virus*	*Capsicum annuum*	Republic of Korea
	Solanum lycopersicum	Netherlands, Thailand, USA

(continued)

Table 6.1 (continued)

Quarantine pests intercepted	Host	Source
15. *Raspberry ring spot virus*	*G. max*	AVRDC (Taiwan), Costa Rica, Sri Lanka, Thailand, USA
16. *Tomato ring spot virus*	*G. max*	AVRDC (Taiwan), Canada, Costa Rica, Sri Lanka, Thailand, USA
17. *Wheat streak mosaic virus*	*T. aestivum*	USA
	Z. mays	Philippines, Puerto Rico, South África, USA
Fungi		
1. *Fusarium oxysporum* f. sp. *cucumerinum*	*Cucumis sativus*	USA
2. *Monographella nivalis*	*Triticum aestivum*	Germany, Hungary, Italy, Mexico, Sweden, Turkey, UK and USA
	Hordeum vulgare	Italy
3. *Peronospora manshurica*	*G. max*	Canada, Belgium, Brazil, Colombia, Costa Rica, Ghana, Hungary, Indonesia, Israel, Italy, Korea, Japan, Nepal, Malaysia, Papua New Guinea, Poland, Romania, Russia, Taiwan, Thailand, USA and Zimbabwe
4. *Phomopsis longicolla*	*Helianthus annuus*	USA
5. *Uromyces beticola*	*Beta vulgaris*	Belgium, Denmark, Holland, Hungary, Germany, The Netherlands, Sweden, UK, USA and SSR
Nematodes		
1. *Aphelenchoides arachidis*	*Arachis hypogea* pods	Nigeria
2. *A. fragariae*	*Fragariae* spp	USA
3. *Ditylenchus destructor*	*Solanum tuberosum* packaging material	Australia
4. *D. dipsaci*	*Humulus lupulus*	Germany and USA
5. *Heterodera humuli*	*H. lupulus*	Australia, Germany
6. *H. schachtii*	*Beta vulgaris*	Denmark and Germany
7. *H. goettingiana*	*Pisum* spp. (in soil clods)	UK
8. *Rotylenchus minutus*	*Hypoxis hemerocallidea*	Swaziland
9. *Scutellonema bradys*	*Dioscorea* spp.	USA
Weeds		
1. *Avena barbata*	*H. vulgare*	Morocco
2. *A. sterilis*	*T. aestivum*	Australia
3. *Bifora testiculata*	*H. vulgare*	Lebanon
4. *Bromus secalinus*	*Cicer arietinum*	Australia
5. *Centaurea maculosa*	*Coriandrum sativum*	Russia

(continued)

Table 6.1 (continued)

Quarantine pests intercepted	Host	Source
6. *C. melitensis*	*H. vulgare*	Morocco
7. *C. solstitialis*	*C. sativum*	Russia
8. *Cichorium pumilum*	*Trifolium alexandrinum*	Uzbekistan
9. *Cichorium spinosum*	*T. alexandrinum*	Uzbekistan
10. *Convolvulus erubescens*	*H. vulgare*	Morocco
11. *Echinochloa crus-pavonis*	*Oryza sativa*	China
12. *Echium plantagineum*	*H. vulgare*	Morocco
13. *Galium tricornutum*	*H. vulgare, Lathyrus sativus, C. sativum*	Lebanon, Uzbekistan
14. *Hordeum pusillum*	*L. sativus*	Lebanon
15. *Ipomoea plebeia*	*Nigella sativa*	Uzbekistan
16. *Salsola vermiculata*	*Lens culinaris*	Canada

1. Those not yet recorded in India;
2. Those with races, biotypes or strains not yet identified in India;
3. Those found on a new host or from a country where they were not previously reported;
4. A completely new pest species not yet described in scientific literature or
5. Those reported in India but with a broad host range.

Interceptions, particularly of pests and their races, biotypes or strains not yet reported in India, highlight the critical role of quarantine in preventing the introduction of harmful exotic pests. The third and fourth categories of pests are not anticipated based on the PRA, which is primarily literature-based. As no reports exist on these pests or hosts, their presence is completely unexpected and significant from a quarantine perspective. The final category 'pests with a broad host range' presents a serious risk, as they could become invasive if they encounter suitable environments.

6.5 Major Biosecurity Threats to Indian Agriculture

6.5.1 Risk Associated with the Import of Plants/ Planting Material

Bulk consignments for sowing or planting pose significant risks due to the challenges of thorough examination, treatment and the extensive planting areas required. Quarantine inspections typically focus on smaller samples from these consignments, with decisions to reject, detain or release the lot based on findings, often following fumigation or chemical treatments. Conversely, bulk consignments for consumption generally present lower risks. Small samples imported for research, such as germplasm, wild relatives or landraces, carry substantial quarantine importance as they often harbour diverse pest biotypes, races or strains. In true seeds, risks stem from deep-seated infections, while vegetative propagules pose greater risks due to higher pest susceptibility.

Quarantine processing has intercepted numerous pests in bulk consignments, germplasm and research materials (http://www.plantquarantineindia.org). Regulated pests have also been detected in imported transgenic materials, such as *Peronospora manshurica* causing soybean downy mildew from the USA and *Cryptolestes ferrugineus* in paddy from Singapore. These examples highlight the critical role of quarantine in preventing pest introductions. A detailed list of intercepted pest categories is provided in Table 6.2.

Table 6.2 Examples of different categories of pests intercepted in quarantine

	Category of pest intercepted	Pest/host/source country
1.	Not known to occur in India	*Peronospora manshurica*/Soybean/USA *Uromyces betae*/Sugarbeet/USA and Italy *Fusarium nivale*/Wheat/UK *Cowpea mottle virus*/Cowpea/Philippines *Tomato black ring virus*/French bean/CIAT (Colombia) *Heterodera schachtii*/Sugarbeet/Denmark *Anthonomus grandis*/Cotton/USA *Quadrastichodella eucalypti*/Eucalyptus/Australia
2.	Known to occur but the race/biotype/strain intercepted is not known to occur	*Bipolaris maydis* Race T/Sorghum/USA Pea seed-borne mosaic virus/Broadbean *Burkholderia solanacearum* Biovar 3/Groundnut/Australia
3.	Intercepted on a host on which it was never reported before	*Alternaria zinniae*/Tobacoo/Japan *Pseudomonas syringae* pv *syringae*/*Hibiscus cannabinus*/Bangladesh *Aphelenchoides besseyi*/*Stylosanthes hamata*/Australia *Merobruchus columbinus*/*Samanea saman*/UK *Bruchus ervi*/*Acacia brachystachya*/Australia *Pachymerus lacerdae*/*Orbignya phalerata*/Italy
4.	Intercepted from a country from where it was never reported before	*Peronospora manshurica*/soybean/Malaysia *Heterodera zeae*/*Vetiveria zizanioides*/Tanzania *Bruchus ervi*/*Acacia brachystachya*/Australia
5.	A new species hitherto unreported	*Drechslera pluriseptata*/*Eleusine coracana*/Zambia *Tylenchorhynchus neoclavicaudatus*/Potato tubers/USA *Polenchus minutus*/Palm plants/UK
6.	Known to occur in India but possess a wide host range	*Bipolaris sorokiniana*/*Carthamus tinctorius*/Italy, USA *Bipolaris siccans*/Soybean/USA *Claviceps purpurea*/*Avena sativa*/USA *Colletotrichum graminicola*, *Bipolaris turcica* and *Gloeocercospora sorghi*/Sorghum/Nigeria

Source: Khetarpal and Gupta (2008)

6.5.2 Transboundary Movement of GMOs/LMOs

The Environment Protection Act regulates the safe use and release of GMOs in India. The Department of Biotechnology (DBT), under the Ministry of Science and Technology, is responsible for reviewing, approving and monitoring GMO experiments and recombinant DNA products. Meanwhile, the Ministry of Environment, Forest, and Climate Change (MoEF&CC) oversees large-scale commercial applications through the Genetic Engineering Appraisal Committee (GEAC). The biosafety guidelines issued by DBT in 1990 require organizations working with GM plants to establish Institutional Biosafety Committees (IBSCs) to liaise with the government. To ensure compliance with these guidelines, DBT has formed committees with specific responsibilities. The Review Committee for Genetic Manipulation (RCGM) assesses R&D projects on GMOs, inspects experimental sites, and approves the import and export of GMOs for research and training. RCGM's Monitoring cum Evaluation Committee supervises limited multi-locational GMO field trials. State and district-level committees also monitor GMO experiments to ensure safety and compliance. ICAR-NBPGR ensures imported transgenic materials are free of pests and terminator gene technology, protecting biosecurity and preventing the introduction of harmful traits or organisms.

6.5.2.1 Cartagena Protocol on Biosafety

LMOs, including GMOs, are regulated under the Cartagena Protocol on Biosafety (CPB), established within the framework of the Convention on Biological Diversity (CBD). Article 19, Paragraph 3 of the CBD provided the basis for this international agreement, aimed at managing the cross-boundary movement of LMOs developed through modern biotechnology (Gupta et al. 2007; Chalam et al. 2018). After extensive negotiations, the Protocol was adopted on 29 January 2000 and came into effect on 11 September 2003. To date, 173 countries, including India, are signatories (https://bch.cbd.int/protocol/).

The CPB seeks to safeguard biological diversity from potential risks posed by the transfer, handling and use of LMOs, while also addressing risks to human health. Its scope covers transboundary movement, transit, handling and use of LMOs that could harm biodiversity or sustainable development. The Protocol emphasizes a precautionary approach to mitigate these risks while promoting biodiversity conservation and sustainable development.

LMOs covered under the Protocol are categorized as:

- LMOs for intentional introduction into the environment (e.g. seeds, live fish)
- LMOs intended for direct use as food or feed, or for processing (e.g. agricultural commodities—corn, canola, cotton)
- LMOs for contained use (e.g. bacteria for laboratory scientific experiments)

However, it does not include products derived from LMOs (e.g. paper from GM trees) and LMOs that are pharmaceuticals for humans (that are addressed by other international agreements or organizations).

The Protocol outlines an Advance Informed Agreement (AIA) procedure (Article 7) for LMOs intended for planned introduction into the environment, which may have adverse effects on biodiversity conservation and its sustainable use. Article 11 applies before the first transboundary movement of LMOs proposed for direct use as food, feed or for processing.

The AIA procedure does not apply to the intentional transboundary movement of LMOs deemed unlikely to have harmful effects on biodiversity conservation and the sustainable use of biological diversity, while also considering potential risks to human health, as determined by a decision of the Conference of the Parties (COP), acting as the meeting of the Parties (MOP) to this Protocol.

Four categories of LMO are exempted from the AIA:

- LMOs in transit,
- LMOs for contained use,
- LMOs identified in a decision of the COP/MOP as not expected to have adverse effects on biodiversity conservation and sustainable use and
- LMOs intended for direct use as food, feed or for processing.

The AIA is meant only for first-time shipments while consecutive shipments are exempt from it.

6.5.2.2 Compliance with the Protocol in India

Article 2 of the Cartagena Protocol on Biosafety (CPB) mandates Parties to adopt legal, administrative and other measures to meet their obligations and achieve the Protocol's objectives. A key requirement is the establishment of a National Biosafety Regulatory Framework. As a CPB signatory, India has actively implemented its commitments, participated in COP-MOP meetings and adhered to their decisions.

Acknowledging the rapid advancements in biotechnology, the Indian government has established a regulatory framework for the development, evaluation and release of biotechnology products. The MoEF&CC acts as India's national focal point for implementing the Protocol, coordinating with the Protocol Secretariat, and serving as the Competent National Authority (CNA). The CNA manages administrative responsibilities under the Protocol on behalf of India.

6.5.2.3 Biosafety Regulatory Framework for GMOs in India

In India, GMOs and their products are regulated under the 'Rules for the Manufacture, Use, Import, Export, and Storage of Hazardous Microorganisms, Genetically Engineered Organisms or Cells, 1989', which were notified under the Environment (Protection) Act, 1986. Known as the Rules of 1989, these regulations govern all activities related to GMOs, including their sale, storage, import, export, production, manufacturing and packaging. They apply to both research and large-scale applications of GMOs and their products, covering:

- The manufacture, import and storage of microorganisms and genetically engineered products

- Genetically engineered organisms, microorganisms and cells, as well as any substances, products or foodstuffs containing such cells, organisms or tissues
- New gene technologies, in addition to cell hybridization and genetic engineering

A set of biosafety guidelines has been issued under these rules to minimize any potential adverse impacts of GMOs and their products on the environment, as well as human and animal health.

Distinct guidelines are provided for various stages of development and use of LMOs, namely, contained use, confined field trials, food safety, and environmental release as indicated:

- Recombinant DNA Safety Guidelines, 1990
- Revised Guidelines for Research in Transgenic Plants, 1998
- Guidelines for generating Preclinical and Clinical Data for rDNA Vaccines, Diagnostics and other biologicals, 1999
- Guidelines for Conduct of Confined Field Trials (CFTs) of Regulated, Genetically Engineered (GE) Plants, 2008
- Standard Operating Procedures (SOPs) for CFTs of Regulated, GE Plants, 2008
- Guidelines for Monitoring of CFTs of Regulated, GE Plants, 2008
- Guidelines for the Safety Assessment of Foods Derived from Genetically Engineered Plants, 2008 (updated in 2012)
- Protocols for Food and Feed Safety Assessment of GE crops, 2008
- Guidelines and Handbook for Institutional Biosafety Committees, 2011
- Guidelines on Similar Biologics: Regulatory Requirements for Marketing Authorization in India, 2012 (updated in 2016).

The rules are implemented by MoEF & CC, the Department of Biotechnology (DBT) and State Governments through six competent committees, viz.

- Recombinant DNA Advisory Committee (RDAC)
- Institutional Biosafety Committee (IBSC)
- Review Committee on Genetic Manipulation (RCGM)
- Genetic Engineering Appraisal Committee (GEAC)
- State Biotechnology Coordination Committees (SBCCs) and
- District Level Committees (DLCs)

The RCGM and GEAC have formed various sub-committees and expert panels, consisting mainly of specialists from public sector institutions, to develop and review biosafety guidelines and data. Central Compliance Committees oversee confined field trials on a case-by-case basis. Several other regulations, such as the PQ Order, 2003; the Food Safety and Standards Act, 2006 and the DGFT Notification on GM policy in the Foreign Trade Policy (2006–09), also apply. India has several guidelines for the risk assessment and management of GMOs, including the 'Guidelines for the Environmental Risk Assessment of Genetically Engineered Plants, 2016' and the 'Environmental Risk Assessment of Genetically Engineered Plants: A Guide for Stakeholders and Risk Analysis Framework, 2016' (MoEF & CC and BCIL, 2017).

6.5.3 Threat from Emerging Pests and Diseases

In the context of biosecurity, the threat from exotic destructive pests is a major concern, especially in the era of liberalized trade under the WTO regime. However, indigenous pests in India are increasingly gaining prominence due to changes in cropping systems. This has resulted in the resurgence of pest problems like necrosis in sunflower and groundnut, and leaf curl in potato, indicating potential threats to crop production and food security. The excessive use of chemical pesticides has also led to the development of resistant pest strains, emphasizing the need for ongoing enhancement of mitigation strategies.

A deeper understanding of resistance genetics is essential, particularly for key crops like rice, maize and grain legumes. This includes identifying resistance genes and ensuring durable resistance through both conventional breeding and molecular techniques. Efforts should also focus on understanding the causes of significant diseases and developing diagnostic protocols for seed certification.

6.5.4 Potential Threat of Biowarfare

The mishandling or misuse of dangerous, previously unreported pests in India, which can severely harm human health and ecosystems, is a major national concern. The agricultural sector is especially vulnerable to threats from exotic pests and diseases, some of which may be used as bioweapons (Table 6.3). Many agricultural

Table 6.3 Major threats from devastating exotic plant pests/diseases

Bacterial and Fungal Pathogens
• Bacterial wilt and ring rot of potato (*Clavibacter michiganensis* subsp. *sepedonicus*)
• Fire blight of apple and pear (*Erwinia amylovora*)
• Black pod of cocoa (*Phytophthora megakarya*)
• Powdery rust of coffee (*Hemileia coffeicola*)
• Sudden death of oak (*Phytophthora ramorum*)
• South American leaf blight of rubber (*Microcyclus ulei*)
• Vascular wilt of oil palm (*Fusarium oxysporum* f sp. *elaeidis*)
• Soybean downy mildew (*Peronospora manshurica*)
• Blue mold of tobacco (*Peronospora hyoscyami* subsp. *tabacina*)
• Tropical rust of maize (*Physopella zeae*)
Viruses, Viroids and Phytoplasma
• *Barley stripe mosaic virus*
• Coconut cadang cadang (Viroid)
• Palm lethal yellowing (Phytoplasma)
Nematodes
• Pine wood nematode (*Bursaphelenchus xylophilus*)
• Red ring nematode of coconut (*Rhadinaphelenchus cocophilus*)
Insects
• Mediterranean fruit fly (*Ceratitis capitata*)
• Cotton boll weevil (*Anthonomus grandis*)
• Russian wheat aphid (*Diuraphis noxia*)

pests, including strains, isolates or biotypes, have bioweapon potential. For example, viruses like rice tungro bacilliform virus, cotton leaf curl virus, groundnut bud necrosis virus and banana bunchy top virus, as well as others like tobacco streak, citrus tristeza and mung bean yellow mosaic viruses, have multiple strains.

Pathogens responsible for serious diseases also exhibit variability, such as cereal rusts caused by *Puccinia triticina* (with numerous virulent pathotypes), rice blast (*Pyricularia oryzae*), *Burkholderia solanacearum* (Race 2, not yet reported in India) and *Xanthomonas campestris* pv. *malvacearum* (with the highly virulent African strain absent in India). Insects like *Bemisia tabaci, Nilaparvata lugens, Orseolia oryzae* and *Tribolium castaneum* also have diverse biotypes. Nematodes such as *Meloidogyne incognita* and *Heterodera avenae* show multiple races.

While balancing scientific and commercial research freedoms, the threat of alien species and potential bioterrorism requires robust policies and enhanced technological capabilities to prevent and respond effectively. In some cases, a lack of genetic diversity weakens natural defences, highlighting the need for proactive measures in managing these threats.

The introduction of naturally occurring or genetically modified viruses, along with other experimental biotechnology products, into weaponry, livestock, crops and medicine is controversial. If not properly regulated, these could pose significant risks to the life. New biological strains of several plant pathogens can easily cause significant harm, as often observed in the case of livestock, because they rapidly infect elements of ecosystems that have not yet developed immunities.

6.6 Salvaging of Infested/Infected/Contaminated PGR

The different approaches used for salvaging germplasm include:

- ***Mechanical cleaning*:** A consignment may be contaminated with plant debris, soil, dead insects/parts thereof, egg masses, pupa, adult insects, fungal bodies like bunt, ergot, sclerotia, smut balls and different kinds of weed seeds. Cleaning by sieving, hand picking, gravitational separators etc. are done to remove contaminants from the consignment.
- ***Hot water treatment*:** It is a very effective tool to save seed material by eradicating many plant pathogenic fungi, bacteria and nematodes. The material is dipped in water and kept at a definite temperature (52 °C for 30 min) for a specified duration. The basic principle employed here is that the treatment temperature should be high to kill the pest but not affect the viability of seed. But, in most of the cases the margin of safety is very narrow, hence, very precise regulation of temperature is needed.
- ***Chemical treatments:*** Chemicals/pesticides could be applied as spray, dip, dust or slurry. It should be ensured that a dose of chemicals should be such as to eradicate the pest but not kill the host seed/propagule. Also, the technical personnel handling the treatment should not be adversely impacted by the chemical used. Upon arrival, the health status of the material should be assessed before any

treatment is administered. The selection of the chemical and its dosage should be based on the specific pest involved. Seed samples infected with various fungi were salvaged through fungicide seed dressing, utilizing a mixture of Carbendazim and Dithane M-45 (1:1 ratio).

- ***Fumigation:*** Fumigation of seed under the atmospheric pressure or lowered pressure is acceptable as a phytosanitary treatment for storage insects. Fumigants like phosphine, EDCT mixture (ethylene dichloride + carbon tetrachloride mixture) and methyl bromide are commonly used fumigants in India. Methyl bromide has critical use exemption for quarantine and pre-shipment purposes while non-feed stock uses of carbon tetrachloride are restricted. Samples are salvaged using aluminium phosphide (Phosphine @ 2 g/cu m for 72 h)/EDCT mixture @ 320 mg/L for 48 h against seed-borne insects.
- ***Alcohol wash:*** An alcohol wash is used to remove seed-borne spores of various fungi. Infected seeds are mixed with ethyl alcohol and a pinch of river sand in a mechanical stirrer for 30 s. The abrasive action of the sand helps dislodge the spores from the seed surface, leaving the seeds clean in the alcohol.
- ***Acid wash:*** To eliminate spores of sugar beet rust (*Uromyces betae*) from sugar beet seeds, concentrated sulfuric acid is used. The seeds are stirred in the acid with a glass rod for 1 min, followed by immediate washing under running tap water until all traces of the acid are removed, and the seeds are dried.
- ***X-ray radiography:*** X-ray radiography is an effective method for detecting and addressing hidden infestations in seeds, such as seed beetles (bruchids), seed wasps (chalcidoids), weevils and other internal feeders. Seeds from over 340 plant genera known to harbour hidden infestations are required to undergo X-ray radiography. The seeds are placed in a single layer and exposed to X-rays. By examining the X-ray images, infested seeds can be manually separated from healthy ones.
- ***Cold treatment*:** Low temperature is employed to kill the insects in seed materials. Most of the insect kills between −10 and 15 °C within hours and this technique is used for disinfestation of seed samples infested with various stages of insects.
- ***Formalin root-dip treatment*:** Several species of plant parasitic nematodes are killed in the root zone of various vegetative propagules. Plant propagule samples were salvaged by root dip in formalin at 0.25% for 10 min.
- ***Healthy harvest through PEQ growing*:** Healthy harvest is made from virus infected seed samples and supplied to the indentors.

6.6.1 Prophylactic Treatments

Suspected seed samples are all given prophylactic fumigation with aluminium phosphide (Phosphine at 2 g/cu m for 72 h)/EDCT mixture at 320 mg/L for 48 h against seed-borne insects while pesticidal dip/spray treatment are given to vegetative propagule samples against insect-pests and pathogens. Prophylactic treatment with hot water (52 °C for 30 min) is given to all samples of paddy against seed-borne pathogens. All seed samples of chilli and tomato are treated with 10% tri-sodium orthophosphate to avoid the entry of tobamovirus strains.

6.7 Issues in the International Exchange of PGR

There are several issues related to quarantine for the safe exchange of plant germplasm, both in terms of legislation and quarantine procedures. The national quarantine legislation has categorized imports as:

(a) Prohibited plant species (Schedule IV);
(b) Restricted plant species where import is permitted only by authorized institutions (Schedule V);
(c) Restricted plant species are permitted only with additional declarations of freedom from quarantine/regulated pests and subject to specified treatment certifications (Schedule VI) and
(d) Plant material for consumption/industrial use is allowed with a Phytosanitary Certificate (Schedule VII).

The PQ Order has made Pest Risk Analysis (PRA) mandatory for all materials being imported into the country, except for those listed in Schedules V, VI and VII. The crops listed in Schedules V, VI and VII of the PQ Order provide a generic PRA, and a detailed PRA is not necessary for their import. In the case of PGR, a several species of cultivated crops (and their wild relatives/landraces) with good traits are being imported. These wild relatives, landraces of germplasm could carry with them new pests/strains as their pest profiles are not adequately documented. This hampers PRA preparation and consequently poses more risk in their import. However, in 2007 the legislation was amended whereby ICAR-NBPGR now undertakes PRA for such germplasm material for pest-free import. This has become more relevant, when access to germplasm is becoming tougher under the Convention on Biological Diversity, 1992.

The technical issues affecting the quarantine processing include:

- Sample size-based non-destructive protocols are required for pest diagnostics.
- Proficiency is also required in the area of taxonomy and biosystematics for identification of unknown/new pests or strains.
- Extremely sensitive and practicable molecular methods are required for the detection of new pests/biotypes/pathotypes/races/strains etc.
- An antisera repository needs to be developed as diagnostic reagents, such as antisera for viruses are in general not easily accessible for exotic pests.
- New reports of pests need to be authenticated by another organization prior to publication.
- Taxonomic information needs to be consolidated and digitized keys should be developed for quick and reliable identification of exotic insect pests, where identification keys are not readily accessible.

- Eco-friendly non-destructive salvaging techniques need to be developed.
- PEQ testing at ICAR-NBPGR and release from indexed virus-free plants may take one crop season.
- Establishment of PEQ facilities for public and private stakeholders.

Websites like http://ppqs.gov.in/, designed by DPPQ&S, provide information on legislation, quarantine procedures, methodologies, pest alerts etc.. However, a web portal for the official exchange of information for facilitating communication among countries needs to be established. Also, easy databases availability related to quarantine and endemic pests could streamline the work of quarantine personnel. At ICAR-NBPGR, lists of potential quarantine and endemic pests of various crop groups have been compiled for various crop groups (Cereals, Grain Legumes, Edible Oilseeds, Tropical and Sub-tropical Fruits, Solanaceous and Cucurbitaceous vegetables) that act as ready reckoners.

Box 6.4 Losses Averted by Interception of Quarantine Pests in India: Few Success Stories

Quarantine pest	Losses caused elsewhere and interception at ICAR-NBPGR
 *Acanthoscelides obtectus*	*A. obtectus* is a serious pest of common beans (*Phaseolus vulgaris*) and lima beans (*P. lunatus*) in tropical South America. It has been recorded on a number of other legumes and has been intercepted in *Phaseolus* seeds imported from countries like Colombia, Italy, Mexico, Netherlands, Peru, Slovakia and United States, including cowpea in Mexico, chickpea and *Vigna subterranea*
 *Peronospora manshurica*	The disease found in 50% of 825 fields surveyed in Iowa and an average yield reduction of 11.8% for susceptible cultivars was estimated. The pest has been intercepted year after year on soybean imported from several countries including United States
 *Ditylenchus dipsaci*	*D. dipsaci* can cause complete failure of host crops and in case of heavy infestation, crop losses of 60–80% were recorded from Italy and >90% from France and Poland. It has been intercepted in *Allium* bulbs, *Humulus* rooted plants, Narcissus bulbs etc. from France, The Netherlands and United States

(continued)

Box 6.4 (continued)

Quarantine pest	Losses caused elsewhere and interception at ICAR-NBPGR
 *Cowpea mottle virus*	*Cowpea mottle virus* is estimated to cause yield losses from 10 to 100%. Yield reductions of >75% are reported in cowpea from Nigeria. The virus is transmitted through seeds of cowpea at a rate of 10% depending on genotype and the time between infection and flowering. It has been intercepted in *Vigna unguiculata* and *Vigna subterranea* from Philippines and Nigeria, and Ghana respectively
 *Polygonum lapathifolium*	*P. lapathifolium* is a damaging weed with great economic impact. In sugarbeet, it has reportedly caused 81% reduction in yield. In addition to competitive effects, *P. lapathifolium* hosts a number of damaging pests. It has been intercepted in *Linum usitatissimum* from Switzerland

6.8 The Way Forward

The WTO's free trade policies have heightened the accountability of quarantine workers to prevent the introduction of alien pests and implementing eradication measures. Pest introduction and spread remain major concerns for India, where inadequate phytosanitary protocols can allow pests to spread and negatively impact agriculture. However, the WTO also offers significant opportunities for stakeholders, including farmers exporting agricultural commodities, to improve economic growth by meeting quality and phytosanitary standards of importing nations. India's active participation in IPPC and Codex Alimentarius negotiations can boost agricultural trade by aligning standards with international norms. Coordinated efforts between the government and relevant institutions are essential to meet these standards while protecting crops and the environment.

Raising awareness about quarantine and biosecurity regulations among the scientific community and other public is vital. Despite quarantine measures for nine key pests, their spread to neighbouring regions underscores the need for comprehensive eradication and surveillance programmes. Prompt action is needed to prevent pest spread and establish domestic quarantine regulations for newly introduced pests. Investing in legal and protective measures is crucial to safeguarding national agriculture. Unlike countries such as Australia and the United States, India lacks sufficient plant quarantine checks at inter-state borders. Establishing check-posts at borders, depots and junctions would strengthen pest control efforts (Gupta and Dubey 2017).

Since the enactment of the DIP Act in 1914, India's plant quarantine system has evolved, but modernization is crucial in the globalized context to prevent foreign pest invasions and protect farm incomes. Quarantine specialists are vital in promoting trade while also protecting the environment. Under the WTO regime, quarantine's importance has grown and strategic pest diagnosis is necessary to ensure pest-free exchange of PGR and prevent harmful pest movement.

6.9 Summary

Plant biosecurity involves managing risks to plant health, addressing emerging pests, regulating GMOs and mitigating biological warfare threats. In India, quarantine laws prevent exotic pest entry during imports and exports, while biosafety for GMOs is managed by the DBT under the Ministry of Science and Technology. Biological weapon concerns are handled by the Ministry of Home Affairs. Key priorities include upgrading quarantine facilities, enhancing risk analysis, prioritizing research, and developing early warning systems for exotic diseases. A strategic approach is needed for coordinating efforts across departments to meet biosecurity goals in a global context.

Multiple-Choice Questions

1. The most widely applied insect used for biological control is
   (a) *Trichogramma chilonis*
   (b) *Metarhizium anisopliae*
   (c) *Beauveria bassiana*
   (d) *Bacillus thuringiensis*
2. The first pesticide used in India was
   (a) DDT
   (b) Endosulfan
   (c) Heptachlor
   (d) Aldrin
3. The word nematode is derived from a …….. word 'Nema'
   (a) Greek
   (b) Latin
   (c) German
   (d) French
4. Sterility mosaic is a disease of
   (a) Pigeon pea
   (b) Tomato
   (c) Chilli
   (d) Wheat

5. Who defines Agro-ecology as 'the study of the relation of agricultural crops and environment'?
   (a) OECD
   (b) FAO
   (c) World Bank
   (d) Indian Ecological Society
6. *Trichoderma viride* has proved a useful microorganism for
   (a) Biocontrol of soil-borne pathogens
   (b) Reclamation of wastelands
   (c) Bioremediation of contaminated soil
   (d) Gene transfer in higher plants
7. The Great Bengal Famine was caused due to
   (a) Helminthosporium disease of rice
   (b) *Sesame leaf spot*
   (c) a & b
   (d) None
8. Third-country quarantine for the safe transfer of groundnut germplasm is located at
   (a) Thailand
   (b) UK
   (c) France
   (d) Italy
9. Destructive Insects and Pests Act was passed in
   (a) 1914
   (b) 1966
   (c) 1912
   (d) 2003
10. How many Regional Plant Quarantine Stations are there in India?
   (a) 5
   (b) 64
   (c) 7
   (d) 46
11. Which is the most widely used technique for the detection of plant viruses in quarantine?
   (a) Infectivity test
   (b) PCR
   (c) ELISA
   (d) EM
12. 10% Trisodium orthophosphate treatment is effective against
   (a) Fungi
   (b) Bacteria
   (c) Phytoplasma
   (d) Tobamoviruses

13. In India, approval of the environmental release of GMOs is given by
    (a) RCGM
    (b) IBSC
    (c) GEAC
    (d) SBCC
14. What are the mandatory requirements for the import of transgenic germplasm into India?
    (a) Import Permit from NBPGR
    (b) Import clearance from DBT
    (c) Phytosanitary Certificate
    (d) All the above
15. Quarantine pest is
    (a) Not reported in the country
    (b) Present but restricted distribution
    (c) under official control
    (d) All

Write brief notes on the following topics:

1. Pesticide residues
2. Invasive alien species
3. SPS Agreement of WTO
4. Methyl bromide is an ideal fumigant and should not be phased out
5. Principle and concept of IPM/IDM

Short Questions

1. What is plant quarantine and how is it different from biosecurity? Write briefly about plant quarantine set-up in India.
2. Are advanced diagnostic techniques for pests more effective than conventional methods in protecting our agriculture from exotic pests?
3. Describe blotter test to determine seed health of seeds of crop germplasm.
4. Write two smut fungi causing disease in wheat and also describe their characteristic symptoms associated with seed.
5. Write 8 important diseases of crops which can be easily detected by visual examination of seed.
6. Differentiate between seed-borne and seed-transmitted diseases

Answers

1. (a)
2. (a)
3. (a)
4. (a)
5. (b)
6. (a)
7. (a)
8. (b)
9. (a)
10. (c)
11. (c)
12. (d)
13. (a)
14. (d)
15. (d)

Acknowledgements This chapter is the outcome of work done by all past and present staff of the Division of Plant Quarantine, ICAR-NBPGR and the authors duly acknowledge their contribution and dedicated service.

Suggested Reading

Bhalla S, Chalam VC, Singh B, Kavita G, Dubey SC (2018) Biosecuring plant genetic resources in India: role of plant quarantine. NBPGR, New Delhi, 216 p +vi (ISBN 978-81-937111-1-8)

Chalam VC, Parakh DB, Maurya AK (2017) Role of viral diagnostics in quarantine for plant genetic resources and preparedness. Indian J Plant Genet Resour 30(3):271–285

Chalam VC, Dubey SC, Murali Krishna C, Bhalla S, Singh K (2018) Transboundary movement of living modified organisms: strengthening capacities of enforcement agencies. ICAR-NBPGR and Ministry of Environment, Forest and Climate Change, New Delhi, vi+159 p. (ISBN 978-81-937111-2-5)

Chalam VC, Kavita G, Singh MC, Khan Z, Jameel A, Pooja K, Gawade Bharat H, Pardeep K, Raj MB, Maurya AK, Meena DS (2023) 75 quarantine pests intercepted ensuring crop biosecurity. ICAR-National Bureau of Plant Genetic Resources, New Delhi, 84 p

Dubey SC, Gupta K, Akhtar J, Chalam VC, Singh MC, Khan Z, Singh SP, Kumar P, Gawade BH, Raj K, Boopathi T, Kumari P (2021) Plant quarantine for biosecurity during transboundary movement of plant genetic resources. Indian Phytopathol 74:495–508. https://doi.org/10.1007/s42360-021-00375-7

Gupta K, Dubey SC (2017) Biosecurity policies influencing international exchange of PGR. Indian J Plant Genet Res 30(3):258–266

Gupta K, Irfan R (2020) Training Manual on Pest Risk Assessment developed for training under TCP/NEP/3702 on Building capacities to improve and sustain forest health to enhance the resilience of forests and livelihoods of forest dependent communities in Nepal, November 23–28, 2020, 100 p

Gupta K, Kapur ML, Khetarpal RK (2007) Safe transboundary movement of transgenic planting material. In: Randhawa GJ, Bhalla S, Chalam VC, Sharma SK (eds) Cartagena protocol on biosafety decisions to diagnostics. National Bureau of Plant Genetic Resources, New Delhi, pp 57–73

Khetarpal RK, Gupta K (2008) Plant biosecurity in India. In: Shetty PK, Parida A, Swaminathan MS (eds) Biosecurity. National Institute of Advanced Studies; M S Swaminathan Research Foundation, Bangalore; Chennai, pp 10–22

Mathur SB, Kongsdal O (2003) Common laboratory seed health testing methods for detecting fungi. International Seed Testing Association, Bassersdorf, p 425

Plant Quarantine (Regulation of Import into India) Order (2003) The Gazette of India Extraordinary, Part II Section 3 Sub-section (ii) No. 1037 Published by Authority, p 105

Suggested Websites

http://bch.cbd.int/protocol/cpb_publications.shtml
http://plantquarantineindia.nic.in/pqispub/docfiles/dip_act.htm
http://ppqs.gov.in/
http://ppqs.gov.in/divisions/plant-quarantine/present-setup
http://www.fao.org
https://www.cabi.org/cpc
https://www.ippc.int/en/
https://pqms.cgg.gov.in/8792aa56-5f28-400f-a202-d46697b590a6

Conservation Strategies for PGR Management

7

Anju Mahendru Singh, Padmavati G. Gore, J. Aravind, Suman Roy, and G. D. Harish

Learning objectives

- An overview of conservation
- Conservation of orthodox seeds
- Processing for long-term conservation of germplasm
- Seed genebank standards
- Related topics

Keywords

Plant genetic resources · Germplasm · PGR · Conservation · Genebank · Ex-situ collection · In-situ collection · Field Genebank · Seed Genebank · Orthodox seeds

7.1 Introduction

Conservation of plant genetic resources (PGRs) aims to maintain and enhance the taxonomic and genetic diversity of plants, the habitats or ecosystems in which they live and the interrelationships between plants, other organisms and their environment. It also aims to check habitat, species and genetic erosion by establishing and implementing conservation programmes. There are two general strategies for

A. Mahendru Singh (✉) · P. G. Gore · J. Aravind · S. Roy
Division of Germplasm Conservation, ICAR-National Bureau of Plant Genetic Resources, New Delhi, India

G. D. Harish
ICAR-NBPGR Regional Station, Shillong, India

K. Tripathi et al. (eds.), *Textbook of Plant Genetic Resources*,
https://doi.org/10.1007/978-981-96-5024-8_7

conservation, namely *in-situ* conservation and *ex-situ* conservation. The two strategies along with their specific techniques are detailed below:

7.1.1 In-situ Conservation Strategy

'Conservation of ecosystems and natural habitats and the maintenance and recovery of viable populations of species in their natural surroundings and, in the case of domesticated or cultivated species, in the surroundings where they have developed their distinctive properties' (Convention on Biological Diversity [CBD] 1992). *In-situ* conservation involves the location, designation, management and monitoring of species at the location where they grow naturally.

The primary techniques of In-situ conservation for plant genetic resources include:

(i) **Protected Area Networks**

National Parks and Nature Reserves: These are large areas where human activities are limited to maintain the natural state of ecosystems and protect biodiversity, for example, Kaziranga National Park, Assam; Gangotri National Park, Uttarakhand.

Biosphere Reserves: These areas comprise multiple ecosystems, focusing on integrating biodiversity conservation with sustainable use, for example, Nanda Devi, Uttarakhand; Sundarban, West Bengal.

Wildlife Sanctuaries: Designated regions protecting specific species or ecosystems from human interference, for example, Bornadi Wildlife Sanctuary, Assam; Chinnar Wildlife Sanctuary, Kerala.

(ii) **Home Gardens and Traditional Agroecosystems**

Home gardens often maintain diverse local varieties. Traditional agroecosystems, where indigenous and local practices are used, often have a high genetic diversity of crops and their wild relatives.

(iii) **Sacred Groves and Forests**

Sacred groves are patches of forests or natural vegetation ranging from a few trees to several acres, often dedicated to local folk deities or tree spirits. These areas are preserved by local communities due to their religious beliefs and traditional rituals passed down through generations. The sanctity of these sacred forests varies widely. In some groves, even dry foliage and fallen fruits remain untouched, as it is believed that disturbing them would offend the local deity, leading to diseases, natural calamities or crop failures, e.g. the Garo and Khasi tribes of northeastern India strictly prohibit any human interference in their sacred groves. In contrast, some communities, allow the collection of deadwood or dried leaves but strictly forbid cutting live trees or their branches, e.g. the Gonds of central India.

(iv) **Gene Management Zones (GMZ)**

Within or adjacent to protected areas, specific zones may be designated as GMZs, where human interventions like controlled burns or thinning might be applied to maintain or enhance genetic diversity.

(v) **On-farm Conservation**

The sustainable management of genetic diversity of locally developed crop varieties (landraces), with associated wild and weedy species or forms, by farmers within traditional agricultural, horticultural or agri-silvicultural systems.

7.1.2 Ex-situ (off-Site) Conservation Strategy

'Conservation of components of biological diversity outside their natural habitats' (CBD 1992). Ex-situ Conservation entails exploration, sampling, transfer and storage of the propagules of the species in an artificially created environment. In 1920, a Russian scientist, Nicholai Ivanovich Vavilov, an avid explorer realized the need to conserve genetic resources. He established the first Genetic Resource Center (GRC) of the world at Leningrad where he brought together his worldwide collections of seeds and propagules of a large number of cultivated crops and their wild relatives. Farming communities all over the world are to be credited with the conservation and utilization of biological diversity. In the last 100 years, the awareness of the need for conservation and the science of conservation have progressed considerably. There are various techniques of ex-situ conservation, such as seed banks, in-vitro genebanks and cryobanks, field genebanks etc. The choice of the particular techniques of conservation is primarily determined by the longevity of the propagule of the species during conservation. The primary techniques of *Ex-situ* conservation for plant genetic resources include:

(i) **Seed Genebanks**

Seeds of many plant species can be dried, up to 3–7% moisture content and stored at low temperatures (−18 °C) for long periods without significant loss of viability (see details in Sect. 7.2).

(ii) **Field Genebanks**

Living plants are grown and maintained in field conditions. These are useful for conserving species that do not produce seeds, produce recalcitrant seeds or represent clonally propagated plants. The field genebanks are spread across the seven regional stations of the NBPGR (see details in Sect. 7.8) as well as the National Active Germplasm Sites across the country (see details in Sect. 7.9).

(iii) **Botanic Gardens and Arboreta**

Botanic gardens and arboreta are institutions that cultivate collections of living plants primarily for scientific, educational and recreational purposes. They play a pivotal role in conservation, especially for ornamental, medicinal and rare plant species.

(iv) **In-Vitro Storage or Tissue Culture** (see details in Chap. 8).

Plant cells, tissues or organs are grown and maintained in sterile containers with a nutrient-rich medium. These techniques are useful for species that are difficult to conserve as seeds or whole plants and allow for the rapid multiplication of plants.

(v) **Cryopreservation** (see details in Chap. 8)

This technique involves the storage of plant tissues (often shoot tips or seeds) at ultra-low temperatures, typically in liquid nitrogen (−196 °C). Under these conditions, all metabolic processes halt, enabling potential indefinite storage of the species.

The difference between in-situ and ex-situ conservation in terms of plant diversity, genetic material to be conserved, strategy, method and place of conservation has been presented in Table 7.1 Relative advantages and disadvantages of in-situ and ex-situ conservation strategies have been presented in Table 7.2.

7.2 Ex-situ Conservation of Orthodox Seeds in the National Genebank of India

The term 'Genebank' is used to describe an organizational unit which has the objective of conserving and managing PGR collections and facilitating their use. The National Genebank (NGB) of India, located at the ICAR-National Bureau of Plant

Table 7.1 In-situ and ex-situ conservation

	Ex-situ conservation		In-situ conservation	
Plant diversity	Orthodox seeded Crops	Non-orthodox seeded crops *viz.*, Root and Tuber, Horticulture, Plantation, Forage, Medicinal and Aromatic crop species	Wild species Keystone species Umbrella species	Land races Wild and weedy relatives of crop species.
Genetic material	Orthodox seed	Recalcitrant seeds, Embryonic axis, Dormant buds, Meristems, shoot tips, Leaf explants, Rhizome buds, *In-vitro* explants, Cell, suspension, Pollen, DNA, Whole Plant	Habitat Ecosystems	Populations
Conservation strategy Conservation method	Low temperature storage	*In-vitro* conservation, Cryopreservation, DNA Libraries, Botanical garden, Arboreta, Field genebank	Biosphere reserves Sacred grooves Heritage sites	On farm Under community agro-ecosystems
Place of Conservation	Seed genebank	In vitro genebank, Cryo genebank, Field genebank, Green house	Conservation of natural habitat or Ecosystem	On-farm conservation
	National Genebank, National Active Germplasm Sites, Regional Genebank		Natural habitat	

Table 7.2 Relative advantages and disadvantages of in-situ and ex-situ conservation strategies (Source: Hawkes et al. 2000)

Conservation strategy	Advantages	Disadvantages
In-situ conservation strategy	Dynamic conservation Permits interactions and co-evolution of species/pathogen Provides easy access to evolutionary and genetic studies Appropriate for recalcitrant or intermediate-seeded species Target taxa conserved alongside associated taxa within a single reserve.	Materials less easily available for utilization Vulnerable to natural and human-mediated catastrophes, for example, fire, vandalism, urban and industrial development, climate change Appropriate management regimes are often poorly understood Require routine active management and monitoring Limited genetic diversity can be conserved in any single location
Ex-situ conservation strategy	Medium and long-term storage is feasible Limitless genetic diversity can be conserved in any single location Easy access for evaluation for resistance to pests and diseases Easy access to plant breeding, farmers and other forms of utilization Little routine maintenance is required once material is conserved (except for field genebanks, in-vitro genebank and botanic garden accessions)	Impossible to store seeds of recalcitrant or intermediate species, unless via in-vitro or field genebanks Practical limitation on storage space for diversity of recalcitrant or intermediate species in field genebanks Freezes evolutionary development in relation to environmental changes Genetic diversity is potentially lost with each regeneration cycle In-vitro storage may result in loss of diversity and mutation

Genetic Resources, is the cornerstone of the Indian National Plant Genetic Resources System (INPGRS). Seeds are the easiest and the most convenient form for long-term conservation and hence the seed genebank forms the major component of a genebank. Plant species possessing orthodox seeds (Box 7.1) are primarily conserved in the seed banks. Conservation of seeds is carried out in controlled environmental conditions that maintain seed viability for longer periods. The duration of viability (longevity) of seeds depends on the initial seed quality, moisture content and temperature during storage. In general, low moisture content and low temperature reduce the loss of seed viability. Different combinations of moisture content and temperature can, therefore, be used to prolong seed viability during storage. The NGB was refurbished and became functional in 2021. The NGB has 12 modules in which temperature is maintained at −18 °C. The NGB is the second largest genebank in the world. The germplasm holding of the National Genebank as of 31st March 2024 is depicted in Table 7.3.

Table 7.3 Status of base collection in National Genebank (−18 °C) (as on 31st March, 2024)

Crop/Crop group	Status of total accessions conserved
Cereals	**177,190**
Paddy	118,393
Wheat	36,253
Maize	11,911
Others	10,633
Millets	**60,435**
Sorghum	26,370
Pearl millet	8398
Minor millet	25,667
Forages	**7522**
Oat	1414
Clover	620
Teff	300
Marvel grass	341
Others	4847
Pseudocereals	**8221**
Amaranth	6494
Buckwheat	1141
Others	586
Legumes	**68,881**
Chickpea	14,876
Pigeon pea	11,964
Mung bean	4438
Pea	4705
Cowpea	4045
French bean	4206
Cluster bean	4334
Horsegram	3174
Ricebean	2239
Others	14,900
Oilseeds	**63,490**
Groundnut	13,890
Oilseed brassica	12,469
Safflower	7500
Sesame	10,518
Soybean	5527
Sunflower	2012
Others	11,579
Fibre	**17,056**
Cotton	11,086
Jute	3526
Mesta	2076
Others	368

(continued)

Table 7.3 (continued)

Crop/Crop group	Status of total accessions conserved
Vegetables	**29,468**
Tomato	2886
Brinjal	4816
Chilli	5343
Okra	4205
Onion	1149
Others	11,069
Fruits and nuts	**302**
Buchanania	97
Others	205
Medicinal and aromatic plants	**9529**
Opium poppy	534
Ocimum	841
Tobacco	2289
Others	5868
Ornamental	**743**
Marigold	375
Others	368
Spices, condiments and flavour	**3697**
Coriander	1206
Fenugreek	1475
Others	1016
Agroforestry	**1710**
Sesbania	669
Others	1041
Duplicate safety samples	**10,235**
Trial material (wheat, barley)	**10,771**
Total	**469,250**

Advantages and Disadvantages of Seed Genebanks:
Advantages

- Handling of germplasm is easy
- A large number of germplasm samples or entire variability can be conserved in a very small space
- Germplasm is conserved under pathogen and insect free environment

Disadvantages

- Seeds of recalcitrant species cannot be stored in seed banks
- Failure of the power supply may lead to loss of viability and thereby loss of germplasm
- It requires a periodical evaluation of seed viability. After some time, regeneration/multiplication is essential to get new or fresh seeds for storage

7.2.1 Seed Longevity

Seeds are subject to losing their viability upon storage. Temperature and moisture content are the two major determinants of seed ageing. Therefore, the effect of these environmental factors on seed viability must be predicted for efficient and proper management during the storage of seeds. The seed longevity equations help us to predict the seed viability after storage for any given time with any combination of temperature and moisture content. Different groups of researchers have given different seed longevity equations in different crops. Among them, the Ellis and Roberts equation is very well accepted as it allows accurate predictions.

Ellis and Roberts Eq. (1980):

$$\log\sigma = K_{E-}C_W \log m - C_H t - C_Q t^2$$

where K_E, C_W, C_H and C_Q are viability constants; m = moisture content; t = temperature in °C.

Seed longevity is an important trait for management aspects of orthodox seed genebanks as different accessions vary among themselves in the context of seed health status. The physiological status of the seed during harvest, processing and handling environments before storage leads to variation in the viability status among the same seed lots. Therefore, testing the seed lots for viability among large accessions can only be done with efficient seed longevity equations. Major factors affecting seed longevity are: (1) Species; (2) Accession heterogeneity; (3) Seed quality and viability; (4) Seed composition (lipid, protein, carbohydrate etc.); (5) Storage temperature; (6) Seed moisture content.

Box 7.1 Classification of Seeds on the Basis of Storage Behaviour

Classification of the seeds on the basis of storage behaviour is vital in terms of selecting the appropriate method of storage. Seeds are being classified in various ways like Ewart's (1908) classification is based on the lifespan of seeds upon normal storage condition. Accordingly, Ewart classified all seed into three categories:

- **Microbiotic:** Lifespan of 3 years or less under ambient condition.
- **Mesobiotic:** Lifespan of 3–15 years under ambient condition.
- **Macrobiotic:** Lifespan of 15–100 years or more under ambient condition.

Elliot (1912) first classified seeds on the basis of their response to desiccation. This classification was done for the forest species only. Accordingly, forest seeds are of three types:

- Species which can withstand drying.
- Species which can withstand partial drying.
- Species which can rarely be dried.

(continued)

Box 7.1 (continued)

Roberts (1973) classified seeds upon their survival response during storage to moisture content and temperature. He introduced the term 'orthodox' and 'recalcitrant'.

- **Orthodox Seeds:** The seeds which can be dried to 2–5% moisture content without any damage are called as orthodox seeds, e.g. Rice, Wheat, Maize, etc. The oldest orthodox seed is Judean date palm seed (about 2000 years old). This seed was discovered during excavations at Herod the Great's palace on Masada in Israel. Preserved in a cool, dry environment, it was successfully germinated in 2005.
- **Recalcitrant Seeds:** The seeds which cannot survive desiccation below a comparatively high moisture content (12–31%) are called as recalcitrant seeds, e.g. Mango, Rubber, Bulbaceous seeds, etc.
- **Intermediate Seeds:** Ellis (1990) added another group named as intermediate which shows desiccation tolerance of 7–10% to 20%. Intermediate seeds, first characterized in crop seeds such as papaya (Ellis et al. 1991) and coffee (Eira et al. 2006), are also found in other species like *Cuphea* and *Salix*.

Seed Viability: It is the ability of seed to germinate under favourable condition. This includes the dormancy breaking before germination.

Seed Dormancy: It is the inability of a viable seed to germinate under favourable environmental conditions.

Seed Longevity: It is the ability of seed to maintain vigour and viability during storage.

7.2.2 Determination of Seed Storage Behaviour

Information on the storage behaviour of a wide range of species is available. Seeds that succumb to desiccation, reaching a moisture content of 15–20% (values in equilibrium with >70% relative humidity at 20 °C), are indicative of recalcitrant characteristics. Seeds capable of enduring desiccation to approximately 10–12% moisture content (values in equilibrium with 40–50% relative humidity at 20 °C) but experience reduced viability with further drying to lower moisture levels are likely to exhibit intermediate seed storage behaviour. Seeds that withstand desiccation, maintaining viability even at 5% moisture content or below (values in equilibrium with 10–15% relative humidity at 20 °C), are likely to demonstrate orthodox seed-storage behaviour. When new species with unexplored storage behaviour are encountered, the protocol outlined in Flowchart 7.1 should be followed.

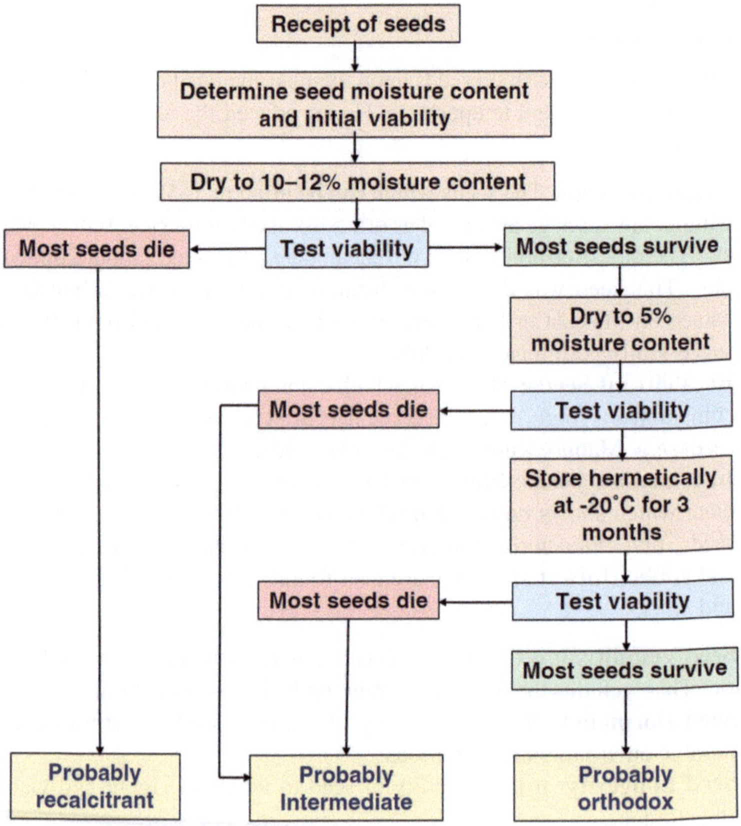

Flowchart 7.1 Determination of seed storage behaviour

7.3 Types of Seed Storage

Two types of seed storage facilities are used for the conservation of genetic resources: those holding seed samples for long-term storage-referred to as *base collections*, and those holding seed samples for immediate use—referred to as *active collections*. The temperature, relative humidity (RH), seed moisture content, containers and distribution arrangements of these stores vary.

(a) **Base collections:** A base collection comprises a selection of accessions, each distinct and conserving the genetic integrity of the original sample as closely as possible. Typically, seeds from base collections are not directly distributed to users. Engels and Visser (2002) introduced the concept of the 'most original sample' (MOS) to describe samples within base collections. The MOS refers to seeds that have undergone the fewest regenerations since acquisition by the genebank; it may constitute a subset of the original seed lot or a sample from the initial regeneration cycle if the original seeds required regeneration before storage. Base collections are stored for extended periods (usually 50–100 years) at temperatures below 0 °C, typically around −18 to −20 °C, with a moisture content ranging from 3% to 7%, to ensure seed viability is maintained.

(b) **Active collections:** Active collections, according to FAO, also referred to as working collections, comprise accessions readily available for distribution. These accessions can be employed for various purposes such as research and breeding. The active collection undergoes regular characterization, evaluation and multiplication to sustain their viability, ensuring a consistent supply of plant genetic resources. Accessions within the active collection are frequently accessed and are maintained to retain at least 65% viability for 10–20 years. For seeds with inherent poor storability, the moisture content should fall between 3% and 8%, while for seeds with good storability, it should be between 7% and 11%, depending on the storage temperature. Opting for a lower moisture content and higher storage temperature can be a practical approach to reduce refrigeration costs. In cases where achieving low moisture contents is impractical, storage with higher moisture at a lower temperature may be considered.

7.3.1 Types of Seed Storage Facilities Depending on the Duration of Storage (Box 7.2)

(a) **Short-term storage:** The period for short-term storage of seeds is from 1 to 2 years. The conditions should be such that the seeds do not lose their viability in this duration. A cool and dry atmosphere at 20–22 °C and 20–30% RH is considered ideal for short-term storage.

(b) **Medium-term storage:** The active collection, typically larger than those designated for the base collection, is preserved in medium-term storage facilities. These accessions are intended for regular distribution, necessitating storage for a maximum period of 5 years. Active collections are maintained at temperatures ranging from 0 to 10 °C with a relative humidity of 15 ± 3%.

(c) **Long-term storage:** The period of storage here is very large spanning 50–100 years. This is the storage facility for base collection, where the seeds that meet all essential criteria for seed bank conservation are maintained at −18 to −20 °C with a relative humidity of 15 ± 3%.

Box 7.2 Types of Facilities for Different Durations of Seed Storage

Condition	Short term	Medium term	Long term
Storage period	12–24 months	0–5 years	50–100 years
Temperature	+18 to +20 °C	0 to +10 °C	−18 to −20 °C
RH	20–30%	15 ± 3%	15 ± 3%
Storage container type	Cloth bags, paper bags, glass bottles	Cloth bags, metal cans, glass bottle and plastic jars	Specially designed tri-layered aluminium foil packets
Uses	Distribution	Evaluation and breeding	Future crop improvement programme

7.4 Processing for Long-Term Storage in Seed Genebank of India's National Genebank

The main objective of long-term storage is to ensure seed viability despite long years of conservation in NGB. Several factors can influence the longevity of seeds in storage. Some of these are mentioned below:

- Environmental Conditions Before Seed Maturation: Suboptimal environmental factors leading up to seed maturation can have detrimental effects on seed quality. For instance, hot and dry conditions may impede proper seed development.
- Seed Maturity at Collection: Seeds collected in an immature state tend to experience a more rapid loss of viability compared to mature seeds.
- Handling Between Collection and Processing: Adverse conditions during handling, such as high temperature, humidity and fungal development, can harm the seeds.
- Processing-Induced Injury: Seed processing, including actions that lead to a cracked seed coat, may diminish the seed's storage life.
- Seed Coat Structure: Seeds with hard seed coats generally exhibit greater resilience compared to those with thin seed coats.
- Seed Chemistry: Oily seeds may pose greater challenges for storage compared to starchy seeds.
- Insects and Fungi: Uncontrolled presence of insects and fungi can lead to the destruction of seeds.
- Storage Conditions: Crucial factors include controlling seed moisture content and storage temperature. Additionally, gaseous environments may also influence the longevity of seeds in storage.

The National Plant Genetic Resources programme in India has developed guidelines for the long-term conservation of germplasm in the National Genebank (NGB), in line with international standards and ensures the long-term viability of the germplasm conserved after due processing. Each of the steps in seed conservation in NGB is depicted in Flowchart 7.2 and explained in detail below:

7.4.1 Germplasm Acquisition

The first step of PGR conservation is germplasm acquisition. The long-term conservation of orthodox seeds in the genebank begins with the receipt of freshly harvested, physically pure seed at the Germplasm Handling Unit (GHU). Upon the arrival of germplasm for genebank conservation, the basic parameters that are checked up for conservation are:

(i) **Uniqueness of the accession:** Maintaining a sample in the genebank is expensive; therefore, the samples are carefully checked for uniqueness to minimize duplication of accessions in the genebanks. Given that each genebank adopts its numbering systems, the same accession may be available under different

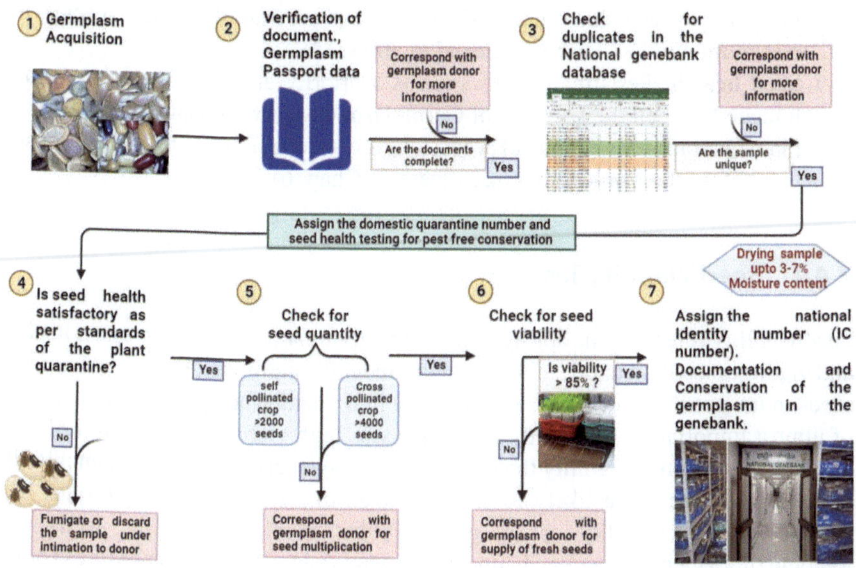

Flowchart 7.2 Steps of processing germplasm for long-term storage in the National Genebank of India

identities. Duplication in the collections is best identified by comparing relevant fields, namely collector number, Indigenous collection (IC), exotic collection (EC) number, passport information on location of germplasm collected in databases using computer programmes.

(ii) **Availability of passport information:** The utilization of the conserved germplasm can be facilitated only if all relevant passport information is available in the database. Hence, only those accessions having passport information are processed further. In the genebank, different types of germplasm are acquired and they require specific passport information as follows: (i) Explored and collected germplasm: basic information on parameters like biological status and collection details (details of passport information provided in Chap. 5); (ii) Released varieties and hybrids: variety identification proposal that contains the details of the variety including pedigree details and the proceedings of the high-level committee that recommends the identification of these varieties; (iii) Registered germplasm: information on the unique trait of the germplasm and pedigree details, if it is a breeding line/genetic stock, are accepted for long-term conservation. If published, the germplasm is reviewed by a high-level Plant Germplasm Registration Committee and recommended for registration and provided a unique INGR number.

(iii) **Seed quantity:** The global genebank standards recommend a minimum of 2000 seeds in self-pollinated crops and 4000 seeds in cross-pollinated crops. In wild germplasm accessions, where the getting required number of seeds is difficult due to seed shattering, and the low seed set trait of the wild species, the minimum number has been relaxed to 500 seeds.

(iv) **Seed health:** Pest-free conservation is a priority at NGB. There is a strong linkage between the Division of Germplasm Conservation and the Division of Plant Quarantine, wherein all accessions are tested for any form of pest infestation before their processing (see details in Chap. 6).

7.4.2 Seed Viability Tests:

Seed viability refers to 'the ability of a seed to germinate and develop into a healthy and vigorous seedling under suitable environmental conditions'. Ensuring the stored seeds in the genebank can successfully grow into plants when planted in the field is of utmost importance. The seeds should possess high viability at the onset of storage and maintain this viability throughout the storage period. Seeds with an initially high viability are more likely to endure for an extended period in storage. The decline in seed viability occurs gradually but accelerates as the seeds age. Recognizing when this decline happens is crucial to implementing measures for the regeneration of the accession (see details in Sect. 7.7). Excessive deterioration could result in the irreversible loss of genetic material. As per the international genebank standards, a minimum seed viability of 85% is essential for conservation. It is essential to assess the viability of accessions: (i) before seeds are packaged and deposited in the genebank; (ii) at regular intervals throughout the storage period, termed as monitoring (see details in Sect. 7.6). Various methods exist for assessing seed viability, but the germination test stands out as the most precise and reliable approach.

The purpose of a germination test is to determine the percentage of seeds within an accession that will germinate under favourable conditions, giving rise to healthy seedlings with essential structures, namely roots, shoots and ample food reserves-capable of progressing into fully developed, reproductive mature plants. Germination test is carried out for each species under specific controlled conditions by following standards given by the International Seed Testing Association (ISTA 2019). The fundamental prerequisites for seed germination are water, oxygen, light, and specific temperature. According to the International Standards for Genebanks (FAO/IPGRI 1994), it is recommended to conduct a minimum of two replicates, each consisting of 100 seeds. If the initial results reveal a germination rate below 90%, an additional 200 seeds should be subjected to the same testing method. For species with large seeds and low seed-multiplication rates or those with challenges in seed regeneration, such as certain wild species, the allocation of 200 seeds for germination testing may pose difficulties. In such instances, two replicates of 50 or 25 seeds each can be employed based on the available quantity. The overall seed viability is determined as the average of these two tests. Germination test, in the genebank, is generally carried out using paper as growing media using either of the two methods: (i) Top of paper (TP): the small-seeded species are germinated on top of one or more layers of paper which are placed into petri dishes, (ii) Between papers (BP): the large-seeded are germinated between two layers of paper. This may be achieved by loosely covering the seeds with an additional layer of filter paper. The method prescribed by the International Seed Testing Association (ISTA 2005) used for germination testing in genebank is depicted in Flowchart 7.3 and the germination testing standardized by ISTA, 2019 is described in Table 7.4.

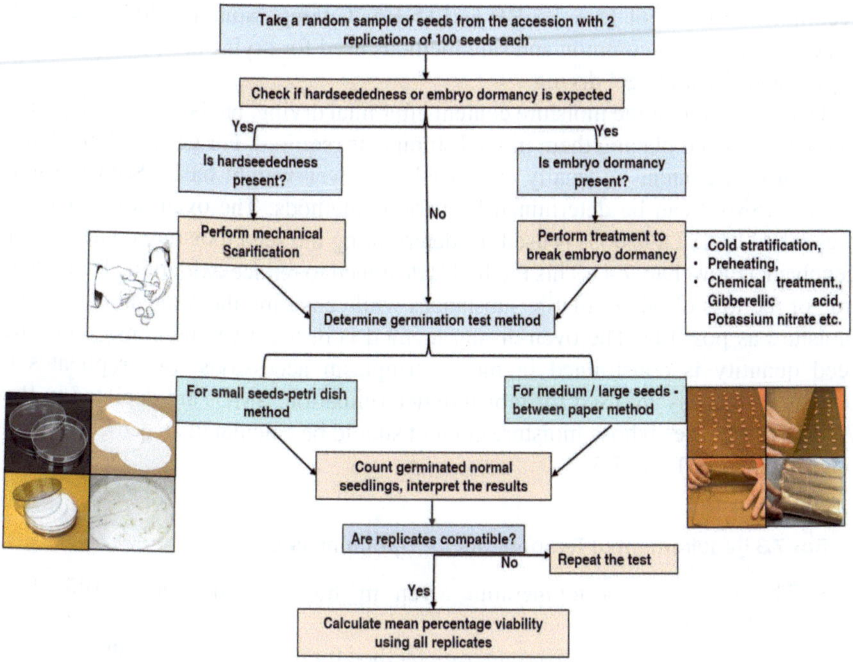

Flowchart 7.3 Protocol for performing germination test

Table 7.4 Germination test standardized by the International Seed Testing Association (some examples)

Crop	Substrate[a]	Temperature (°C) 16 h/8 h	First count	Final count	Pre-conditioning
Rice	TP, BP, S	20/30	5	14	Preheat at 50 ± 2 °C, HNO_3 for 24 h
Maize	TP, BP, S	20/30	4	7	–
Wheat	TP, BP, S	20	4	8	Preheat at 30–35 °C, GA_3
Pulses	BP, S	20/30	5	7	–
Brinjal	TP, BP, S	20/30	7	14	–
Sorghum	TP, BP,	20/30	5	14	Prechilling

[a] TP = Top of paper, BP = Between paper, S = Sand

7.4.3 Seed Moisture Content

'The moisture content is the amount of free water in the seed' and is usually expressed as a percentage. Even small changes in the moisture content have a large effect on the storage life of seeds. The moisture content should be brought down to the range of 3–7% for most orthodox seeds. The FAO/IPGRI genebank standards recommend the use of 15 ± 5% RH and 15 ± 5 °C temperature for drying seeds. In genebank, the most common and safe methods used for drying are (i) dehumidified drying and (ii) silica gel drying.

Determination of the moisture content after final drying, but before packing them in containers and placing them in seed storage, is required. For genebank purposes, the moisture content is usually expressed on a wet–weight basis. Seed moisture content (SMC) can be determined by various methods. The oven drying method prescribed by ISTA (2005) is used for determining the seed moisture content in the genebank (Flowchart 7.4). This method is designed to reduce oxidation, decomposition or the loss of other volatile substances while ensuring the removal of as much moisture as possible. The oven-drying method is destructive and considering that seed quantity is constrained in most germplasm accessions, two replicates of 0.5–1.0 g of seeds are used for moisture determination. SMCs are depicted in Box 7.3 and the species whose moisture content should be calculated as per ISTA methods are listed in Table 7.5.

Box 7.3 Requirement of Temperature for Optimum SMC

- The low constant temperature oven method for oily seeds: **103 °C for 17 ±1 h**
- The high constant temperature oven method for non-oily seeds: **130–133 °C for 1 h ±3 min, 2 h ±6 min or 4 h ±12 min (depending on species)**

Calculation of SMC:

The moisture content as a percentage by weight must be calculated to three decimal places for each replicate using the following formula (on a fresh weight basis):

$$\textbf{Moisture\%} = M2 - M3 \,/\, M2 - M1^{*}\ 100$$

Where,

- M1 is the weight in grams of the container and its cover
- M2 is the weight in grams, of the container, its cover and its contents before drying
- M3 is the weight in grams of the container, its cover and its contents after drying

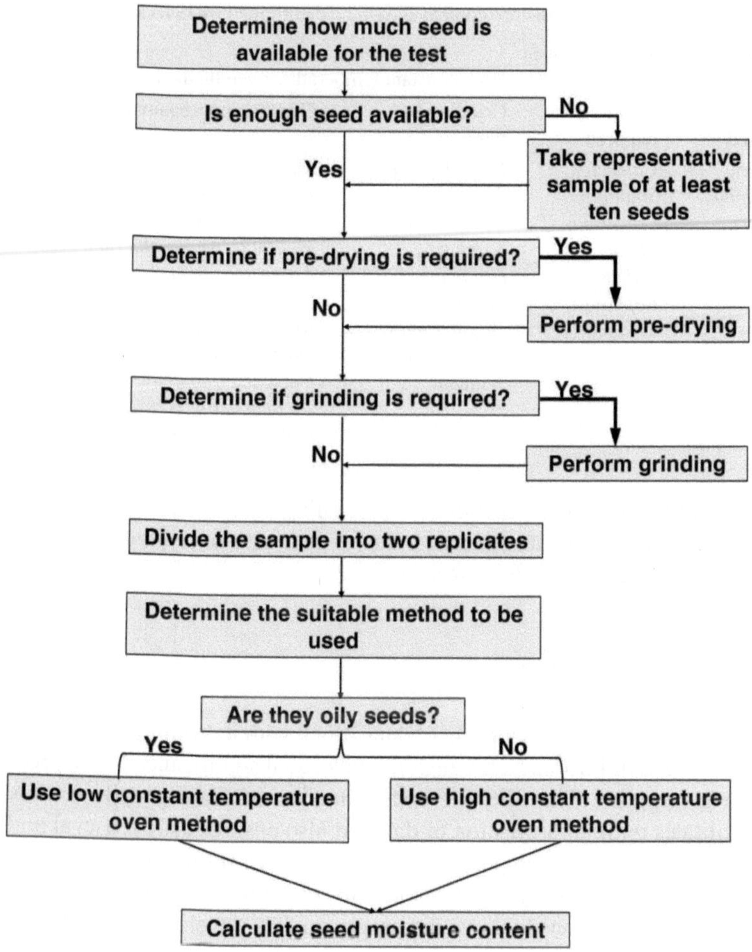

Flowchart 7.4 Protocol for determining moisture content

7.4.4 Packaging

Once the desirable moisture is achieved depending on the species, the seeds are packed immediately in the container for storage. This is done to prevent moisture absorption from the air, averting any mix-up of individual accessions, and safeguard against contamination by insects and diseases. Various types of containers are available for packaging, and the selection depends on both storage conditions and the specific species involved. The packaging material must be impermeable to water and suitable for prolonged use. When the relative humidity (RH) of the storage room is uncontrolled, it becomes crucial to opt for moisture-proof containers. Commonly utilized containers in genebanks include glass bottles, aluminium cans, aluminium foil packets and plastic bottles. Glass bottles are effective but fragile,

Table 7.5 Species whose moisture content should be determined by ISTA methods

Low constant temperature oven method		High constant temperature oven method	
Grinding is required	Grinding not required	Grinding is required	Grinding not required
Ricinus, Gossypium, Arachis, Glycine	*Capsicum, Solanum, Camelina, Linum, Allium, Raphanus, Sesamum*	*Lupinus, Zea, Avena, Pisum, Oryza, Secale, Sorghum, Hordeum, Phaseolus, Fagopyrum, Cicer, Vicia, Vigna, Triticum*	*Alopecurus, Lactuca, Panicum, Petroselinum, Chloris, Lolium, Onobrychis, Ornithopus, Cucurbita, Melilotus, Arrhenatherum, Medicago, Asparagus, Beta, Agrostis, Cynodon, Scorzonera, Phalaris, Carum, Daucus, Anthriscus, Cichorium, Trifolium, Dactylis, Lepidium, Cynosurus, Cucumis, Cuminum, Paspalum, Festuca, Lycopersicon, Lotus, Deschampsia, Holcus, Citrullus, Phleum*

while aluminium cans pose challenges in resealing once opened. Among these, aluminium foil packets hold an advantage due to their ease of sealing and resealing, as well as their space-saving characteristics.

7.4.5 Documentation

The accurate and reliable recording of data, along with its comprehensive documentation and efficient information transfer, holds equal significance to the proper handling of germplasm. The effective documentation of plant genetic resources not only facilitates optimal utilization of data but also ensures easy retrieval and applicability. This documentation plays a crucial role in disseminating information about germplasm holdings, enhancing the ability to search for specific traits and enabling the selection of accessions for breeding programmes to develop new traits. Moreover, it serves as a valuable source of information for planning and executing genebank activities. In the National Genebank (NGB), the documentation of genebank holdings involves two types of information files: passport data descriptors and genebank management descriptors. Passport data descriptors encompass details like the Name of the Crop, Taxonomic Code, Cultivar Name, National Number, Collector No, Other ID and Location in the genebank. On the other hand, genebank management descriptors include information such as seed quantity, seed germination percentage, seed moisture percentage, source of material and date of storage. As the central component of the National Genebank, this documentation process is efficiently managed through a well-organized computerized network.

Box 7.4 Harrington Thumb Rule for Seed Storage

Seed ageing and loss of germination cannot be stopped, but can be minimized by proper seed storage conditions. Harrington thumb rule on seed storage is a useful measure for assessing the effect of moisture and temperature on seed storage.

According to the *Harrington's rules:*

- For every **decrease of 1% seed moisture content, the life of the seed doubles**. This rule is applicable when moisture content ranges between 5% and 14%.
- For every **decrease of 5.6 °C in storage temperature, the life of the seed doubles.** This rule applies between 0 and 50 °C.
- Good seed storage is achieved when the % of relative humidity in the storage environment and the storage temperature in degrees/Fahrenheit adds up to hundred but the contribution from temperature should not exceed 10 °C.

7.5 International Genebank Standards

Ex-situ conservation in genebanks is recognized as the safest and most reliable method for long-term conservation of plant genetic resources for food and agriculture. The most up-to-date scientific and technical practices in conservation need to be followed for safe, efficient as well as cost-effective management of seed genebanks. Recognizing this need, based on the recommendation of the Commission on Plant Genetic Resources of the United Nations Food and Agriculture Organization (later expanded as Commission on Genetic Resources for Food and Agriculture), 'Genebank Standards' were developed and published in 1994 in collaboration with the erstwhile International Plant Genetic Resources Institute (reorganized as International Board of Plant Genetic Resources and subsequently as Bioversity International). This iteration dealt only with the storage of seeds with orthodox seed storage behaviour. Subsequently, this was revised and expanded to include non-orthodox seeds and vegetatively propagated crops and published as 'Genebank Standards for Plant Genetic Resources for Food and Agriculture' in 2014. This has proven to be a valuable reference document for key genebank activities such as acquisition, storage, documentation, characterization, evaluation, distribution, exchange, safety duplication, security and personnel. To facilitate the implementation of these standards more effectively, a series of 'Practical Guide for Application of Genebank Standards for Plant Genetic Resources for Food and Agriculture' for seed, in-vitro and field genebanks were published in 2022 with technical and operational guidelines. Genebank standards for the conservation of orthodox seeds in seed genebanks (FAO 2014) have been presented in Table 7.6.

Table 7.6 Genebank standards for conservation of orthodox seeds in seed genebanks

Activity	Genebank standards
Acquisition of germplasm	• All seed samples added to the genebank collection have been acquired legally with relevant technical documentation. • Seed collecting should be made as close as possible to the time of maturation and before natural seed dispersal, avoiding potential genetic contamination, to ensure maximum seed quality. • To maximize seed quality, the period between seed collecting and transfer to a controlled drying environment should be within 3–5 days or as short as possible, bearing in mind that seeds should not be exposed to high temperatures and intense light and that some species may have immature seeds that require time after harvest to achieve embryo maturation. • All seed samples should be accompanied by at least a minimum of associated data as detailed in the FAO/Bioversity multi-crop passport descriptors. • The minimum number of plants from which seeds should be collected is between 30–60 plants, depending on the breeding system of the target species.
Drying and storage	• All seed samples should be dried to equilibrium in a controlled environment of 5–20 °C and 10–25% of relative humidity, depending upon species. • After drying, all seed samples need to be sealed in a suitable airtight container for long-term storage; in some instances where collections that need frequent access to seeds or are likely to be depleted well before the predicted time for loss in viability, it is then possible to store seeds in non-airtight containers. • Most-original-samples and safety duplicate samples should be stored under long-term conditions (base collections) at a temperature of −18 ± 3 °C and relative humidity of 15 ± 3%. • For medium-term conditions (active collection), samples should be stored under refrigeration at 5–10 °C and relative humidity of 15 ± 3%.
Seed viability monitoring	• The initial seed viability test should be conducted after cleaning and drying the accession or maximum within 12 months after receipt of the sample at the genebank. • The initial germination value should exceed 85% for most seeds of cultivated crop species. For some specific accessions such as wild and forest species that do not normally reach high levels of germination, a lower percentage could be accepted. • Viability monitoring test intervals should be set at one-third of the time predicted for viability to fall to 85% of initial viability or lower depending on the species or specific accessions, but no longer than 40 years. If this deterioration period cannot be estimated and accessions are being held in long-term storage at −18 °C in hermetically closed containers, the interval should be 10 years for species expected to be long-lived and 5 years or less for species expected to be short-lived. • The viability threshold for regeneration or other management decisions such as recollection should be 85% or lower depending on the species or specific accessions of initial viability.

(continued)

Table 7.6 (continued)

Activity	Genebank standards
Regeneration	• Regeneration should be carried out when the viability drops below 85% of the initial viability or when the remaining seed quantity is less than what is required for three sowings of a representative population of the accession. The most original sample should be used to regenerate those accessions. • The regeneration should be carried out in such a manner that the genetic integrity of a given accession is maintained. Species-specific regeneration measures should be taken to prevent admixtures or genetic contamination arising from pollen gene flow that originated from other accessions of the same species or other species around the regeneration fields. • If possible, at least 50 seeds of the original and the subsequent most-original-samples should be archived in long-term storage for reference purposes.
Characterization	• Around 60% of accessions should be characterized within 5–7 years of acquisition or during the first regeneration cycle. • Characterization should be based on standardized and calibrated measuring formats and characterization data follow internationally agreed descriptor lists and are made publicly available.
Evaluation	• Evaluation data on genebank accessions should be obtained for traits that are included in internationally agreed crop descriptor lists. They should confirm to standardized and calibrated measuring formats. • Evaluation data should be obtained for as many accessions as practically possible, through laboratory, greenhouse and/or field analysis as may be applicable. • Evaluation trials should be carried out in at least three environmentally diverse locations and data collected over at least 3 years.
Documentation	• Passport data of 100% of the accessions should be documented using FAO/Bioversity multi-crop passport descriptors. • All data and information generated in the genebank relating to all aspects of conservation and use of the material should be recorded in a suitably designed database.
Distribution and exchange	• Seeds should be distributed in compliance with national laws and relevant international treaties and conventions. • Seed samples should be provided with all relevant documents required by the recipient country. • The time span between the receipt of a request for seeds and the dispatch of the seeds should be kept to a minimum. • For most species, a sample of a minimum of 30–50 viable seeds should be supplied for accessions with sufficient seeds in stock. For accessions with too little seed at the time of request and in the absence of a suitable alternative accession, samples should be supplied after regeneration/multiplication, based on a renewed request. For some species and some research uses, smaller numbers of seeds should be an acceptable distribution sample size.
Safety duplication	• A safety duplicate sample for every original accession should be stored in a geographically distant area, under the same or better conditions than those in the original genebank. • Each safety duplicate sample should be accompanied by relevant associated information.

(continued)

Table 7.6 (continued)

Activity	Genebank standards
Security and personnel	• A genebank should have a risk management strategy in place that includes inter alia measures against power cuts, fire, flooding and earthquakes. • A genebank should follow the local occupational safety and health requirements and protocols where applicable. • A genebank should employ the requisite staff to fulfil all the routine responsibilities to ensure that the genebank can acquire, conserve and distribute germplasm according to the standards.

Source: FAO (2014)
Source: Genebank Standards for Plant Genetic Resources for Food and Agriculture, FAO (2014)

7.6 Monitoring of Conserved Germplasm

Ex-situ conservation of 'orthodox seed' germplasm is not limited to the acquisition and storage of the material. It also includes guaranteeing the continued survival of the stored material with the original genetic makeup of the population and ensuring that enough seeds are available for distribution to end users for utilization. To ensure this, periodic monitoring of both seed viability and seed quantity is a key activity in seed genebanks to identify those accessions that require regeneration.

Despite orthodox seed storage behaviour ensuring long-term survival under low moisture and temperature conditions, several other factors such as the genotype, pre-harvest conditions, post-harvest handling etc. also affect the longevity of accessions. Moreover, genebanks depend on germination test data and not vigour data which is proven to be an accurate early indicator of decrease in viability for management decisions. There is also a lack of information on seed viability loss described by a survival curve in several species. Moreover, even if available, such data are based on only a few accessions which may not capture the between-accession diversity in longevity. Therefore, the seed viability of the conserved accessions needs to be tested at periodic intervals. As monitoring is costly both in terms of resources as well as using the valuable conserved seeds, monitoring should be planned in such a way as to ensure seeds do not deteriorate more than the threshold level and that it is not unnecessarily wasting precious seeds.

For long-term conservation of most of the crop species, the initial viability should exceed 85%. This threshold is relaxed for specific accession and species such as crop wild relatives which usually do not achieve high levels of germination. It is recommended that viability monitoring test intervals are set at one-third of the time predicted for viability to fall to 85% (or lower for specific accessions or species) of the initial viability, but no longer than 40 years. But in practice, due to a lack of accurate longevity information, this is set at 10 years for species with good storability and 5 years for those with poor storability. Germination tests can be performed with either a fixed sample size (200 seeds as 2 replicates of 100 seeds each or 4 replicates of 50 seeds each) if sufficient seeds are available or using sequential germination tests if the seed quantity is limited.

Sequential germination tests: The sequential germination test employs a reduced seed count of 40 seeds per replicate. It involves testing replicates sequentially until conclusive results are obtained, determining whether regeneration is required. However, all other procedures and conditions for sequential germination adhere to those for normal germination tests (see details in Sect. 7.4.2). The procedure of sequential germination test has been given in Flowchart 7.5. Table 7.7 is given for referring the result and making the informed decision.

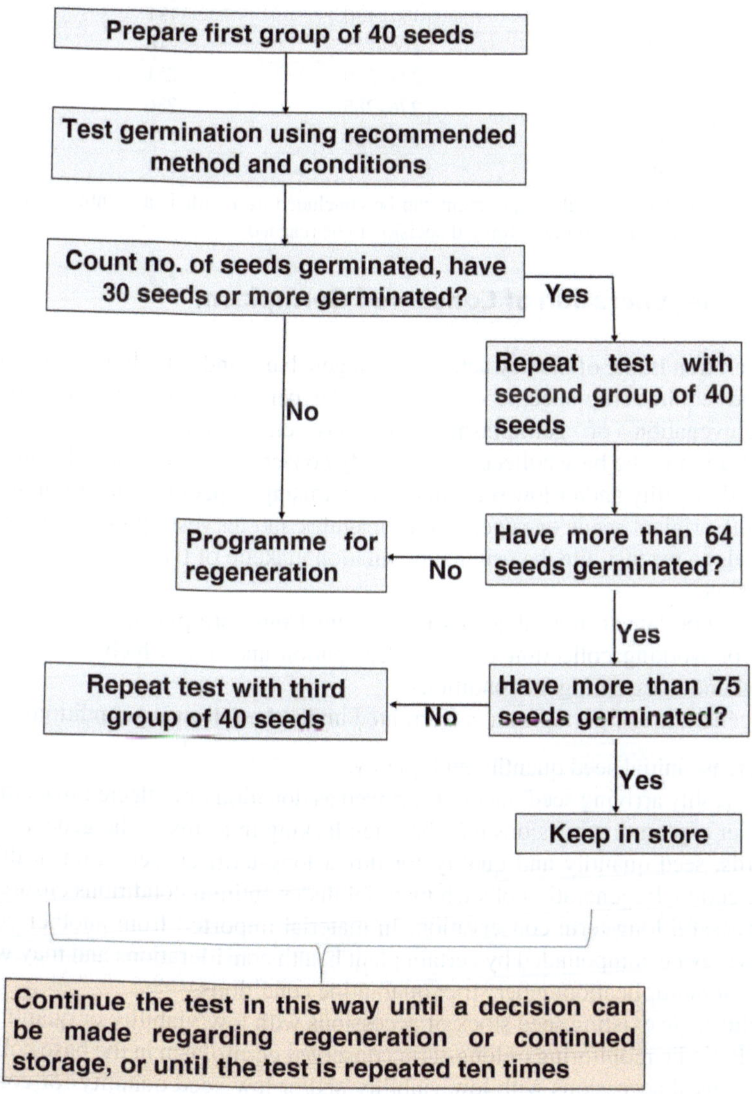

Flowchart 7.5 Procedure for sequential germination test

Table 7.7 Sequential germination test plan for 85% regeneration standard when testing seeds for germination in groups of 40[a] (adopted from Manual of Seed Handling in Genebanks by NK Rao et al. 2006)

Number of seeds tested	Regenerate if the number of seeds germinated is less than or equal to	Repeat the test if number of germinated seeds is in the range of	Store if number of seeds germinated is more than or equal to
40	29	30–40	–
80	64	6575	76
120	100	101–110	111
160	135	136–145	146
200	170	171–180	181
240	205	206–215	216
280	240	241–250	251
320	275	276–285	286
360	310	311–320	321
400[a]	340	–	341

[a] Upon testing 400 seeds, the experiment can be concluded as a sufficient number of tests have been conducted to allow for an informed decision to be reached

7.7 Regeneration of Conserved Germplasm

Regeneration is one of the key activities in a genebank and is highly resource intensive and requires crop-specific knowledge and expertise. It is the process of restoration/rejuvenation of germplasm accessions either intended for long-term conservation in the base collection or already conserved in the base collection with low seed viability and/or low seed quantity by raising a specific population of plants from the original seeds or vegetative propagules and harvesting the seeds or plant materials to reconstitute the original population makeup of the accession as closely as possible.

It is important to note that this is different from germplasm multiplication in active or working collection for only distribution and use, which is usually performed under less stringent conditions.

Regeneration of germplasm is warranted under the following conditions:

1. Increase initial seed quantity and quality.

   Freshly arriving seed material received as donation or collected from the field under exploration missions may be often lacking in terms of the genebank standards, seed quantity and quality for direct long-term conservation in the base collection. Regeneration of such material under optimal conditions ensures their successful long-term conservation. In material imported from another country, this may be compounded by certain plant health considerations and may warrant initial multiplication under strict quarantine conditions.
2. Rejuvenate existing seed stock of accessions with low viability or quantity.

   Periodic monitoring of long-term conserved germplasm in the base collection can reveal accessions with low viability and/or low seed quantity for conservation and distribution. Generate additional samples for active collection and safety duplication.

In addition to the base collection, accessions need to be also replicated for other collections.

To ensure the security of collection, 'safety duplicates' of accessions need to be conserved at a location far from the originating genebank. Likewise, for ensuring distribution and use, an active collection needs to be also maintained under medium-term storage conditions.

Owing to the resource-intensive and time-consuming nature of regeneration, subsamples to fulfil multiple requirements may be generated from the same regeneration cycle.

Regeneration is to be performed very cautiously and under optimum conditions as this exposes the valuable germplasm to the following risks, leading to a drastic change in its composition or even the complete loss of the accession in the worst case.

1. Mechanical mixture with any other seeds or contaminant material.
2. Exposure to biotic and abiotic stresses leading to low yield or loss of material.
3. Inadvertent hybridization particularly in cross-pollinated or often cross-pollinated crops leading to genetic contamination.
4. Genetic erosion and drift due to insufficient sampling and/or variable seed production among population fractions leading to change in gene frequencies.

As the goal of regeneration is to produce the maximum quantity of the highest quality seeds most efficiently with minimal loss in genetic integrity and with minimal cost.

The following factors are to be considered during regeneration.

1. Type of collection

   Accessions from the base collection should be prioritized over those from the active collection for regeneration. Active collections should ideally be regenerated from the seed samples from the base collections. Alternatively, for up to three regeneration cycles, seeds from the active collection may be used. Base collections should always be regenerated from the most-original seed samples in the base collection.
2. When to regenerate

   Regeneration being a resource-intensive and time-consuming process, it must be carried out only when essential to save resources and to avoid unnecessary exposure to risks. Usually, it is undertaken when the seed viability falls below 85% of the initial viability during periodic monitoring or when the number of viable seeds in accession is <1500 for populations and < 250 for inbred lines/pure lines. Low viability accessions should be prioritized over low quantity accessions for regeneration.
3. Sample size

   A random sample should be drawn from the original accessions to capture within-accession diversity and avoid loss of low-frequency alleles. For 10–150 loci, to retain 90–95% of alleles with a frequency of 0.003–0.05, it is estimated that a plant population of 90–210 is required. Cross-pollinated species require more individual plants compared to self-pollinated species to maintain the within-accession diversity. The number of seeds required also depends on the viability of the accession and can be estimated as follows.

$$\text{Number of seeds required for regeneration} = \frac{\text{Desired plant population for regeneration}}{\text{Germination\%}^{*} \times \text{Expected field establishment\%}}$$

* Expressed as decimals

4. Preparation of seeds

   For seeds stored in genebank, the containers need to be retrieved and left overnight at room temperature to avoid seed deterioration due to rapid uptake of moisture from the atmosphere.

5. Maintaining effective population size

   Effective population size is the number of breeding individuals in an ideal population that would have the same amount of random genetic drift or inbreeding as the actual population. It is one of the key parameters to be considered during regeneration for reducing genetic drift. The level of heterozygosity and within-accession diversity affects the effective population size. The contribution of each plant in the regeneration plot to the next generation has to be carefully considered as it can have a drastic effect on reducing effective population size. Practices such as bulking of seeds and allowing open pollination of cross-pollinated species can lead to this. The best strategy to ensure high effective population size is to sample an equal number of seeds from the maximum number of maternal plants separately.

6. Choice of environment

   Ideally, germplasm accessions should be regenerated in an ecological region as close as possible to that from which they have been collected. In addition, to avoid undue selection pressures leading to a change in the population composition of accessions, any biotic or abiotic stresses should be avoided and it should be ensured that the plots are as uniform as possible. Fields where the same species has been previously raised should be avoided.

7. Pollination control

   Proper isolation distances, temporal isolation, bagging, pollination cages, use of pollinators etc. should be used particularly in cross-pollinated and often cross-pollinated species.

8. Crop management

   The best crop management practices should be adopted to keep the regeneration plots free of all pests. Weeds and soil-borne pests are of particular concern and should be managed by practices such as herbicide sprays, soil sterilization, periodic weeding, deep ploughing etc. Proper monitoring for common pests and diseases should be periodically done with the help of plant protection specialists to ensure the production of the best quality seeds.

9. Maintaining accession identity

   The plots should be diligently labelled and filed maps are to be maintained to keep track of the accession identity. After regeneration, the seeds can be compared with the original samples or the characterization data can be compared to verify their identity.

10. Harvesting

    Harvesting should be done at the most optimum maturity of the plant. This ensures that the maximum number of seeds is harvested before natural dispersion and at physiological maturity, which are desiccation tolerant and can be handled with minimal mechanical damage.

7.8 Field Genebank

Many of the horticultural and forestry species are difficult to conserve as seeds due to one or the other reasons. Some species have short-lived seeds known as recalcitrant, for example, cocoa, coconut, oil palm, rubber and many tropical fruits like mango, mangosteen, jackfruit, durian and rambutan while some are sexually sterile and vegetatively propagated, for example, yam, taro, cultivated banana, potato, sweet potato, cassava, pineapple and sugarcane. Some reproduce seeds readily like fruit trees but are highly heterozygous and clonally propagated to maintain the genotypes. Seeds of recalcitrant species can only be stored without desiccation for a few days, weeks or months (Roberts et al. 1984). Even if the technology for conserving recalcitrant seed is developed, there will still be a problem with the long regeneration cycle of perennial species (Hawkes 1982). The genetic resources of such plant species are generally conserved in field genebanks (FGBs) where germplasm is maintained in the form of plants as a permanent living collection. FGBs are often established to maintain working collections of living plants for experimental purposes. It is an ex-situ method of conservation where genetic variation is maintained away from its original location. FGBs are different from gene sanctuary which is generally established in the centre of diversity or microcentre.

7.9 National Active Germplasm Sites (NAGS)

NAGS is designated by the ICAR-NBPGR where active collections of PGRs are maintained and made available to researchers, breeders and other interested parties. The purpose of NAGS is to conserve, evaluate and utilize genetic diversity in plants for the benefit of agriculture and the environment in close collaboration with the regional stations of the NBPGR as well as other research institutes and universities. The number of NAGS has been recently updated to 49 centres located throughout India each of which specializes in specific plant groups or regions (Fig. 7.1). The Bureau is gradually developing and strengthening the national plant genetic resources system by linking up the National Base Collection (kept under long-term storage at ICAR-NBPGR) with National Active Germplasm Sites (NAGS).

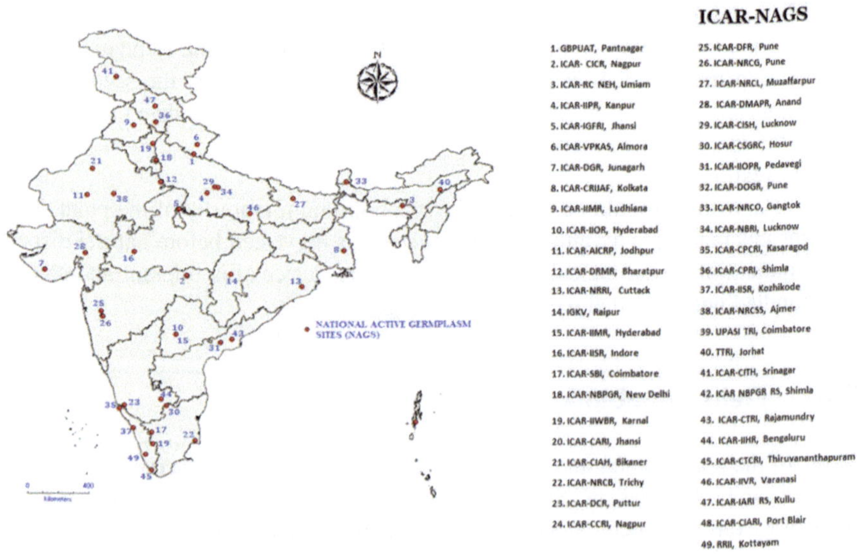

Fig. 7.1 National active germplasm sites in India

7.10 Summary

Conservation of Plant Genetic Resources (PGRs) aims to maintain the genetic diversity of plants. It also aims to halt habitat, species and genetic erosion by establishing and implementing conservation programmes. There are two general strategies for conservation, namely in-situ conservation and ex-situ conservation. In-situ conservation involves the management and monitoring of species at the location where they grow naturally. National Parks, nature reserves, biosphere reserves, wildlife sanctuaries, sacred groves and forests are some examples of in-situ conservation. Ex-situ conservation on the other hand entails exploration, sampling, transfer and storage of the propagules of the species in an artificially created environment. The sources of PGRs brought for Ex-situ conservation are from explorations, products developed by breeders and those protected or used by farmers. Thus, old and new varieties, landraces, crop wild relatives, vegetative propagules, buds, pollen and embryo all can be conserved. Seed genebanks, field genebanks, botanic gardens and arboreta, in-vitro tissue repositories and cryopreservation are the predominant methods of ex-situ conservation of different types of PGRs. Seeds of all seed-bearing species can be divided into orthodox, recalcitrant and intermediate types based on their tolerance to reduction in seed moisture without loss of their viability. More than 90% of the species possess orthodox seeds that can be dried up to 3–7% moisture content and stored at low temperatures (−18 °C) for long periods without significant loss of viability. Orthodox seeds are typically conserved in the long-term storage modules of a seed genebank after proper processing, quarantine testing, drying and packaging. Seeds require regeneration, when the quantity falls below a threshold in the seed bank. The recalcitrant and intermediate types of seeds are highly sensitive to desiccation and, therefore, need to be preserved at higher moisture and in ultra-low temperatures achieved by using liquid nitrogen through the process called cryopreservation. The species that are predominantly vegetatively propagated can be conserved through in-vitro conservation techniques, while those, such as tree species that are highly heterozygous are best conserved in field genebanks.

Questions

1. **Short questions**
   - (i) Define ex-situ conservation
   - (ii) At what interval is the monitoring of germplasm required to be done in genebank?
   - (iii) What is Harrington's thumb rule on seed storage?
   - (iv) Differentiate between short-, medium- and long-term storage.
   - (v) What are the seed viability and seed moisture standards for base collection?
   - (vi) What is the seed quantity requirement for long-term conservation of germplasm?
   - (vii) Write a short note on sequential germination
   - (viii) How to determine seed storage behaviour?
   - (ix) Write short notes on active and base collection.
   - (x) Write a short note on NAGS.
   - (xi) Write a short note on the estimation of the moisture content of the seed by using constant oven temperature.

Multiple-Choice Questions

1. The seeds which can be dried at low moisture content (2–5%) for storage without getting damaged are called as
   (a) Recalcitrant seeds
   (b) (b) Dry seeds
   (c) Orthodox seeds
   (d) Dicot seeds
2. 2. The inability of viable seeds to germinate under favourable environmental conditions is known as
   (a) Seed longevity
   (b) Dormancy
   (c) Both (a) & (b)
   (d) None of the above
3. The time during storage when the viability of the seed lot has declined to 50% (p50) is a trait of
   (a) Longevity
   (b) Germination
   (c) Diffusion
   (d) All of these
4. The variables of Ellis and Roberts's eq. (1980) of seed longevity are
   (a) Moisture content
   (b) Temperature
   (c) Both (a) & (b)
   (d) None of the above

5. Mature fertilized ovules are
   (a) Fruits
   (b) Seeds
   (c) Berries
   (d) Flowers
6. An intermediate category of seed storage behaviour was first reported in
   (a) Coffee
   (b) Tea
   (c) Cocoa
   (d) Black pepper
7. The National Genebank located at ICAR-NBPGR, New Delhi conserves seeds using the following strategy
   (a) Long-term conservation at −20 °C
   (b) In vitro-conservation and cryopreservation
   (c) Field Genebank
   (d) (a and b)
   (e) (a, b and c)

Answers

1. (c)
2. (b)
3. (a)
4. (c)
5. (b)
6. (a)
7. (d)

Suggested Reading

Annonymous. ISTA (2019) International rules for seed testing. International Seed Testing Association, Bassersdorf, 276 pp. https://doi.org/10.15258/istarules

Breese EL (1989) Regeneration and multiplication of germplasm resources in seed genebanks: the scientific background. International Board for Plant Genetic Resources, Rome, 69 pp. ISBN: 92-9043-186-5. https://hdl.handle.net/10568/104939

Ellis RH, Hong TD, Roberts EH (1985) Handbook of seed technology for genebanks: Volume I. Principles and methodology. International Board for Plant Genetic Resources, Rome, 210 pp. ISBN: 92-9043-118-0. https://hdl.handle.net/10568/104237

FAO (2022) Practical guide for the application of the Genebank Standards for Plant Genetic Resources for Food and Agriculture: Conservation of orthodox seeds in seed genebanks. Commission on Genetic Resources for Food and Agriculture. Rome. https://doi.org/10.4060/cc0021en

Harrington JF (1960) Thumb rules of drying seed. Crops and Soils Publisher 13:16–17.

Rao NK, Hanson J, Dulloo ME, Ghosh K, Nowell D, Larinde M (2006) Manual of seed handling in genebanks. Bioversity International, Rome, 147 pp. ISBN: 978-92-9043-740-6, https://hdl.handle.net/10568/2945

Tissue Culture and Cryopreservation Strategies

8

Sandhya Gupta, Vartika Srivastava, Subhash Chander, Era Vaidya Malhotra, R. Gowthami, and Kanchit Thammasiri

Learning objectives

- Plant tissue culture and its role in plant genetic resource (PGR) management.
- Tissue culture techniques in germplasm collection, exchange, disease eradication etc.
- Tissue culture strategies and methods for the conservation of germplasm.
- Cryopreservation techniques for long-term conservation of germplasm.
- And related topics.

Keywords

Plant genetic resources · Germplasm · PGR · Tissue culture · Cryopreservation · Micropropagation · In vitro conservation · Hardening · Acclimatization · Embryo rescue

S. Gupta (✉) · V. Srivastava · S. Chander · E. V. Malhotra · R. Gowthami
Division of Germplasm Conservation, ICAR-National Bureau of Plant Genetic Resources, New Delhi, India
e-mail: sandhya_gupta87@yahoo.com; sandhya.gupta@icar.org.in; vartika.srivastava@icar.org.in; subhash.chander3@icar.org.in

K. Thammasiri
Department of Gardening and Horticulture, Xishuangbanna Tropical Botanical Garden, Chinese Academy of Sciences, Menglun, Mengla, Yunnan, China
e-mail: kanchitthammasiri@xtbg.ac.cn

K. Tripathi et al. (eds.), *Textbook of Plant Genetic Resources*,
https://doi.org/10.1007/978-981-96-5024-8_8

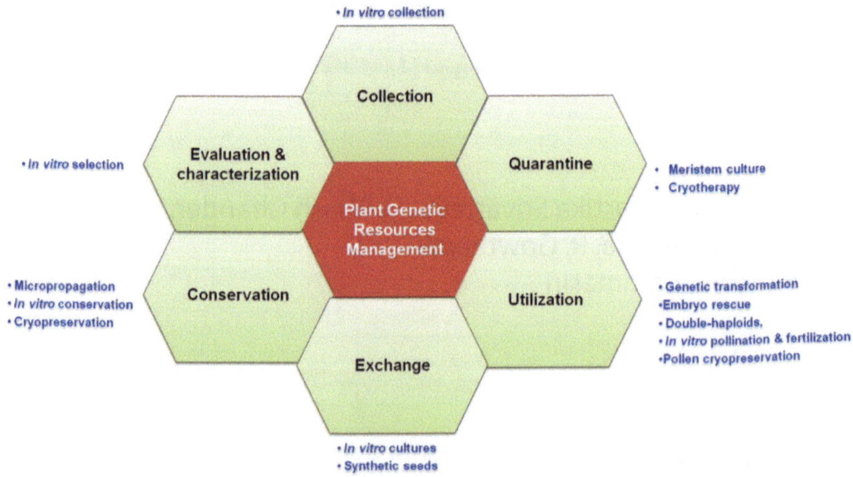

Fig. 8.1 Application of plant tissue culture and cryopreservation techniques for PGR management

8.1 Introduction

Plant genetic resources play a crucial role in food, nutrition and livelihood security. In the majority of the species, seeds can be used to conserve plant genetic resources. Many species, with recalcitrant seeds, clonally propagated species, vegetatively propagated species, and short lived seeds can't be conserved using conventional storage method. An alternative method of conservation, i.e., field genebanks is being followed for these crops. However, Genetic resources conserved in field genebanks are prone to several threats, namely pests and diseases, natural hazards and high maintenance costs for the maintenance of collections. Plant tissue culture plays a significant role in plant genetic resource management, namely collection, conservation, exchange, evaluation, quarantine, and utilization (Fig. 8.1). This chapter provides an overview of plant tissue culture and cryopreservation techniques for plant genetic resources management.

8.2 In vitro Collecting

In vitro collecting is the process of collection of different explants, namely seed, embryo, vegetative tissues (such as apical bud, nodal segment and leaf tissue), from the plants in the field, after surface sterilization, inoculate on sterilized minimal media supplemented with antimicrobial agents, further subsequent processing in the laboratory. In vitro collecting supplements the process of germplasm collection of clonally propagated species, seed species and exceptional plant species (threatened) where the plant genetic resources collection is a challenging task.

8.3 In vitro Germplasm Exchange

There is a significant risk of disease and pest dissemination during germplasm exchange. Germplasm can be exchanged either through (a) non-aseptic conditions (seeds, cuttings, tubers, bulbs etc.) or through (b) aseptic conditions (in vitro) (in vitro shoot cultures, in vitro micro cuttings, in vitro microtubers, somatic embryos, callus, cells suspension etc.). There are laws and policies in practice in several counties to reduce the entry and spread of plant diseases and pests. International guidelines on the safe movement of germplasm recommend the use of in vitro cultures for the safe transfer of germplasm across various nations for vegetatively and clonally propagated crops.

8.4 Micropropagation

Micropropagation is a method of propagation of small plant parts called explants using in vitro techniques through axillary branching or adventitious branching followed by regeneration. The term micropropagation means micro/miniature shoots and their propagation. It provides a rapid and reliable system for the production of many genetically uniform disease-free plantlets and contributes towards commercial plant propagation. The potential of tissue culture for micropropagation was first used by George Morel in 1960 for the propagation of orchids (*Cymbidium*) using shoot-tip explants.

8.4.1 Stages of Micropropagation

Murashige (1978) suggested four unique steps that may be used in commercial in vitro propagation. Stages I, II and III are carried out in vitro, while stage IV is completed ex vitro in a greenhouse/polyhouse. Later, in 1981, Debergh and Maene proposed an additional stage called stage 0 for the micropropagation of various plant systems (Fig. 8.2). The different stages of micropropagation are explained in detail in separate headings as given below:

Stage 0: Selection of Mother Plants and Their Maintenance (Pre-propagation Stage)
The pre-propagation stage (also called stage 0) is the initial step in micropropagation and involves the selection and maintenance of mother plants under controlled conditions in glass houses under disease- and insect-free conditions with minimal duration for 3 months. Mother plants are grown at relatively low humidity and watered either with irrigation tubes or by capillary sand beds or mats; also explants for in vitro propagation are done after appropriate pretreatment of the mother plants with fungicides and pesticides to minimize contamination in the in vitro cultures.

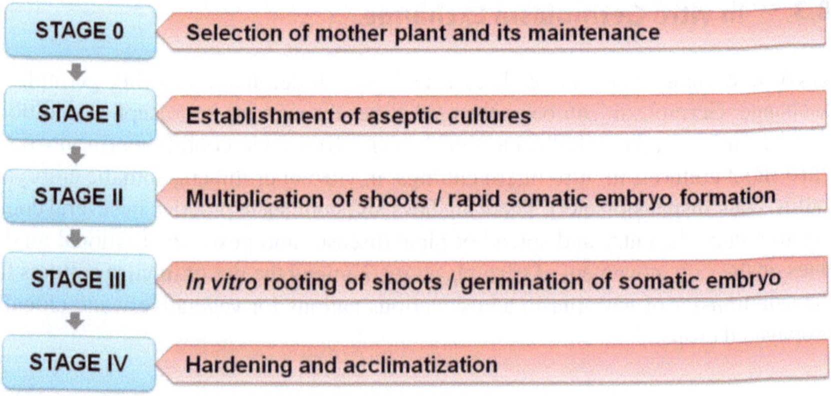

Fig. 8.2 Stages of micropropagation

Stage I: Establishment of Aseptic Cultures
Stage I refers to the initiation and establishment of aseptic cultures using suitable explants. This stage involves a series of steps, namely selection of suitable explants, sterilization of explants and selection of suitable medium. The selection of suitable explants for micropropagation depends on the purpose and the mode of regeneration. The apical buds or axillary buds/nodal segments are used for clonal propagation to maintain true to type; apical meristems are used for meristem-tip culture to eliminate viruses; and nucellus explants are used for somatic embryo induction etc. After the selection of suitable explants, the explants are washed in autoclaved distilled water, and rinsed in ethanol. Surface sterilization is achieved by using chemicals such as sodium hypochloride and mercuric chloride. Stage I requires at least four passages of subculture in different media and cultures were maintained at standard culture room conditions.

Stage II: Multiplication of Shoots or Rapid Somatic Embryo Formation
Stage II refers to the propagation phase in which the explants are cultured on the defined culture medium for shoot multiplication or rapid embryo formation. The main objective is to achieve propagation without losing genetic stability. This stage takes place by following any one mode of regeneration (somatic embryogenesis or organogenesis) and the regeneration pathway may be either direct or indirect (callus mediated). In this stage, repeated subcultures of axillary or adventitious shoots, nodal segments, somatic embryos and other explants from stage I on a defined culture medium to stimulate maximum proliferation of regenerated shoots lead to the multiplication of propagules in large numbers. The propagation media generally contains a basal media (Murashige and Skoog, Gamborg B5 media, Woody plant media etc.) and cytokinin at the concentration of 1–2 mg/L but up to 30 mg/L has been used either alone or in combination with an auxin (0.1–1 mg/L) depending on the crop species. BAP (6-Benzylaminopurine) is the most commonly used

cytokinin. Among auxins NAA, IAA and IBA are generally used. 2,4-D is not used as it promotes callusing and is used during indirect organogenesis.

Stage III: In vitro Rooting of Shoots and/or Germination of Somatic Embryo
Multiple shoots induced or formed through regeneration from callus, axillary branching and direct adventitious bud formation must undergo a process of rooting for complete plantlet formation. In a few cases, the shoots need an additional step of shoot elongation before rooting, as due to exposure to cytokinins shoots remain short. Rooting is generally done by transferring individual shoots to a medium supplemented with a suitable auxin in vitro or in a few cases ex vitro where shoots are directly established in soil as microcuttings for developing roots. If somatic embryos are obtained from stage II, then somatic embryos should be germinated to get plantlets in stage III. This stage may require 1–6 weeks and in general, rooting can be obtained on reduced concentrations of salt (½ or ¼) of the basal medium without any growth regulators, reduced sugar levels (usually 1 g/L) or supplemented with auxins (0.1–1 mg/L NAA or IBA) to produce sturdy well-formed rooted plants for high survival ex vitro.

Stage IV: Hardening and Acclimatization of Plantlet
The success of micropropagation is determined by the in vitro-propagated plants' establishment ex vitro through hardening. In this stage, in vitro plants with well-formed roots are transferred from test tubes to external soil conditions. This stage may take 4–6 weeks and for acclimatization, the individual plant is taken out of the medium followed by washing of the roots with running tap water to remove the agar and then individual plants are planted in pots with a suitable potting mixture (peats, vermiculite, soil or sand in different proportions). Transplanted plantlets or shoots are immediately irrigated with an inorganic nutrient solution and maintained under high humidity (90–100%) for 10–15 days initially. For a high humidity built up around transplanted plants, the plants should be covered with clean transparent plastic bags having small holes for air circulation.

8.5 In vitro Conservation

In vitro conservation refers to the conservation of plant genetic resources in relatively stable form under a defined artificial environment and nutrient conditions in the in vitro genebanks for short- to medium-term conservation (under normal and slow growth conditions), and for long-term (under suspended growth) using cryopreservation techniques (Fig. 8.3). Varied explants are used for in vitro conservation, namely, shoot tips, shoot meristems, nodal segments, dormant buds, cell suspension cultures, callus, hairy root cultures, seeds, zygotic embryos, embryonic axes and pollen. However, cell and callus culture conservation is more prone to somaclonal variations. The basic goal of developing an in vitro conservation protocol is to reduce the frequency of subculture and ensure the genetic stability of conserved germplasm.

8.5.1 Normal Growth

Under normal growth, in vitro cultures are maintained under standard culture room conditions with regular subculturing to avoid loss. This strategy is widely applicable for naturally slow-growing culture systems or for those for which there are no reported methods. However, maintenance of cultures under normal growth conditions is cost and labour-intensive, cultures are exposed to microbial contamination, somaclonal variations or loss of regenerative capacity with time. The main advantage of this strategy is that germplasm is readily available for multiplication and distribution.

8.5.2 Slow Growth Conservation

The slow growth conservation strategy is based on optimizing the subculture period without risking germplasm loss and compromising genetic stability through different strategies (depending on the species) for short- to medium-term conservation. In this strategy, in vitro cultures are subjected to one or combinations of strategies such as low temperature, low light or photoperiod cycle, minimal containment, reduced oxygen concentration, osmotic adjustment, modification of gaseous environment, growth regulator reduction, growth retardants, minimal media, synthetic seeds, induction of in vitro storage organ (see Fig. 8.3). The best slow-growth conservation strategy should have high viability/regeneration (%) after conservation, cost-effective, technically less demanding strategy and generic protocol (wide and easy applicability) to several genotypes.

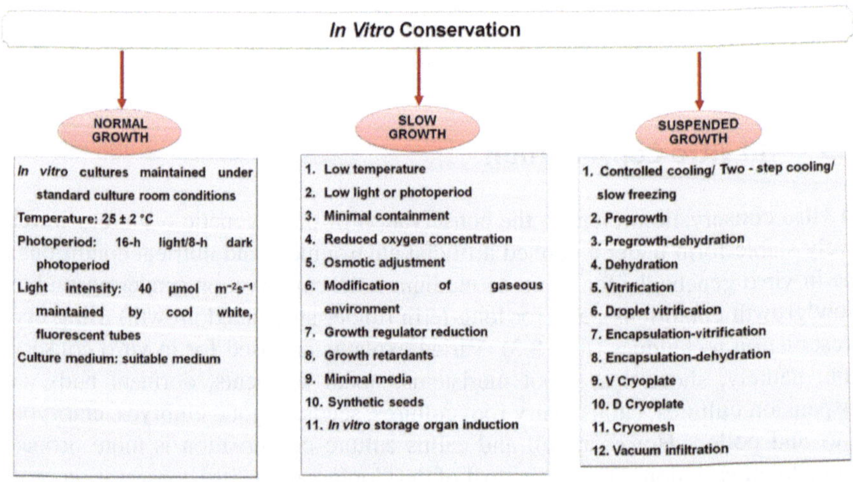

Fig. 8.3 In vitro conservation of plant genetic resources under normal, slow and suspended growth

8.5.3 Suspended Growth-Cryopreservation

Cryopreservation is the only available safe and cost-effective technique for the long-term conservation of diverse germplasm. Cryopreservation is a process where biological material can be stored at ultra-low temperature either in the liquid phase of nitrogen (−196 °C) or vapour phase of nitrogen (−150 °C), without losing its viability. This technique works on the principle of vitrification. The term vitrification is derived from the Latin word vitreum, which means glass. Vitrification is a physical process, which encompasses the phase transition of a concentrated aqueous solution from the liquid phase into an amorphous glassy non-crystalline solid or glass. Conversely, ice formation is the phase transition of a liquid to ice crystals. Water is difficult to vitrify; hence, different cryoprotectants are used as they are easily supercooled (< −70 °C). For the viability of any cell or tissue, it is necessary to escape from ice crystallization by the process of physical/osmotic dehydration or by substituting the water molecules with various cryoprotective agents (CA). Vitrification is the only freeze avoidance mechanism that enables hydrated tissues to survive exposure to liquid nitrogen (LN) temperatures. There are different dehydration procedures available and used for the preparation of the explant to cryopreservation such as controlled freezing, air desiccation (laminar airflow or silica gel dehydration) and cryoprotective agents (CA). Germplasm can be cryopreserved in the form of whole seed, embryo/embryonic axes, pollen, somatic embryo, meristem, shoot or root tips, axillary buds, nodal segments, bulbils etc.

The first report on the survival of plant cells/tissue with scientific evidence was witnessed by Prof. Akira Sakai in pre-frozen (−30 °C) cortex cells of mulberry on immersion in LN in 1956. He successfully applied the same technique in *Salix koriyanagi* and *Populus sieboldii*. Later, other workers successfully reported the cryopreservation in flax cells, and cell cultures of *Ipomoea* sp. and *Daucus carota*. These scientific reports laid the foundation pillars for the cryobiology and development of classical technique (freeze-induced dehydration) for cryopreservation in 1968 (Fig. 8.4). Later in the 1990s, the development of vitrification techniques revolutionized the long-term conservation of germplasm of vegetatively propagated crops and non-orthodox seeded species. In the vitrification method, intra and intercellular viscosity is attained by replacing the freezable water through osmotic and physical dehydration followed by ultrafast freezing in LN. With this combination of dehydration and ultra-fast freezing, the remaining intra/intercellular water molecule does not have adequate time to form ice crystallization and phase transition to an amorphous glassy non-crystalline solid. Thus, several vitrification-based techniques have been developed, namely dehydration, vitrification, encapsulation-dehydration, pregrowth dehydration, pregrowth, encapsulation–vitrification and droplet–vitrification protocols. Later during 2011 and 2013, V and D cryoplate-based techniques, respectively, were developed. More recently vacuum infiltration vitrification and cryomesh techniques were developed. These techniques have been regularly used as a complementary strategy to conventional seed genebanks, in vitro genebanks and field genebanks. Detailed generic protocols for all the cryopreservation techniques are given in Fig. 8.4.

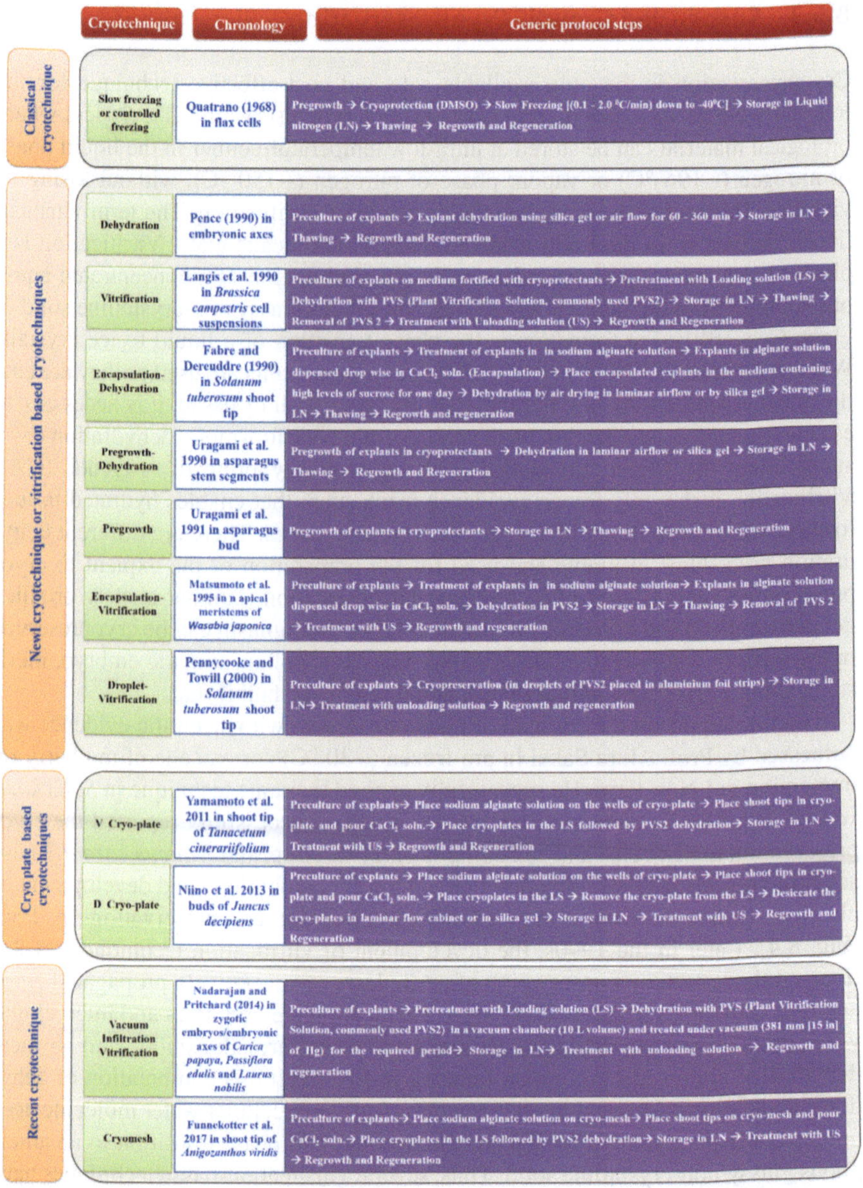

Fig. 8.4 Cryopreservation techniques and procedures

8.5.3.1 Cryoprotective Agents (CPAs)

CPAs are constituents that are used to protect biological cells/tissues/organs from freezing injury. These are featured as water-soluble and low/high molecular weight substances. It simply operates on the principle of osmotic dehydration and increasing the solute concentration of the cells/tissues. After the dissolving of CPAs in a cryoprotective solution, it reduces the melting point of the cryoprotective solution so that cells/tissues/organs can freeze within their surroundings but not within their cells. Chemically CPAs interfere with the hydrogen bond of water molecules and prevent ice crystallization and cryogenic injury on LN immersion. CPAs should be easily penetrable to cells and non-toxic to cells/tissues/organs. Standardization of CPAs is a challenging task, as higher concentration may lead to cytotoxicity. CPAs are of two types based on their permeability.

Penetrating CPAs: CPAs which penetrate the cell membrane and enter the cytosol are called penetrating CPAs. Such CPAs are exclusively small molecules and to prevent ice crystallization, they form hydrogen bonds with water. They act by replacing water, and thus control cell size changes as well as prevent intracellular ice formation and dehydration during cryopreservation. The toxicity of such CPAs is an obstacle to successful cryopreservation, such as dimethyl sulfoxide (DMSO), glycerol, ethylene glycol, proline and acetamide.

Non-penetrating CPAs: Non-penetrating CPAs do not penetrate the cell membrane as they are generally larger molecules. Such CPAs are performed by desiccating the cell/tissues/organs before cryopreservation; thus, the available water content in the cell/tissues/organs is minimal to cause ice crystallization during cryopreservation. They prevent ice formation by a similar mechanism as penetrating CPAs, but they do not penetrate the cells, such as, polyethylene glycol (PEG), mannose, sucrose, propylene, glucose and ethylene glycol.

8.5.3.2 Cryoprotective Solution

Preculture medium: It is basal media with 0.1–1.0 M sucrose without any hormones. It enhances the freeze tolerance of the explant.

Loading Solution: It is a mixture of 2 M glycerol and 0.4 M sucrose; wherein sugar concentrations can vary. The loading treatment either neutralizes the osmotic stress and/or induces the physiological adaptation of tissues and cells (including membranes), to both dehydration and freezing.

Plant Vitrlfication Solution 2 (PVS2): This solution was identified by Sakai et al. (1990). It comprises 30% (w/v) glycerol, 15% (w/v) ethylene glycol and 15% (w/v) DMSO in MS medium containing 0.4 M sucrose. DMSO when added to the cell media, increases the permeability of the cellular membrane, which allows water to flow more freely through the membrane. Glycerol exerts an extracellular effect by osmotic stimulation of cell dehydration as well as an intracellular effect by permeating the membrane and restricting the dehydration effect. Ethylene glycol has high permeation into cells compared with other cryoprotectants due to its low formula weight.

Plant Vitrification Solution 3 (PVS3): This solution was identified by Nishizawa et al. in 1993. It comprises 50% glycerol (w/v) and 50% sucrose (w/v).

Unloading solution: This solution was suggested by Sakai et al. in 1990. It consists of 1.2 M sucrose. The unloading step is mainly to avoid the detrimental osmotic shock, which might occur to the cryopreserved materials, if they were transferred directly on to recovery medium.

8.6 Somaclonal Variations

Micropropagation offers rapid clonal multiplication of vegetatively propagated crops in relatively smaller spaces under controlled environmental conditions. Mass production of a population of genetically identical plants is the major advantage of plant tissue culture; however, there are chances of the emergence of variations among cultured cells owing to several factors such as culture conditions, re-differentiation and de-differentiation of cells, and ploidy of the cultured cells. Some of these variations are transient physiological or developmental changes, while some variations are a result of epigenetic changes, both not transmitted to the off-spring. On the other hand, the third type of variation includes alterations in the genetic makeup and mutations, which are genetically stable and can be inherited by the progeny. The first documented report on variations observed in in vitro cultures was in sugarcane (Heinze and Mee 1971). Larkin and Scowkraft (1981) coined the term somaclonal variation for genetic and epigenetic variations occurring in cells regenerated from somatic cells. Variations observed in cultures derived from gametic cells are termed gametoclonal variation (Evans et al. 1984).

Somaclonal variations include morphological, cytological, biochemical, genetic and epigenetic changes. These variations can be considered useful as well as harmful, depending on the objective of clonal multiplication. In the context of using clonal multiplication for elite genotypes with known characteristics and in commercial establishments, it is undesirable to have off-types as it is important to maintain the original genetic makeup of that particular genotype. Somaclonal variants can also be used for the generation or selection of plants with superior/desirable traits and as a source of genetic variability. Over the past years, somaclonal variations have been reported in several tissue culture systems of different crops across the globe and somaclone-derived new cultivars have been released in several crops.

8.6.1 Sources

Somaclonal variations are a result of mutations induced by various stress factors to which plants are subjected during the tissue culture process. Consequently, plant cells get exposed to oxidative stress resulting in the generation of reactive oxygen species (ROS) such as $O_2^{\cdot-}$, H_2O_2 and OH^- which cause changes in DNA methylation, chromosomal aberrations and induced point mutations. Several factors affect the frequency of development of somaclonal variations under in vitro conditions:

Tissue Culture Environment

In vitro cultures are subjected to artificially controlled conditions of light, temperature and humidity. All these factors affect the cell cycle. An inefficient control of the cell cycle tends to allow cell division before the completion of replication, thereby inducing chromosomal breakage, which can result in chromosomal aberrations, such as duplication, inversion, translocation, or deletion, in the in vitro cultured cells (Phillips 1994).

Parallelly, the addition of exogenous growth hormones affects (stimulates or inhibits) cell division and growth. Plant growth regulators tend to induce rapid cell division and are used for inducing multiple shoots, both axillary and adventitious. Some of these growth regulators, above certain concentrations, tend to act as mutagens. Prolonged culture in a medium containing high growth regulator concentrations has been reported to influence the ploidy levels of cells, by stimulating rapid cell division and induction of callus. Growth regulators have also been observed to affect chromosome remodeling and methylation patterns of cellular DNA leading to the induction of epigenetic changes in cells.

Explant Source

When differentiated tissues such as leaves, roots or stems are used as explants for culture initiation and multiplication, the cells must first undergo de-differentiation followed by re-differentiation to form shoot primordia and buds with or without an intervening callus phase. This whole process of reorganization of the cellular structures alters the cell cycle controls with high chances of induction of mutations in the regenerated plants. The de-differentiation and re-differentiation processes involve both qualitative and quantitative changes in the genome. The more the organizational structure of cells is broken down, the greater the chances of occurrence of variations. On the contrary, explants with pre-existing meristems such as shoot tips and axillary buds regenerate into new plantlets with low chances of any variation amongst them. In vitro plants raised through axillary bud induction and proliferation are generally free from any induced mutations. However, cultures that are generated through an intervening callus phase are more prone to induced mutations. De-differentiation of well-organized cells to an unorganized cellular mass of callus followed by rapid cell division may lead to chromosomal instability.

Culture Duration and Subculture Interval

Systems with extended duration of in vitro culture are more prone to the emergence of somaclonal variations. This may be attributed to induction of ploidy instability, accumulation of ROS species, depletion of nutrients in the culture medium or restriction of space to the dividing cells. In general, a reduction in growth potential and an increase in the accumulation of variants has been observed after eight subcultures. Based on several studies, it can be concluded that in general the expected rate of induction of variants increases as an exponential function of the number of subculture cycles, and, after several multiplication cycles different mutants can be expected in the population of subcultured plants.

Activation of Transposable Elements

Transposable elements (transposons) or mobile DNA elements can induce gene mutations or genome rearrangements. It is believed that disruption of the normal cell cycle during organogenesis and rapid cell division phases of tissue culture provides conducive conditions for the activation of transposable elements. These can act as insertional mutagens in genomes and can cause a wide spectrum of chromosomal rearrangements, in turn interfering with gene structure, expression and regulation.

8.6.2 Somaclonal Variations in Crop Improvement

Somaclonal variations provide a source of genetic variability which can be used for the development of plants with desirable traits. Several cultivars with different traits have been developed in different crops using somaclonal variations; a few examples have been provided in Table 8.1.

Table 8.1 Some commercially viable varieties developed through somaclonal variation in different crops

Sl. no.	Crop	Somaclone name	Characteristic of somaclone
1.	Banana	CIEN-BTA-03	Resistant to Yellow Sigatoka disease
2.	Banana	Tai-Chiao No. 1	Fusarium wilt-resistant and high yielding
3.	Sugarcane	Phule Savitri (Co 94012)	Early, high sucrose and high yielding
4.	*Geranium* spp.	CIM Pawan	Higher herbage and essential oil yield
5.	*Java citronella*	CIMAP/Bio-13	37% more oil and 39% more citronellon
6.	Tulip	Bs6, source	Red-violet coloured longer flower and stem
7.	Potato	Somaclones IBP-10, IBP-27 and IBP-30	Higher resistance to *Alternaria solani* and *Streptomyces scabiei*
8.	Potato	White Baron	Reduced tuber browning
9.	*Anthurium* sp.	Orange Hot	Light orange spathes and spadices
10.	Wheat	HeZu 8	Improved yield and agronomy
11.	Strawberry	Serenity	Pale skin-colored, late season, resistant to powdery mildew
12.	Sugarcane	Ono	Disease resistance and higher yield
13.	Brown mustard	Pusa Jai Kisan	High Yield
14.	Sweet potato	Scarlet	Colour, shape and baking quality of roots
15.	Barley	AC Malone	Improved yield and Downey Mildew resistance
16.	Amaryllus	Suryakiran	Red and white double colour flowers

8.6.3 Somatic Embryogenesis

The development of an embryo from a fertilized egg through a predetermined pattern of cell division and differentiation is known as embryogenesis. The development of embryos from somatic cells is known as somatic embryogenesis. This process was reported for the first time in carrot by Reinert and Steward, independently in 1958. Here, embryos are generated without the process of meiosis and fertilization. Somatic embryogenesis exploits the ability of plant cells to de-differentiate and re-differentiate into new cell types. The plant regenerating from a somatic embryo is genetically identical to the parent plant. A somatic embryo develops from a non-zygotic cell, has a bipolar structure that resembles a zygotic embryo and does not have vascular connections with the original tissue. Somatic embryos differentiate either directly from the explant without callus or indirectly via an intervening callus phase. Somatic embryogenesis involves several steps starting with the generation of PEMs (proembryogenic masses), somatic embryo formation, maturation and culminates in plant regeneration.

Induction of Embryogenic Cells

To acquire embryogenic competence, the foremost requirement for cells is to undergo de-differentiation. The induction of somatic embryogenesis involves the repression of the existing developmental gene expression pattern within the explant tissue and the activation of the embryogenic gene expression pathway. Auxins are commonly used to induce de-differentiation in plant cells. In vitro formation of embryogenic cells is preceded by asymmetric cell division and cell elongation. Auxins promote rapid cell division and alter cell polarity leading to asymmetric cell division and unorganized callus growth.

Proliferation of Embryogenic Cultures

Embryogenic cells once generated continue to proliferate, giving rise to proembryogenic masses (PEMs). Auxin is essential for the proliferation of PEMs, but it inhibits the transition of PEMs into somatic embryos. Embryogenic callus can be maintained and further proliferated on the high auxin medium like that used for initiation. PEMs are made up of small embryogenic cells. They have angular shapes and are connected to adjacent cells through numerous plasmodesmata. These cells contain several small vacuoles, as well as numerous starch grains, and exhibit a high density of ribosomes, numerous rough endoplasmic reticulum and spherosome-like vesicles. The embryogenic cells are held together by larger non-embryogenic cells.

Development and Maturation of Somatic Embryos

PEM-to-embryo transition is an important step in somatic embryo formation. To stimulate the continued growth of somatic embryos, it is essential to transfer the embryogenic cultures to a medium that does not contain auxin as the continued presence of auxin after the induction of embryogenesis results in cell elongation and separation of previously attached cells. These enlarged cells lose their ability to develop into somatic embryos permanently. The development of somatic embryos closely resembles that of zygotic embryos.

Somatic embryos undergo several morphological and biochemical transformations. Maturation includes the accumulation of storage substances and the acquisition of desiccation tolerance. The initial stage, known as the globular stage, is characterized by small cell clusters that appear within 5–7 days after transfer to auxin-free medium. After 2–3 days of isodiametric growth, the embryos begin to grow in a bilaterally symmetrical manner and enter the heart stage. In this stage, the hypocotyl begins to elongate, and the radicle is initiated. This stage is followed by the torpedo stage, finally growing into a plantlet with well-developed shoots and roots.

Plantlet Development

Morphologically normal embryos which have accumulated storage material develop into healthy plants on further culturing onto the suitable medium.

8.6.3.1 Factors Affecting Somatic Embryogenesis

Explant

Generally immature or less differentiated tissues, including hypocotyl segments, young leaves, embryonic shoot tips, young floral parts and immature zygotic embryos, are selected as explants for somatic embryo production. Zygotic embryos contain pre-embryogenic determined cells (PEDCs) which possess embryogenic competence. On the contrary, cells in differentiated explants have to undergo induction to acquire embryogenic potential, and these cells are termed induced embryogenic determined cells (IEDCs).

Genotype

Somatic embryogenesis is a genetically controlled trait and differences in response to induction of somatic embryos have been observed among different genotypes. Variations in genotype can arise due to differences in the endogenous levels of growth regulators.

Culture Medium

Somatic embryogenesis is mostly induced on an auxin-enriched medium, the most commonly used auxin being 2,4-D. Callus differentiates into meristematic cells called proembryogenic masses (PEMs) on auxin-rich medium. The PEMs proliferate on the same medium but do not mature into embryos. For embryo transition and maturation, auxin-free medium is used. Cytokinins have been reported to have a stimulating effect on embryo development.

8.6.3.2 Application of Somatic Embryogenesis

Somatic embryos are used for several applications in plant tissue culture:

Large Scale Propagation

Unlimited numbers of embryos can be generated from a single explant if all conditions are favourable. Somatic embryos have a greater potential for large-scale propagation in comparison to the other methods of propagation. The scale of propagation can be further enhanced if bioreactors are applied to produce embryogenic systems.

Synthetic Seeds

Somatic embryo-derived synthetic seeds have been developed in several crop plants. They are more useful in crops which do not produce seeds and their vegetative propagation is cumbersome, where very few seeds are produced, and in crops where the zygotic embryos are recalcitrant.

Germplasm Conservation

Somatic embryos can be encapsulated in alginate beads to generate synthetic seeds. These synthetic seeds can be conserved for medium- or long-term using different in vitro and cryopreservation approaches.

Genetic Transformation

Due to their high multiplication rates and potential for scale-up via bioreactor, somatic embryos can be used as a suitable target for gene transfer. Further, regeneration of the transformed cells can be initiated from single cells in somatic embryos enabling the regeneration of non-chimeric plants.

8.6.4 Genetic Stability of In vitro Raised Germplasm

Plants maintained in vitro have a greater chance of containing genetic alterations in the form of somaclonal variations. As a result, it is essential in tissue culture-based propagation to verify the genetic uniformity of regenerated plantlets using various types of markers, including phenotypic, biochemical, physiological and molecular markers.

Timely identification and elimination of variants are necessary for maintaining the genetic uniformity of tissue culture-raised plants. Moreover, efficient detection can help in the identification of variants harbouring beneficial agronomic characteristics. Somaclonal variants can be detected using morphological, physiological/biochemical, and molecular techniques as elucidated in Table 8.2.

8.7 Tissue Culture Techniques for Pre-breeding

Tissue culture techniques that can be of use in breeding programmes include protoplast culture, anther/microspore culture, embryo rescue, ovary and ovule culture etc.

Table 8.2 Techniques for the detection of somaclonal variations

Morphological detection	Biochemical detection	Molecular detection	Cytological detection
Differences in plant morphology, such as plant structure, stature, leaf anatomy and morphology, pigmentation	Based on the detection of variation in isozymes	Changes in genome	Variation in ploidy or chromosomal number and structure
Influenced by environment; not true representation of genetic composition	Sensitive to ontogenic changes and other environmental factors	Genetically stable, environment-independent	Genetically stable, environment-independent
Morphological traits are limited in number.	Limited in numbers	Several molecular markers are available—RAPD, ISSR, RFLP, AFLP, SSR and SNPs to name a few	Karyological analysis for chromosomal aberrations, flow cytometry for chromosome counting
Usually applicable for field/greenhouse-established plants	Tissue-specific expression	Can be conducted at an early growth stage, while plants are still in tissue culture and prior to full regeneration	Can be conducted at an early growth stage, while plants are still in tissue culture

8.7.1 Protoplast Culture

Protoplast refers to the entire cell, excluding the cell wall. In plants, the cell wall can be removed by mechanical, chemical or enzymatic means. Protoplasts are single cells and could be utilized to raise genetically identical plants, especially in horticultural crops where clonal mass multiplication is required for quality planting material production. In addition, this also could provide an opportunity to raise somatic hybrids as the cell wall is absent and the protoplast fusion may take place under controlled conditions. The absence of a cell wall in the isolated protoplasts also facilitates the transfer of genes between different species via uptake of DNA enabling transformations related research. Subsequently, this transformed protoplast may lead to the formation of the complete plant using a suitable tissue culture procedure, which later may be used in breeding programmes.

8.7.2 Anther/Microspore Culture

Anther culture is an in vitro technique to develop homozygous plants through microspores. These homozygous plants may be very important in plant breeding. The development of haploids is important because of their potential applications in mutation research, selection, genetic analysis, transformation and the production of homozygous cultivars.

8.7.3 Ovary and Ovule Culture

The aseptic culture of unfertilized ovaries and ovules to obtain haploids from egg cells or other haploid cells comprising embryo sac is called ovary culture. This in vitro technique of gynogenesis is effective with the appropriate use of growth regulators in the culture media. The process of gynogenesis is often divided into two stages, the first is induction and the second is regeneration. During the induction phase, a liquid medium usually contains low auxin content and explants are cultured in dark conditions. Later these are transferred into semi-solid media with higher auxin content and explants are kept in light for regeneration of plants through embryogenesis.

8.7.4 Embryo Rescue

In breeding programmes, to increase the genetic variability in germplasm collections, and/or to introduce desirable genes from wild species in commercial cultivars, wide hybridization is often performed. However, embryos obtained by hybridization between related species or genera are often impeded by interspecific and intergeneric incompatibility. The embryo rescue technique is the most effective in vitro procedure utilized to save the embryos derived from distant crosses as it is found beneficial in overcoming the natural barriers to fertility. This is particularly about the post-fertilization barriers which end up either in the slow degradation of the embryos after reaching a certain stage of maturity or hindering in the initial developmental phases of the embryo. In case of pre-fertilization incompatibilities between the species, the genomes of both species may be combined using the somatic hybridization technique (Fig. 8.5). The embryo rescue technique is simple to use and has a high potential for saving desirable embryos that would otherwise degenerate.

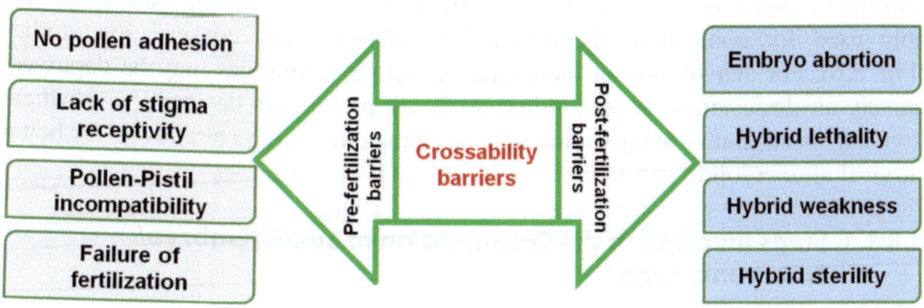

Fig. 8.5 Pre-and post-fertilization barriers during wide hybridization

8.8 Cryopreservation of Orthodox Seeds

Seeds are the best propagules to be used for the conservation of most of the agriculturally important crops including some horticultural crops due to ease in handling and regeneration. In general, it is much easier to process and conserve orthodox seeds because they tolerate desiccation without significant reduction in germination. These seeds are also tolerant to low-temperature storage as low as cryogenic temperatures (−170 to −180 °C) and the regeneration potential remains the same in most species. Contrary to this, the difficult-to-conserve seeds which fall in the category of non-orthodox (intermediate and recalcitrant) types mostly belonging to wet humid tropics and sub-tropical areas need special processing techniques for conservation. Therefore, in most of the genebanks worldwide, maximum conserved crop diversity includes seeds of orthodox types.

8.8.1 Cryopreservation of Non-orthodox Seeds

The non-orthodox seeds are sensitive to desiccation and freezing and based on their level of tolerance, they are classified as intermediate and recalcitrant types. The intermediate types of seeds (e.g. *Carica papaya*, *Citrus* spp., *Piper nigrum*, *Musa* spp.) are more tolerant to desiccation than the recalcitrant types (e.g. *Mangifera* spp., *Litchi chinensis*, *Artocarpus heterophyllus*, *Madhuca longifolia*, *Syzygium cumini*, *Garcinia* spp.). Depending on the desiccation tolerance, the type of storage, that is, short-, medium- and long-term conservation, can be recommended for a particular species. For the conservation of these difficult seeds, it is, therefore, important to record the change in germination percentage values with successive desiccation levels and identify the critical moisture content (CMC) for that particular species beyond which the germination percentage reduces significantly. In desiccation-tolerant seeds, the germination of seeds after desiccation and cryopreservation remains unchanged; however, for certain freezing-sensitive species, the germination reduces significantly and hence alternate cryopreservation strategy must be followed. In case of larger seed size, embryo or embryonic axes are preferably used for conducting the desiccation tolerance and conservation studies (Fig. 8.6). For dehydration of the explants, various techniques may be deployed, namely air desiccation (under laminar air flow), pre-growth desiccation, vitrification and encapsulation-dehydration to achieve the critical moisture levels for better survival after cryopreservation.

8.8.1.1 Steps Involved in the Conservation of Seeds/Embryo/Embryonic Axes

- **Harvest of fruits and seeds:** Whole fruits of the same maturity should be collected from the parent tree to ensure uniform quality. To develop a cryopreservation protocol in non-orthodox seeds ample number of seeds would be required; hence, an adequate number of fruits should be collected. Passport data should be recorded at the time of collection.

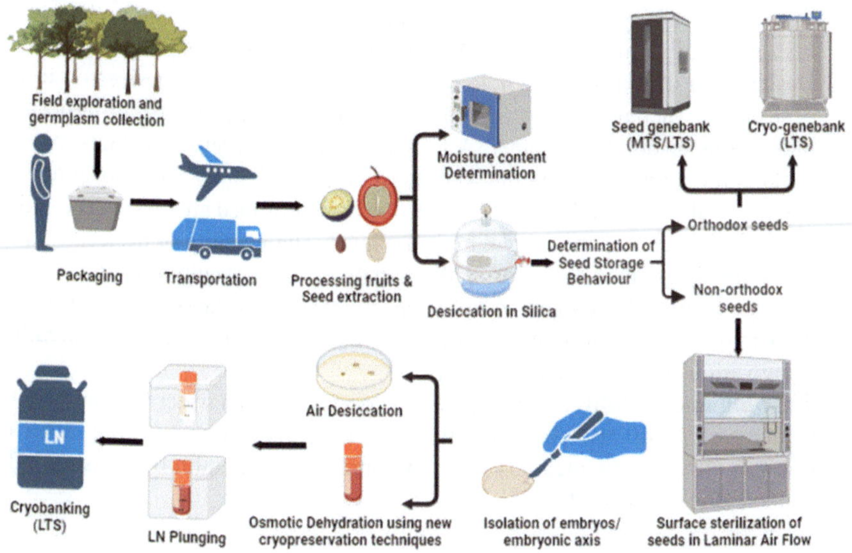

Fig. 8.6 Steps followed for processing seeds for medium- and long-term conservation

- **Post-harvest handling and transport:** As most of the non-orthodox seeds lose viability quickly, rapid transport of fruits instead of seeds is preferred. This process of seed extraction and experimentation should be carried out within a week. Delay in transport and handling adversely affects the viability. The soft and succulent fruits often decay fast; hence, seeds from half of the fruits may be extracted at the site of collection, coated with Bavistin powder, packed in moistened sawdust/charcoal/peatmoss and should be transported to reach the laboratory within 48 h of extraction.
- **Processing in the laboratory:** After recording the fruit characters, the developmental stage of the seed needs to be identified to understand the maturity level. The water content of fresh seeds is to be determined using a minimum of ten seeds. Depending on the initial data on seed size and moisture levels, the ideal explants (either seed/embryo/axis basis) were selected for processing. In species like coconut and oil palm, where large seeded fruits are produced, explants like, excised embryos/ plumules must be isolated aseptically and in litchi, jackfruit, *Madhuca* spp. and other similar types, embryonic axes prove to be the explants of choice.
- **Extraction and excision of explants in the laboratory:** For excision of embryo/embryonic axes, the seeds are surface sterilized under aseptic conditions using sodium hypochlorite (2–2.7%) for 8–10 min followed by washings with double distilled water three to four times. The concentration of the disinfectant and the duration of the treatment may be altered as required. The explants were then carefully excised using sterilized forceps and scalpels.
- **Determination of moisture content and air-desiccation:** Moisture content can be measured using the low constant or high constant temperature oven method.

In the case of very small explants like embryonic axes, buds and pollen, small vials made of aluminium foil can be used instead of glass weighing bottles.

- **Desiccation of explants:** Whole seeds and embryos can be desiccated in the desiccators/ vacuum desiccators filled with charged silica gel or air desiccated under laminar air flow cabinet. After each desiccation level, explants may be packed in the cryovials and plunged into LN after labelling. The moisture content and viability of embryonic axes need to be determined after each desiccation interval to determine the critical moisture level. The viability of desiccated axes should be determined by culturing them in the optimized culture media.
- **Storage in liquid nitrogen:** A sufficient number of explants must be used for immersion in liquid nitrogen for at least 24 h. The freezing rate can be modified using various methods and containers.
- **Thawing and viability:** Thawing is to be performed in a water bath at the temperature of 37–40 °C. The viability of the explants can be assessed either by using in vitro culture methods for embryonic axes and embryos or using ISTA methods for whole seeds.

8.8.2 Cryopreservation of Pollen

Among the various propagules for long-term conservation, pollen grains are conserved to mainly support breeding programmes. However, pollen (haploid male gamete) cryopreservation is the only option available to conserve a part of the gene pool in some of the important recalcitrant horticultural species, for example, *Mangifera* spp., *Litchi chinensis* where only conservation in field genebanks (FGBs) is possible. Pollen grains can be stored in viable conditions for a sufficiently long time and can be used at any time throughout the year in breeding programmes. Pollen banks are generally established as supplements to the genebanks where conservation of certain difficult recalcitrant seeded species is not possible.

8.8.2.1 Steps Involved in Conservation of Pollen

- **Pollen collection:** The collection of pollen is done depending on the pollen type, that is, dry or sticky pollen. For dry pollen, the flowers are collected at peak anthesis, and the pollen grains are dusted on butter paper. Pollen from other debris is sieved on clean butter paper. For sticky pollen, the anthers after dehiscence are suspended in an organic solvent (e.g. cyclohexane) for a few seconds and the solution is quickly passed through a filter paper through a syringe to obtain the pollen on the surface of the paper which can then be stored in the cryovial.
- **Desiccation of pollen grains:** Sometimes when the pollen has moisture of more than 10–12%; it needs to be desiccated before conservation in liquid nitrogen to avoid any injury due to ice crystal formation and recovery of maximum viable pollen after storage. For this, the collected samples of pollen are either left open

in the lab to dehydrate under normal conditions or they are kept inside a silica desiccator for quick dehydration.

- **Cryopreservation:** As soon as the pollen grains are collected/ after dehydration to optimal moisture content, they should be quickly plunged in liquid nitrogen by either packing in aluminium foils, gelatin capsules or cryovials.

8.8.3 Cryopreservation of Dormant Buds

In plant systems, Sakai (1960) provided one of the first studies showing that winter twigs of poplar (*Populus sieboldii*) and willow (*Salix koriyanagi*) could survive low temperatures of liquid nitrogen if cooled slowly followed by LN plunging. This technique follows protocols which induce a freeze-dehydration process using a slow freezing regime. This helps the vegetative apical meristematic tissues to dehydrate and survive the low-temperature stress during cryopreservation. Vegetative dormant buds in most of fruit crops have been utilized for the propagation of identical clones for centuries and hence could be ideal for long-term conservation of selected woody crops. This technique is being profusely utilized in the case of conservation of fruit trees, namely *Malus* spp., *Pyrus* and *Morus*. Besides having the advantage of clonal integrity, this method is relatively simple, space-efficient, amenable to long term conservation, and costs effective when compared to new cryopreservation techniques. Bud sticks of the desired length are first desiccated to the optimum moisture content at which its viability is intact and then cryopreserved followed by step-wise freezing.

8.8.3.1 Steps Involved in Conservation of Dormant Buds

- **Step wise freezing:** Nodal sections, pre-desiccated to optimal moisture contents, are kept in a refrigerator followed by sequentially lowering the temperature at −5 °C/day till the terminal temperature of −25 °C (holding for 48 h) followed by rapid immersion in LN.
- **Thawing and rehydration of buds:** This is the most critical step of recovery in dormant bud cryopreservation. The protocol of thawing must be standardized for each species. Mainly two types of thawing are followed, that is, slow thawing and rapid thawing. In slow thawing, the cryovials are simply removed from LN storage tank and placed in a refrigerator (4 °C) for 24 h before subjecting to rehydration, while in rapid thawing cryovials are suspended in lukewarm water (temperature at 37 °C) for about 20–30 min followed by rehydration.
- **Rehydration of buds:** Buds/nodal sections must be rehydrated in moist peat moss for 10–15 days at 4 °C temperature (in case of temperate species) or room temperature (for tropical/subtropical species).
- **Recovery of thawed buds:** For in vivo recovery, chip budding can be done on a suitable rootstock in the field and bud emergence and re-growth should be recorded. In the case of a test of in vitro recovery, bud pieces must be surface sterilized first followed by removal of four to seven outer scales before culturing on a suitable nutrient media.

8.9 Tissue Culture and Cryopreservation as Tools for Virus Eradication

Viruses are causing diseases to almost all the important vegetatively propagated crops and are responsible for major yield losses as well as product quality. Vegetatively propagated material may be infected with one or several viruses which can be detected by various techniques, namely double antibody sandwich—enzyme-linked immunosorbent assay (DAS-ELISA), immunosorbent electron microscopy (ISEM), phage display, lateral flow immune assay (LFIA), tissue blot immunoassay (TBIA), quartz crystal microbalance (QCM) immunosensors, PCR-based techniques (RT-PCR, multiplex PCR) and next-generation sequencing (NGS). Traditionally meristem culture, and heat-/thermotherapy were used for virus elimination and several scientific evidence are available. However, there are several bottlenecks in using traditional techniques, namely required technical expertise to isolate such as tiny meristem (0.1–0.3 mm), low regeneration capacity, expensive if combined with heat-/thermotherapy (required expensive equipment), some tissues are heat/chemosensitive and not all viruses are eliminated. Cryotherapy is a comparatively new technique used to eliminate viruses from infected plant tissue based on cryopreservation. Recently, silver nanoparticles have been successfully used to obtain virus-free (Grapevine fan leaf virus and grapevine leaf roll-associated virus-1) explants in combination with tissue culture technique in Asgari, Peykani and Shahani cultivar of grape (Dolatabadi et al., 2023). Following tissue culture techniques are used for virus elimination.

8.9.1 Meristem Culture

A localized region of actively dividing cells in plants, that is, tips of stems and roots, is known as meristem. Culturing of meristem in the culture media under aseptic condition to get plantlet is called meristem culture. The success of plantlet development and virus elimination is solely dependent on the size of meristem. In general, the apical meristems are free from pathogen infection or carry very low concentrations of virus (apex to downward) infection due to following reasons (1) Absence of vascular tissue in the meristems through which viruses readily move in the plant body. (2) Rapidly dividing meristematic cells with high metabolic activity do not allow viruses to multiply. (3) Virus replication is inhibited by a high concentration of endogenous auxin in shoot apices.

Meristem tip culture includes the following steps.

- Selection of suitable/healthy stock plant.
- Isolation of meristem tip: Explant of 0.1–0.3 mm (length) with 1 mm diameter is suitable for virus elimination. Meristem includes an apical dome along with 1–2 leaf primordia.
- Culturing of meristem in suitable media combinations.
- Development of plantlets: Shoot and root development.
- Transplanting of rooted plantlets.
- Virus indexing of every regenerated plant.

8.9.2 Cryotherapy

Cryotherapy can be defined as the eradication of plant viruses from infested plant material using cryopreservation techniques. Brison et al. 1997 first time successfully used a cryotherapy technique for the elimination of Plum pox virus from interspecific *Prunus* rootstock. Since the first report, cryotherapy has been effectively applied to eliminate the number of viroids, viruses and phytoplasmas in numerous economically important crop species growing in wide climatic variation from subtropical to temperate regions. Cryotherapy comprises numerous features compared to conventional technique (meristem culture), namely (1) cryotherapy of shoot tips has very high efficiency for virus eradication; (2) though the size of shoot tip has a major influence on the virus elimination frequency, size of shoot tip required in cryotherapy is much larger than needed in meristem culture; therefore, this technique avoids the difficulty in isolating extremely small meristem; (3) different cryopreservation techniques are equally effective in virus elimination; (4) time duration for the development of virus-free plant is almost equal to, or even lesser than meristem culture; (5) plantlet regeneration by cryotherapy is usually lower than that of meristem culture; however, it is adequate for yielding nuclear stock plants; (6) species/genotype-specific effect is common, which is also present in meristem culture for virus elimination. Cryotherapy comprises seven principal steps, namely (1) in vitro establishment of infested planting material, (2) shoot tip isolation, (3) cryoprotection of shoot tip, (4) liquid nitrogen immersion for at least 1 h, (5) thawing and regeneration in multiplication media, (6) ex vitro establishment of root plantlets in mist chamber and (7) virus indexing of each regenerant for specific virus.

8.10 Summary

In plant genetic resources management, tissue culture and cryopreservation are alternative methods for conserving the species which produce recalcitrant seeds or vegetatively propagated. In addition, they help to solve the problems of field genebanks. In vitro conservation is applied for short- to medium-term conservation under normal and slow growth conditions using tissue culture techniques and for long-term under suspended growth using cryopreservation techniques.

Although somaclonal variation in in vitro conservation is not desirable, however it provides a source of genetic variability for development of desirable traits. Somatic embryos have a greater potential for large-scale propagation and production of synthetic seeds. Synthetic seeds produce seedlings for several crop plants which do not produce seeds, or their seeds are not viable for a long time. Its techniques support pre-breeding, protoplast culture, anther/microspore culture, embryo rescue and a suitable target for gene transfer.

The technique of cryopreservation, may be applied in orthodox seeds, non-orthodox seeds, pollen, dormant buds, meristem etc for their long term conservation. Several techniques *viz.* slow freezing, vitrification, encapsulation-dehydration, encapsulation-vitrification, droplet-vitrification, V cryo-plate, D cryo-plate, vacuum infiltration vitrification and cryomesh, have been developed for cryopreservation of orthodox seeds of rare, endangered or threatened species, non-orthodox seeds, pollen, dormant buds, in vitro explant for their long term conservation.

Virus eradication before plant conservation is necessary for successful plant genetic resource management. Using tissue culture techniques for meristem culture and cryopreservation techniques for cryotherapy will help to eradicate viruses from in vitro conservation of plant material.

Multiple-Choice Questions

1. Non-orthodox seeds include
   (a) Mango
   (b) Jackfruit
   (c) Citrus
   (d) All of the above
2. Critical moisture content is
   (a) Moisture content of seed at time of harvesting
   (b) Moisture content below which seed germination decreases significantly
   (c) Moisture content below which seed germination decreases slightly
   (d) None of the above
3. In case of larger seeds, the ideal propagule for conservation is
   (a) Seed
   (b) Embryo
   (c) Embryonic axes
   (d) Both b & c
4. For long-term conservation of pollen in liquid nitrogen, organic solvents are used for pollen collection in
   (a) Dry pollen
   (b) Sticky pollen
   (c) Both a and b
   (d) None of the above
5. For long-term conservation of true-to-type genotype, the ideal technique is
   (a) Seed cryobanking
   (b) Pollen cryobanking
   (c) Dormant bud cryobanking
   (d) All of the above
6. For the gelling agent, agar is obtained from
   (a) Red algae
   (b) Green algae
   (c) Blue algae
   (d) All of the above
7. Which is the commonly used antioxidant to prevent browning due to wounding of explants in vitro?
   (a) Citric acid
   (b) Ascorbic acid
   (c) Activated charcoal
   (d) Polyamines

8. _______ is the naturally occurring auxin in plants.
   (a) 1-napthaleneacetic acid (NAA)
   (b) 2,4-Dichlorophenoxyacetic acid (2,4-D)
   (c) Indole 3-acetic acid (IAA)
   (d) 1-napthoxyacetic acid (NOA)
9. Use of _______ as gelling agent forms a clear gel and facilitates easy detection of any contamination.
   (a) Agar
   (b) Agarose
   (c) Phytagel
   (d) All of the above
10. High cytokinin and low auxin induces _______
    (a) Root formation
    (b) Shoot formation
    (c) Both a and b
    (d) None of the above
11. In vitro active genebank conserves germplasm for
    (a) Short-term conservation
    (b) Medium-term conservation
    (c) Short- to medium-term conservation
    (d) Long-term conservation
12. Physical growth limitations during in vitro conservation does not include
    (a) Reduced oxygen concentration
    (b) Osmotic adjustment
    (c) Growth regulator reduction
    (d) None of the above
13. Small culture vessels minimized the growth and development of plants by limiting
    (a) Gaseous exchange
    (b) Nutrient supply
    (c) Both a and b
    (d) None of the above
14. Reduced oxygen concentration during in vitro slow growth is achieved by
    (a) Mineral oil overlay
    (b) Low temperature
    (c) Low light
    (d) All of the above
15. Inclusion of _______ in the medium is used for in vitro organ formation.
    (a) High concentration of BA
    (b) High concentration of sucrose
    (c) High concentration of GA_3
    (d) High concentration of 2,4-D

16. Pusa Jai Kisan is
    (a) Somaclone
    (b) Hybrid
    (c) Bred through MAS
    (d) Selection from wild
17. Chances of somaclonal variations increase in
    (a) Shoot tip cultures
    (b) Axillary bud cultures
    (c) Meristem cultures
    (d) Callus cultures
18. Somaclonal variations are undesirable for germplasm conservation.
    (a) True
    (b) They do not affect any conservation strategy
    (c) False
    (d) Somaclonal variations and germplasm conservation are unrelated
19. Morphological markers are not preferred because
    (a) They are influenced by the environment.
    (b) They are limited in number.
    (c) They may not be associated with important economic traits.
    (d) All of the above.
20. Somatic embryogenesis is
    (a) Germ line cells developing into embryos
    (b) Non-germ line cells developing into embryos
    (c) Embryos developing from zygotes
    (d) Embryonic tissue becoming somatic
21. What is the boiling point (temperature) of liquid nitrogen at which long-term conservation is possible for non-orthodox and vegetatively propagated species?
    (a) −195.8 °C
    (b) −115 °C
    (c) +195.8 °C
    (d) +115 °C
22. Which of the following is the penetrating type of cryoprotectant?
    (a) Dimethyl sulfoxide
    (b) Polyethylene glycol
    (c) Sucrose
    (d) Mannose
23. ………… is a physical process, which encompasses the phase transition of a concentrated aqueous solution from the liquid phase into an amorphous glassy non-crystalline solid or glass.
    (a) Liquid nitrogen
    (b) Cryoprotection
    (c) Vitrification
    (d) Tissue culture

24. Which one is a more efficient technique for virus elimination?
    (a) Meristem culture
    (b) Thermotherapy
    (c) Cryotherapy
    (d) Chemotherapy
25. Which part of plant is used for cryotherapy?
    (a) Nodal segment
    (b) Leaf segment
    (c) Shoot tip
    (d) Stem segment
26. Cryoprotective agents work on the principle of
    (a) Osmotic dehydration
    (b) Vitrification
    (c) Totipotency
    (d) Meristem culture

Answers

1. (d)
2. (b)
3. (d)
4. (b)
5. (c)
6. (a)
7. (b)
8. (c)
9. (c)
10. (b)
11. (c)
12. (d)
13. (c)
14. (a)
15. (b)
16. (a)
17. (d)
18. (a)
19. (d)
20. (b)
21. (a)
22. (a)
23. (c)
24. (c)
25. (c)
26. (a)

Suggested Reading

Agrawal A, Sharma N, Gupta S, Bansal S, Srivastava V, Malhotra EV, Singh K et al (2021) Biotechnological applications for plant germplasm conservation at ICAR-National Bureau of Plant Genetic Resources, India–recent achievements. In: IX International scientific and practical conference on biotechnology as an instrument for plant biodiversity conservation 1339, pp 29–42

Bairu MW, Aremu AO, Van Staden J (2011) Somaclonal variation in plants: causes and detection methods. Plant Growth Regul 63:147–173

Bettoni JC, Costa MD, Gardin JPP, Kretzschmar AA, Pathirana R (2016) Cryotherapy: a new technique to obtain grapevine plants free of viruses. Revista Brasileira de Fruticultura 38:e-833

Bhojwani SS, Dantu PK (2013) Plant tissue culture: an introductory text. Springer, London

Biswas P, Kumar N (2023) Application of molecular markers for the assessment of genetic fidelity of in vitro raised plants: current status and future prospects. In: Kumar N (ed) Molecular marker techniques. Springer, Singapore

Brison M, de Boucaud MT, Pierronnet A, Dosba F (1997) Effect of cryopreservation on the sanitary state of a cv Prunus rootstock experimentally contaminated with Plum Pox Potyvirus. Plant Sci 123(1-2):189–196

Debergh PC, Maene J (1981) A scheme for commercial propagation of ornamental plants by tissue culture. Sci Horticult 14(4):335–345

Dolatabadi K, Davarynejad G, Safarnejad MR, Ghayoor Z (2023) Developing virus-free grapevine explants by using silver-nanoparticles and its comparison with chemo and thermotherapy-based approaches. J Crop Prot 12(1):15–27

Engelmann F (2004) Plant cryopreservation: progress and prospects. In Vitro Cellular Dev Biol-Plant 40:427–433

Evans DA, Sharp WR, Medina-Filho HP (1984) Somaclonal and gametoclonal variation. Am J Bot 71:759–774

Funnekotter B, Whiteley SE, Turner SR, Bunn E, Mancera RL (2015) Evaluation of the new vacuum infiltration vitrification (VIV) cryopreservation technique for native Australian plant shoot tips. CryoLetters 36(2):104–113

Grout BW (1999) Meristem-tip culture for propagation and virus elimination. Plant Cell Cult Protocols:115–125

Heinz DJ, Mee GWP (1971) Morphologic, cytogenetic, and enzymatic variation in Saccharum species hybrid clones derived from callus tissue. Am J Bot 58:257–262

Larkin PJ, Scowcroft WR (1981) Somaclonal variation – a novel source of variability from cell cultures for plant improvement. Theor Appl Genet 60:197–214

Morel G (1960) Producing virus-free cymbidiums. Am Orchid Soc 29:495–497

Murashige T (1978) Plant tissue culture: history, current status and prospects. Tissue Cult Assoc Rep 12:41–47

Nishizawa S, Sakai A, Amano Y, Matsuzawa T (1993) Cryopreservation of asparagus (*Asparagus officinalis*) embryogenic suspension cells and subsequent plant regeneration by vitrification. Plant Sci 91(1):67–73

Phillips RL, Kaeppler SM, Olhoft P (1994) Genetic instability of plant tissue cultures: Breakdown of normal controls. Proc Natl Acad Sci USA 91:5222–5226

Ramírez-Mosqueda MA (2022) Somatic embryogenesis: methods and protocols. Springer, New York

Sakai A (1956) Survival of plant tissue of super-low temperatures. Low Temp Sci B 14:17–23

Sakai A (1960) Survival of the twigs of woody plants at -196°C. Nature 185:393–394

Sakai A, Kobayashi S, Oiyama I (1990) Cryopreservation of nucellar cells of navel orange (Citrus sinensis Osb. var. brasiliensis Tanaka) by vitrification. Plant Cell Rep 9(1):30–33

Slack SA, Tufford LA (1995) Meristem culture for virus elimination. In: Plant cell, tissue and organ culture: fundamental methods. Springer, Berlin, pp 117–128

Wang MR, Chen L, Zhang Z, Blystad DR, Wang QC (2018) Cryotherapy: a novel method for virus eradication in economically important plant species. Plant Cell Cult Protocols:257–268

9 Germplasm Characterization and Evaluation

Jyoti Kumari, Kuldeep Tripathi, Sneha Adhikari, Priya Pardeshi, Lekshmy Sathee, Mahesh Kumar, Rinky Resma Panda, S. B. Suby, K. S. Hooda, Sebastian Raubach, Paul D. Shaw, and Raj Kumar Gautam

Learning objectives

- To learn about the characterization and evaluation of plant genetic resources
- To gain knowledge about types of descriptors
- To know about efficient germplasm utilization
- Development and importance of core and mini-core sets
- And related topics

The original version of the chapter has been revised. A correction to this chapter can be found at https://doi.org/10.1007/978-981-96-5024-8_18

J. Kumari · K. Tripathi · P. Pardeshi · K. S. Hooda · R. K. Gautam (✉)
Division of Germplasm Evaluation, ICAR-National Bureau of Plant Genetic Resources, New Delhi, India
e-mail: jyoti.kumari@icar.org.in; kuldeep.tripathi@icar.org.in; ks.hooda@icar.org.in

S. Adhikari
Division of Genetics, ICAR-Indian Agricultural Research Institute, New Delhi, India
e-mail: sneha.adhikari@icar.org.in

L. Sathee · M. Kumar
Division of Plant Physiology, ICAR-Indian Agricultural Research Institute, New Delhi, India

R. R. Panda
Graduate School, Division of Plant Genetic Resources, ICAR-Indian Agricultural Research Institute, New Delhi, India

S. B. Suby
ICAR-Indian Institute of Maize Research, Ludhiana, India
e-mail: sb.suby@icar.org.in

S. Raubach · P. D. Shaw
Department of Information and Computational Sciences, The James Hutton Institute, Dundee, UK
e-mail: sebastian.raubach@hutton.ac.uk; paul.shaw@hutton.ac.uk

K. Tripathi et al. (eds.), *Textbook of Plant Genetic Resources*,
https://doi.org/10.1007/978-981-96-5024-8_9

Keywords

Plant genetic resources · Germplasm · PGR · Characterization · Evaluation · Descriptors · Pre-breeding · Utilization · Biotic stress · Abiotic stress · Core set · Minicore · Genetic stocks · Genetic diversity · High throughput phenotyping

9.1 Introduction

Plant Genetic Resources (PGR) comprise a diverse range of collections, including landraces, farmers' varieties, breeding materials, genetic stocks, obsolete and modern varieties, as well as wild and weedy relatives of cultivated plants, mutants and potential domesticates such as wild species. The efficacy of PGR in enhancing crop improvement programs requires identifying promising accessions. Collected or introduced germplasm needs thorough characterization and evaluation to assess its potential variability and utility. After a comprehensive evaluation of germplasm, documentation of data on agronomic traits, quality and resilience to both biotic and abiotic stresses is warranted for increasing awareness and proper utilization. Additionally, molecular tools are employed to scrutinize the germplasm for novel traits, and underlying genes that could be exploited in the development of new varieties tailored to effectively meet global food requirements.

9.2 Understanding Genetic Diversity

Germplasm are essential and valuable resources for genetic diversity. When this genetic diversity is assessed and identified for specific beneficial traits, it becomes genetic resource ready to be used in breeding programs. Crop diversity on the other hand, refers to the differences among crop species, their varieties and individual plant characteristics. We usually think about crop diversity in three main ways: the different species, the different varieties within each species and the genetic differences within those varieties. To understand and estimate the genetic diversity within and between plant populations, different markers such as morphological, biochemical, or molecular markers are used.

9.2.1 Statistical Procedures to Measure Population Genetic Variation

One of the primary statistical methodologies employed in assessing population genetic variation is called 'Genetic Diversity Analysis.' This analytical approach aims to quantitatively evaluate the genetic variation both within and between populations. Several statistical techniques and metrics are conventionally utilized for this purpose.

F-statistics (F_{ST}): F_{ST} gives the extent of genetic differentiation among populations by comparing the genetic variation within populations to the total genetic variation across the entire sample. Its values, ranging from 0 to 1, reflect the degree of divergence between populations, with higher values indicating greater differentiation. F_{ST} equals 0 when the subpopulations are identical in allele frequencies, and 1 when they are fixed for different alleles. F_{ST} is a measure of genetic differentiation over subpopulations and is always positive.

Allelic Diversity: This metric estimates the diversity of alleles present at individual genetic loci within a population. Indicators such as allelic richness and heterozygosity are commonly employed to quantify this diversity, with higher values suggesting increased genetic variation.

Heterozygosity: Heterozygosity evaluates the frequency of heterozygotes within a population, wherein individuals possess two different alleles at a given genetic locus. Expected heterozygosity (He) and observed heterozygosity (Ho) are typically computed and compared to assess genetic diversity.

Genetic Distance Measures: Various distance measures, such as Nei's genetic distance, Euclidean distance, Jaccard or pairwise F_{ST}, quantitatively express the genetic divergence among populations by assessing allele frequencies.

Neighbour-Joining Trees: These phylogenetic trees are constructed based on genetic distance metrices, illustrating the evolutionary relationships among populations. Through depicting genetic similarities and differences, these trees offer insights into population genetic structures.

Principal Component Analysis (PCA): PCA represents a multivariate statistical technique utilized to visually depict and analyse patterns of genetic variation across populations. By reducing the dimensionality of genetic data while retaining its variability, PCA facilitates the identification of population structure and genetic relationships.

Population Structure Analysis: Techniques such as STRUCTURE and ADMIXTURE employ Bayesian clustering algorithms to infer population structure and assign individuals to distinct genetic groups based on their genotypic data.

Effective Population Size (Ne): Ne estimates the hypothetical size of an idealized population experiencing equivalent rates of genetic drift or inbreeding as the actual population under investigation. It provides valuable insights into the genetic robustness and evolutionary potential of populations.

9.2.2 Concept of Genetic Integrity, Genetic Shift, Genetic Drift and Optimum Environment

Genetic Integrity

It refers to the maintenance of the genetic diversity in terms of allelic diversity and genotype frequencies, and overall genetic structure within a population over time. It specifies the preservation of the natural genetic variation that exists within populations, which is crucial for their long-term viability and evolutionary potential.

Genetic integrity is essential for ensuring entire spectrum of genetic variation within a species or population for effective use of PGR in research and breeding program.

Genetic Shift

The genetic shift is the changes in the genetic composition of a population over time, which can occur due to various evolutionary processes like natural selection, mutation, genetic drift and non-random mating.

Genetic Drift

Genetic drift refers to the random fluctuations in allele frequencies within a population due to sampling error in finite populations. Over time, genetic drift can lead to changes in the genetic makeup of a population, particularly in small populations, and can result in the fixation of certain alleles or loss of genetic variation.

Optimum Environment

In population genetics, the concept of an 'optimum environment' typically refers to the ideal ecological conditions under which a population experiences maximal fitness and maintains optimal genetic diversity. Considering that the environment supports high genetic diversity and resilience to natural selection, imposes balanced selection pressure and minimizes the effect of genetic drift.

9.3 Crop Descriptors

Crop descriptors are essential for documenting and managing plant genetic resources. They represent the attributes, characteristics or measurable traits observed in genebank accessions, which aid in classifying, storing, retrieving, exchanging and utilizing data. Over time, descriptor lists have progressed from basic to comprehensive, intending to achieve international harmony. Descriptors marked with asterisks (*) are highly discriminating descriptors particularly important for assessing diversity. The development of descriptors reflects global knowledge and expertise, with collaboration among scientists from various organizations at both national and global levels ensuring the creation of comprehensive and standardized lists. Compatibility and common standards facilitate the efficient exchange of data.

9.3.1 Brief History of Crop Descriptors

Crop descriptors are vital for documenting plant genetic resources (PGRs). They became prominent with the establishment of the International Board for Plant Genetic Resources (IBPGR) in the 1970s, leading to the creation of the first descriptor list in 1977. Initially, these lists aimed to offer a minimum set of characteristics for crops like soybean. However, a major issue arose: many descriptors lacked universally accepted definitions and states, hindering data exchange between institutions. A brief history of descriptor development is depicted in Fig. 9.1.

Fig. 9.1 Brief history of descriptors

9.3.2 From Minimum to Comprehensive Descriptors

In 1990, descriptor lists evolved to become more comprehensive, covering all relevant traits for characterization and evaluation. Collaborative efforts with organizations like AVRDC and CIP led to the development of comprehensive lists such as the 'Descriptors for Sweet Potato' in 1991. These lists incorporated standard sections like site, environment and management, improving compatibility between documentation systems and facilitating information exchange.

9.3.3 Highly Discriminating Descriptors

In 1994, revisions were made to descriptor lists to accommodate curators' preferences for essential descriptors. The goal was to offer comprehensive lists while emphasizing a minimum set of highly discriminating descriptors, marked with asterisks (*). These descriptors effectively distinguished between accessions and played a key role in harmonizing international plant genetic resources documentation. Users, such as curators and researchers, could supplement these descriptors with additional ones based on their collection objectives.

9.3.4 Development of Crop Descriptors

Descriptors are required for proper resource management, accurate documentation, increasing access to diversity and efficient information exchange. The development of crop descriptors is a meticulous process involving extensive research on specific crops and collaboration among scientists worldwide. Key facilitators in this process are organizations like the Consultative Group on International Agricultural Research (CGIAR) and Bioversity International. These organizations act as international coordinators, ensuring that the global knowledge and expertise are reflected in the descriptor lists. Maintaining a standard format and seeking consensus among experts are crucial steps in descriptor development. When dealing with mandate crops of CGIAR, Bioversity seeks scientific advice from the relevant CGIAR centres. These centres possess the expertise required to create high-quality descriptor lists, owing to their work with diverse collections of specific crops.

9.3.5 Role of Bioversity International in Descriptor Development

Bioversity International assumes a pivotal role in the documentation of plant genetic resources. The international standing of descriptor lists is guided by collaborative efforts with both international and national organizations, which include esteemed entities such as the International Union for the Protection of New Varieties of Plants (UPOV), Organization Internationale de la vigne et du vin (OIV), the World Vegetable Centre (WVC), CGIAR centres, Instituto Nacional de Investigación Agropecuaria (INIA), French Agricultural Research Centre for International Development (CIRAD), Institut national de la recherche Agronomique (INRA) and numerous international universities and research organizations.

9.3.6 Concept of Descriptor Lists

Descriptor lists are the backbone of standardized documentation. When a species name is identified and listed along with its accession number, different attributes, characteristics, or traits are described using descriptors. The descriptor list aims to include information relevant to different genebank operations, from initial registration to characterization, evaluation and management of accessions.

Descriptor Elements

Each descriptor consists of three essential elements and their examples are mentioned in Box 9.1.

1. **Descriptor Name:** This is a descriptive, unambiguous and concise name for the attribute, characteristic or trait. It often comprises an object or item and an attribute name. It is crucial to ensure that technical terms are correct and widely accepted.
2. **Descriptor State:** A descriptor state defines a state of expression to characterize a particular trait. It represents the variation observed or measured for that descriptor. Each descriptor is associated with a numeric code for data recording and exchange.
3. **Descriptor Method**: This method outlines the procedures and conditions for measuring or scoring descriptors, ensuring accurate and consistent interpretation of results.

Box 9.1 Examples of Descriptor Name, States and Code

Descriptor name	Descriptor states	Descriptor code
Plant growth habit	Prostrate	1
	Spreading	2
	Semi-erect	3
	Erect	4
Leaflet shape	Linear	1
	Lanceolate	2
	Ovate-lanceolate	3
	Ovate	4
	Remarks	99
Flower colour	White	1
	White blue	2
	Blue	3

Reference Materials

Reference materials are essential for defining descriptor states. These materials include:

1. **Drawings:** Simple line drawings or pictures help users select states of expression, especially when access to standard references is limited.
2. **Check Cultivars and Standard References:** These provide objective baselines for measurements and comparisons and ensure consistency in the performance of accessions.
3. **Colour Charts:** In cases where descriptors involve colour descriptions, colour charts or reference standards are invaluable for precise and consistent scoring.

4. **Parameters:** Parameters are used to clarify descriptor states, especially when interpretation is challenging. They provide an objective baseline for observations.

Descriptor Method Elements

A descriptor method comprises three elements:

1. **Object:** It defines the specific part of the plant to be observed or measured. Clear measurement points are crucial for consistency.
2. **Condition:** Condition specifies the circumstances under which observations are to be recorded, including factors like plant growth stage, temperature and humidity.
3. **Sampling Procedure:** This indicates the number of samples used for observations and the method for sample selection, ensuring data accuracy.

9.3.7 Descriptors and Derived Standards

Compatibility in documentation systems is vital for effective data exchange. Three types of standards are used in this context:

1. **Crop Descriptors:** These provide internationally recognized guidelines for describing crop accessions. They standardize documentation and promote uniformity.
2. **Multi-Crop Passport Descriptors (MCPD):** Developed by FAO and IPGRI, these descriptors are used for passport data and enhance data exchange across multiple crops.
3. **Descriptors for Genetic Marker Technologies:** These standards are specific to genetic marker technologies and enable consistent data recording and exchange in genetic research.

9.3.8 Categories of Crop Descriptor

To streamline the maintenance, retrieval and updating of information, descriptors are organized into five primary categories as depicted in Fig. 9.2:

Thumb Rules for Crop Descriptor Use

To effectively use crop descriptors, several rules and guidelines are followed:

1. Data from the first four categories (passport, management, environment and site, characterization) should be available for every accession.
2. Number of descriptors selected in each category depends on the crop's characteristics and objectives.
3. Descriptor ordering and numbering should be followed closely, using recommended descriptor states.
4. Management descriptors can be defined by referring to lists of crops with similar maintenance requirements.

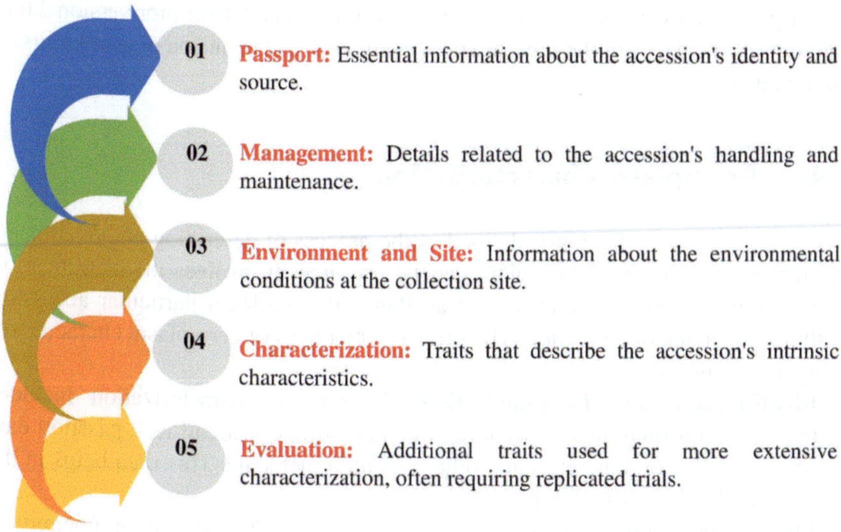

Fig. 9.2 Categories of crop descriptor

5. Characterization and evaluation descriptors are highly crop-specific and must be defined uniquely for each crop.

9.3.9 Descriptor 2.0

In the contemporary landscape, descriptor lists for crops and gene pools have evolved to be comprehensive, providing internationally recognized references for most known descriptors related to a particular crop or gene pool. However, the use of all listed descriptors is not mandatory. Curators can select those deemed relevant to their collections' objectives. For instance, a fruit tree collection aiming to represent ornamental diversity might emphasize different descriptors than a general diversity collection. Additionally, temperate and tropical species may employ different descriptors to describe environmental conditions during collection, regeneration or evaluation. Recognizing the importance of value-added traits and nutritional traits, there is a need of concerted effort to develop comprehensive and inclusive descriptor lists. It is crucial to have illustrative, self-explanatory and clear descriptors in these lists. For example, instead of simply categorizing nutritional content as low, medium and high, the specific ranges of these parameters should be provided. For protein content, this could mean defining low as less than 20%, medium as 20–25% and high as greater than 25%. Descriptor lists should be prepared with a view toward all possible uses of a particular crop. For example, a cowpea descriptor list should include traits related to its use as a pulse, vegetable and fodder. This holistic approach ensures that the descriptors are relevant and useful for various stakeholders, ranging from researchers and breeders to farmers and consumers.

Descriptors are dynamic in nature, so the need for updated descriptor version 2.0 is felt to meet the needs of the present era of evaluation and inclusion of novel traits of significance.

9.4 Germplasm Characterization

Germplasm characterization is defined as the process of describing and documenting the essential features of a plant genetic resource. It involves the identification and documentation of key traits and attributes that define a particular accession within a genebank or collection. The primary objectives of germplasm characterization are as follows:

1. **Identification and Documentation:** Germplasm characterization includes recording information such as accession names, unique identifiers, passport data (origin, collector, date etc.) and other essential metadata. This data helps in the precise identification of genetic resources.
2. **Phenotypic Traits:** Characterization involves the description of phenotypic traits, including morphological, agronomic and physiological characteristics. These traits provide valuable information about the appearance, growth and development of plant materials.
3. **Molecular Characterization:** With advancements in molecular biology, DNA-based methods are increasingly used for germplasm characterization. Molecular markers and genetic profiling techniques can reveal genetic diversity, relatedness and population structure within and among accessions.
4. **Disease Resistance and Tolerance:** Classification also includes evaluating accessions for resistance or tolerance to harmful pests and diseases.
5. **Nutritional and Quality Traits:** Evaluating nutritional content, flavour and other quality traits is essential, especially for crops with diverse culinary and end-use applications.

9.4.1 Methods of Germplasm Characterization

Several methods and tools are employed in germplasm characterization, depending on the nature of the plant's genetic resources and the objectives of the study:

1. **Field Observations:** It involves the direct observation of plants in their natural or cultivated environment. Researchers record characteristics such as plant height, leaf shape, flower colour, fruit size and other visible traits. Generally chracterization should be performed under non stress favorable environments to allow the manifestation of the expression of all traits.
2. **Laboratory Analysis:** Laboratory techniques include DNA fingerprinting, genotyping and biochemical assays to assess genetic diversity, disease resistance and nutritional traits.

3. **Phenotyping Platforms:** High-throughput phenotyping platforms, such as automated imaging systems and sensors, allow for the rapid and precise measurement of various plant traits under controlled conditions.
4. **Traditional Knowledge:** In some cases, local and indigenous knowledge can provide valuable information about the uses, adaptations and characteristics of specific plant genetic resources.

9.4.2 Germplasm Evaluation

Germplasm evaluation is the systematic assessment of plant genetic resources to determine their potential utility for breeding, research or other useful purposes. It involves a more comprehensive analysis of accessions, often extending beyond simple characterization. Evaluation of germplasm covers the whole range of activities starting from the receipt of the new samples by the curator and growing these for regeneration, seed increase, characterization, preliminary evaluation and also detailed evaluation and documentation. It requires systematic characterization and evaluation plans and techniques to understand various agro-morphological, physiological and quality traits including stress tolerance, insect pest and disease resistance. The detailed process of characterization and evaluation of germplasm is mentioned in Fig. 9.3. The basic differences between characterization and evaluation are mentioned in Box 9.2.

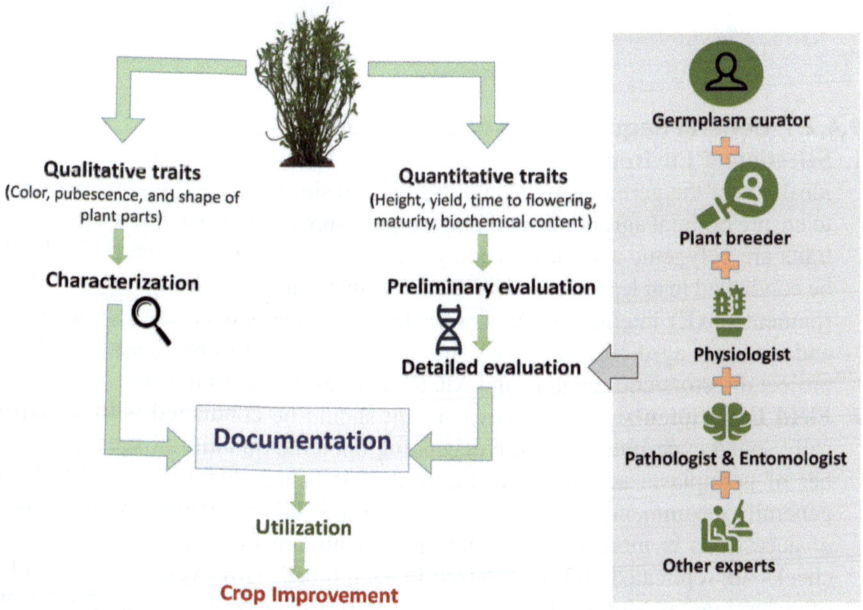

Fig. 9.3 Process of characterization and evaluation of germplasm

Box 9.2 Difference Between Characterization and Evaluation

	Characterization	Evaluation
Definition	Germplasm characterization is the recording of distinctly identifiable and heritable characteristics	Germplasm evaluation refers to the agronomic description of materials in a genebank, focusing on traits crucial for crop improvement as deemed significant and practically useful by breeders and researchers
Nature of traits	Usually qualitative traits, with monogenic or oligogenic inheritance, environmentally stable and easily measurable and usable for classification	Commonly, quantitative traits, which are governed by multiple genes, affected by environments and are relatively less easy to measure and utilize for classification
Descriptor traits	Colour, pubescence and shape of plant parts etc.	Traits such as plant height, yield and its components, time to flowering, maturity, biochemical quality traits such as, protein or starch, as well as tolerance to both biotic and abiotic stresses
Purpose	To describe accessions and establish distinguishable characteristics among the accessions	To identify potentially useful variability for further use in the genetic enhancement of crops
Environment and design	Characterization may be done in a single non stress environment with an augmented design	Evaluation process requires multi-location trials in replicated randomized block design for stability and estimating genotype x environment interaction

9.4.2.1 General Requirements of Evaluation

1. **Selection of Environment**: Evaluation should occur in a suitable environment, ideally near the germplasm collection site or in similar agro-climatic conditions to ensure optimal agronomic and phenotypic expression. Since many agronomic traits are polygenic and influenced by the environment, evaluation trials should be conducted in at least three diverse environments to minimize genotype x environment (GxE) interaction. A coordinated approach across multiple locations and different agro-climatic zones, in collaboration with crop-based institutes, project directors/coordinators and AICRP centres, is recommended.
2. **Field Experiments:** The field experiment should be conducted with a statistically sound experimental design depending upon the quantity of seed and number of germplasm accessions in the trial. Augmented block design (ABD) is generally recommended if the planting material is less in quantity or the number of accessions is more than 50. Here treatments are unreplicated whereas the checks are replicated and randomized in each block. However, for less number of accessions, randomized block design (RBD) may be followed. The blocks should be laid out across the soil fertility gradient of the experimental plot. The number of checks depends upon the crop and the parameters under study and is

representative of the type of germplasm. Three or more checks, in which one national as well as one locally adapted check, should be used for comparative assessment of germplasm.

The experimental plot should have at least three rows of 3–5 m length for each accession with a recommended gap between the accessions. For cross-pollinated species, it is recommended to have a greater number and longer rows compared to self-pollinated crops to accommodate adequate population size. Standard agronomic practices should be followed depending on the objectives of the experiments.

3. **Validation of Promising Accessions:** Following evaluation, promising accessions should undergo further validation via assessments conducted across multiple locations, seasons and years. Adhering to standard agronomic practices, including appropriate row spacing, fertilizer application, weeding, irrigation and plant protection measures, is crucial for cultivating a successful crop. Nowadays, advanced digital technologies such as drone-based phenotyping are increasingly utilized for accurate, precise and efficient data collection. Selection of hotspot locations/creation of artificial epiphytotic conditions for evaluation and validation of disease-resistant genotypes is essentially required. Similarly, controlled environmental condition/growth chamber/phenomics facility is required for evaluation and validation of abiotic stress tolerant genotypes.
4. Standard laboratory infrastructure facility and efficient non-destructive protocol for precise estimation of quality traits.
5. **Participatory Approaches:** Involving local communities and stakeholders in germplasm evaluation can provide valuable insights into the suitability and acceptance of specific accessions.

Knowledge of the scale of measurement is essential for both data recording and analysis when evaluating traits. Examples of descriptor scales and types are provided in Box 9.3.

Box 9.3 Scale of Measurement Data

Scale	Measurement	Typical descriptor
Ratio	Direct measurement, ratio may be estimated.	Height, yield components, days to flowering, maturity and harvest, protein contents
Interval	Differences or intervals can be estimated, but CV cannot be estimated	Temperature, date
Ordinal	Relative value assigned within a standard scale	Host plant resistance (measured by disease scoring from 1: highly tolerant to 9: highly susceptible)
Nominal/ Categorical	Multistate: Qualitative state assigned into arbitrarily number-scale and different categories	Colour, seed pattern and growth habit
Binary	Trait presence or absence	Pubescence, anthocyanin colouration

Significance of Germplasm Characterization and Evaluation

1. **Conservation:** Germplasm characterization and evaluation help to protect agricultural biodiversity, which is a prerequisite for long-term food security. Germplasm characterization ensures the accurate identification and documentation of plant genetic resources, facilitating their conservation for future generations.
2. **Genetic Resource Management:** It helps genebanks and repositories to manage their collections effectively, preventing duplication and ensuring the availability of diverse genetic materials through core set development.
3. **Crop Improvement:** Germplasm evaluation for the desirable qualities, yield and resilience traits results in promising donors for utilization in breeding programs.
4. **Resilience and Adaptation:** Understanding the response of different accessions toward stresses can help in developing resilient and adaptable cultivars.

9.5 Markers and their Use in PGR Characterization

Plant genetic resources (PGRs) are vital for food security, biodiversity and sustainable agriculture. Their effective utilization relies on characterization and conservation efforts. Markers play a crucial role in PGR management by enabling accurate identification, diversity assessment and efficient genetic conservation strategies. They help in assessing genetic variability, authenticate accessions, monitor changes over time and facilitate efficient utilization of genetic materials. This chapter explores the diverse applications and significance of markers in PGR characterization and evaluation.

9.5.1 Morphological Markers

Morphological markers are easily scorable plant traits such as leaf shape, flower colour and seed size. Characterization using morphological markers enables the identification and classification of different germplasm, facilitating the conservation and utilization of plant genetic resources. Morphological characterization is carried out by raising germplasm in a particular experimental design. It is easy to measure and does not require expensive technology. However, it has certain disadvantages such as less in number, environmental sensitivity and subjective characterization when compared to other methods of characterization. Different sets of morphological descriptors are taken into consideration for different groups of crop plants.

9.5.2 Biochemical Markers

Biochemical marker analysis reveals genetic diversity and identifies unique accession within a population of crop species, helping understand genetic relationships among germplasm accessions. Biochemical markers or traits may include various proximate compositions, micronutrients and other nutritional parameters, in addition to proteins and isozymes. This technique detects genetic differences and assesses the relatedness of PGR accessions, making it useful for diversity assessment in crop species.

Isozymes

Isozymes are variants of enzymes that catalyse the same biochemical reactions but differ in their genetic origin or structure. They are valuable biochemical markers used in PGR characterization. Isozyme analysis can reveal genetic diversity and assist in identifying unique alleles or variants within a population. For example, different varieties of a crop may exhibit distinct isozyme patterns, providing insights into their genetic relationships. However, isozyme polymorphism is often limited within cultivated species, restricting their full potential in genetic mapping for plant breeding.

Protein

Protein electrophoresis involves separating proteins based on their charge and size using an electric field. This technique allows for the visualization of protein variations among different accessions. By analysing protein profiles, researchers can detect genetic differences and assess the relatedness of PGR accessions. Protein electrophoresis is particularly useful for diversity assessment in crop species.

9.5.3 Cytological Markers

Cytological markers are related to chromosomes and nuclear characteristics such as, chromosome size, secondary constriction in chromosomes, position of centromere, arm ratio, constitutive heterochromatic patterns, banding characteristics (G, Q, R and N banding), total genomic chromosome length, chromosome volume and DNA content. However, cytological markers have limited applications for characterization due to their high cost, complex measurement, limited number and low resolution.

9.5.4 Molecular Markers

Molecular markers are DNA sequences or variations used to identify and distinguish individuals, accessions or populations within a species. They are stable, heritable and abundant across the genome, serving as valuable tools for genetic analysis and diversity. Various types of molecular markers are employed in PGR characterization and conservation.

Microsatellites (SSRs): Microsatellites are short, tandemly repeated DNA sequences that exhibit high levels of polymorphism. They are widely employed for assessing genetic diversity within and among populations.

Single nucleotide polymorphisms (SNPs): SNPs are single base-pair variations in the DNA sequence. They are abundant in the genome and serve as highly informative markers for genetic analysis.

Sequence-related amplified polymorphisms (SRAPs): SRAPs are PCR-based markers that amplify specific regions of the genome. They are valuable for assessing genetic diversity and population structure.

Random amplified polymorphic DNA (RAPD): RAPD markers are PCR-based markers that generate DNA fragments of varying lengths. They are used for fingerprinting and diversity analysis.

Genotyping-by-sequencing (GBS): GBS is a high-throughput genotyping method that combines DNA sequencing with SNP discovery. It allows researchers to analyse large portions of the genome, making it an efficient choice for PGR characterization. GBS is particularly useful when dealing with large and diverse collections of PGRs.

9.5.4.1 Applications of Molecular Markers

Molecular markers help scientists measure how much genetic diversity exists in PGR collections including several other applications as given below.

Genetic Diversity Assessment

Molecular markers are instrumental in assessing the genetic diversity present within PGR collections. By analysing the molecular profiles of different accessions, researchers can quantify the extent of diversity, identify unique genotypes and prioritize accessions for conservation and breeding programs.

Identification of Germplasm

Molecular markers allow for the clear identification of plant materials, reducing the risk of mislabelling or contamination in genebanks and seed repositories, and maintaining the integrity of accessions.

Phylogenetic Analysis

Molecular markers can uncover patterns of genetic differentiation, gene flow and historical relationships among populations of the same or different species through population structure and phylogenetic analysis.

Monitoring Genetic Erosion Molecular markers help track changes in genetic diversity and genetic erosion over time.

Development of Core Collections

Molecular markers can help to select core samples, ensuring that they capture the maximum genetic diversity with minimal redundancy. These core collections simplify management and enhance the accessibility of genetic resources to researchers and breeders.

Facilitating Ex situ Conservation
Molecular markers assist in optimizing the conservation of PGRs in genebanks and seed vaults. They are used in the prioritization of accessions for conservation and duplicate identification.

Conservation of Endangered Species
For rare or endangered plant species, molecular markers are crucial for identifying and conserving unique genetic variants. These markers guide conservationists in preserving genetic diversity within small populations, reducing the risk of extinction.

9.6 Principles and Practices of Germplasm Regeneration and Maintenance, Breeding Systems and Mode of Reproduction

Germplasm regeneration and maintenance are essential for conserving and managing plant genetic resources. Regeneration involves renewing germplasm accessions by growing a random sample of seeds under optimal conditions to ensure they retain the same characteristics as the original population. This process is performed for various purposes: initial seed increase for newly collected germplasm, long-term conservation, voucher samples in medium-term storage (MTS) and replenishing seed stocks in base and active collections. For clonal crops conserved in field genebanks, re-establishment may occur at the same site or a different site to ensure security or avoid diseases and pests.

9.6.1 Methods of Germplasm Regeneration

1. **Seed Regeneration:** The most common method involves growing plants from seeds, ensuring genetic stability, and obtaining a new set of seeds for storage and distribution.
2. **Vegetative Propagation:** For clonally propagated plants, regeneration involves taking cuttings and vegetative propagules in the field or using tissue culture techniques to create new individuals.
3. **In Vitro Regeneration:** For recalcitrant or difficult-to-regenerate species, in vitro techniques offer a controlled environment where plantlets can be generated from small tissue samples.
4. **Cryopreservation:** Cryogenic storage of plant tissues, especially for recalcitrant seeds or tissues, allows long-term conservation at ultra-low temperatures.

9.6.2 General Guiding Principles of Regeneration Method

Assuring Genetic Integrity

Germplasm regeneration poses great risks to the genetic integrity of germplasm, due to selection, out-crossing or mechanical mixing. The risk of losing genetic integrity during regeneration is higher in the case of out-crossing crops of a genetically heterogeneous nature. Regeneration aims to maintain the original genetic diversity and structure of the accession or collection as well as the allelic frequencies. The effectiveness of germplasm bank in collecting, conserving and regenerating PGR depends upon sampling procedures, effective population size, genetic drift and seed viability.

Assuring Efficiency and Quantity of Regenerated Material

The reproductive mechanism, rate of reproduction of the species and suitable regeneration site should be considered to produce sufficient seeds or planting materials required after regeneration.

Assuring the Quality of Regenerated Material

The regeneration process should ensure the production of sufficient quantities of healthy, viable seeds, tubers or other vegetative plant propagules. Before distributing germplasm, genebanks must screen it for the presence of seed-borne or vegetative-borne pathogens and pests to prevent the spread of diseases and pests.

General regeneration guidelines about the type of collection, regeneration time, sample size and site of regenerations are given in Sect. 7.3.

9.6.3 Breeding Methods of Crop

Open-Pollinated (OP) Populations

Open-pollinated plants are naturally pollinated by various pollinators, including wind, insects and other agents. These populations consist of a mixture of genotypes. These populations tend to be diverse, and they can be maintained over time through sibbing, bagging, caging, cluster bagging etc.

Pure Lines and Inbreds

Pure lines and inbreds are developed through successive generations of self-pollination and selfing in the case of cross-pollinated species, resulting in a population with highly uniform genotypes. Pure lines and inbreds are genetically stable and ideal for creating homozygous lines with consistent traits. They are often used in hybrid seed production.

Hybrid Varieties

Hybrid varieties are the outcome of crossing between two distinct, inbred pure lines, resulting in a hybrid with heterozygous vigour. Hybrids often exhibit improved

Table 9.1 Genetic structure and variability

Homozygous and homogeneous	Genetically similar and non-segregating population	• Pure line cultivars of autogamous crops (wheat, barley) • Inbred lines of allogamous crops (maize, sunflower and sorghum) • Population of autogamous wild and weedy plants
Homozygous and heterogeneous	Genetically dissimilar but non-segregating population	• Traditional cultivars of autogamous crops • Landraces of autogamous crops • Multiline variety of autogamous crops
Heterozygous and homogeneous	Genetically similar but segregating on selfing	• Specialized genetic stock • Single cross hybrids (F_1) • Asexually propagated derivative of sexually propagated, allogamous species • Vegetative propagated species (Sugarcane)
Heterozygous and heterogeneous	Genetically dissimilar but segregating population	• Allogamous cultivated species (maize, sunflower) • Wild and weedy allogamous crops • Traditional landraces of often cross-pollinated species

performance, but their seeds cannot be saved for replanting because they do not produce true-to-type progeny due to segregation in later generations.

Clonal Varieties
Clonal varieties are propagated through vegetative methods, such as cuttings or tissue culture, preserving the genetic makeup of the parent plant. Clonal varieties are genetically identical to the parent plant, ensuring uniformity in traits. They are common in crops like potatoes, grapes and bananas.

Understanding the breeding methods of crops, along with their genetic structure and variability (Table 9.1), is crucial to determining their uniformity and breeding behaviour.

9.7 Core and Minicore Collections

The large size of germplasm accessions of any crop species accumulated and conserved in genebanks globally poses challenges in the evaluation, conservation and utilization for crop improvement. Therefore, the development of a smaller, diverse and manageable subset of genebank collections, such as core, minicore, reference set and trait customized core set, is the essential requirement for efficient management of genebank collection.

Core Collections Core collections are smaller subsets of larger germplasm collections, aimed at efficiently capturing the maximum genetic diversity of a crop species making it more accessible for breeders, researchers and other users. A genebank core set is a smaller subset of a larger collection of genetic resources that are selected to represent the maximum genetic diversity of the entire collection with the minimum redundancy (Frankel and Brown 1984; Brown 1989). This core set aims to capture the genetic variability present in the entire set while being more manageable and cost-effective for conservation, characterization and utilization. A good core collection should have the following characteristics: representativeness, low redundancy, manageability, data completeness and usability. Those accessions not retained in the core collection are maintained as reserve collections to preserve rare alleles. Core collections are not necessarily static (Brown et al. 1987) and accessions may be added over time to represent new geographical areas or taxons, or with novel traits/genes. For the development of the core, various workers have suggested sampling proportions ranging from 5 to 30% (Brown 1989; Bisht et al. 1998; Hintum et al. 2000). Brown (1989) using the sampling theory of selectively neutral alleles, emphasized that the entries in a core subset should be ~10% of the total collection with a ceiling of 3000 per species. This level of sampling is effective in retaining 70% of the alleles of the entire collection.

Core sets can be developed using a combination of qualitative data, quantitative data and passport data and extended to include biochemical data and molecular data. Core collections have been developed in various crops conserved in genebank globally, for example: wheat, barley, lentil, linseed and chickpea at ICAR-NBPGR and all mandated crops at ICRISAT.

Mini Core Collections When the number of core collections is large, precise evaluation under replicated multilocations/environments would be costly and ultimately not be precise, reducing its utility in a breeding program. Therefore, a smaller subset of the core collection, known as a 'mini-core', is created to facilitate more practical management, and comprehensive evaluation for trait discovery and utilization. The mini core concept was postulated by Upadhyaya and Ortiz (2001) which is 10% of the core or 1% of entire germplasm accessions of the species and is often used for more focused breeding programs.

Reference Set A 'reference set' is a set of existing diverse germplasm collections that facilitates and enables access to existing crop diversity for specific desired traits, such as drought tolerance or resistance to disease or pests. This set is sampled based on preliminary evaluation followed by advanced evaluation and validation.

Trait Customized Core Set The loss of rare alleles, especially in plant and resistance breeding, is one of the main concerns when working with core collections (Odong et al. 2012). To overcome this, a subset called, trait-customized core collections aim to maximize diversity for that particular trait, which would be a better option for finding rare alleles than a core collection that is designed to represent the original population.

9.7.1 Principles for Formulating Core Collections

1. **Maximize Genetic Diversity:**
   Ensure the core collection represents the maximum genetic diversity present in the entire germplasm collection, considering genetic, morphological and phenotypic traits.
2. **Objective-Driven Selection:**
   Clearly define objectives for the core collection, such as capturing all genetic groups, focusing on specific traits, or representing different geographic regions.
3. **Determine Collection Size:**
   Balance the size of the core collection to manageability while capturing sufficient diversity. Smaller collections are easier to manage; larger collections may better represent diversity.
4. **Ensure Representative Samples:**
   Represent each genetic subgroup in the germplasm collection. Use clustering techniques, geographic information, and molecular markers to achieve a balanced representation.
5. **Sampling strategies:**
   The first step is to determine the desired size of the core collection. Then, accessions can be grouped or stratified based on their characteristics, such as passport data, phenotypic data and genetic distance. Next various sampling strategies, such as proportional (P), logarithmic (L), constant (C), square root (S) or genetic diversity-dependent allocation (G), are applied to determine the number of entries that are to be chosen from each group. Finally, one can either randomly select entries within each group or use a stepwise clustering approach to make the final selection.

9.7.2 Software for Core Set Development

PowerCore (Kim et al. 2007)**:** It was developed by the Rural Development Administration (RDA), in South Korea, and uses the advanced M (maximum) strategy with a heuristic search for establishing core sets.

CORE Hunter (Thachuk et al. 2009)**:** It is based on an optimization tool for selecting core subsets from germplasm collections. It uses criteria like maximizing genetic diversity and minimizing redundancy.

MSTRAT (Gouesnard et al. 2001): Implements sampling strategies to maximize genetic diversity, including hierarchical clustering.

GenoCore (Jeong et al. 2017)**: GenoCore is an advanced bioinformatics tool designed for the efficient selection of core collections based on large genotyping data.**

Other Statistical Softwares for Classification and Validation—SAS, SPSS, R.

9.7.3 Establishing a Core Collection

Creating a core collection involves a systematic procedure, which can be tailored to different situations. A detailed overview of the core collection establishment has been given in Fig. 9.4. Core set development follows different strategies as given below for extracting representative diverse sets from the entire collection

- ***Identifying the material to be represented*:** The selection of material depends on the crop species, the genebank's collection and the core collection's objectives. It may include wild relatives, landraces, modern cultivars, or specific trait-focused subsets.
- ***Collection of data***: Agro-morphological descriptors, utilization descriptors, biochemical traits, molecular data, taxonomic data and geographical passport data.
- ***Grouping of accessions***: Based on different factors such as plant taxonomy, geographical origin, phenotypic traits and genetic distance and clustering.
- ***Determination of sampling strategies:*** Sampling methods for core set selection can be divided into simple random sampling and stratification sampling. Various sampling strategies, such as maximization (M), proportional (P), logarithmic (L), constant (C), square root (S) or genetic diversity-dependent allocation (G),

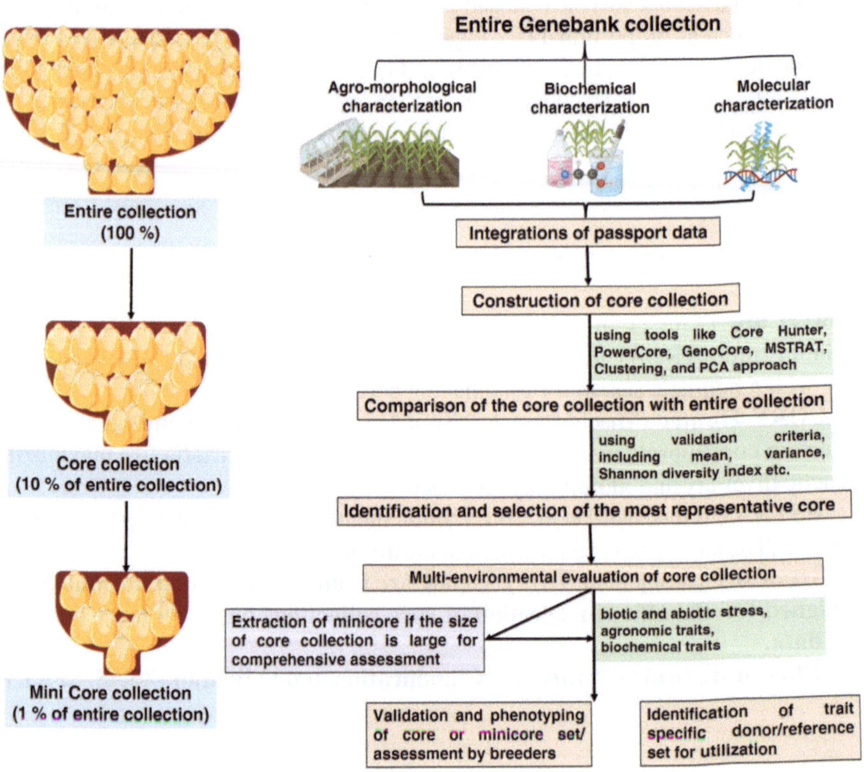

Fig. 9.4 Establishment of core collection for trait discovery

are applied to determine the number of entries from each group. Both stratified sampling method (wherein all accessions are first clustered into different groups and then accessions are drawn from each cluster following appropriate allocation methods) and non-stratified procedure (accessions are directly selected from the entire collection following the maximization approach) have been used to develop core collections.

- ***Sampling proportion***: A suitable sampling proportion is another critical factor in the process of constructing core collections. A high sampling ratio may result in the inclusion of redundant samples in the core collection, while a low sampling ratio may lead to the loss of key materials. In general, size of the core collection is restricted to 10% of the whole collection to represent the maximum possible allelic diversity, although it may retain 5–30% of accessions of the entire collection. While sampling, we may also exercise preferential selection for trait-specific unique germplasm or rare variants in the core collection.

9.7.4 Quality Evaluation of Core Collections

Core collections are compared with the entire set for representativeness and capturing a whole range of diversity. Core collections should undergo thorough validation to ensure their representativeness and utility.

Evaluation of genetic diversity: Compare the diversity captured in the core to that in the entire germplasm collection, using molecular markers/phenotypic traits. The Shannon-diversity index is most commonly used for the diversity estimation of any traits.

Statistical criteria: Core collections and entire collections are compared using statistical parameters such as mean, range, variance and coefficient of variation (%). For categorical data or qualitative traits, the Class coverage, Shannon-Weaver diversity index and Chi-square goodness of fit may be tested. Comparative analysis between the two collections for quantitative trait means can be done by Newman-Keuls test and t-test, variance homogeneity with Levene's test (Levene 1960), frequency distribution by Wilcoxon rank, sum non-parametric test and correlation by Mantel test.

Other evaluation parameters include the mean difference percentage (MD%), variance difference percentage (VD%), coincidence rate of range percentage (CR%) and variable rate percentage (VR%) are used to compare the mean and variance ratio between core and entire collections. The percentage of the significant difference between the core sets and the entire collection was calculated for the mean difference percentage (MD %) and the variance difference percentage (VD %) of traits. Coincidence rate (CR %) and variable rate (VR %) were designed to evaluate the properties of the core set against the entire collection. Their estimation formula is given below:

Mean Difference (MD%)—which is estimatedas: $MD(\%) = \frac{1}{m}\sum_{j=1}^{m} \frac{M_e - M_c}{M_c} \times 100$

where M_e = mean of entire collection; M_c = mean of core collection, and m = number of traits.

Variance Difference (VD%)—which is estimated as: $VD(\%) = \frac{1}{m}\sum_{j=1}^{m}\frac{V_e - V_c}{V_c}\times 100$

where V_e = variance of the entire collection, V_c = variance of the core collection, and m = number of traits.

Coincidence Rate (CR%)—which is estimated as: $CR(\%) = \frac{1}{m}\sum_{j=1}^{m}\frac{R_c}{R_e}\times 100$

where R_e = range of the entire collection, R_c = range of the core collection, and m = number of traits.

CR% indicates whether the distribution ranges of each variable in the core set are well represented.

Variable Rate of CV (VR%)—It is estimated as $VR(\%) = \frac{1}{m}\sum_{j=1}^{m}\frac{CV_c}{CV_e}\times 100$

where CV_e = coefficient of variation of the entire collection, CV_c = coefficient of variation of core collection and m = number of traits. VR% allows a comparison between the coefficient of variation values existing in the core collections and the entire collections and determines how well it is represented in the core sets.

A lower value in MD (<20%) and a higher value in VD, CR and VR (>80%) could be considered to indicate a more representative core collection. In addition, principal component analysis (PCA) plots have been widely used to compare the distribution characteristics between the core collection and the initial population. Moreover, correlation analysis is commonly conducted to infer whether the inherent relationship between traits in the original collection is well retained in the core group.

Agronomic Performance and User Feedback: Assess the core's performance in field trials to ensure that it maintains essential agronomic traits. Obtain feedback from researchers and breeders to understand if the core meets their needs.

9.7.5 Limitations of Core Set Development

Genebank core sets are valuable tools for managing genetic resources; however, they have certain limitations as mentioned below:

Loss of rare alleles and biasedness in selection: Core sets may exclude rare alleles, which may be an important trait for future breeding programs. Further, core sets might not fully capture the genetic diversity of the entire collection.

Dynamic nature: Germplasm collections are not static and number grows regularly with the addition and augmentation of new collections. Therefore, a core set established may become outdated, necessitating periodic evaluation of germplasm and updating the core set, which can be time and resource-intensive.

Sampling method: Different sampling methods can favour certain traits or places of origin, impacting the overall representativeness of the core set.

Traits used for phenotyping: Core sets development uses mostly the easily measurable traits; hence, difficult-to-score traits including root structure, and nutrition quality etc are neglected. Hence, development of trait-specific core sets and reference sets are also warranted.

Resource allocation: Developing and maintaining core sets require significant resources, including time, funding and expertise. Limited resources might hinder the creation or update of core sets, particularly for genebanks with a large number of collections.

Advanced software and data quality: Advanced core development tools and good quality data are essential to develop the most reliable and representative core sets. Future research should focus on choosing appropriate sampling strategies and software, integrating multiple data types.

9.8 Evaluation of PGR for Various Stresses

Stress in plants refers to any external factor that negatively impacts their well-being. This stress can arise from both abiotic and biotic sources (Box 9.4). Biotic stress, caused by living organisms like viruses, bacteria, fungi, nematodes, insects, arachnids and weeds, can lead to nutrient deprivation and hinder crucial physiological processes, ultimately reducing productivity. With over 8000 fungal species and approximately 14 bacterial genera known to induce plant diseases, viruses also pose a significant threat. Microorganisms can trigger plant wilt, leaf spots, root rot and seed damage, while insects can cause substantial physical harm to various plant parts and serve as vectors for spreading diseases.

Box 9.4 Different Types of Stresses

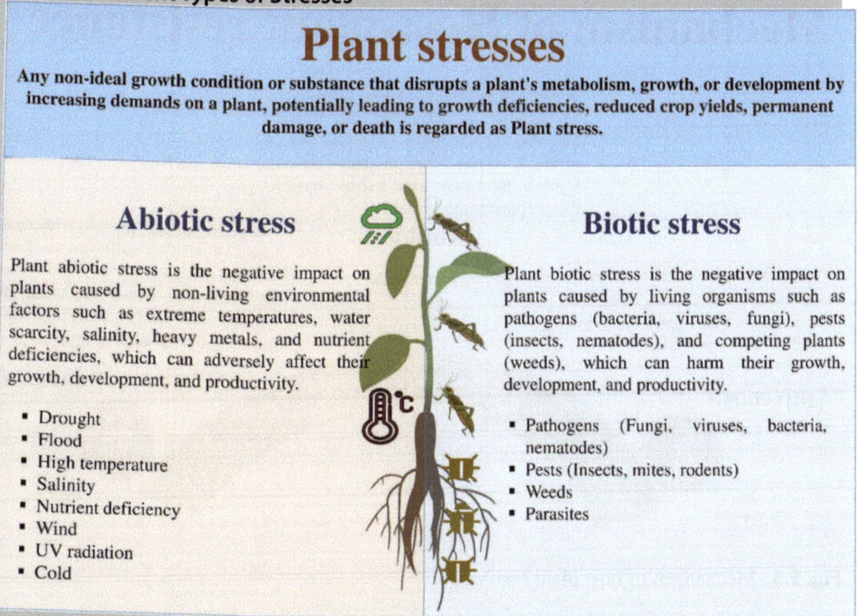

To combat stresses, plants have evolved intricate mechanisms to modify their physiology, metabolism, gene expression and developmental activities. For instance, in response to diseases, plants employ strategies such as establishing physical barriers like thick cuticles, leaf waxes and robust cell walls, producing toxic compounds for chemical defence, activating defence mechanisms through induced systemic resistance, recognizing pathogen-associated molecular patterns (PAMPs) via pattern recognition receptors (PRRs) and deploying resistance (R) genes that encode proteins capable of directly detecting pathogen effectors and eliciting defence responses. However, the effectiveness of these mechanisms varies among different plant species.

Moreover, plants employ various strategies to resist insect pests, including non-preference, antibiosis, tolerance and avoidance (Fig. 9.5 and Table 9.2). Researchers working with plant genetic resources should be aware of these resistance and tolerance mechanisms in crops as they seek to understand and address pest challenges effectively.

However, intensive modern breeding efforts in the crops for higher yields have led to the narrowing of the gene pool, thus increasing the genetic uniformity of crop varieties. Consequently, vulnerability to various biotic stresses is increased. Therefore, evaluation of germplasm for the development of promising cultivars with pest resistance is warranted.

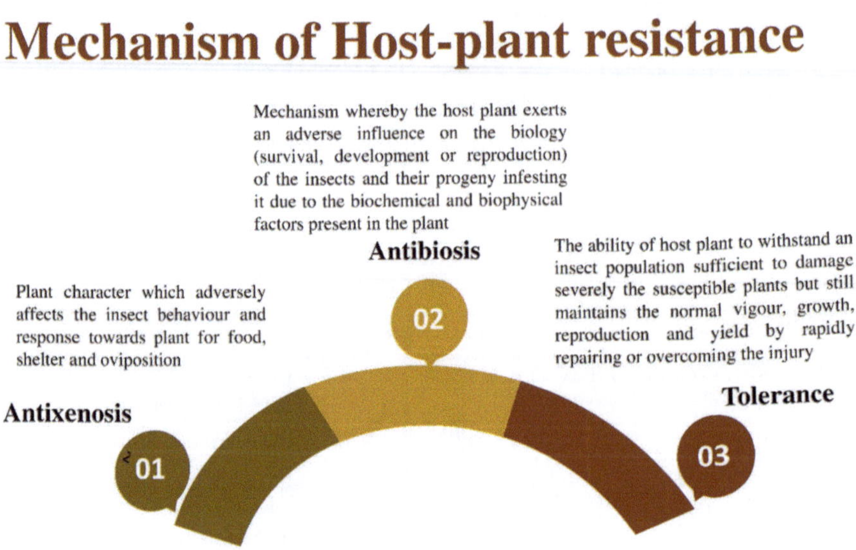

Fig. 9.5 Mechanism of host plant resistance

Table 9.2 Mechanism of resistance in different crop plants

Host crop	Pest	Mechanism of resistance
Insect pests:		
Wheat	Hessian fly	Antibiosis
Barley	Cereal leaf beetle	Tolerance and antibiosis
Rice	Rice stem borer	Antibiosis, non-preference, tolerance
	Brown plant hopper	Non-preference
Maize	European corn borer	Antibiosis
	Maize stem borer	Antibiosis
	Corn earworm	Non-preference
	Corn leaf aphid	Non-preference and antibiosis
Cotton	Bollworms	Antibiosis, non-preference and escape
	Boll weevil	Non-preference, escape
	Jassids	Non-preference
Brassica spp.	Cabbage aphid	Non-preference and antibiosis
Potato		Non-preference and antibiosis
Diseases:		
Tomato	*Pseudomonas syringae*	Incompatible response
Barberry	*Puccinia graminis*	The pathogen can infect young barberry leaves with thin cuticles; however, the germ tubes emerging from basidiospores do not penetrate thicker cuticles on mature leaves (*Adult plant resistance*).
Citrus fruits	*Xanthomonas campestris* pv. *citri*	Mandarins are resistant because their stomata are too small to allow entry of the bacterium whereas grapefruit is susceptible because of open stomata. (*Physical barrier*).
Onion	*Colletotrichum circinans*	Brown onion skins contain quinones, catechol and protocatechuic acid, which inhibit the germination of spores of pathogen whereas white onions do not produce these compounds and are therefore, susceptible (*Chemical barriers*).
Groundnut	*Cercospora arachidicola*	Early maturing varieties of groundnut escape early leaf spot infection (*Disease escape*).

9.8.1 Screening of Germplasm for Biotic Stresses

Germplasm is evaluated for resistance to biotic stresses using standard protocols. Initial screening is conducted in natural disease hotspots for one year. Promising accessions identified in this phase undergo detailed evaluation, including the identification of specific pests or pathogens. During advanced evaluation, artificial disease epiphytotic conditions are created, and inoculations are performed to ensure that the disease does not escape. This approach helps us understand the true resistance or tolerance response in plants. Additionally, efforts should be made to understand the number and nature of resistance genes and the molecular basis of resistance by employing genetic and genomic approaches under natural conditions, disease hotspots, or artificially created epiphytotic conditions.

9.8.1.1 Methods Used for Screening Germplasm against Diseases

(i) *Field trials:* A large number of accessions are grown in the presence of the pathogen and evaluated for disease symptoms using standard methodology.
(ii) *Greenhouse trials:* Plants are grown in controlled environments, inoculated with the pathogen, and evaluated for disease symptoms using standard methodology.
(iii) In vitro *tests:* Plant tissues are grown in laboratory conditions and challenged with the pathogen to evaluate resistance.
(iv) *Molecular markers:* DNA markers are used to identify plants with resistance genes, allowing for the rapid selection of resistant individuals. These molecular markers are gaining importance because of the lack of environmental influence on them.

The screening method relies on factors such as the specific pathogen, the target crop and available resources. Screening for disease resistance is an ongoing endeavour, as new pathogen races evolve and existing ones develop novel strategies to overcome plant defences. To document observations regarding fungal, bacterial and viral diseases, it is recommended to adhere to the standard evaluation system (SES) endorsed by CGIAR Institutes or All India Coordinated Crop Improvement Programs. For instance, The International Rice Research Institute (IRRI) in Los Baños, Laguna, Philippines, has developed the SES for rice, facilitating uniform evaluation practices for rice scientists worldwide. According to the SES for rice, a panicle blast is characterized by dark, necrotic lesions that partially or completely cover the panicle base (node) or the uppermost internode, or the lower part of the panicle axis. Infected panicles are assessed based on the given rating scale (Box 9.5). Rice landraces have been identified with blast resistance through SES for example: Blast: Tetep (Vietnam); Leaf blast: CM 76, Champe, Chirakey, Bada Atte, Akola, Jyothi, Birun, Jhapaca, Jeeram (India); Leaf and neck blast: Meghalaya lakang, Chingphourel, Manuikhamei, Kemenyakepeyu, Wainem, Thekruhla, Koyajang (North East Asia).

Box 9.5 Scale (Based on Symptoms) for Panicle Blast Evaluation in Rice

0	No visible lesions or observed lesions on only a limited pedicel
1	Lesions on numerous pedicels or secondary branches
3	Lesions on a few primary branches or the middle part of the panicle axis
5	Moderate lesions partially around the base (node) or the uppermost internode or the lower part of the panicle axis near the base
7	Complete lesions around the panicle base or uppermost internode or panicle axis near base with>30% of filled grains
9	Complete lesions around the panicle base or uppermost internode or the panicle axis near the base with <30% of filled grains

9.8.1.2 Methods Used for Screening Against Insect Pests

(i) *No-choice assays*: No-choice assays involve exposing test genotypes directly to constant insect pressure, such as introducing 6–10 neonates of stem borer or fall armyworm into the whorl of each maize plant. This method identifies genotypes highly resistant to specific insect pests and is generally used for antibiosis testing. No-choice assays can be conducted under protected or field conditions and may involve exposure to multiple pest species, as seen in the development of multi-borer resistant in maize genotypes screening.

(ii) *Multi-choice assays:* In multi-choice assays, the test genotypes are exposed to a fixed number of insects released in a common place, where the insects are free to choose among the genotype. This is used generally in antixenosis assessment and tested under enclosures, such as moths being released to assess preference for egg laying.

Selection criteria: The criteria that are used to select plants for insect resistance can vary depending on the objectives of the screening program. Some of the most common selection criteria include:

Level of insect damage: Plants that have a low level of insect damage suggest less preference and/or antibiosis are typically selected for further evaluation. Visual rating methods for stem borer resistance screening in maize typically involve the following steps:

Raising genotypes: The test genotypes are planted along with resistant and susceptible checks under a suitable statistical design. In maize, two rows of three to five-meter length are used.

Infestation and timing: Release 6–10 neonate larvae or 15–20 black-headed stage eggs into the whorl using a soft brush. Stem borers typically prefer 3–5 leaf stage plants (10–20 days after germination), so it is important to infest at the right stage to avoid overestimating resistance. Screening can be done under natural insect pressure if the location is a hotspot, but it takes at least three seasons for reliable results. This method is essential for pests that cannot be mass-reared, such as shoot fly resistance screening in sorghum and maize, where planting is aligned with the peak infestation period.

Visual assessment: Allow the infestation to progress for a specific period of 14–28 days, to allow the stem borer larvae to develop and cause damage to the plant. Visual assessments are usually done on the basis of damage caused by stem borers on a scale, often from 1 to 9 or in percentages. The visual scoring system categorizes germplasm into three categories such as 1–3: Little or no visible damage (Resistant); 4–6: Some damage, but the plant can recover and produce reasonably well (Moderately Resistant); 7–9: Severe damage, often resulting in a significant reduction in yield (Susceptible).

9.8.2 Evaluation for Abiotic Stresses

Abiotic stress in plants is usually defined as an adverse impact on the physiology of a plant, induced by a sudden change from optimal environmental conditions. In most cases, stress is measured in terms of plant survival, crop productivity or the primary assimilation processes, which are correlated to overall growth. The primary abiotic environmental parameters that affect plant growth are light, water, carbon dioxide, oxygen, soil nutrient content and availability, temperature, salts and heavy metals.

Responses to abiotic stress depend on the extremity and time duration of the stress, developmental stage, tissue type and interactions between multiple stresses. Exposure to stresses typically promotes alterations in gene expression and metabolism, and reactions are frequently centred on altered patterns of secondary metabolites.

Types of Abiotic Stresses

Depending on the kind of causative agent involved, abiotic stresses may be divided into physical or chemical types of stress. Physical stress includes salinity (due to its osmotic component), water scarcity, temperature extremes, excessive or insufficient irradiation, anaerobiosis caused by waterlogging or flooding and mechanical stress caused by wind or excessive soil compaction. Chemical stress can be due to salinity (due to its ionic or toxic components), a lack of mineral nutrients and environmental contaminants.

Effect of Abiotic Stresses

Drought stress: Reduces water availability, causes stomatal closure, decreases photosynthesis and alters hormone levels.

Salinity stress: Disrupts ionic balance and osmotic pressure, impairs nutrient uptake and leads to ionic toxicity.

Extreme temperatures: Denature proteins, disrupt cellular functions and affect membrane fluidity and metabolic processes.

Cold stress: Affects membrane fluidity, disrupts metabolic processes and impairs photosynthesis.

Heavy metals: Induce oxidative stress, damage cellular components through reactive oxygen species and interfere with enzyme activities.

In addition to the particular damage caused by each of these stresses, they all facilitate the overproduction of reactive oxygen species (ROS) such as superoxide anions, hydrogen peroxide or hydroxyl radicals associated with oxidative stress. Here, we discuss phenotyping for salt, waterlogging, heat and drought stress.

9.8.2.1 Salt Tolerance

Salinity refers to the detrimental effects caused by the presence of salts in the soil. Under extreme salt stress conditions, plants may die. However, moderate-to-low salt stress levels can hinder plant growth rates and lead to various symptoms. These symptoms may include morphological, physiological, or biochemical alterations in the plants.

Types of salt-affected soils: There are three main types of salt-affected soils: saline, sodic (or alkali) soils and saline-sodic (Box 9.6).

Box 9.6 Characteristics of Different Salt Affected Soils

Salt-affected type	Electrical conductivity EC (dSm^{-1})	Exchangeable sodium percentage ESP (%)	Sodium absorption ratio SAR	pH
Saline	>4	<15	<13	<8.5
Sodic	<4	>15	>13	8.5–10.5
Saline—Sodic	>4	>15	Variable	>8.5

- Saline soils: These are characterized by high levels of sodium cations, with an electrical conductivity (EC) exceeding 4 dSm^{-1}. The exchangeable sodium percentage (ESP) is typically below 15, and the pH of saline soils is relatively lower compared to sodic soils.
- Sodic or alkaline soils: These soils contain carbonate, bicarbonate and sodium ions. They have a high pH, usually exceeding 8.5 and a high ESP, typically above 15.
- Saline sodic soils: These soils contain large amounts of total soluble salts and greater than 15% exchangeable sodium. The pH is generally less than 8.5.

Effect of Salt Stresses on Plants

- **Leaf chlorosis**: Yellowing of leaves due to nutrient imbalances and reduced chlorophyll production.
- **Leaf necrosis**: Browning and death of leaf tissue, particularly at the tips and margins.
- **Stunted growth**: Reduced plant height and overall biomass due to inhibited cell division and elongation.
- **Leaf abscission**: Premature dropping of leaves as a stress response.
- **Wilting**: Loss of turgor pressure leading to drooping leaves.
- **Reduced yield**: Lower reproductive success due to impaired metabolic processes resulted in a decrease in tillering, florets per panicle, grain weight and ultimately yield.
- **Root damage**: Shortened and thickened roots with fewer lateral roots, limiting water and nutrient uptake.
- **Ion toxicity**: Accumulation of toxic ions like sodium (Na+) and chloride (Cl-) in plant tissues.

Mechanism of Salt Tolerance

Most crop plants are sensitive to salt during seedling and flowering stages. To cope, crops may accelerate maturity, but this can reduce yield. Traditional tall rice and wheat varieties exhibit high seedling vigour, helping tolerate growth reduction and dilute accumulated salt. The following are the mechanisms of salt tolerance in plants:

- **Ion homeostasis:** Plants use transporters like SOS1, HKT1 and NHX to regulate ion uptake and compartmentalization, maintaining low cytosolic sodium levels.

Na+/H+ antiporters sequester excess sodium into vacuoles, reducing its cytotoxic effects in the cytoplasm.

- **Osmotic adjustment:** Accumulation of osmolytes such as proline, glycine betaine and sugars to maintain cell turgor and enzyme function under osmotic stress.
- **Antioxidant defence through Reactive Oxygen Species (ROS) Scavenging**: Enhanced production of antioxidants like superoxide dismutase, catalase and glutathione peroxidase neutralize ROS generated under salt stress.
- **Hormonal regulation:** Increases in ABA levels promote stomatal closure to reduce water loss and activate stress-responsive genes. Ethylene and cytokinins modulate root architecture and enhance stress tolerance.
- **Gene expression and signal transduction:** Activation of transcription factors such as DREB, MYB and NAC, which regulate genes involved in ion transport, osmoprotection and antioxidant defence.
- **Salt exclusion:** Mechanisms that limit the uptake of sodium at the root level, reducing its translocation to shoots and leaves. Variation in sodium ion transport from the rhizosphere is a key comparison for identifying salt-tolerant plants. Salt tolerance is profoundly associated with the ability to restrict Na^+ ion uptake while maintaining the uptake of K^+. Often the tolerant varieties maintained lower Na concentrations besides maintaining K concentrations under high salt concentrations. For example, salt-tolerant rice genotypes like CSR1 display significant control over the partitioning of salt within plant parts as the reproductive organs are excluded efficiently. Tissue tolerance to salt stress is measured as the concentration of leaf sodium, which results in a 50% loss of leaf chlorophyll content (Yeo and Flowers 1983) also designated as LC50.
- **Ion sequestration:** Active transport of excess ions into vacuoles to prevent ionic toxicity in the cytoplasm.
- **Transport:** Salt-tolerant rice genotypes limit salt entry into flag leaves and panicles. Older leaves and culms near senescence store toxic salts, salvaging actively growing regions.

Scoring System of Salt Tolerance

High germination percentage and maintenance of coleoptile and radicle length after salt treatment are an indicator of salt tolerance. In the seedling stage, plant survival or mortality is a good selection criterion. The tip burning of seedling leaves, spikelet sterility and retarded growth are major indicators of salt stress. The comparative reduction in yield is a major criterion for the selection/rejection of the genotypes. IRRI (1988) has developed the Standard Evaluation System for salinity and sodicity. The genotypes are scored from 1 to 9 wherein the lower score of 1 states tolerant and the higher score indicates sensitivity (Box 9.7).

Box 9.7 Visual Scoring for Salt Tolerance in Hydroponically Grown Plants at the Seedling Stage (Adapted from IRRI 1988)

Scale	Symptom
1	Normal growth
3	Nearly normal growth, some discolouration of leaves
5	Majority of leaves are discoloured/whitish
7	Complete growth arrest, majority of leaves dry
9	Almost all plants are dead or drying

9.8.2.1.1 Screening for Salt Tolerance

- **Field screening**: Field screening helps to expose genotypes to salt stress at varying soil stress conditions. The layout of field screening is usually augmented block design by including repeated checks in each block for a large number of accessions.
- **Microplots**: Screening in microplots helps in avoiding the issues of soil heterogeneity and spatial variability. In micro plots within concrete boundaries, naturally occurring saline soil is filled and used for the evaluation of genotypes. Microplots offer the advantage of simulation of fields like salinity level and are generally used to screen mapping populations or confirmation of tolerant germplasm.
- **Pot screening**: Pot screening offers further control over variations and soil heterogeneity and allows periodic destructive sampling of plants for root studies.
- **Hydroponic screening**: Hydroponic screening offers the advantage of handling a large number of lines at a time. A hydroponic solution suitable for the crop under consideration with periodic maintenance of optimum pH is required. For rice, Yoshida medium is used. Approximately 3 and 6 g/L NaCl is needed to raise the salinity to 6 and 12 dSm^{-1} respectively. Surface-sterilized seeds are kept in petri dishes lined with moistened filter paper for 48 h to germinate. Adequate temperature for the crop needs to be maintained during this period. Pregerminated seedlings are transplanted to hydroponic support system and after 3–4 days of establishment, the salinized nutrient solution is provided. Half the level of required salinity is maintained initially and slowly salinity level is increased to the desired level by adding NaCl. Sensitive and tolerant check genotypes are planted in each tray for comparative evaluation (e.g. for rice, IR29 is a sensitive check and FL478 is a tolerant check). Genotypes are scored for visual symptoms after 15–20 days of treatment.

9.8.2.2 Waterlogging Tolerance

Among the different abiotic stresses, complete submergence and waterlogging are constraints for crop production around the world. The majority of crop plants are strictly aerobic for optimal growth and yield. Crops like rice develop aerenchyma and adapt to low oxygen conditions while other crops like maize do not possess such a system for gas exchange from aerated parts to submerged roots. Hence, the submerged roots progressively succumb to low oxygen stress, nutrient imbalance and, eventually, irreversible plant damage and yield losses. Water logging at the pre-emergence stage, early seedling stage, at vegetative growth stage and or flowering/grain filling stage can be detrimental to crop fitness in different crops at varying magnitudes.

The crop and stage-specific phenotyping is required for identifying tolerant germplasm. The selection of soil type according to the target soil environment, and the selection of genotypes according to crop duration are also essential to generate useful phenotyping data. In the waterlogging phenotyping trials, maintenance of uniform stress intensity and duration under field conditions are difficult. For evaluating a large number of lines, pot screening with uniform stress intensity is advantageous. Lines selected from pot screening can then be evaluated in the field in smaller plots for confirmation and validation.

Mechanism for Waterlogging Tolerance

When there is an abundance of water surrounding the plants, they will wilt due to oxygen deprivation, which affects nutrient and water uptake and is the primary cause of damage during waterlogging. When there is not enough oxygen, the energy metabolism switches from aerobic to anaerobic mode. The higher soluble sugar availability, enhanced glycolytic pathway activity, higher activity of fermentation enzymes helps in survival. Waterlogging stress curtails photosynthesis due to stomatal closure along with the accumulation of abscisic acid (ABA) and gaseous plant hormone ethylene. The stomata closure induced by waterlogging limits CO_2 levels and drives the accumulation of reactive oxygen species (ROS). The ROS induces severe cellular damage to membranes and organelles.

9.8.2.2.1 Traits for Measurement of Water Stress Tolerance

Seed germination and days to emergence are important traits to determine pre-emergence waterlogging of sensitive crops. Genotypes with high germination percentage (number of coleoptiles emerged*100/number of seeds sown) are selected. The count of coleoptiles is recorded daily from the 4th day up to the 15th day after planting in maize (Zaidi et al. 2016). Traits such as days to 90% of emergence, seedling biomass, seedling mortality ratio (number of dead seedlings*100/total number of seedlings), root/stem lodging, days to 50% flowering and grain yield per plant. In maize, additional parameters like the number of surface roots, number of brace roots and anthesis to silking interval are also considered. Tissue tolerance to waterlogging can be assessed by the measurement of RWC, MSI and antioxidant enzyme(s) activities (Sairam et al. 2008).

9.8.2.3 Heat Tolerance

Heat stress can be defined as exposure to temperatures (maximum and/or minimum) above a threshold level for a period of time that causes irreversible damage to crop growth and development and is a function of the intensity, duration and rate of increase in temperature. For different crops, the optimal temperatures for growth can vary between day and night, as well as over the entire growing season; for example, in maize, the daytime, optimal temperatures range between 25 and 33°C, while the optimal night temperatures range between 17 and 23°C. There are different mechanisms of heat stress tolerance in plants listed in Box 9.8.

Box 9.8 Different Mechanisms of Heat Stress Tolerance in Plants

Mechanism	Characters
Heat avoidance Plants use this mechanism to reduce the harm from high temperatures	**Transpirational cooling:** Transpiration reduces canopy temperature **Leaf orientation:** Orientation of leaves to minimize exposure to direct sunlight **Refractive properties:** Absorption of light radiation by leaves is influenced by leaf characteristics such as pubescence and glaucousness
Heat tolerance	**Membrane stability:** Certain lipids and protein compounds maintain the integrity of cell membranes. **Osmotic adjustment:** Solutes or osmoprotectants like proline and glycine betaine maintain cell turgor and hydration under heat stress **Stem reserve mobilization**
Photosynthetic enzyme stability/pigment	Some heat-resistant plants have evolved stable RuBisCO enzymes and specific photosynthetic pigments that are more heat resistant

Effect of Heat Stress

1. **Physiological effects:** High temperatures affect processes like respiration and photosynthesis, reducing the plant's ability to capture light and convert it into energy. Heat stress also increases water loss through transpiration.
2. **Cellular damage:** High temperatures can damage cellular structures, including proteins and membranes, compromising overall plant health.
3. **Oxidative stress:** Heat stress triggers the production of reactive oxygen species (ROS) in plant cells, which can damage cellular structures.
4. **Impact on plant growth and yield:** High temperatures during critical stages like flowering and grain filling can significantly reduce yields and quality.
5. **Adaptations and tolerance:** Some plants have developed adaptive mechanisms, such as changes in leaf anatomy and increased production of heat shock proteins, to cope with heat stress.

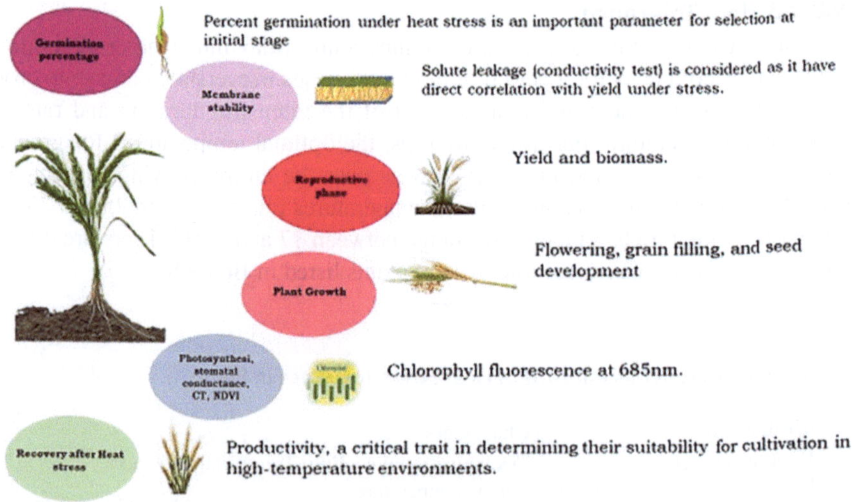

Fig. 9.6 Selection criteria for heat resistance

Screening for Heat Resistance

Field environment: By growing germplasm accessions under different dates of sowing so as to expose plants to high/low temperatures. It is the most convenient approach but its effectiveness depends upon the repeatability of heat over the years. Also, we cannot maintain the critical temperature required, as it depends on weather conditions.

Heat-prone environment: Certain locations where heat stress appears more frequently based on climatic analogue data are used for screening purposes.

Controlled environment: A greenhouse environment or growth chamber with controlled temperature, humidity and irrigation facility provides more accurate results.

Selection criteria for heat resistance: Different criteria for selecting heat stress-tolerant plants are depicted in Fig. 9.6.

9.8.2.4 Drought Tolerance

In agriculture, the term drought refers to a meteorological condition in which the amount of water available through rainfall and/or irrigation is insufficient to meet the crop needs for optimal growth and development; this eventually affects overall productivity'. Drought in plants refers to a situation in which the available water supply is inadequate to meet the requirements of the plants, which can result in various physiological and morphological changes. To develop drought-tolerant genotypes, precise phenotypic data at different environmental conditions are required. The foremost trait of interest is high yield potential. A genotype with good inherent yield potential often performs well under drought conditions.

9.8.2.4.1 Plant Drought Resistance Mechanism

I. **Drought escape**

Early maturing varieties demonstrate resistance by avoiding the drought period. This mechanism is predominantly observed in late-season drought stress. Early varieties exhibit a low evaporation rate, possess a reduced leaf area index, and have lower yield potential, however, avoid the drought through fast maturity.

II. **Dehydration avoidance and tolerance**

Plants employ dehydration avoidance to maintain water levels under stress conditions, protecting various physiological, biochemical and metabolic processes. Such plants are also called water savers. Water turgor potential under stress is commonly measured, and plants exhibit reduced transpirational losses. Mechanisms involved in drought stress are given in Box 9.9 and Fig. 9.7.

Box 9.9 Different Mechanisms of Drought Stress Tolerance

Mechanism	Characteristics
Dehydration avoidance	
Leaf characteristics	Leaf pubescence reduces leaf temperature and leaf angle affects net radiation, favouring erect leaves over horizontal ones. Leaf rolling is a common feature observed in drought-tolerant crops
Reduced transpiration through cuticle and stomatal regulation	A waxy layer on the surface of leaves and stems, helps reduce water loss through transpiration. Stomatal closure is another response to water scarcity, minimizing water vapour loss from the leaf surface
Osmotic adjustment	Plants can adjust the concentration of solutes within their cells to regulate water movement
ABA (abscisic acid)	Under stress conditions, there is an increase in abscisic acid concentration synthesized in the roots of stressed plants. ABA reduces leaf expansion and promotes root growth
Root architecture	Plants with deeper and well-branched root systems can explore a larger soil volume, accessing water stored at greater depths
Dehydration tolerance	
Cellular protective mechanisms	Some plants produce specific proteins and compounds that protect cellular structures from damage during dehydration
Accumulation of osmoprotectants	Osmoprotectants, such as sugars and amino acids, help maintain cell structure and function during dehydration
Efficient stem reserve mobilization	ABA accumulation triggers the stem reserve mobilization

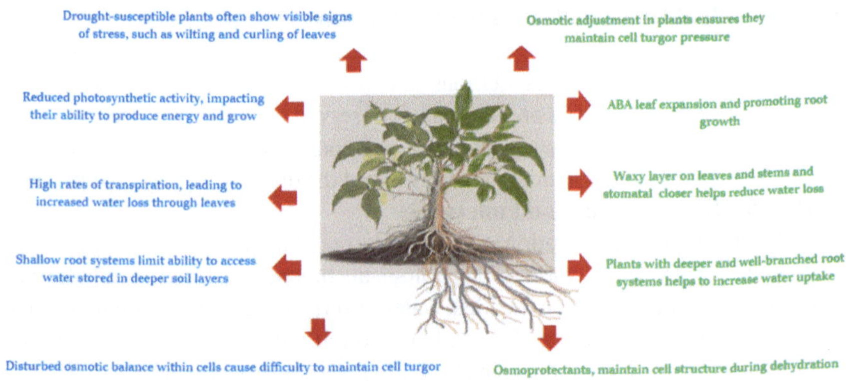

Fig. 9.7 Drought stress mechanisms in plants

Traits Used for Screening of Drought Tolerance

- Leaf rolling: Assessed on a 0–5 scale, where 0—no rolling and 5—tightly rolled leaves.
- Leaf firing: Refers to the drying of leaves due to insufficient cooling.
- Canopy temperature
- Leaf water retention (LWR)
- Survival and recovery of seedlings
- Seedling growth rate
- Stem reserves mobilization
- Proline and ROS compound measurements
- Water use efficiency
- Photosynthesis, radiation use efficiency
- Carbon isotope discrimination
- Staygreen/delayed senescence
- Yield and its components

Screening of Drought Tolerance

A drought environment is created for studying plant responses, classifying drought-tolerant varieties, and understanding the mechanisms underlying drought stress.

Controlled irrigation under field conditions: This can be done by reducing irrigation frequency or decreasing the volume of water applied.

Rainout shelters: Rainout shelters are arrangements created to protect specific land areas or experimental plots from receiving precipitation, enabling the imposition of drought stress.

Drip irrigation with sensors: Drip irrigation systems with soil moisture sensors should be used to monitor and control water delivery in experimental pots/plots. This ensures that plants receive a specific and limited amount of water.

Greenhouse experiments: Under greenhouses conditions, humidity and temperature can be controlled with adjusted irrigation practices to induce drought stress.

Use of absorbent materials: Introduce absorbent materials, such as hydrogels/PEG, into the soil or water to reduce water availability. These materials can absorb and retain water, creating a drier environment around plant roots.

Natural drought-affected area: Choose regions or seasons with naturally occurring drought conditions for field experiments. This allows researchers to study the responses of crops to environmental stress under natural conditions.

Soil moisture at different depths should be measured to understand water availability in the field using a soil moisture sensor, or by employing the volumetric method of moisture measurement. Alternatively, it can be measured by using a soil moisture neutron probe, tensiometers, thetaprobe, Diviner-2000 etc. at regular intervals.

9.8.2.5 Physiological Techniques for Screening of Abiotic Stress Tolerance

a. **Relative Water Content (RWC) and Water Deficit (WD)**

Seedlings are harvested from control and stress treatments, and leaf fresh weight is measured. Leaf samples are then immersed in de-ionized water for 4 h till they attain full turgidity and turgid weight is measured. Samples are then oven-dried at 60 °C till they attain constant dry weight, and the dry weight is recorded. Finally, the relative water content (RWC) and water deficit (WD) are measured by the formula given below (Weatherley 1950).

$$\text{Water deficit} = \frac{\text{Turgid weight} - \text{Field weight}}{\text{Turgid weight} - \text{Oven dry weight}} \times 100$$

$$RWC = \frac{\text{Field weight} - \text{Oven dry weight}}{\text{Turgid weight} - \text{Oven dry weight}} \times 100$$

b. **Membrane Stability Index (MSI)**

The membrane stability index (MSI) is determined by following the method described by (Sairam et al. 1997). Leaf samples weighing 100–500 mg are washed with de-ionized water and placed in test tubes containing 20 mL of double distilled water. Tubes containing samples are incubated in a water bath at 40 °C for 30 min. The conductivity of the solution was measured and recorded as C1. Another test tube containing the leaf sample is subjected to 100 °C for 10 min. The conductivity of the solution was measured and recorded as C2. The membrane stability index of the leaf samples is calculated by using the following formulae:

$$MSI = \left(1 - \left(\text{C1}/\text{C2}\right)\right) * 100$$

c. **Canopy Temperature by Thermal Imaging**

Thermal images of the plants are captured by the methodology described by (Bayoumi et al. 2014). A portable thermal infrared camera like Fluke TiX620 can be used. The camera is positioned 1.5 m above the plants, and images are captured in the direction of sunlight. The IR thermal images are analysed using SmartView 4.3 software that converts the images from IRB files to Fluke thermal images. From each treatment, replicated thermal images were captured, and the temperature values were averaged to ensure accuracy in the measurements. The recording should be done between 12 noon to 2 PM during the daytime.

d. **Estimation of Photosynthetic Pigment Content (mg g^{-1} FW)**

Chlorophyll and carotenoid contents in leaves are estimated according to Hiscox and Israelstam (1979). The freshly collected uppermost fully expanded leaves are cut into small pieces. 50 mg of fresh leaves sample was put into the test tube containing 10 mL of dimethyl sulphoxide (DMSO). Then, the test tubes were incubated in the dark for overnight to facilitate the extraction of chlorophyll into the solution. The absorbance of the solution was measured at 470, 645 and 663 nm using a UV-visible spectrophotometer (model Specord Bio-200, AnalytikJena, Germany). Total chlorophyll and total carotenoids were calculated according to Arnon (1949) and expressed as mg g^{-1} FW.

$$\text{Chlorophyll a}(Chl\,\text{a}) = \left\{\left[\left(12.7 \times OD_{663}\right)^{+}\left(2.69 \times OD_{645}\right)\right] \times \text{V}\right\} / \left(1000 \times \text{W}\right)$$

$$\text{Chlorophyll b}(Chl\,\text{b}) = \left\{\left[\left(22.9 \times OD_{645}\right)^{+}\left(4.68 \times OD_{663}\right)\right] \times \text{V}\right\} / \left(1000 \times \text{W}\right)$$

$$\text{Total chlorophyll} = \left\{\left[\left(20.2 \times OD_{645}\right)^{+}\left(8.02 \times OD_{663}\right)\right] \times \text{V}\right\} / \left(1000 \times \text{W}\right)$$

$$\text{Total carotenoids} = \left[1000 \times OD_{470} - \left(3.27 \times Chl\,\text{a}^{+}\,104 \times Chl\,\text{b}\right)\right] / 229$$

where.

OD_{470} = Absorbance values at 470 nm
OD_{663} = Absorbance values at 663 nm
OD_{645} = Absorbance values at 645 nm
W = Weight of the sample in mg
V = Volume of the solvent used (mL).

e. **Chlorophyll Content Index (CCI)**

Non-destructive measurement of chlorophyll content is done by Opti-Science CCM-200 plus Chlorophyll Content Meter (CCM). The CCI of 10 leaf samples from each treatment is averaged to calculate the mean CCI.

f. **Chlorophyll Fluorescence**

It is a non-destructive measurement to evaluate photosynthetic efficiency in plants under stress conditions. It involves measuring the fluorescence emitted by chlorophyll molecules to assess the efficiency of photosystem II (PSII) by measuring the parameters, **Fv/Fm**, maximum quantum efficiency of PSII, **ΦPSII (YII)**, actual quantum yield of PSII during illumination, indicating photochemical efficiency and **NPQ**, non-photochemical quenching. For measuring chlorophyll fluorescence, plants or leaf parts are dark-adapted to stabilize PSII. There are several instruments for measuring it: Mini-PAM, FluorPen, OS5p + etc. It generally varies from 0.78 to 0.84 in healthy plants.

g. **Normalized Difference Vegetation Index (NDVI)**

It is a commonly used non-destructive technique based on the reflectance values of the red and near-infrared (NIR) wavelengths captured by sensors. NDVI values range from −1 to +1, where higher positive values indicate healthier and denser canopy. It is measured by a handheld instrument, green seeker or by tractor-mounted/drone-mounted instrument.

h. **Measurement of Photosynthetic-related parameters (IRGA: Infrared Gas Analyser)**

The LI-6400XT Portable Photosynthesis System, developed by LI-COR in Lincoln, Nebraska, USA, is employed to measure photosynthetic parameters listed below:

Main variables	Abbreviations		Unit
	Common	*LI-6400*	
Photosynthetic rate	PN	*Phot*	µmol (CO_2) m^{-2} s^{-1}
Transpiration rate	TR	*Trmmol*	mmol (H_2O) m^{-2} s^{-1}
Stomatal conductance to H_2O	GS	gsw	mol (H_2O) m^{-2} s^{-1}
Intercellular CO_2 concentration	Ci	*Ci*	µmol (CO_2) m^{-2} s^{-1}
Vapour pressure deficit	VPD	*vpd*	

9.9 High Throughput Phenotyping (HTP)

Imaging and Image Processing Concepts for Automated Germplasm Characterization (Phenotyping)

Numerous initiatives were launched over the past few decades for the establishment of precise phenotyping for large numbers of plants. These efforts resulted in the emergence of plant phenomics, a technique facilitating high-throughput phenotyping by using multidimensional phenotypic data and non-destructive technologies to characterize plant responses to environmental stimuli. Traditional phenotyping methods are laborious and time-consuming, making it challenging to evaluate large plant populations. However, with the advent of high-throughput phenotyping platforms, the process has become more efficient and cost-effective. These platforms extensively use imaging and image processing in addition to automation to move sensors to plant or plant to sensors thus making the phenotyping accurate, repeatable and rapid.

Features of HTP: Use of precise sensors, imaging systems and automation to monitor environmental factors such as radiation, ambient temperature and humidity and soil moisture throughout the plant growth Image acquisition protocols, image analysis configuration and feature extractions and interpretations are key components that result in extraction of features of plants for interpretation of their responses.

9.9.1 Imaging in HTP

Imaging systems to record plant responses are equipped with sensors designed to capture electromagnetic radiation across different wavelengths according to specific needs. They incorporate high-resolution cameras capable of capturing radiation in the visible, infrared, near-infrared, X-ray and terahertz ranges of the electromagnetic spectrum. Moreover, non-destructive measurement techniques, such as chlorophyll fluorescence, NMR and PET-based imaging systems, are also utilized for assessing various plant traits. Different imaging sensors and their applications are mentioned in Box 9.10.

Box 9.10 Imaging Sensors and Applications

S. No	Imaging system	Description	Phenotypic trait measured
1	Visible light	Camera-sensitive imaging and produce Grey or colour scale image	Prediction of biomass, dynamic growth, colour, shape, seed morphology, root architecture, panicle traits etc.
2	Infrared imaging	Includes NIR, multispectral	Shoot or leaf temp, leaf and canopy water status, disease severity, nutrient stress, insect infestation etc.
3	Fluorescence imaging	Detect chlorophyll fluorescence signal	Performance of photosynthesis, plant health, quantum yield
4	Hyper-spectral imaging	Produce continuous and desecrate spectral data	Water status, quality, pigment, nutrients status etc.
5	Magnetic resonance imaging	Suitable for the internal structure and physiological processes of plants	Root system architecture, water content, dynamic studies of physiological processes

RGB Imaging: The widespread application of visible light imaging in plant science is attributable to its affordability and ease of use. By employing this imaging system, which operates at a wavelength range of 400 to 700 nanometres an extensive array of phenotypic attributes can be analysed in two-dimensional (2D) images. This methodology acquires colour photographs of vegetation, enabling scientists to examine attributes including the colour, size and shape of leaves and plants. The disadvantage of visible images is that they only provide physiological information. In addition, overlapping foliage and the soil background can occasionally present obstacles during the segmentation phase of image processing.

Infrared Imaging: In thermography, the blackbody law is used to determine the temperature. Thermal imaging visualizes infrared radiation, which represents the surface temperature of an object. The spectral range of thermal cameras is 7–14 μm. The measurement of temperature is a crucial physical element in the study of plants. Leaf orientation, heat capacity, surface characteristics, infrared (IR) absorption and transpiration rate are all structural and functional aspects that are indirectly related to the apparent temperature indicated by thermal imaging. Infrared cameras detect temperature differences, aiding in the identification of plant stress, disease and other physiological responses.

Spectroscopic Imaging: The use of spectroscopic imaging is particularly promising for plant phenotyping. It was developed as part of the study of plant growth in remote locations using satellite imagery. Multispectral or hyperspectral cameras, which can scan wavebands of interest at high speeds, can be used to produce spectral measurements of the electromagnetic spectra. Canopy water content can be estimated using multispectral and hyperspectral measurements of the absorption band in the infrared region. Reflectance vegetation indices are excellent examples of practical applications of spectral measurements. Hyperspectral imaging in high-throughput phenotyping (HTP) revolutionizes plant and biological studies by recording comprehensive spectral information across a wide range of wavelengths (350–2500 nm).

Near-Infrared Imaging: NIR imaging is based on the idea that various materials have distinct spectral fingerprints in the NIR range. Organic substances including water, cellulose and starch absorb and reflect NIR light very well. Important biochemical components including water, cellulose, lignin and other organic substances can be determined with NIR imaging. To capture a wider range of spectrum information, it is frequently combined with other imaging techniques such as multispectral and hyperspectral imaging. This multi-modal method improves the richness and quality of phenotypic data, allowing for a more thorough examination of plant properties.

Chlorophyll Fluorescence Imaging: Chlorophyll fluorescence is a phenomenon that arises when chlorophyll molecules capture light energy during the process of photosynthesis and then release a portion of this absorbed energy as fluorescence. The utilization of chlorophyll fluorescence imaging serves as a technique for quantifying the photosynthetic efficacy of plants, hence elucidating alterations in several parameters such as the maximum quantum yield of photosystem II (PSII).

Image Processing in High Throughput Phenotyping: Image processing is essential in high throughput phenotyping (HTP), which is a technology-driven strategy for measuring plant or biological properties on a wide scale. HTP analyses a vast number of photographs using various imaging techniques and computational tools to extract significant information for scientific study, plant breeding, and other uses. The major processes involved are given below:

Pre-processing: Images obtained from imaging systems may contain background noise and inconsistencies. Preprocessing includes tasks like noise reduction, image enhancement and background removal to improve the quality of the images.

Segmentation: Separating the regions of interest (e.g. plants, leaves, roots) from the backdrop in images is one of the most important tasks in HTP. The regions in question are identified and delineated using image segmentation algorithms.

Feature extraction: Following segmentation, several attributes such as size, shape, texture, colour and others are extracted from the segmented parts. These characteristics can be used to quantify and describe the properties of biological samples.

Object tracking: Image processing can be used to follow the movement and growth of biological things over time in longitudinal research or dynamic trials. This is necessary for determining how features change during the life cycle.

Machine learning and statistical analysis: Based on visual data, advanced machine learning algorithms can be used to categorize, predict and assess phenotypic features. This is extremely useful when working with large, complex feature sets. Statistical analysis helps in quantifying variations among different germplasm samples. Image processing procedures comprise batch processing, robotics and workflow management systems for efficiently handling the enormous amount of photographs. Large populations of plants or biological samples can be analysed simultaneously by researchers due to automation, which drastically cuts down on the amount of time needed for investigations. HTP reduces the possibility of subjectivity and human mistakes by automating the phenotyping process, guaranteeing the consistency and repeatability of experimental outcomes.

9.9.2 Applications of HTP

Crop improvement: Researchers can discover plants with desirable features such as high yield, disease resistance and stress tolerance by precisely assessing and quantifying phenotypic attributes of large genotypes for use in crop improvement.

Precision agriculture: Engineers associated with designing precision agriculture tools can utilize insights developed from the plant phenomics platform for optimizing resource management practices such as water and fertilizer use, pest control and other inputs.

Disease and stress detection: The HTP platform can facilitate the development of tools for advanced detection and monitoring of diseases and stress factors in plants. Image analysis and other HTP methods can reveal changes in phenotypic traits that can be signs of diseases or environmental stresses. Early detection enables farmers to act quickly to prevent or reduce crop losses.

Functional genomics: Functional genomics study is made easier by HTP, which connects genetic information to traits that can be visualized. Scientists can look at how changing genes affects the looks of plants and learn more about how genes work and how they are controlled.

Climate change research: Phenomics helps researchers to study the impact of climate change by looking at how plants react to changes in their surroundings. The ability of plant species to respond to different climates can be learned by studying phenotypic traits in different climates.

Nutrient and water use efficiency: HTP technologies can be used to figure out how well plants use water and nutrients. By looking at how different phenotypes react to changes in water and nutrients, scientists can find the best ways to farm, make better use of resources and solve problems related to water shortages and managing nutrients.

9.10 Germplasm Utilization

Utilization refers to the use of potential donor germplasm with an identified trait of economic importance in the breeding programme for the development of improved cultivars. Knowledge of gene pool and breeding methodologies is required for the utilization of specific germplasm, through breeding, varietal development, marker-assisted utilization etc. Surprisingly, despite large collections available across the world, only a few germplasm accessions (<1%) have been utilized in crop improvement programs such as in wheat, maize, spring barley, soybean and other grain legumes (Sharma et al. 2013).

9.10.1 Major Bottleneck for Limited Use of Germplasms in Crop Improvement Programme

Large size of germplasm collections: Large size of germplasm collections is difficult to characterize systematically. The establishment of core collection, minicore sets, trait specific subsets of genebank accessions is the essential requisites for enhancing utilization through the creation of smaller sets easy to maintain, evaluate and search for novel and unique traits of economic importance.

Lack of infrastructure facility and trained manpower: Lack of infrastructure facility and trained manpower to carry out high throughput phenotyping in controlled environments are other important issues posing hindrances in PGR utilization.

Documentation: Documentation and easy access to characterization and evaluation data of germplasm are also very important for increasing awareness of the users.

Limited use of advanced tools and techniques: In general, the use of modern concepts and tools is lacking in germplasm characterization and evaluation, which hinders getting precise outcomes and searching for potential donors.

9.10.2 Effective Utilization of Germplasm

For effective utilization of germplasm, the important criteria and tools to be adopted are given in the following paragraphs.

9.10.2.1 Documentation

Plant genetic resources crucial for agriculture and food security are conserved in genebanks worldwide. They provide genetic material for breeding, research and conservation. Precise documentation, including passport and characterization data, is vital for the effective management of PGRs. These data help categorize PGRs for breeders' evaluation, ensuring accurate selection based on desired traits. Regular data collection maintains quality control during regeneration. Catalogues and digital databases facilitate breeders' access to specific traits. The exchange of data between national programs optimizes efficiency, minimizes duplication and informs priority setting for germplasm activities. Further, to enhance the management and utilization of plant genetic resources, several steps must be taken such as:

1. **Accurate documentation:** Emphasize the accurate documentation of germplasm information regarding origin, characterization and performance.
2. **Promotion of documentation:** Promote the documentation of plant genetic resources and encourage data exchange.
3. **Uniform guidelines:** Develop uniform guidelines for documenting PGR information through the production of descriptor lists, allowing better communication between scientists and institutions and ultimately leading to increased use of conserved genetic resources.
4. **Common standards:** To achieve seamless data exchange, it is crucial to establish compatible documentation systems. This can be accomplished through the adoption of common standards for information exchange, with descriptor lists serving as well-established standards.

9.10.2.2 Focused Identification of Germplasm Strategy (FIGS)

Mining huge genebank collections to identify useful traits is a gigantic task. Germplasm under natural selection pressure in specific environments undergo gene evolution and possess potential sources of novel genes for which selection pressure exists. Geographic information systems (GIS) have been applied to genetic resource conservation as a database management system that can simultaneously handle digital spatial data and attached non-spatial attribute data (Gepts 2006).

Definition: A recent approach 'Focused Identification of Germplasm Strategy (FIGS)'combines agro-ecological information with plant traits and characteristics. A prototype system that combines the biological information managed by a genebank with environmental information, using geographic information system (GIS) technologies, is being developed to identify candidate accessions for targeting specific breeding objectives.

Origin of concept: The focused identification of germplasm strategy (FIGS) was used to create virtual collections by combining thousands of bread wheat landraces from different institutions such as the International Centre for Agricultural Research in the Dry Areas (ICARDA), the N.I. Vavilov Research Institute of Plant Industry (VIR) and the Australian Winter Cereals Collection (AWCC). These collections were augmented with geographical data including collecting site coordinates, long-term temperature and rainfall records, agro-ecological zoning schemes and estimates of potential salinity incidence.

Application: FIGS is employed to develop subsets of landrace accessions, called FIGS sets, which have a higher likelihood of possessing specific desired traits. These sets are then used for rapid and precise screening to identify donor germplasm and discover new traits. For example, FIGS has been instrumental in identifying sources of resistance to various biotic stresses in wheat, such as powdery mildew, Russian wheat aphid, and stem rust, as well as resistance to net blotch disease in barley. It has also aided in identifying tolerance to abiotic stresses like drought tolerance in *Vicia faba*. One notable application of FIGS is its use in prioritizing the evaluation of wheat and rice germplasm from regions prone to drought, heat and salinity, as well as genebank collections in India. By combining FIGS with climate analogue data, targeted exploration efforts can be directed toward areas with specific environmental challenges, facilitating more efficient germplasm evaluation and trait discovery.

9.10.2.3 Data Collection Tools

Accurate and efficient data collection is the foundation of plant breeding and crop research (Pieruschka and Schurr 2019). It forms the base upon which vital advancements are made, ensuring food security for future generations. Therefore, implementing robust mechanisms to facilitate and improve data collection is critical and a required step in any modern research.

Traditionally, recording plant data was undertaken using handwritten records, complex data meticulously inscribed on printed field plans or researchers' notebooks. These served an important purpose but were prone to errors like typos, misplaced numbers and confusion due to unclear field layouts. Transcribing these records, once a researcher was back at their desk added another layer of potential mistakes, further highlighting the limitations of this method.

The introduction of spreadsheets marked a natural progression from handwritten notes that most researchers were comfortable with. They offered a familiar and widely used platform for data entry, equipped with basic error-checking functionalities. However, utilizing spreadsheets on mobile devices outdoors presented significant challenges. The complexity of the applications, coupled with sunlight glare and cumbersome input methods on smartphones and tablets, hampered the accuracy of data entry. Moreover, these spreadsheets lacked vital features like image tagging and geolocation, limiting their usefulness in capturing comprehensive trial data.

Today, innovative mobile applications (apps) have emerged as the current frontier tools which help researchers to save time and, work more efficiently in comparison to manual plant phenotyping (Table 9.3). These apps, specifically designed for low-cost mobile devices, aim to address the shortcomings of previous methods. They prioritize user-friendliness and incorporate functionalities crucial for comprehensive data capture. These features often include:

- Intuitive interfaces: User-friendly interfaces and simplified data entry options minimize errors and streamline the data collection process.
- Offline functionality: The ability to operate independently of internet connectivity ensures that data can be captured even in remote locations with unreliable internet access.
- Data validation and error checking: Built-in safeguards like drop-down menus, pre-defined values, and range checks help prevent errors during data entry.
- Image and audio capture: Integrating camera and microphone capabilities allows for real-time image and audio documentation of observations, enriching the data collected.
- Geolocation tagging: Automatic location recording provides valuable spatial context for the collected data, aiding in further analysis and data visualization.

Table 9.3 Description of data collection software

	Software/App	Description	Developer
1.	Field Book App (Rife and Poland 2014) 	• Customized for eight crops. Reduces paper usage and streamlines data collection, analysis and storage	Kansas State University and CIMMYT for data collection in plant genetic resources (PGR).
2.	PhenoTop Mobile App 	• Used for field data collection, including observations and selections with images • Allows offline and online data collection, automatic synchronization and features sorting, filtering and barcode reader • Streamlines data capture, analysis and evaluation	Phenome Nerwork LTD
3.	X Farm and Plantae Software 	• Use for data management, monitoring, analysis, mapping, data acquisition, planning, accounting, weather data analysis, statistical analysis, counting and crop phenotyping • Special applications in precision agriculture, vehicle fleet management, sensor integration and pest control	Agri Expo company for various agricultural tasks
4.	Digital Passport Data Sheet (NBPGR-PDS) 	• App for Android phones facilitating data collection in the field of plant genetic resources • Works offline, acts like a digital book and allows easy organization of data with pictures • Users can transfer data to MS Excel, operate offline for data protection, and is free on the Google Play Store Designed for researchers, students, plant experts and business owners	ICAR -National Bureau of Plant Genetic Resources, New Delhi, India

5.	ClimMob	• Online tool designed for farmers • Instead of researchers conducting complex field trials, this software makes it easy to manage and simplify the handling of data for small trails on farmers' fields	Food and Agricultural Organization of United Nations
6.	GridScore NEXT (Raubach et al. 2022) 	• Cross-platform app for collection of phenotypic observation data • Allows offline and online data collection, data synchronization, image tagging, GPS georeferencing and barcode support • Offers advanced data visualizations for pattern detection and error highlighting	Information and Computational Sciences Department, The James Hutton Institute, UK
7.	PhenoApp (Röckel et al. 2022) PHENOAPPS	• Can include sample images of phenotype descriptors • Predefined BBCH scales for some crops	Julius Kühn Institute, Germany

The development of these advanced features empowers researchers to gather a broader spectrum of data with increased speed and accuracy. This, in turn, fuels advancements in plant research and breeding, ultimately paving the way for a more sustainable and resilient future for agriculture.

In this section, we will describe the common types of errors that are seen in plant phenotyping experiments along with some possible solutions to help in data collection.

9.10.2.3.1 Field Trial Data Collection Error Types

Different types of errors can be introduced during phenotypic data collection. Many of these errors can be avoided by simple data validation measures. We have defined these errors into six main classifications, which cover most of the commonly seen data recording errors.

1. **Number swaps:** One of the easiest mistakes to make is introducing number swaps into recorded data, so instead of writing down or entering 19, a 91 might be recorded. Depending on the value range of a trait, an error like this can be almost an order of magnitude of its expected value.
2. **Decimal point issues:** Like number swaps, recording the decimal point of a decimal number in the wrong place can skew the data distribution. Recording a value of 20.7 instead of 2.07 introduces an error with one order of magnitude. Both number swaps and decimal point issues can be prevented by setting restrictions on allowed trait values. Data visualizations can also highlight these errors as clear outliers so that they can be corrected in time.
3. **User differences:** While the previous two error types take place at the point of entering the data, differences in the perception of the person collecting the data take place even earlier. People perceive things differently based on their physical properties like how their eye sees colours or based on personal experience. Traits that require a perceptive measurement like flower colour or leaf shape can therefore be scored differently by two people for the same plant. This can be mitigated by using technical equipment to measure the colour or by providing reference sheets that clearly show the different options for a trait.
4. **Different units:** When data are collected in different locations around the world, people may be using different units to measure the same characteristic. While centimetres are widespread for height measurements, it may be common practice in some countries to measure in feet and inches. Weights could also be measured in grams or pounds. Ensuring everyone uses the same units when recording observations is crucial to ensure data integrity down the line.
5. **Different categories:** Often traits are scored on a discrete scale. Disease severity, for example, might be scored as 1, 3, 5, 7, or 9 with 1 being no disease and 9 high disease occurrence. Similarly, the same trait could be scored with 'very low', 'low', 'medium', 'high' and 'very high'. While these two sets of categories represent the same discrete scale, data that are collected using different categories are difficult to combine and therefore prone to errors. Sharing the same trait definitions with their data type, scale and restrictions between all parties collecting data can help prevent these issues.

6. **Plot mix-ups:** Even when the data are collected correctly, there is another type of error that can arise. The data could have been collected for the wrong plant/ plot altogether. Plot/plant mix-ups are a common error that can be prevented using barcodes or QR codes to identify each individual or by utilizing technologies available in some data collection applications for a structured walk order through the trial.

Collecting data for managing plant genetic resources is a crucial task, especially when dealing with a large amount of complex information. It can be very challenging. However, field data collection software can provide valuable tools to help make this job much easier and reduce the potential for errors to be introduced into datasets. These platforms help organize, present and ensure the accuracy of the data in a systematic and easy-to-understand way. These are requirements for plant breeding experiments.

There is often resistance to using digital technology for data collection amongst some groups, but the benefits such systems bring far outweigh the disadvantages. From previous work, we estimate that we see somewhere between 5 and 10% error rate while recording trait data using traditional paper and pen-based methods when compared to electronic data collection (and that assumes that the germplasm being scored is what you think it should be and not something else). It is worth remembering that if data are scored electronically, it is available immediately in digital format whereas recording data on paper provides the opportunity for errors to be introduced while recording and then while transcribing from paper to digital format—a process that is also time-consuming. We would advocate against writing anything down on paper and that the use of electronic data collection along with the use of barcodes is a minimum requirement. In addition, simple measures must be taken such as agreeing on protocols/ontologies for data collection so that phenotypes are recorded in a way that minimizes an individual's interpretation of characteristics and so that equivalence can be established. Examples on how to achieve this include the use of standard operating procedures on how a trait is scored and the assignment of characters against standard ontologies (Jaiswal et al. 2005). This information can then be shared with collaborators and the wider community so that it is transparent and clear how measurements were taken.

9.11 Summary

The very basis of any crop improvement programme is the extent of variability available for different economically important traits in the germplasm. Characterization of genetic resources conserved in the genebank is a prerequisite for gaining knowledge about their preliminary information. Further, a comprehensive evaluation of conserved germplasm for economically important traits such as agronomic, biotic, abiotic stress and quality traits will provide information about potential donors for target traits. Knowledge of characterization and evaluation descriptors, hotspot locations and methodology of evaluation are essential requirements for precisely identifying unique trait-specific germplasm. Establishment of core, mini-core, reference set of large germplasm collection and identification of potential donor germplasm for agronomic, stress-related and quality traits are critical for enhanced utilization of germplasm in the breeding programme. Further, novel, high precision and robust high throughput phenotyping tools and data collection and analysis software enhance the efficiency of trait discovery.

Multiple-Choice Questions

1. What is the primary goal of creating a core collection of plant genetic resources?
   (a) To increase the number of samples in a genebank
   (b) To maximize genetic diversity with minimal redundancy
   (c) To reduce the genetic diversity in a population
   (d) To improve seed germination rates
2. Which software is commonly used for developing core collections in large sets of genotyping data?
   (a) Excel
   (b) GenoCore
   (c) Gen Stat
   (d) SPAR
3. What does NDVI stand for in remote sensing?
   (a) Normalized Difference Vegetation Index
   (b) National Data Verification Initiative
   (c) New Development in Vegetation Imaging
   (d) Normalized Data Value Index
4. How do plants defend themselves from the pests attack?
   (a) Physical barriers (Thick cuticles, waxes on leaves and rigid cell walls)
   (b) Chemical defences (Production of toxic compounds)
   (c) Resistance (R) genes
   (d) All of the above
5. What is one of the main applications of chlorophyll fluorescence measurement in plants?
   (a) Estimating soil moisture levels
   (b) Assessing photosynthetic efficiency
   (c) Measuring plant height
   (d) Calculating seed germination rates
6. Which of the following is a key parameter measured in chlorophyll fluorescence?
   (a) Fv/Fm
   (b) pH
   (c) Soil texture
   (d) Temperature

7. What does the acronym PAM stand for in the context of chlorophyll fluorescence measurement?
   (a) Pulse-Amplitude Modulation
   (b) Plant Analysis Method
   (c) Photo-Acoustic Monitoring
   (d) Phyto-Assay Measurement
8. In the context of plant breeding, what is the significance of using a core collection?
   (a) To enhance the nutritional value of crops
   (b) To increase the size of the plants
   (c) To utilize diverse genetic traits for improving crop yield
   (d) To reduce the water requirements of plants
9. What type of platform can be used to capture NDVI data for agricultural monitoring?
   (a) Satellite sensors
   (b) Ground-penetrating radar
   (c) Ultrasound machines
   (d) X-ray imaging
10. Which hormone is commonly associated with plant stress responses and is often analyzed in studies involving genetic resources?
    (a) Insulin
    (b) Abscisic acid (ABA)
    (c) Auxin
    (d) Cytokinin
11. How do deep root systems contribute to drought resistance in plants?
    (a) By increasing the rate of photosynthesis
    (b) By accessing deeper water sources in the soil
    (c) By reducing the need for stomatal closure
    (d) By storing carbohydrates during drought periods
12. Which of the following mechanisms is NOT associated with drought resistance in plants?
    (a) Stomatal closure
    (b) Deep root systems
    (c) Increased leaf surface area
    (d) Accumulation of osmolytes
13. Which of the following is a physiological response of plants to drought stress?
    (a) Increased transpiration rate
    (b) Reduced photosynthesis
    (c) Enhanced leaf expansion
    (d) Increased chlorophyll content
14. What is the role of proline in drought-resistant plants?
    (a) It serves as a structural component of cell walls
    (b) It acts as a signal molecule for stomatal opening
    (c) It functions as an osmoprotectant and helps in osmotic adjustment
    (d) It enhances root growth by promoting cell division

15. How does heat stress affect the photosynthetic apparatus of plants?
    (a) It increases chlorophyll concentration
    (b) It causes ion leakage from leaf cells
    (c) It enhances photosynthetic efficiency
    (d) It stabilizes membrane structures
16. What type of proteins do plants produce in response to high-temperature stress?
    (a) Storage proteins
    (b) Transport proteins
    (c) Heat shock proteins (HSPs)
    (d) Structural proteins
17. The primary effect of high temperatures on plant photosynthesis is
    (a) Increased photosynthetic rate
    (b) Enhanced production of chlorophyll
    (c) Reduced activity of enzymes like Rubisco
    (d) Increased energy reserves
18. Common physiological injury observed in plants under elevated temperatures
    (a) Increased root growth
    (b) Scorching of leaves and stems
    (c) Enhanced chlorophyll production
    (d) Delayed leaf senescence

Answers

1. (b)
2. (b)
3. (a)
4. (d)
5. (b)
6. (a)
7. (a)
8. (c)
9. (a)
10. (b)
11. (b)
12. (c)
13. (b)
14. (c)
15. (b)
16. (c)
17. (c)
18. (b)

Short Questions

Q1. Enumerate different methods of germplasm evaluation.
Q2. Which statistical design should be adopted for field evaluation of germplasm?
Q3. How observation of pest severity is recorded for identification of promising accession(s)?

Suggested Reading

Alercia A, Diulgheroff S, Metz T (2001) FAO/IPGRI multi-crop passport descriptors [MCPD]. 4 p. https://hdl.handle.net/10568/105205

Arnon DI (1949) Copper Enzymes in Isolated Chloroplasts. Polyphenoloxidase in Beta Vulgaris. Plant Physiol 24:1–15. https://doi.org/10.1104/pp.24.1.1

Bayoumi TY, El-Hendawy S, Yousef MS, Emam MA, Okasha SA (2014) Application of infrared thermal imagery for monitoring salt tolerant of wheat genotypes. J Am Sci 10(12):227–234

Bisht I, Mahajan R, Loknathan T, Agrawal R (1998) Diversity in Indian sesame collection and stratification of germplasm accessions in different diversity groups. Genet Resour Crop Evol 45(4):325–335. https://doi.org/10.1023/a:1008652420477

Bioversity International (2007) Developing crop descriptor lists: guidelines for developers. Bioversity Technical Bulletin no. 13:72 p. ISBN: 978-92-9043-729-1. https://hdl.handle.net/10568/74489

Brown AHD (1989) Core collections—a practical approach to genetic-resources management. Genome 31(2):818–824

Brown AHD, Grace JP, Speer SS (1987) Designation of a “core” collection of perennial Glycine. Soybean Genet. Newsl 14:59–70

De Vicente C, Metz T, Alercia A (2004) Descriptors for genetic markers technologies. International Plant Genetic Resources Institute, Rome. https://cropgenebank.sgrp.cgiar.org/images/file/learning_space/descriptors_genetic_markers.pdf

Frankel OH, Brown AHD (1984) Plant Genetic Resources Today: A Critical Appraisal. In: Holden JH, Williams JT (eds) Crop Genetic Resources: Conservation and Evaluation, George Allen and Unwin, London 249–257

Gautam RK, Kaur V (2023) Characterization and evaluation of plant genetic resources: principles and techniques. In: Gautam RK, Kumari J, Tripathi K et al (eds) Plant genetic resources: theory and practice e-book. ICAR-NBPGR, pp 58–63

Gepts P (2006) Plant Genetic Resources Conservation and Utilization. Crop Sci 46:2278–2292. https://doi.org/10.2135/cropsci2006.03.0169gas

Gotor E, Alercia A, Ramanatha Rao V et al (2011) The scientific information activity of Bioversity International: the descriptor lists. Impact Assessment Discussion Paper 3. Bioversity International 16 p. https://hdl.handle.net/10568/73451

Gouesnard B, Bataillon T, Decoux G, Rozale C, Schoen D, David J (2001) MSTRAT: an algorithm for building germ plasm core collections by maximizing allelic or phenotypic richness. J Hered 92(1):93–94. https://doi.org/10.1093/jhered/92.1.93

Hiscox JD, Israelstam GF (1979) A Method for Extraction of Chlorophyll from Leaf Tissue without Maceration. Can J Bot 57:1332–1334. https://doi.org/10.1139/b79-163

Hintum van Th.JL, Brown AHD, Spillane C, Hodgkin T (2000) Core collections of plant genetic resources. IPGRI Technical Bulletin No. 3. International Plant Genetic Resources Institute, Rome, Italy

IRRI (1988) Standard evaluation system for rice, 3rd edn. International Rice Testing Programme, Los Banos, p 43

Jaiswal P, Avraham S, Ilic K et al (2005) Plant ontology (PO): a controlled vocabulary of plant structures and growth stages. Comp Funct Genomics 6(7–8):388–397. https://doi.org/10.1002/cfg.496

Jeong S, Kim J-Y, Jeong S-C, Kang S-T, Moon J-K, Kim N (2017) GenoCore: A simple and fast algorithm for core subset selection from large genotype datasets. PLOS ONE 12:e0181420. https://doi.org/10.1371/journal.pone.0181420.

Kim K, Chung H, Cho G, Ma K, Chandrabalan D, Gwag J, Kim T, Cho E, Park Y (2007) PowerCore: a program applying the advanced M strategy with a heuristic search for establishing core sets. Bioinf 23(16):2155–2162. https://doi.org/10.1093/bioinformatics/btm313

Kumar A, Balamurugan A (2023) Screening of plant genetic resources (PGR) for disease resistance. In: Gautam RK, Kumari J, Tripathi K et al (eds) Plant genetic resources: Theory and practice e-book. ICAR-NBPGR, India, pp 120–126

Kumar A, Kaur V (2010) Characterisation and evaluation of PGR: principles and techniques. Division of Germplasm Evaluation, ICAR-NBPGR, New Delhi, pp 139–145

Levene H (1960) Robust Tests for Equality of Variances. In: Olkin I (ed) Contributions to Probability and Statistics. Stanford University Press, Palo Alto 278–292

Mahajan RK, Sapra RL, Srivastava U et al. (2000) Minimal descriptors (for characterization and evaluation) of Agri-Horticultural Crops (Part-I). ICAR-National Bureau of Plant Genetic Resources, Pusa Campus, New Delhi, p 230

Meena VK, Shekhawat HVS, Chand S (2023) Advances in molecular marker technology and their significance in plant improvement strategies. IntechOpen. https://doi.org/10.5772/intechopen.1002773

NBPGR (2000) Manual on Characterization and evaluation of plant genetic resources Jai Vigyan National Science Technology Mission on Conservation of Agro-biodiversity (Plant Genetic Resources) NATP (Plant Biodiversity). National Bureau of Plant Genetic Resources, New Delhi

Odong TL, Jansen J, Van Eeuwijk FA, Van Hintum TJL (2012) Quality of core collections for effective utilisation of genetic resources review, discussion and interpretation. Theor Appl Genet 126(2):289–305. https://doi.org/10.1007/s00122-012-1971-y

Patrik JW, Tyerman SD, van Bel AJE (2015) Long distance transport. In: Buchanan BB, Gruissem W, Jones RL (eds) Biochemistry & molecular biology of plants. Wiley-Blackwell, Chichester, pp 658–701

Pieruschka R, Schurr U (2019) Plant phenotyping: past, present, and future. Plant Phenomics 7507131:1–6. https://doi.org/10.34133/2019/7507131

PPV&FRA (2022) Guideline for the conduct of tests for distinctness uniformity and stability on teak (Tectona grandis L.f.). Protection of Plant Varieties and Farmers' Rights Authority Department of Agriculture Cooperation and Farmers Welfare Ministry of Agriculture and Farmers Welfare, Government of India, NASC Complex, DPS Marg, New Delhi-110 012. https://plantauthority.gov.in/sites/default/files/teakguideline.pdf

Raubach S, Schreiber M, Shaw PD (2022) GridScore: a tool for accurate cross-platform phenotypic data collection and visualization. BMC Bioinformatics 23(1):214. https://doi.org/10.1186/s12859-022-04755-2

Rife TW, Poland JA (2014) Field book: an open-source application for field data collection on Android'. Crop Sci 54(4):1624–1627. https://doi.org/10.2135/cropsci2013.08.0579

Röckel F, Schreiber T, Schüler D, et al (2022) PhenoApp: a mobile tool for plant phenotyping to record field and greenhouse observations. F1000Res 11:12. https://doi.org/10.12688/f1000research.74239.2

Sairam RK, Deshmukh PS, Shukla DS (1997) Tolerance of drought and temperature stress in relation to increased antioxidant enzyme activity in wheat. J Agron Crop Sci 178(3):171–178. https://doi.org/10.1111/j.1439-037x.1997.tb00486.x

Sairam RK, Kumutha D, Ezhilmathi K et al (2008) Physiology and biochemistry of waterlogging tolerance in plants. Biol Plant 52(3):401–412. https://doi.org/10.1007/s10535-008-0084-6

Sharma S, Upadhyaya HD, Varshney RK, Gowda CLL (2013) Pre-breeding for diversification of primary gene pool and genetic enhancement of grain legumes. Front Plant Sci, 4. https://doi.org/10.3389/fpls.2013.00309

Singh RK, Mishra B, Chauhan MS et al (2002) Solution culture for screening rice varieties for sodicity tolerance. J Agric Sci Cambridge (UK) 139:327–333. https://doi.org/10.1017/S0021859602002447

Taiz L, Zeiger E (2010) Plant physiology, 5th edn. Sinauer Associates, Sunderland, pp 85–105

Thachuk C, Crossa J, Franco J, Dreisigacker S, Warburton M, Davenport GF (2009) Core Hunter: an algorithm for sampling genetic resources based on multiple genetic measures. BMC Bioinf 10(1). https://doi.org/10.1186/1471-2105-10-243

Tripathi K, Bhardwaj R, Bhalla S et al. (2018) Plant genetic resources evaluation: principles and procedures. Indian Council of Agricultural Research -National Bureau of Plant Genetic Resources. New Delhi vi + 50 p. https://www.researchgate.net/publication/324942043_Plant_Genetic_Resources_Evaluation_Principles_and_Procedures/citations

Upadhyaya HD, Ortiz R (2001) A mini core subset for capturing diversity and promoting utilization of chickpea genetic resources in crop improvement. Theor Appl Genet 102(8):1292–1298. https://doi.org/10.1007/s00122-001-0556-y

UPOV Test Guidelines - English Index: http://www.upovint/en/publications/tg_rom/tg_indexhtml

USDA ARS National Genetic Resources Program Germplasm Resources Information Network - (GRIN) [Online Database] National Germplasm Resources Laboratory Beltsville Maryland. http://www.ars-gringov/cgi-bin/npgs/html/croplistpl

Weatherley P (1950) Studies in water relations of cotton plants I The field measurement of water deficit in leaves. New Phytol 49:81–87. https://doi.org/10.1111/j.1469-8137.1951.tb05168.x

Yang L, Chen J, Hu W et al (2016) Population genetic structure of *Glycyrrhiza inflata* B (Fabaceae) is shaped by habitat fragmentation water resources and biological characteristics. PLoS One 11(10):e0164129. https://doi.org/10.1371/journal.pone.0164129

Yeo AR, Flowers TJ (1983) Varietal differences in the toxicity of sodium ions in rice leaves. Physiol Plant 59(2):189–195. https://doi.org/10.1111/j.1399-3054.1983.tb00756.x

Yu Q, Liu Q, Xiong Y et al (2019) Genetic diversity and population divergence of a rare endemic grass (*Elymus breviaristatus*) in the southeastern Qinghai-Tibetan plateau. Sustainability 11(20):5863. https://doi.org/10.3390/su11205863

Zaidi PH, Vinayan MT, Seetharam K (2016) Phenotyping for abiotic stress tolerance in maize: waterlogging stress A field manual CIMMYT: Hyderabad India. https://repository.cimmyt.org/server/api/core/bitstreams/b099d162-443f-4089-8ab7-59c9b7a821c5/content

Zaman-Allah M, Zaidi PH, Trachsel S et al (2016) Phenotyping for abiotic stress tolerance in maize: drought stress A field manual. http://hdl.handle.net/10883/17716

Biochemical Tools for PGR Management 10

Antil Jain, Racheal John, Simardeep Kaur, Maharishi Tomar, Nand Lal Meena, Sapna Langyan, and Rakesh Bhardwaj

Learning objectives

- Analytical tools for assessing food composition and nutritional quality
- Strategies for method selection in quality control and assurance
- Chromatographic approaches to studying bioactive compounds
- Comprehensive analysis using NIR and Raman spectroscopy: Non-Destructive Approaches
- Chemometric prediction modelling: Techniques and applications

Keywords

Plant genetic resources · Germplasm · PGR · Nutritional quality · Biochemical tools · Proximate analysis · NIR · Chromatography

A. Jain · R. John · N. L. Meena · S. Langyan · R. Bhardwaj (✉)
Division of Germplasm Evaluation, ICAR-National Bureau of Plant Genetic Resources, New Delhi, India
e-mail: Sapna@icar.org.in; Rakesh.Bhardwaj1@icar.org.in

S. Kaur
Division of Crop Science I ICAR-Research Complex for North Eastern Hill Region (NEH), Umiam, Meghalaya, India

M. Tomar
ICAR-Indian Grassland and Fodder Research Institute, Jhansi, Uttar Pradesh, India

K. Tripathi et al. (eds.), *Textbook of Plant Genetic Resources*,
https://doi.org/10.1007/978-981-96-5024-8_10

10.1 Introduction

Germplasm resource diversity encompasses a vast array of genetic variations that are crucial for agricultural sustainability and food security. Genetic resource diversity not only helps crops adapt better to varied environments, but also provides resilience against environmental challenges. In addition, these resources have rich nutritional value, with some germplasm having particularly high concentrations of essential nutrients for human health. It is crucial to tap into the genetic resources available to develop breeding programs that can strengthen crops by adding desirable traits and improving their nutritional value. Therefore, using germplasm as a source of vital genes, especially those that are rich in specific nutrients, has enormous potential to combat nutritional deficiencies and ensure food sustainability in the face of global challenges. To ensure accuracy in identifying and using essential genes, it is of utmost importance to generate high-quality and reliable data on the nutrient profiles of germplasm resources. This data is essential for capturing invaluable genetic traits from germplasm collections, which can drive targeted breeding efforts, thereby enhancing crop resilience, nutritional value, and overall agricultural sustainability. Good Laboratory Practices (GLP) serve as the foundation for ensuring the integrity, reliability, and reproducibility of scientific data generated in laboratory settings. Central to GLP is a commitment to precision and accuracy, starting with correct measurements and adherence to standardized protocols. Understanding the principles of designing and validating analytical methods is crucial, as it ensures the robustness and validity of experimental outcomes. Experimental design plays a pivotal role in GLP, guiding researchers in systematically planning and executing experiments to minimize bias and maximize the reliability of results. Calibration, validation, and optimization of laboratory instruments and techniques are essential steps in ensuring the accuracy and consistency of measurements. Therefore, it is essential to follow these procedures to maintain the quality and integrity of the research and experimentation.

10.2 Understanding Important Terminologies in Nutritional Quality Evaluation and Data Generation

- **Quantification vs. Qualification:** Quantification involves the determination of the exact amount or concentration of a particular substance present in a sample. This process typically uses calibration curves or standard reference materials to establish a relationship between the measured signal (e.g., absorbance and fluorescence intensity) and analyte concentration. Quantitative analysis provides precise numerical data, allowing for accurate comparisons, calculations, and assessments of the quantity of the substances present. For example, by measuring the weight of any sample, concentration, use of scale, and SI units to represent the measurement can be quantified. Qualification is a critical aspect of ensuring the reliability, accuracy, and consistency of analytical methods, processes, and personnel in laboratory settings. Here, we provide a brief overview of each of the following:

- **Analytical Method Qualification:** Analytical method qualification involves assessing and validating the performance characteristics of an analytical method to ensure its suitability for its intended purpose. This procedure commonly involves assessing several validation parameters, including specificity, accuracy, precision, linearity, range, and robustness. Through method qualification, laboratories can establish confidence in the method's ability to consistently produce accurate and reliable results.
- **Process Qualification:** Process qualification encompasses the validation and verification of laboratory processes to ensure that they consistently produce results that meet predefined specifications and requirements. This includes validation of sample preparation procedures, instrument calibration, data analysis protocols, and quality control measures. Process qualification aims to identify and mitigate the sources of variability or error within laboratory workflows, thereby enhancing the reliability and reproducibility of analytical outcomes.
- **Personnel Qualification:** Personnel qualification involves ensuring that laboratory staff possess the necessary skills, knowledge, training, and competency to perform their assigned tasks effectively and accurately. These may include training programs, proficiency testing, ongoing competency assessments, and documentation of qualifications. Qualified personnel are essential for maintaining the integrity of analytical processes, adhering to quality standards, and minimizing the risk of errors or deviations.
- **Analysis:** Analysis entails delving into the intricacies of a complex substance or system and breaking it down into its fundamental components through myriad analytical techniques. The primary aim of the analysis is to unravel the composition and behavior of the material, method or result, shedding light on its properties and structure. For example, consider the analysis of an extracted wheat sample using spectroscopy. This technique enables researchers to identify and quantify mineral composition, offering valuable insights into its nutrient content and facilitating a deeper understanding of its nutritional profile.
- **Determination:** When a precise measurement of the quantity or concentration of a specific component within a sample is performed using calibrated instruments, this process yields a numerical value for the analyte, primarily focusing on quantification. For example, in the determination of the fat content in a food product, the Soxhlet extraction method is commonly used, in which fat is extracted from the sample using a solvent such as hexane. After extraction, the solvent is evaporated, leaving behind fat, which can then be weighed to determine its mass. The fat content is calculated based on the weight of the extracted fat and the initial mass of the sample.
- **Measurement:** Measurement involves assigning a numerical value to a physical quantity, such as length, mass or time. Precise quantification is essential for comprehending the composition, structure, and quantity of any sample in food chemistry. The measurement facilitated both the analysis and determination processes, enhancing our understanding of the characteristics of the sample. For example,

when measuring the weight of a sample using a weighing balance and representing it in kilograms or grams, we obtain a numerical value that accurately reflects the mass of the sample. This measurement aids in assessing the quantity of the analyte present, providing valuable information for various applications, such as formulation, analysis, and quality control.

- **Laboratory Techniques:** These tools are used to look at or change something in our tests. These may include instrumental techniques, such as chromatography and spectroscopy, as well as non-instrumental techniques, such as titration or gravimetric analysis. Each technique is chosen based on its suitability for the type of analysis required and the properties of the sample being analyzed.
- **Laboratory Methods:** Methods outline the overall approach or strategy for conducting a particular analysis. They provided a systematic framework for the standardized implementation of the chosen techniques.
- **Laboratory Procedures** are similar to detailed instructions. If methods are big plans, the procedures are a step-by-step guide. They tell us exactly what to do at each stage of our experiment. These may include sample handling and preparation, instrument setup and calibration, data acquisition, analysis, and interpretation, as well as quality-control checks and documentation practices.
- **Laboratory Protocols:** These are similar to super-detailed guidelines. They not only tell us what tools and steps to use but also cover everything else, such as how to prepare samples, how to make sure our machines are working correctly, and how to check if our results are accurate. Protocols are similar to a complete manual for the entire experiment.
- **Standard Operating Procedures (SOPs):** This is the most fundamental QA method of quality assurance. This involves committing step-by-step instructions established by the laboratory or technique so that consistency is maintained and variability can be minimized.
- **Calibration and Maintenance:** Regular calibration and maintenance of the laboratory equipment are essential to ensuring accurate and precise measurements. Instruments such as balances, pipettes, and spectrophotometers must be calibrated to the recognized standards. Calibration is the process of adjusting and confirming the accuracy of the measurement instrument. This is similar to tuning the equipment before use. In the laboratory, the weighing balance, pipettes, laboratory instruments, and pH meters need to be calibrated so that reliable and precise measurements can be made. For example, to measure the acidity of any solution, the pH meter must be calibrated by adjusting it to a known standard, which has a defined pH value that ensures that the meter can provide accurate readings.
- **Proficiency Testing:** In this procedure, laboratories participate in testing programs where samples with known and unknown characteristics are received from external sources. Laboratories often participate in external proficiency-testing programs. Once the samples have been analyzed, they are compared with the expected values to assess their performance through various statistical tests.

- **Validation and Verification:** Validation confirms that a method is suitable for its intended purpose, which is a documented process that ensures consistent results within specified acceptance criteria. New instruments undergo validation during installation, commissioning or significant changes, guaranteeing reliable performance under diverse conditions such as relocations or alterations in environmental settings. Verification ensures that equipment, methods, and processes operate as intended while confirming the accuracy of activities or data before sharing them with relevant parties. For example, pH meter verification may involve checking that the meter displays the correct pH value when measuring a solution with a known acidity. The validation involves conducting experiments to ensure that the method is accurate and precise, and provides consistent results when applied to various samples.
- **Traceability:** For traceability, a documented chain is established by referencing known standards or certified materials. This ensured that the measurements and data generated in the laboratory could be linked back to recognized standards, providing confidence in the accuracy and reliability of the results.
- **Reference Materials:** These are known as standard reference materials (SRMs) or certified reference materials (CRMs). These substances have standardized, extensively tested, and certified materials used to calibrate and validate analytical instruments and methods. For example, for the determination of minerals in food samples through atomic absorption spectroscopy, certified reference materials are used to draw the calibration curve and to obtain a linear graph for verification of the results. Similarly, food reference materials (FRMs) are also reference materials, which are well-characterized samples with known properties. FRMs may include a variety of food matrices, such as fruits, vegetables, grains, meats, dairy products, and processed foods. These materials are prepared and certified by specialized organizations or institutions, such as national metrology institutes, international standards organizations, and commercial providers. FRMs are characterized using validated analytical methods to determine their composition, analyte concentration, and other relevant properties.
- **Internal Quality Control (IQC):** In IQC, controls, samples, and standards are used within the laboratory to accurately and precisely monitor the method and technique involving the use of control samples and standards within the laboratory to monitor the accuracy and precision of analytical methods. For example, a control sample with a known concentration of starch is run along with the rest of the samples. Any deviations from the expected results are investigated through identification and rectification, ensuring the quality of the testing process.
- **External Quality Control (EQC):** This is an inter-laboratory comparison (ILC) that involves a laboratory's participation in an external proficiency-testing program. In this scenario, a laboratory receives samples with known characteristics from an external provider. The laboratory analyzed these samples, and the results are then compared to the expected values. This external assessment helps the laboratory gauge its performance compared to other laboratories.

10.3 Key Considerations in Method Selection for Analysis

The selection of an appropriate method is crucial for determining the particular characteristics of any analyte in a given matrix. This method refers to the application of any technique. These criteria are defined as follows:

- **Accuracy:** Accuracy measures how closely the results of an experiment align with the true or expected values. Accuracy can be represented in terms of absolute error or percentage relative error. For example, when determining the concentration of salt in a solution, accuracy refers to how close one can get to the actual concentration.
- **Precision:** Precision refers to the consistency or reproducibility of results when a sample is analyzed multiple times. A precise result can or cannot be accurate; it is simply a representation of how close individual results are to each other. For example, when titrating the acidity of a solution, repeating the titration yields consistent results, but it might or might not be close to the actual value.
- **Sensitivity:** This is the ability of any method/technique to determine differences in the amount of analyte. The smallest detectable difference in the analyte amount is determined based on the sensitivity of the method. For example, for measuring the difference in the analyte amount of a colored compound, a method with high sensitivity, like UV-Vis spectroscopy, can distinguish subtle color variations indicative of concentration changes.
- **Specificity and Selectivity:** Specificity refers to how exclusively a method or technique responds to a particular analyte—it reacts only with the target substance and not with others. Selectivity, on the other hand, describes the ability of a method to accurately measure an analyte even in the presence of potential interfering substances. It is often quantified using a selectivity coefficient, which indicates how well the method can distinguish the analyte from other components. Example: Suppose you are analyzing a mixture of metal ions and want to determine the concentration of copper. A method that uses a reagent reacting only with copper ions demonstrates high specificity. The selectivity coefficient in this case indicates how effectively the method can differentiate copper from other metal ions that might interfere with the measurement.
- **Robustness and Ruggedness:** Robust methods provide reliable results across various sample matrices, whereas rugged methods are insensitive to variations in experimental conditions. For example, in a reaction that is sensitive to changes in temperature, a robust method would still provide reliable results, even if the temperature fluctuates slightly. Ruggedness is evident when the method remains reliable despite minor variations in the experimental conditions.
- **Scale of Operation:** Analyzing the scale of operation involves considering the amount of sample, the number of samples, the expected analyte concentration, and the minimum detectable amount. For example, for a limited number of samples, proteins can be estimated by quantifying individual amino acids using high-pressure liquid chromatography (HPLC). If more samples are to be analyzed, estimating nitrogen and multiplying by a specific Jones factor is an appropriate

method. However, if a very large number of samples are evaluated, then rapid and non-destructive NIR/Raman spectroscopy-based methods are appropriate.

- **Limit of Detection (LOD):** LOD is the lowest concentration or amount of analyte that can be reliably detected but not necessarily quantified. At this level, the signal is distinguishable from the background noise with a specified level of confidence. For example, consider a method for detecting the Pb concentration in wheat. The LOD is the concentration at which the instrument can reliably identify the presence of Pb, even if the exact quantity cannot be accurately measured.
- **Limit of Quantification (LOQ):** LOQ is the lowest concentration or amount of analyte that can be reliably measured and can be accepted with an ensured level of accuracy and precision. The main idea of LOQ is that the sample concentration should not only be detected but also be quantified. For example, the LOQ could be the concentration at which the instrument provides a reliable and accurate measurement of lead. If the LOQ is determined to be 0.05 ppm, it means that concentrations below this level may be detected, but accurate quantification is challenging.
- **Limit of Blank (LOB):** The limit of blank refers to the highest measured value that can be obtained while analyzing a sample with no analyte of interest. Imagine it as finding the noise in a completely silent room. The significance of the blank limit is to set a baseline (any instrument/technique) for detecting an actual analyte, ensuring accurate and reliable measurements against noise or signals. For example, to detect the concentration of cineole in citrus fruit using gas chromatography, before analyzing the actual samples, a series of tests are performed with pure, cineole-free methanol: hexane solution.

10.4 Proximate Analysis

Proximate analysis provides a comprehensive understanding of the basic composition of foods. The quantification of key components, viz., moisture, ash, protein, fat, and dietary fiber, gives invaluable insights into the nutritional content, quality, and safety of foods. This information is essential for assessing dietary adequacy, planning balanced diets, and addressing nutritional deficiencies. Moreover, the proximate analysis serves as a cornerstone for quality control and regulatory compliance in the food industry, ensuring that consumers receive safe, authentic, and accurately labelled food products. Additionally, the data obtained from proximate analysis guides product development and formulation processes, enabling food manufacturers to create innovative products with specific nutritional profiles and sensory characteristics tailored to consumer preferences. Key proximate components (Fig. 10.1) in food are given below:

- **Moisture:** Moisture content is measured to assess the amount of water present in a food sample. This is crucial, as water content affects the food's texture, stability, and shelf life. Methods for moisture determination include oven drying, Karl Fischer titration, and moisture analyzers.

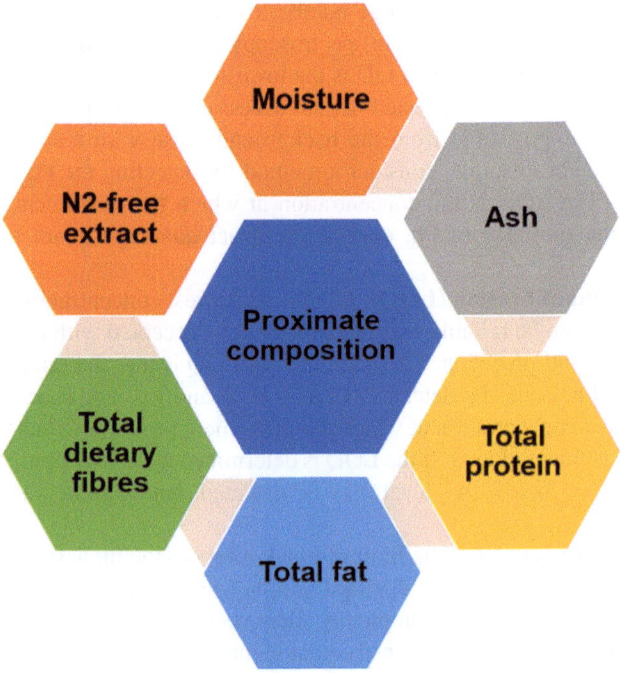

Fig. 10.1 Proximate composition parameters

- **Ash:** Ash content represents the inorganic mineral content remaining after the incineration of organic matter in a food sample. It provides insights into the mineral composition of the food. Ash content is determined by burning the sample at high temperatures, followed by weighing the residue.
- **Crude protein:** Crude protein content indicates the total nitrogen content in a food sample, which is then converted to protein using conversion factors. The protein content is important for assessing nutritional quality and labelling requirements. Protein determination methods include Kjeldahl nitrogen analysis, the Dumas method and quantification through amino acids profiling.
- **Crude fat:** Crude fat, or lipid content, represents the total fat content in a food sample. It is determined by extracting lipids from the sample using solvents and then weighing the extracted fat. Soxhlet extraction and solvent extraction methods are commonly used for fat determination.
- **Dietary fiber:** Dietary fiber is a crucial component of proximate composition, representing the portion of plant-derived foods that resists digestion in the human gastrointestinal tract. While not a nutrient in the traditional sense, dietary fiber plays a vital role in maintaining digestive health, regulating bowel movements, and reducing the risk of chronic diseases such as heart disease, diabetes, and certain cancers.
- **Nitrogen-free extract (NFE):** NFE, also known as available carbohydrates, represents the portion of carbohydrates in a food sample that is readily digestible

and absorbed by the human body. It includes sugars, starches, and some types of fiber that provide energy to the body upon digestion. NFE is calculated by subtracting the sum of moisture, crude protein, crude fat, dietary fiber, and ash content from one hundred. NFE content is crucial for assessing the nutritional value and glycemic impact of foods.

10.4.1 Moisture Estimation Methods

In general, the choice of method depends on the type of plant material, the required accuracy, and the available equipment. Accurate moisture content estimation is essential for ensuring the quality, safety, and suitability of plant-based products for various applications. There are several methods for estimating the moisture content of plant materials, such as the following:

10.4.1.1 Oven/Vacuum Oven Drying Method

Principle: This method is based on the principle of weight loss due to the evaporation of moisture from the sample during drying in an oven/vacuum oven to a constant weight. The sample is weighed before and after drying, and the loss in weight represents the moisture content.

$$\text{Calculation :}$$
$$\text{Moisture content } (\%) = \left[(\text{Initial weight} - \text{Final weight}) / \text{Initial weight} \right] \times 100$$

Advantages:

- Widely accepted and used as a standard reference method.
- Suitable for a wide range of sample types.

Disadvantages:

- Time-consuming, as it may take several hours to obtain accurate results.
- Requires a dedicated drying oven and balance.

Samples with high moisture should not be dried at temperatures above 60 °C, as it may cause oozing of tissue fluids and loss pf nutrients. Further drying of any sample at a temperature beyond 70 °C affects the extractability of membrane lipids. Samples with high sugar may result in charring and/or Maillard reactions and alteration in composition/structure, which can be avoided in vacuum oven. This method is not suitable for samples with high content of volatile constituents like essential oils.

10.4.1.2 Infrared Moisture Analyzers

Principle: These analyzers work by measuring the absorbance of infrared radiation by water molecules in the sample. The instrument provides direct moisture content readings.

$$\text{Calculation:}$$
$$\text{Moisture content}\,(\%) = \text{Readings from the analyzer or instrument display.}$$

Advantages:

- Rapid results, typically in seconds.
- Non-destructive and suitable for routine testing.

Disadvantages:

- Limited to samples with high moisture content.
- A clean and uniform surface is required for accurate readings.

10.4.1.3 Karl Fischer Titration

Principle: The Karl Fischer reagent reacts with water in the sample, and the volume of reagent consumed is proportional to the moisture content. The Karl Fischer reagent is a specialized solution containing iodine, sulfur dioxide, a base, and a solvent. It is used in Karl Fischer titration for the accurate determination of water content in various substances. The redox reaction with water facilitated the precise measurements.

$$\text{Calculation:}$$
$$\text{Moisture content}\,(\text{mg}/\text{g}) = \left(\begin{array}{l}\text{Volume of Karl Fischer reagent used}\,(\text{mL})\,/\\ \text{Sample weight}\,(\text{g})\end{array}\right) \times \text{Correction factor}$$

Advantages:

- Highly accurate, sensitive, and suitable for trace moisture analysis.
- Used on a wide range of sample types, including non-aqueous samples.

Disadvantages:

- Requires specialized equipment (Karl Fischer titrator) and reagents.
- May not be suitable for all sample types.

10.4.1.4 Dielectric Moisture Measurement

Principle: Changes in the dielectric properties of the sample as the moisture content varies. The instrument is calibrated to provide moisture readings.

Calculation: Calibration curves are used to convert instrument readings into moisture content values.

Advantages:

- Non-destructive, suitable for in-line monitoring.
- Works well for many sample types, including solids and liquids.

Disadvantages:

- Calibration is essential for obtaining accurate results.
- The instrument may not function with any type of material.

10.4.1.5 Near-Infrared Spectroscopy (NIRS)

Principle: NIRS analyzes the near-infrared absorbance spectrum with water-content-specific peaks. A calibration model is created to relate the spectral data to the moisture content.

Calculation: A calibration model is used to convert the spectral data into moisture content.

Advantages:

- Rapid analysis without sample preparation.
- Suitable for a wide range of sample types.

Disadvantages:

- A well-established and maintained calibration model that requires samples in large numbers is required.
- Instrument costs and expertise required for calibration.

10.4.2 Ash Content

Ash content estimation in plant materials involves determining the mineral residue remaining when the organic matter is completely combusted. This helps assess the purity, nutritional value, and presence of impurities or contaminants in plant samples. Ash content estimation is based on the principle of heating a known weight of plant material to a high temperature (generally 450–500 °C) to combust its organic content, leaving behind inorganic ash. The weight of ash is expressed as a percentage of the initial sample weight.

Methods of Ash Content Estimation

(i) ***Direct ignition method***

Principle: The plant material is ignited in a muffle furnace at a high temperature (starting from 100 °C to a gradual increase till 450–500 °C) until complete combustion occurs, leaving only the ash.

$$\text{Calculation}: \text{Ash content}\,(\%) = (\text{weight of ash} / \text{weight of the sample}) \times 100.$$

Advantages: Simple, widely used, and applicable to a variety of plant materials.

Disadvantages: Loss of some volatile compounds and possible alteration of some minerals.

(ii) ***Wet digestion method***

Principle: The plant material is digested with concentrated acids to dissolve the organic matter, leaving behind the inorganic content which is subsequently heated to determine ash content.

$$\text{Calculation}: \text{Ash content}\,(\%) = (\text{weight of ash} / \text{weight of the sample}) \times 100.$$

Advantages: Retains a larger portion of mineral content, useful for samples with high organic matter.

Disadvantages: Time-consuming, potential loss of some volatile minerals, the possibility of incomplete digestion.

10.4.3 Total/Crude Fats

Crude fat estimation is based on the principle of extracting lipids from the plant material using a non-polar solvent, followed by evaporation of the solvent and weighing the residual fat. There are several methods for crude fat estimation, including solvent extraction methods such as Soxhlet extraction, Randal extraction and cold percolation; and nuclear magnetic resonance (NMR) direct estimation based on hydrogen activity.

(i) ***Solvent Extraction Methods: Soxhlet/Randal/Cold Percolation***

(a) **Soxhlet Method**

Principle: The solid sample is placed inside a porous extraction thimble, which is then inserted into the Soxhlet extractor. A nonpolar solvent, such as hexane, is introduced into the reservoir flask located beneath the Soxhlet extractor and heated. The solvent vaporizes and ascends through the Soxhlet extractor before condensing back into liquid form through a reflux condenser. This condensed solvent then drips onto the solid sample. As the solvent interacts with the sample, it dissolves the fat present. Once the solvent level within the extractor reaches the height of the siphon tube, it is automatically siphoned back into the reservoir flask. This cyclic process of solvent percolation, dissolution, and siphoning continues until the extraction process is completed. The solvent is then evaporated, leaving the crude fat behind.

$$\text{Calculation}: \text{Crude fat}\,(\%) = (\text{weight of extracted fat} / \text{weight of the sample}) \times 100.$$

Advantages: Reliable, widely used, suitable for most sample types.

Disadvantages: Time-consuming, uses toxic solvents, and may result in loss of heat-sensitive compounds.

(b) **Randal Method**

Principal: In the Randal method, the principle involves boiling the sample in a solvent, followed by refluxing to extract fat efficiently. Initially, the sample is immersed and boiled in a suitable solvent, which is often a mixture of diethyl ether and petroleum ether. This boiling process allows the solvent to penetrate the sample and dissolve the fat content. Subsequently, refluxing is used to facilitate the thorough extraction of fat from the sample. Overall, the combination of boiling and refluxing in the Randal method maximizes the fat extraction efficiency, yielding accurate results in fat estimation analyses.

(c) **Cold Percolation Method**

The cold percolation method for fat estimation involves the extraction of fat from a food sample using a nonpolar solvent at ambient or low temperatures. This method relies on the ability of the solvent to dissolve fat components without requiring high temperatures, thereby minimizing the risk of thermal degradation. In this method, anhydrous sodium sulfate (Na_2SO_4) plays a crucial role in ensuring accurate fat estimation. Before extraction, dry samples are mixed with anhydrous sodium sulfate. This serves two purposes: first, it aids in homogenizing the sample and ensuring consistent extraction; second, it helps to bind any moisture present in the sample, preventing it from interfering with the extraction process.

(ii) ***NMR Method***

The principle of using proton nuclear magnetic resonance spectroscopy (^1H NMR) for estimating fat content involves exploiting the unique chemical environment of the hydrogen atoms in fat molecules. In ^1H NMR, the hydrogen nuclei in fat molecules absorb energy from radiofrequency radiation and resonate at characteristic frequencies, generating a spectrum that provides information about the chemical composition and structure of fat. To calibrate ^1H NMR for fat estimation, pure oil samples with known fat contents are analyzed using spectroscopic techniques. By comparing the spectral signals from these pure oil samples with their known fat concentrations, a correlation between spectral data and fat content is established. This calibration curve or relationship allows for the quantification of fat content in unknown samples based on their ^1H NMR spectra. The advantages of using ^1H NMR for fat estimation include its nondestructive nature, rapid analysis, and high sensitivity to fat molecules. Additionally, ^1H NMR does not require sample preparation or the use of hazardous chemicals, making it the preferred method for fat analysis in research and quality control settings.

10.4.4 Total/Crude Protein

The principle behind crude protein estimation is the quantification of nitrogen content in a sample. Proteins contain nitrogen; therefore, the protein content can be estimated by measuring the nitrogen content. The standard method for crude protein estimation is the Kjeldahl method, which involves digesting a sample and converting nitrogen to ammonium ions.

Methods of Crude Protein Estimation

(i) ***Kjeldahl Method***

Principle: The nitrogen estimation using the Kjeldahl method comprises multiple procedural stages. First, the sample is digested with concentrated sulfuric acid in the presence of a catalyst such as copper sulfate or selenium. The addition of potassium sulfate or lithium sulfate increases the boiling point of the solution, facilitating digestion. During digestion, organic nitrogen compounds are converted into ammonium sulfate. After digestion, the ammonia released from the ammonium sulfate is distilled by alkali distillation. The distillate containing ammonia gas is trapped in a solution of boric acid with a mixed indicator such as bromocresol green and methyl red. The mixed indicator changes color in response to the pH change caused by ammonia. Finally, the trapped ammonia is titrated with a standard acid solution, typically sulfuric or hydrochloric acid, to determine the amount of nitrogen present in the sample. The endpoint of the titration is reached when the indicator changes color, indicating the completion of the reaction between the ammonia and the acid.

Calculation :

$$\text{Crude protein}\,(\%) = (\text{nitrogen content}\,(\text{g}) \times \text{protein factor} / \text{sample weight}\,(\text{g}),$$

where the protein factor is specific for each crop plant.

Advantages: Widely accepted, accurate, and applicable to a wide range of sample types.

Disadvantages: Time-consuming, requires the use of hazardous chemicals and specialized equipment, and may underestimate or overestimate true protein content in certain situations.

(ii) ***Dumas Combustion Method***

Principle: The principle of nitrogen estimation by the Dumas method involves the combustion of a sample in the presence of excess oxygen to convert nitrogen in the sample to nitrogen oxides (NOx). These nitrogen oxides are then passed over a catalyst, typically composed of copper and platinum, which facilitates the conversion of NOx to nitrogen gas (N_2). Subsequently, moisture, carbon dioxide, and other gases are removed from the nitrogen gas, and the purified nitrogen is passed to the thermal conductivity detector (TCD) to ensure accurate quantification. The TCD detects changes in thermal conductivity caused by the presence of nitrogen gas, allowing for the precise determination of nitrogen content in the sample. Overall, the Dumas method with NOx conversion, removal of moisture and other gases, and TCD detection provides a robust and efficient technique for nitrogen estimation in various samples. The total protein content is determined by multiplying the measured quantity of N_2 by a conversion factor, typically 6.25, and expressing the result as a percentage.

$$\text{Calculation :}$$
$$\text{Crude protein}(\%) = (\text{Nitrogen content}(g) \text{ x } 6.25) / \text{Sample weight}(g).$$

Advantages: Faster than Kjeldahl, less risk of contamination, widely applicable.

Disadvantages: Requires specialized equipment, somewhat expensive, may slightly overestimate protein content.

Crude protein (CP) estimation plays a pivotal role in various fields and applications. In nutritional analysis, it is indispensable to assess the protein content of foods and feed ingredients, thereby aiding dietary planning and ensuring quality control. In agriculture, the estimation of protein content is essential for evaluating the nutritional value of crops, influencing breeding and cultivation strategies, and making informed decisions about plant varieties. The food industry relies on this analysis to maintain product quality and to ensure compliance with labelling regulations. Furthermore, in environmental science, crude protein estimation has applications in studying soil and water samples, helping researchers assess nitrogen pollution and environmental impact. This versatile analytical technique has far-reaching implications, ranging from sustaining agriculture to safeguarding environmental health. The standard conversion factor of 6.25 is used to estimate protein content from nitrogen analysis in many crops. However, some crops may have different conversion factors (Box 10.1).

Box 10.1 Is the conversion factor of 6.25 consistently applied for estimating protein content in all crop types?

No, 6.25 is not always the same factor used to estimate the protein content in all crops. The factor 6.25 is a commonly used conversion factor, and it is based on the assumption that proteins contain about 16% nitrogen. To estimate protein content using the factor 6.25, you typically measure the nitrogen content in a sample and then multiply it by 6.25 to get an estimate of the protein content. This factor works for many plant materials and food products, but the actual nitrogen-to-protein conversion factor can vary depending on the specific composition of the proteins in the sample.

In some cases, different conversion factors may be used for specific crops or products. For example, in wheat, the factor may be closer to 5.7, in rice, it is closer to 5.95, and in sorghum, it is 6.25. It's important to consider the specific conversion factor for the crop or product of interest to obtain a more accurate estimation of protein content.

10.4.5 Total Dietary Fibers

The principle behind dietary fiber estimation is to quantify the non-digestible components of food that contribute to its physiological benefits, such as aiding in digestion and providing bulk food. These non-digestible components include cellulose, hemicellulose, pectin, lignin, and other components. These methods aim to separate and quantify the components.

Total Dietary Fibers Estimation Method

Dried and defatted (if the fat content is less than 10%) material can be used to estimate the total dietary fiber (TDF). Selected samples are incubated at 100 °C with heat stable α-amylase to gelatinize, hydrolyze and depolymerize the starch. The samples are then incubated with protease at 60 °C (to depolymerize protein) and amyloglucosidase (to hydrolyze starch to glucose). This is followed by treatment with four volumes of ethanol to precipitate soluble fiber and remove depolymerized proteins and glucose. The residue is filtered, washed with 75% ethanol, dried, and weighed again. One part is analyzed for protein, and the other part is incubated at 525 °C to determine the ash content. The TDF is then determined by the weight of the filtered and dried residues, which reduces the weight of the protein and ash.

10.5 Mineral Profile

Minerals play a crucial role in human nutrition, contributing to various physiological functions that are essential for health and well-being. Iron, zinc, selenium, and calcium are of particular importance because of their critical roles in metabolism, immunity, and overall health. Profiling mineral content, particularly of these key elements, in food crops and germplasm diversity is essential for several reasons. First, it helps ensure the nutritional adequacy of food crops, allowing for the development of nutrient-rich varieties that can help combat malnutrition and deficiencies. Second, understanding the mineral composition of different germplasm varieties enables researchers to identify and select cultivars with naturally high levels of these essential minerals, thereby contributing to biofortification efforts. Moreover, mineral profiling aids in the evaluation and improvement of agricultural practices and breeding strategies to enhance mineral uptake and accumulation in crops.

10.5.1 Flame Photometry

Flame photometry, also known as emission spectroscopy, is a widely used method for mineral profiling. It operates by measuring the emission of light by atoms in an air-natural gas flame when heated to temperatures ranging from 2000 to 2200 degrees Kelvin (K). However, it is limited in that it can only analyze a few minerals, such as sodium, potassium, calcium, lithium, and barium.

10.5.2 Atomic Absorption Spectroscopy (AAS)

AAS is another popular technique for mineral profiling that uses absorption spectroscopy. It operates at high temperatures, typically between 2400 and 2700 degrees Celsius (approximately 2673–2973 degrees Kelvin) in an air-acetylene flame. A nitrous oxide-acetylene flame with temperatures between 2900 and 3100 degrees Kelvin is employed for applications requiring higher sensitivity. AAS relies on the absorption of specific wavelengths of light typically emitted by cathode lamps for specific elements, offering high sensitivity and providing reliable limits of detection (LOD) and quantification (LOQ) values.

10.5.3 Graphite Furnace Atomic Absorption Spectrometry (GFAAS)

GFAAS, also known as electrothermal atomic absorption spectrometry (ETAAS), is a well-established technology used for quantifying elements at trace and ultra-trace levels while requiring only small sample volumes, typically less than 100 μL. In GFAAS, a known amount of the sample solution is injected into a graphite- or pyrolytic carbon-coated graphite tube, which is then heated to vaporize and atomize the analyte. The atoms absorb ultraviolet or visible light with element-specific wavelengths and transition to higher electronic energy levels. The degree of absorption correlates with the concentration of the element in the sample solution, allowing for concentration determination after calibration with standards of known concentrations.

10.5.4 Microwave Plasma Atomic Emission Spectroscopy (MP-AES)

MP-AES is a cutting-edge atomic emission technique that is used for elemental analysis. It capitalizes on the characteristic emission spectra emitted by the excited atoms as they return to their ground state. MP-AES uses high-temperature sources such as microwave plasma (MP), reaching temperatures up to 5000 K, ensuring robust atomic emission with excellent detection limits and a linear dynamic range for most elements. In the MP-AES instrument, microwave energy forms plasma from nitrogen extracted from compressed air, generating a highly stable plasma suitable for analyzing diverse sample types. An optimized microwave waveguide concentrates electromagnetic fields at the torch, whereas a nebulizer and spray chamber create aerosols from liquid samples introduced into the hot plasma. The emitted light from the plasma is directed into a fast-scanning monochromator and imaged onto a high-efficiency Charge Coupled Device (CCD) detector for simultaneous spectra and background measurements, allowing for precise quantification of element concentrations by comparing emissions to known standards. MP-AES offers significant advantages, including a low cost of ownership, enhanced safety by eliminating flammable gases, superior performance compared to conventional flame atomic absorption spectroscopy (AAS), and the ability to handle difficult matrices with ease.

10.5.5 Inductively Coupled Plasma—Optical Emission Spectrometry (ICP-OES)

ICP-OES is a powerful technique used for determining the elemental composition of mostly water-dissolved samples. Owing to its reliability, multi-element options, and high throughput, it has been widely applied in both routine research and specific analysis purposes. The analysis process involves passing the sample solution through a nebulizer into a spray chamber, where it forms an aerosol introduced into an argon plasma. This plasma, generated at the end of a quartz torch by a cooled induction coil through which a high-frequency alternating current flows, reaches extremely high temperatures of 6000–10,000 K. Ionization occurs as a result of collisions between argon atoms and electrons, creating a stable and hot plasma. In the torch, desolvation, atomization, and ionization of the sample occur, with excited electrons emitting light as they return to ground level. Each element has a characteristic emission spectrum, which is measured using a spectrometer. A wavelength separation mechanism, often via an optical grating device, is employed to address overlapping signals from multiple elements. The detector and signal processor correlates the wavelengths of light to the identity of the elements, determining the final sample composition after calibration.

10.5.6 X-ray Fluorescence (XRF)

XRF is a versatile analytical method used to determine the mineral profile of various materials, including solids, liquids, powders, and more. It is fast, accurate, and non-destructive, requiring minimal sample preparation. XRF spectrometer systems can be classified into two main groups: energy dispersive systems (EDXRF) and wavelength dispersive systems (WDXRF). The elemental range and detection levels depend on the spectrometer system used. EDXRF covers elements from sodium to uranium, whereas WDXRF covers elements from beryllium to uranium. The detection limits generally improve with increasing atomic number, and the precision and reproducibility of XRF analysis are high, allowing for accurate results even without specific standards. The measurement time varies depending on the number of elements and the required accuracy, ranging from seconds to 30 minutes, with analysis times after measurement taking only a few seconds. X-rays, which have wavelengths and energies between those of γ-rays and ultraviolet light, interact with matter through fluorescence, Compton scattering, and Rayleigh scattering. Fluorescent radiation is produced when an incoming photon expels an electron from an atom, creating a hole that is filled by an outer shell electron, emitting characteristic X-ray photons. The energy of the emitted X-rays depends on the difference in energy between the initial hole and the filling electron, producing characteristic emission lines that are specific to each element. Fluorescence yield, the ratio of emitted photons to initial vacancies, varies with the atomic number, with a low yield for light elements. Different notations, such as Siegbahn and IUPAC, are used to indicate the emission lines, specifying the elements and shells involved in the transition. Qualitative analysis determines the elements present by measuring their energies, whereas quantitative analysis measures the intensities of the emitted energies to determine the element concentrations.

10.6 Chromatographic Techniques

Chromatography serves as a technique for segregating a mixture of molecules based on their distribution between a mobile phase and a stationary phase. The mobile phase, alternatively termed the solvent, can take the form of either a liquid or a gas. As for the stationary phase, also known as the sorbent, it can exist in either the solid state or the liquid state, with a liquid stationary phase being anchored by a solid support or matrix. In this process, the solid support effectively immobilizes the liquid stationary phase.

Types of Chromatography

• Column chromatography	• Gas chromatography
• Gel-permeation chromatography	• Ion-exchange chromatography
• Paper chromatography	• Thin-layer chromatography
• Affinity chromatography	• High-pressure liquid chromatography (HPLC)

10.6.1 Partition Chromatography

The process of solute distribution between two immiscible phases, known as partitioning, relies on the solute's solubility in each phase. Thin-layer chromatography and paper chromatography are the two most common forms of partition chromatography, using a liquid stationary phase bound to a matrix in both instances. In paper chromatography, this stationary phase consists of water molecules bound to a cellulose matrix.

Distribution coefficient: The foundational concept underlying all chromatography methods is the distribution or partition coefficient (K_d), representing how a compound (analyte) distributes between two immiscible phases, A and B. The K_d value, a constant at a given temperature, is expressed as:

$$K_d = \frac{\text{Concentration in phase } A}{\text{Concentration in phase } B^{1/4}}$$

Fig. 10.2 Paper chromatography

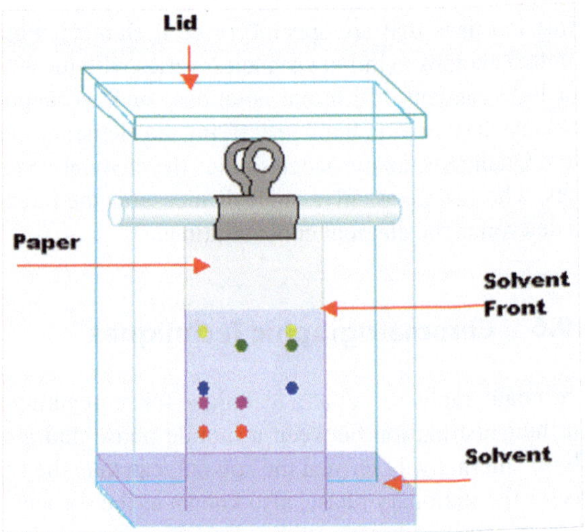

10.6.2 Paper Chromatography

Paper chromatography operates on the principles of capillary action and solubility, where the cellulose support, rich in bound water, serves as the stationary phase (Fig. 10.2). The experimental procedure involves depositing a solution or mixture onto a paper strip, allowing it to dry, immersing one end in a solvent, and drawing the solvent through the paper by capillary action. The resulting chromatogram, developed through ascending or descending solvent flow, facilitates the detection and marking of separated components. The ratio of the distance moved by a solute to the distance moved by the solvent, denoted as Rf, is a critical parameter, always less than one, and is unitless.

$$R_f = \frac{\text{Distance travelled } by \text{ solute}}{\text{Distance travelled } by \text{ solvent}}$$

In the paper chromatography process

(a) The sample is deposited at the paper's bottom line, which is further placed in a solvent-filled tank.
(b) Solvent migration in the paper elutes the solutes, and their movement depends on their affinity for the solvent.
(c) Different development techniques with various flow directions are illustrated in Fig 10.2.
(d) A chromatogram is the term given to a sample applied on TLC or paper, and paper chromatograms can be developed through either ascending or descending solvent flow.

Identification of spots in paper chromatograms can be accomplished through distinct methods. In the process, the spots are commonly identified by comparing them to standards with known Rf values, ensuring accurate and reliable characterization.

Natural Color Observation: Spots can be detected based on their inherent color.

Fluorescence: Detection is possible through the fluorescence emitted by the spots.

Chemical Reactions: After the paper is sprayed with different reagents, chemical reactions can reveal spots. For instance, in amino acid paper chromatography, chromatograms are sprayed with ninhydrin to induce reactions.

Radioactivity: Detection can also be achieved by identifying spots exhibiting radioactivity.

10.6.3 Thin-Layer Chromatography

In Thin-Layer Chromatography (TLC), the stationary phase is prepared by applying the solvent to the support to create the thin layer, resulting in the increased binding of the solvent to the matrix (Fig. 10.3). Partition chromatography primarily serves the separation of small molecular weight molecules. For paper chromatography, cellulose-derived paper is exclusively employed, whereas in TLC, any substance that can be finely divided and formed into a uniform layer is suitable. Both organic and inorganic materials, such as cellulose, polyamide, polyethylene, silica gel, aluminum oxide, and magnesium silicate, can be used for TLC.

Advantages: TLC functions as a method for substance identification and compound purity testing, offering the advantages of rapidity and minimal sample requirements. Notably, TLC surpasses paper chromatography with greater resolving power due to reduced spot diffusion, accelerated separation speed, and a broad selection of sorbent materials.

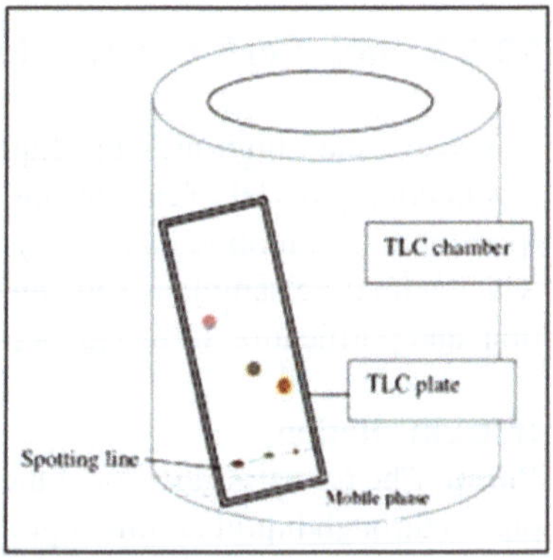

Fig. 10.3 Thin layer chromatography (TLC)

Preparation:

(e) To prepare the chromatography chamber, an appropriate developing liquid is added to a jar, creating a depth of 0.5–1 cm.

(f) The closed jar is left to stand for approximately 30 minutes to saturate the atmosphere inside with the solvent.

(g) The partition of a solute between a moving solvent phase and a stationary aqueous phase determines the solute's rate of movement in the direction of solvent flow.

(h) Factors influencing this movement include the solubility of the solute in the moving phase, an ion exchange effect and temperature, as temperature affects the solute's solubility in a given solvent.

10.6.4 Column Chromatography

In column chromatography, a glass or metal column is filled with the stationary phase. Subsequently, the mixture of analytes is introduced, and the mobile phase, commonly known as the eluent, is propelled through the column, facilitated either by a pumping system or by applied gas pressure. The stationary phase is implemented by either coating discrete small particles (the matrix), which are then packed into the column, or by applying it as a thin film to the inner wall of the column. As the eluent traverses the column, the analytes get separated based on their distribution coefficients, emerging individually in the chromatogram upon exiting the column. This section will delve into two forms of column chromatography: liquid chromatography (LC), particularly high-performance liquid chromatography (HPLC), and gas chromatography (GC).

10.6.5 High-Performance Liquid Chromatography

It is also called high-pressure liquid chromatography or high-performance liquid chromatography (HPLC). The stationary phase may be solid (adsorption) or liquid (partition). The mobile phase may be gas (GC) or liquid (LC). It is a technique by which a mixture sample is separated into components for identification, quantification, and purification of mixtures (Fig. 10.4).

Instrumentation

Pump: The primary function of the pump is to propel a liquid (known as the mobile phase) through liquid chromatography at a specified flow rate. The pump can maintain a consistent composition of the mobile phase, known as isocratic, where the flow per hour (f.ph) remains constant throughout the analysis. Alternatively, it can operate in a gradient mode, where the f.ph changes throughout the analysis.

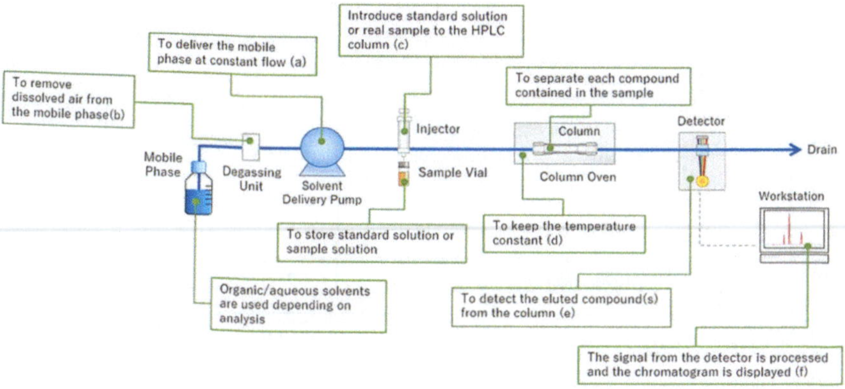

Fig. 10.4 HPLC flow diagram

Injector: The injector is responsible for introducing the liquid sample into the flowing mobile phase. It can be either an auto-sampler or a manual injector.

Stationary Phases in HPLC: High-performance liquid chromatography (HPLC) offers a diverse range of stationary phases:

Normal Phase: Involves a polar stationary phase and a non-polar solvent, such as silica gel.

Reverse Phase: Uses a non-polar stationary phase and a polar solvent, like silica gel C18.

Ion Exchange: The **stationary** phase contains ionic groups, and the mobile phase consists of an aqueous buffer.

Size Exclusion: It is characterized by the lack of interaction between the sample compounds and the column. Larger molecules elute first, followed by smaller molecules.

Quantitative Analysis

The measurement of the concentration of compounds in a sample is shown in Fig 10.5.

1. Determination of the peak height
2. Determination of the peak area Resolution (RS) of a column provides a quantitative measure of its ability to separate two analytes.

$$RS = 2\left(\mathrm{tR2} - tR1\right) / \mathrm{W2} + \mathrm{W1}$$

3. Theoretical plates (N): The number of theoretical plates characterizes the efficiency of a column.

$$\mathrm{N} = 16\left(tR / \mathrm{W}\right)^2$$

where tR1, tR2 and W1, W2 are retention times and peak widths of analyte 1 and analyte 2, respectively.

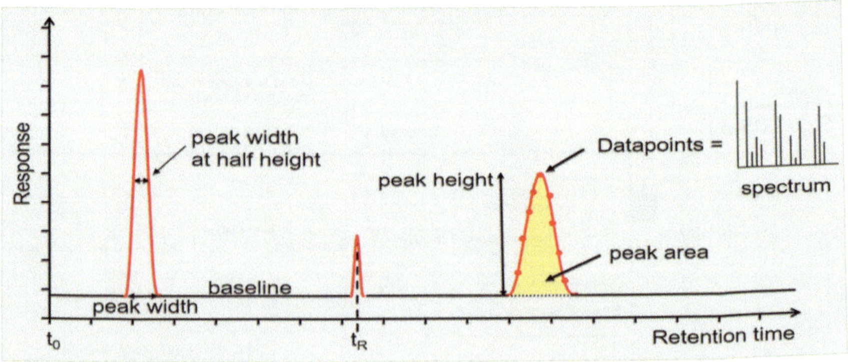

Fig. 10.5 HPLC Chromatogram

Uses of HPLC

HPLC finds application in both chemistry and biochemistry research, serving purposes such as the analysis of intricate mixtures, purification of chemical compounds, formulation of processes for synthesizing chemicals, isolation of natural products, and prediction of physical properties. Additionally, it plays a crucial role in quality control by verifying the purity of raw materials, enhancing process yields, quantifying assays of final products, and assessing product stability while monitoring degradation.

10.6.6 Gas Chromatography

Gas chromatography (GC) is a chromatographic method employed for the separation of volatile organic compounds. This technique relies on the solubility and boiling points of organic liquids to effectively isolate them from a mixture. Illustrated in Fig. 10.6, it serves both qualitative (identifying substances) and quantitative (measuring quantities) purposes. The key components include a flowing mobile phase, an injection port, a separation column (acting as the stationary phase), an oven, and a detector. The underlying principle of the instrumentation lies in the separation of organic compounds based on their distinct partitioning behavior between the mobile gas phase and the stationary phase within the column.

1. The injection port features a rubber septum through which a syringe needle is introduced for sample injection.
2. The injection port is maintained at a temperature higher than the boiling point of the least volatile component in the sample mixture.
3. To control the partitioning behavior, the separation column is typically housed in a thermostat-controlled oven.
4. The separation of components with varying boiling points is achieved by initiating at a low oven temperature and gradually increasing it to elute high-boiling-point components over time.

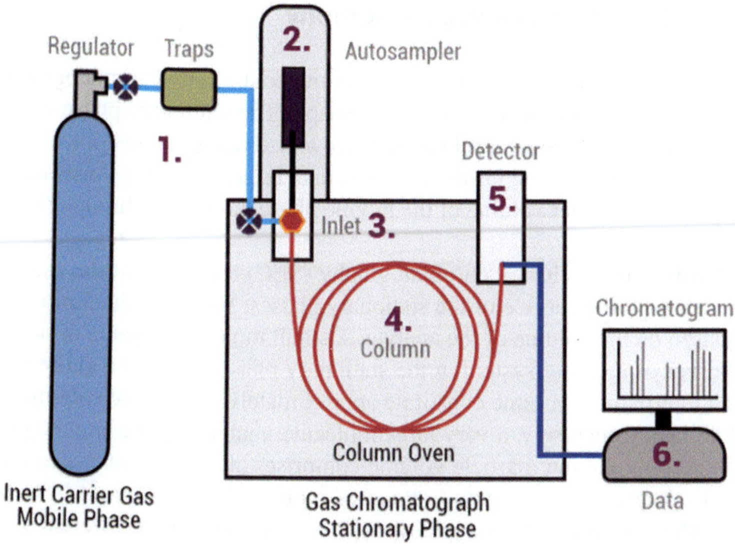

Fig. 10.6 Gas chromatography instrumentation and flow diagram

Theory: As the gas moves the solute (analyte) through and over the stationary phase, the solute will be in equilibrium with the gas and the solid phase. Since there is a mobile phase, the separation will appear as a chromatogram showing the separation of the analytes.

GC Applications

Food Analysis: The analysis of food involves the examination of lipids, proteins, carbohydrates, preservatives, flavors, essential oils, aroma constituents, colorants, texture modifiers, as well as vitamins, steroids, drugs, pesticide residues, and trace elements.

Quality Control Analysis: GC plays a vital role in confirming the presence and quantities of analytes in food products, such as fruits, fruit-derived foodstuffs, vegetables, soft drinks, tea, and coffee.

Food and Cancer: GC is instrumental in identifying chemicals with carcinogenic properties, including hydrocarbons, amines, certain drugs, some metals, and naturally occurring substances in plants and molds. Nitrosamines, for example, can be detected in trace quantities.

Plant Secondary Metabolites and Drugs: The NIST/EPA/NIH Mass Spectral Library contains more than 300,000 mass spectra for over 267,000 different compounds and more than 400,000 chemical structures and retention indices. Over 600 drugs, poisons, and metabolites have published analytical procedures, chromatographic methods, and retention data. These resources are valuable in profiling plant secondary metabolites.

10.6.7 Gel Permeation Chromatography

Gel filtration chromatography, also known as molecular sieve chromatography, separates molecules based on their size and shape. The stationary phase consists of beads with pores that cover a relatively narrow size range. In this process, smaller molecules spend more time inside the beads than larger molecules, causing them to elute later, after a larger volume of the mobile phase has passed through the column (Fig. 10.7).

Principle: Size exclusion chromatography (SEC) requires that the analyte does not interact with the surface of the stationary phases. Variations in elution time are solely based on the volume of the analyte. A small molecule capable of penetrating every corner of the pore system in the stationary phase, where the entire pore volume and interparticle volume constitute approximately 80% of the column volume, will elute late. Conversely, a very large molecule unable to penetrate the pore system, where only the interparticle volume comprises about 35% of the column volume, will elute earlier after this volume of mobile phase has passed through the column. The fundamental principle of SEC is that particles of different sizes will elute (filter) through a stationary phase at different rates, while particles of the same size should elute together. Cross-linked polymers serve as molecular sieves in the form of gels, and these polymers are uncharged and inert, meaning they do not bind or react with the substances under analysis.

Three distinct types of gels are commonly used:

Dextran: This is a homopolysaccharide consisting of glucose residues. Prepared with varying degrees of cross-linking to regulate pore size, it is available in dry bead form that swells upon the addition of water. Sold under the trade name Sephadex, it is primarily used for separating small peptides and globular proteins with small to average molecular masses.

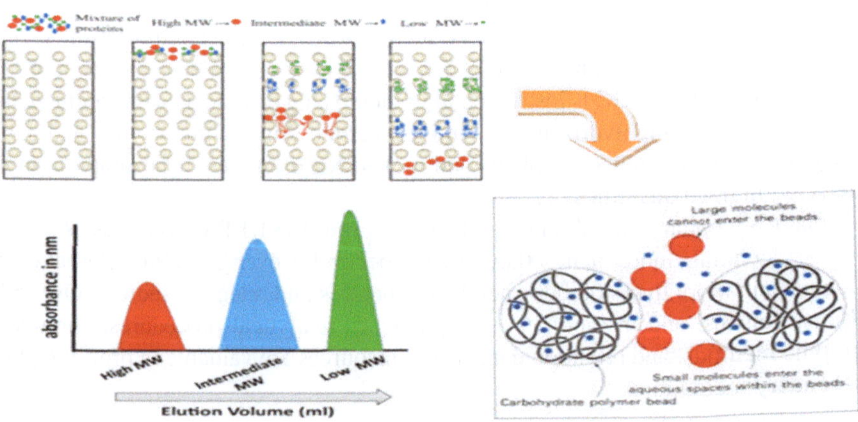

Fig. 10.7 Gel permeation chromatography depicting the separation of proteins; working

Polyacrylamide: These gels are created by cross-linking acrylamide with N, N-methylene bisacrylamide. The pore size is determined by the extent of cross-linking. Polyacrylamide gels exhibit separation properties similar to those of dextrans and are marketed as Bio-Gel P, with a diverse range of available pore sizes.

Agarose: Agarose is a linear polymer made of D-galactose and 3,6-anhydro-L-galactose. It dissolves in boiling water and forms a gel upon cooling, with hydrogen bonds maintaining the gel structure. The pore size of the gel depends on the agarose concentration—lower concentrations create larger pores. Compared to Sephadex or Bio-Gel P, agarose gels have significantly larger pores, making them ideal for separating large globular proteins or long, linear molecules like DNA.

Applications of Gel Filtration

1. Purification of enzymes and other proteins.
2. Estimation of molecular weight, primarily for globular proteins: This involves running several proteins with known molecular weights on the column and determining their elution volumes. By plotting the elution volumes against the logarithm of the molecular weight of the corresponding proteins, a straight line is obtained for the specific separation range of the gel in use. If the elution volume of a protein with an unknown molecular weight is identified, it can be compared to the calibration curve to determine its molecular weight.

Advantages

1. Unlike in ion exchange or affinity chromatography, molecules do not adhere to the medium, so changes in buffer composition do not directly impact resolution.
2. Well-suited for biomolecules sensitive to variations in pH, concentration of metal ions or co-factors, and harsh environments.
3. Conditions can be adjusted to accommodate the specific sample type or meet requirements for further purification, analysis or storage without compromising the separation.

It can be employed following any chromatography technology since components of the elution buffer do not influence the final separation.

10.6.8 Ion Exchange Chromatography

Similar to other chromatography methods mentioned earlier, the Ion Exchange Chromatography (IEC) technique uses both a mobile and stationary phase for the separation of molecules based on their charge. In this method, the stationary phase

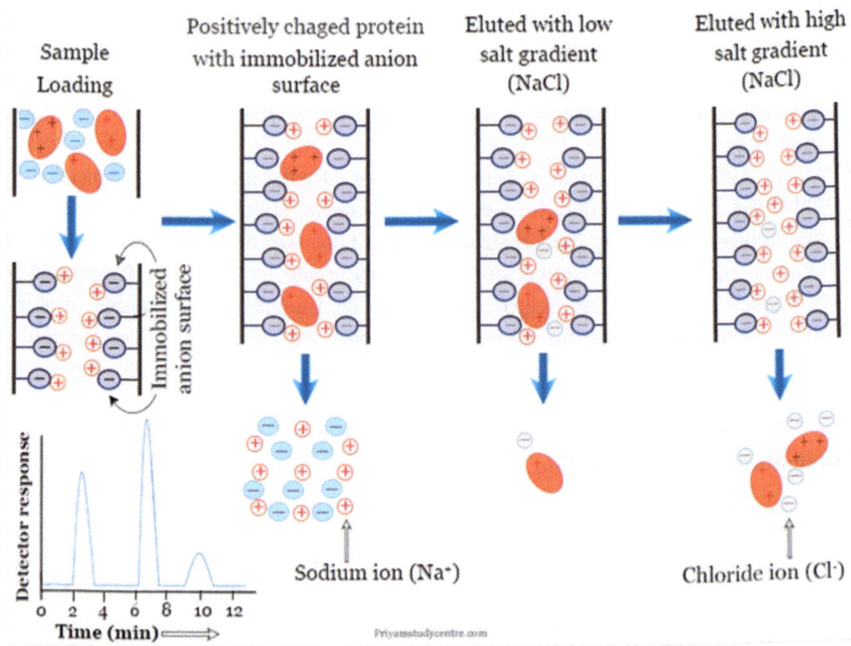

Fig. 10.8 A schematic diagram of ion exchange chromatography

consists of a column filled with ion exchange resins, while the mobile phase typically comprises an aqueous buffer system into which the mixture to be separated is introduced. Within the system, counter ions exist in equilibrium between the two phases, leading to two potential IEC formats: anion-exchange and cation-exchange (refer to Fig. 10.8). Resins that exchange cations are referred to as "catex," while those exchanging anions are termed as "annex." Exchangeable matrix counter ions encompass protons (H+), hydroxide groups (OH−), singly charged monoatomic ions (Na+, K+, Cl−), doubly charged monoatomic ions (Ca2+, Mg2+), and polyatomic inorganic ions (SO4 2−, PO4 3−), as well as organic bases (NR2 H+) and acids (COO−).

Advantages and Uses

Ion exchange chromatography is employed in plant genetic resources for its ability to separate and purify molecules based on their net charge. This technique is particularly valuable in the context of plant genetic resources for several reasons:

(a) *Separation of Charged Biomolecules:* Plant genetic resources often involve the study of charged biomolecules such as proteins and nucleic acids. Ion exchange chromatography allows for the separation of these molecules based on their charge characteristics, enabling researchers to isolate and analyze specific components.

(b) *Purification of Genetic Material:* Often due to the need to purify genetic material, including DNA and RNA, ion exchange chromatography facilitates the purification of these nucleic acids by exploiting differences in their charge properties, leading to cleaner and more concentrated samples.
(c) *Isolation of Specific Proteins:* Many genetic studies focus on understanding the functions of specific proteins. Ion exchange chromatography enables the isolation of proteins based on their net charge, allowing researchers to study individual proteins in more detail.
(d) *Analysis of Metabolites:* A number of metabolites that play crucial roles in plant development and responses to environmental factors are analyzed by ion exchange chromatography for separating these charged compounds, providing insights into plant metabolic pathways.
(e) *Enzyme Purification:* Ion exchange chromatography is a powerful tool for enzyme purification, helping to obtain highly pure and active enzymes from plant tissues.

Questionnaire

(a) What are the basic principles underlying chromatography techniques?
(b) Explain the key components of an HPLC system and their roles in the separation process?
(c) What factors influence the choice of ion exchange resins in a given application?
(d) What is the fundamental principle underlying ion exchange and how does it facilitate the separation of ions?
(e) What methods are commonly used for visualizing separated compounds on TLC plates?

10.7 Rapid and Non-destructive Methods in Germplasm Evaluation

10.7.1 NIR and IR Spectroscopy

Near-infrared (NIR) spectroscopy and infrared (IR) spectroscopy are indispensable tools in food chemistry, enabling rapid, non-destructive analysis of complex food matrices.

Principle: Both techniques operate on absorption spectroscopy principles, wherein molecules absorb specific wavelengths of electromagnetic radiation corresponding to their vibrational frequencies.

Spectrum Range

- NIR spectroscopy covers infra-red regions close to the visible spectrum (800 nm–2500 nm) and relies on molecular vibrations, specifically oscillatory alterations in bond lengths (stretching vibrations) and angles between these bonds (deformation vibrations), particularly involving polar bonds such as -C=O, -C-SH, C-OH, and C-NH. Unlike FTIR spectroscopy, which primarily detects fundamental vibrational transitions, NIR spectroscopy detects overtones and combination bands.

- IR spectroscopy, on the other hand, covers the mid-infrared region from 400 to 4000 cm^{-1} (2500–25,000 nanometres). It provides detailed information about the molecular structure and functional groups by detecting fundamental vibrational transitions.

Advantages

- This technique can be applied in various modes, including reflectance, transmission, interactance, and transflectance, offering flexibility in sample analysis. Its advantages include being chemical-free, cost-effective, and environmentally friendly.
- Chemical-free, cost-effective, and environmentally friendly, making it suitable for both laboratory and on-site applications.
- IR spectroscopy is widely used in food chemistry for qualitative and quantitative analysis, offering high sensitivity and spectral precision.

Disadvantages

- NIR spectroscopy faces constraints such as background interference, complex spectral data, light scattering effects, and limited sensitivity to minor constituents.
- More extensive sample preparation may be required in IR than in NIR spectroscopy.

Advancements and Applications

The above technologies have undergone significant advancements, offering expeditious, precise, and cost-effective analysis, with probe adaptability for online applications and multivariate data analysis.

The development of compact, handheld autonomous spectrometers has enabled NIR/IR spectral analysis in both laboratory environments and the field. NIR spectroscopy is particularly effective in process monitoring and quality control, involving stages like acquiring spectra, preprocessing, recognizing and excluding outlier samples, formulating and validating models, and projecting target parameters of unknown samples.

Chemometrics is essential to extract chemically pertinent insights and devise calibration models correlating spectral characteristics with sample parameters. The amalgamation of NIR and chemometrics is crucial for quality control and prompt detection, facilitating the understanding of composition, structure, and relevant information about substances.

NIRS and IR have revolutionized germplasm screening, facilitating the assessment of multiple plant genetic traits. NIRS expedites the selection of superior germplasm, supporting breeding and research while conserving plant genetic diversity.

10.7.2 Raman Spectroscopy

Raman spectroscopy is a powerful analytical technique that provides detailed information about molecular vibrations and structural characteristics of materials, discovered by Indian physicist Chandrasekhara Venkata Raman in 1928.

Principle

Raman spectroscopy is based on the Raman effect. When incident light interacts with molecules in a sample, most of the photons undergo elastic scattering, known as Rayleigh scattering, A small fraction of incident photons undergoes inelastic scattering, known as Raman scattering, where the scattered photons have different energies (frequencies) due to energy exchange with the molecules. Therefore, the shift reveals the sample's molecular structure, chemical composition, and interactions.

Light Sources

Light sources include lasers with wavelengths ranging from ultraviolet (UV) to near-infrared (NIR), such as argon ion (488.0 and 514.5 nm), krypton ion (530.9 and 647.1 nm), helium-neon (632.8 nm), and diode lasers (630 and 780 nm). Longer-wavelength lasers, particularly in the near-infrared region (e.g., 1064 nm), are preferred due to reduced fluorescence interference. Molecules with varying degrees of polarizability exhibit different Raman scattering intensities. Functional groups such as –C–X, –CN, –S–H–, –N=N–, –C=S–, –C=C–, –S–S–, –C–S–, –C–NO2 exhibit significant polarizability changes, yielding robust Raman signals.

Types of Raman Scattering

Stokes Scattering: The most common type of Raman scattering. It involves the emission of photons with lower energy (longer wavelength) compared to the incident photons. This phenomenon occurs when a molecule absorbs energy from incident photons and undergoes a transition to a higher vibrational energy state. As a result, the scattered photons lose energy, corresponding to the energy gained by the molecule during the vibrational transition.

Anti-Stokes Scattering: This involves the emission of photons with higher energy (shorter wavelength) compared to the incident photons. This phenomenon occurs when a molecule loses energy to incident photons and undergoes a transition from a higher vibrational energy state to a lower one. As a result, the scattered photons gain energy, corresponding to the energy lost by the molecule during vibrational relaxation.

Applications

- Agriculture and Plant Breeding: Raman spectroscopy enables rapid and accurate identification of genetic variations and desirable traits in plants, facilitating breeding programs for improved crop varieties.
- Food Industry: It is used to analyze the composition of food products, detect adulteration, and assess nutritional content.
- Seed Quality and Germination: Raman spectroscopy helps in the identification of species, assessment of seed quality, and monitoring of germination processes.
- Quality Control: This technique distinguishes genuine products from adulterated or counterfeit ones by analyzing characteristic molecular signatures, ensuring product quality and consumer safety.

10.7.3 Near-Infrared (NIR) Spectroscopy and Raman Spectroscopy for Comprehensive Non-destructive Evaluation

Aspect	NIR spectroscopy	Raman spectroscopy
Principle	Sensitive to molecular vibrations in the near-infrared region (overtones and combinations of fundamental modes)	Based on inelastic scattering of monochromatic light, with frequency shifts revealing molecular vibrations
Information provided	Probes bulk properties, providing overall chemical composition and molecular structure	Offers detailed insights into specific chemical bonds, functional groups, and molecular conformations
Applications	Well-suited for samples with organic compounds, such as food, pharmaceuticals, and agricultural products	Ideal for identifying and characterizing individual compounds, differentiating similar molecules, and detecting subtle changes like polymorphism
Depth of penetration	High depth of penetration, suitable for analyzing larger sample volumes	Generally lower penetration depth, focusing on surface and near-surface layers
Sensitivity vs. selectivity	High sensitivity to overall sample composition	High selectivity, with the ability to distinguish specific molecular features
Limitations	May lack specificity when analyzing samples with similar chemical compositions	Can be less effective with samples that fluoresce under laser illumination
Data analysis techniques	Uses chemometric modelling and multivariate analysis to extract and correlate spectral data with sample properties	Similar data analysis techniques are used to enhance interpretation and quantitative analysis
Synergistic advantages	Combining NIR's bulk property sensitivity with Raman's molecular specificity provides a comprehensive understanding of complex samples	Enhanced accuracy and reliability in analyses, particularly for diverse chemical components

10.8 Chemometric Techniques for Prediction Modelling and Classification

Chemometric techniques for prediction modelling and classification involve using mathematical and statistical methods to analyze complex data sets. These techniques help in building models that predict outcomes or classify data into different categories based on patterns and relationships identified in the data.

10.8.1 Linear Regression Methods

Near-Infrared (NIR) Chemometrics heavily relies on linear regression methods, such as Partial Least Squares (PLS), Modified PLS (mPLS), Principal Component Regression (PCR), Orthogonal Partial Least Squares (OPLS), and Iteratively Reweighted Partial Least Squares (IPLS), to establish predictive models between spectral data and analyte concentrations. These methods extract relevant information from complex NIR spectra while mitigating multicollinearity and noise. Essential preprocessing steps include spectral data normalization, derivatization to enhance analyte absorption features, and noise reduction via smoothing techniques like moving average and Savitzky–Golay filtering. Scatter correction methods, including Multiplicative Scatter Correction (MSC) and Standard Normal Variate (SNV), address unwanted spectral variations caused by scattering effects and baseline shifts. Detrending methods help remove systematic trends, while inverse MSC and weighted MSC refine scatter correction approaches for better model accuracy. The PLS and IPLS techniques are particularly effective in handling multicollinearity by orthogonalizing and reweighting spectral variables. Rigorous model validation using metrics such as R-Squared (RSQ), Standard Deviation (SD), Standard Error of Cross Validation (SECV), Ratio of Performance to Deviation (RPD), and Root Mean Square Error (RMSE) are crucial for assessing model reliability, predictive accuracy, and performance. Thus, integrating linear regression methods with spectral preprocessing and thorough validation is essential for developing robust NIR Chemometrics models for diverse analytical applications.

10.8.2 Non-linear Regression Modelling

Non-linear modelling in Near-Infrared (NIR) Chemometrics for food and agriculture has revolutionized analytical techniques, enabling more accurate predictions and a better understanding of complex spectral data relationships. Advanced methodologies such as Support Vector Machines (SVM), Artificial Neural Networks (ANN), Deep Learning (DL), Convolutional Neural Networks (CNN), Deep Neural Networks (DNN), Recurrent Neural Networks (RNN), Long Short-Term Memory (LSTM), Inception modules, and Hybrid models have significantly affected this field. SVM effectively handles non-linear relationships by mapping data into higher-dimensional space and finding optimal hyperplanes for classification or regression, while ANN captures intricate patterns and correlations within spectral data, enhancing prediction accuracy for various traits.

DL techniques, including CNNs, have shown remarkable success in feature extraction from NIR spectra, particularly in food authentication and quality control. CNNs efficiently capture spatial dependencies in spectral data, enabling precise classification based on their NIR signatures. RNNs and LSTM networks have emerged as powerful tools in food chemometrics and NIR spectroscopy due to their ability to model sequential data and capture temporal dependencies within spectral information, allowing for the prediction of changes in food properties over time. For example, they can be used to predict changes in chemical composition during food processing or to monitor the ripening process of fruits and vegetables and the ageing of seeds. By analyzing the relationship between process variables and spectral information, these networks can suggest optimal settings for parameters such as temperature, pressure or mixing time, leading to improved efficiency and product quality. Inception modules, known for their multi-scale feature extraction capabilities, have been integrated into CNN architectures to enhance the discriminative power of models in NIR Chemometrics. Hybrid models, combining linear techniques like Principal Component Regression (PCR) with non-linear approaches such as CNNs or LSTM networks, offer a comprehensive solution for NIR Chemometrics in food and agriculture. Combining CNNs with LSTM networks and Inception modules allows for the integration of spatial and temporal information from NIR spectra, leading to superior performance in predicting complex agricultural outcomes. In summary, non-linear modelling techniques, including SVM, ANN, DL, CNN, RNN, LSTM, Inception modules, and Hybrid models, have transformed NIR Chemometrics in food and agriculture, enabling more accurate predictions and deeper insights into agricultural processes and food quality assessment.

10.9 Summary

Analytical tools and techniques play vital roles in nutritional research, food quality control, and regulatory compliance, providing valuable insights into the nutritional content. Each method offers unique advantages in terms of sensitivity, specificity, speed, and applicability to different types of nutritional analysis. Proximate composition analysis techniques determine the major components of food and provide insights into its nutritional quality and characteristics. Each technique plays a crucial role in assessing the nutritional profile of food, helps evaluate its energy content and nutrient composition, and forms the foundation for understanding and interpreting the nutritional value and properties of various food products. Chromatographic techniques are fundamental to nutritional research for analyzing and quantifying a wide range of compounds present in food. High-performance liquid chromatography (HPLC) has been extensively used to separate and quantify vitamins, amino acids, fatty acids, antioxidants, and other bioactive compounds in complex food matrices. Gas chromatography (GC) is valuable for analyzing volatile compounds, such as fatty acids and flavors, and provides insights into food aroma and lipid composition. Rapid and non-destructive indirect assessment methods like NIR and Raman spectroscopy are used for nutrient profiling and sample categorization. These techniques use variations in spectra, influenced by the sample chemical composition of organic molecules, to accurately determine the composition of nutrients. Their applications span various industries, including food science, agriculture, and pharmaceuticals, where they enable rapid and precise assessments without altering samples, ensuring effective quality control and process monitoring.

Multiple-Choice Questions

1. What is the purpose of adding a catalyst such as copper sulfate or selenium during the digestion step of the Kjeldahl method?
   (a) To increase the boiling point of the solution
   (b) To facilitate the conversion of organic nitrogen compounds into ammonium sulfate
   (c) To trap the released ammonia gas
   (d) To titrate the ammonia with a standard acid solution
2. Which of the following substances is added to increase the boiling point of the solution during digestion in the Kjeldahl method?
   (a) Copper sulfate
   (b) Selenium
   (c) Potassium sulfate or lithium sulfate
   (d) Boric acid
3. What is the purpose of alkali distillation in the Kjeldahl method?
   (a) To convert organic nitrogen compounds into ammonium sulfate
   (b) To trap ammonia gas in a solution of boric acid
   (c) To facilitate the digestion process
   (d) To determine the amount of nitrogen, present in the sample
4. Which indicator is typically used in the distillate solution to detect the presence of ammonia in the Kjeldahl method?
   (a) Bromocresol green
   (b) Methyl red
   (c) Phenolphthalein
   (d) Litmus
5. In the Kjeldahl method, what is titrated with a standard acid solution to determine the amount of nitrogen present in the sample?
   (a) Copper sulfate
   (b) Ammonium sulfate
   (c) Boric acid
   (d) Ammonia trapped in the distillate
6. How is the endpoint of the titration determined in the Kjeldahl method?
   (a) By the change in color of the mixed indicator
   (b) By the addition of a catalyst
   (c) By the boiling point of the solution
   (d) By the presence of ammonia gas

7. What serves as the stationary phase in paper chromatography?
   (a) The solvent
   (b) The cellulose support
   (c) The mixture to be separated
   (d) The solvent front
8. What principle drives the movement of the solvent through the paper in paper chromatography?
   (a) Evaporation
   (b) Capillary action
   (c) Gravity
   (d) Osmosis
9. In paper chromatography, what does Rf stand for?
   (a) Retention factor
   (b) Relative flow
   (c) Ratio of fractions
   (d) Retention front
10. What does the Rf value represent in paper chromatography?
   (a) The rate of solvent flow
   (b) The rate of solute migration
   (c) The ratio of the distance moved by a solute to the distance moved by the solvent
   (d) The ratio of the distance moved by the solvent to the distance moved by the solute
11. What is the typical range of Rf values in paper chromatography?
   (a) 0 to 1
   (b) 1 to 10
   (c) −1 to 1
   (d) 0 to infinity
12. Which of the following statements about the cellulose support in paper chromatography is true?
   (a) It serves as the solvent.
   (b) It is hydrophobic.
   (c) It is rich in bound water.
   (d) It is immiscible with the solvent.
13. What is the primary application of HPLC in chemistry and biochemistry research?
   (a) DNA sequencing
   (b) Cell culture techniques
   (c) Analysis of intricate mixtures
   (d) Electron microscopy
14. How does HPLC contribute to the purification of chemical compounds?
   (a) By measuring electrical conductivity
   (b) By separating components based on size
   (c) By using high-pressure pumps and columns
   (d) By inducing radioactive decay

15. Which of the following is not a purpose served by HPLC in the formulation of processes for synthesizing chemicals?
    (a) Monitoring degradation
    (b) Verifying the purity of raw materials
    (c) Enhancing process yields
    (d) Accelerating chemical reactions
16. In quality control, how does HPLC contribute to verifying the purity of raw materials?
    (a) By measuring the absorbance of ultraviolet light
    (b) By determining the viscosity of the material
    (c) By analyzing the chemical composition of the material
    (d) By assessing the material's color change
17. What role does HPLC play in quantifying assays of final products?
    (a) It measures the pH of the final product.
    (b) It determines the concentration of specific components in the product.
    (c) It records the temperature at which the product is stored.
    (d) It observes the product's crystalline structure.
18. How does HPLC contribute to assessing product stability while monitoring degradation?
    (a) By measuring the product's fluorescence
    (b) By analyzing changes in the product's molecular weight
    (c) By monitoring changes in the product's chemical composition over time
    (d) By determining the product's electrical conductivity
19. What is the primary principle underlying both NIR spectroscopy and IR spectroscopy?
    (a) Absorption spectroscopy
    (b) Fluorescence spectroscopy
    (c) Emission spectroscopy
    (d) Raman spectroscopy
20. What type of molecular vibrations does NIR spectroscopy primarily rely on?
    (a) Rotational vibrations
    (b) Symmetric stretching vibrations
    (c) Overtone and combination bands of vibrational transitions
    (d) Translational vibrations
21. Which of the following is NOT a mode of NIR spectroscopy?
    (a) Reflectance
    (b) Transmission
    (c) Fluorescence
    (d) Interactance
22. In which region of the electromagnetic spectrum does NIR spectroscopy operate?
    (a) Ultraviolet
    (b) Visible
    (c) Near-infrared
    (d) Far-infrared

23. What type of molecular vibrations does IR spectroscopy primarily detect?
    (a) Overtone and combination bands
    (b) Symmetric bending vibrations
    (c) Fundamental vibrational transitions
    (d) Rotational-vibrational transitions
24. Which of the following regions of the electromagnetic spectrum does IR spectroscopy cover?
    (a) Near-infrared
    (b) Mid-infrared
    (c) Far-infrared
    (d) Ultraviolet-visible
25. What advantage does NIR spectroscopy offer over traditional analytical techniques?
    (a) Requires extensive sample preparation
    (b) Slow analysis speed
    (c) Minimal sample preparation
    (d) Limited ability to analyze multiple constituents simultaneously
26. What is the disadvantage of IR spectroscopy when compared to NIR spectroscopy?
    (a) Limited spectral precision
    (b) Requires sophisticated data analysis techniques
    (c) Lower sensitivity
    (d) More extensive sample preparation

Answers

1. (b)
2. (c)
3. (d)
4. (a)
5. (d)
6. (a)
7. (b)
8. (b)
9. (a)
10. (c)
11. (a)
12. (c)
13. (c)
14. (c)
15. (d)
16. (a)
17. (b)
18. (c)

19. (a)
20. (c)
21. (c)
22. (c)
23. (c)
24. (b)
25. (c)
26. (d)

Suggested Reading

Burgess C (2007) The basics of spectrophotometric measurement. In: Techniques and instrumentation in analytical chemistry, vol 27, pp 1–19. Elsevier

Greenfield H, Southgate DA (2003) Food composition data: production, management, and use. Food Agric Org:1–288

Guyot D, Westad F, Houmøller LP (2004) Multivariate data analysis–in practice. Camo Process AS, Norway

Latimer GW (2023) Official methods of analysis of AOAC International, 22nd Edition Published by Oxford University Press, Oxford

Robards K, Ryan D (2021) Principles and practice of modern chromatographic methods. Academic Press, pp 1–515

Genomics Tools for PGR Management

11

Debjani Roy Choudhury, Laxmi Sharma, and Rakesh Singh

Learning objectives

- Understand the basis of polymorphism and its utilization in the generation of genomic resources.
- Understand the types and role of genomic tools for plant genetic resource (PGR) management.
- Explain the uses and application of various genomic tools in pre-breeding and breeding programs and related topics.

Keywords

Plant genetic resources · Germplasm · PGR · Genomics · Markers · Trait discovery · GWAS · Transcriptome · Transgenic · Fingerprinting · QTLs · Mapping · Pangenome · Allele mining

11.1 Introduction

Plant genetic resources (PGRs) represent the total pool of heredity that includes various alleles of different genes in the form of cultivated crop species or their wild relatives as useful resources for improving farming resilience and enhanced profitability. Since PGRs form the fundamental inputs for genetic improvement, deploying these resources in different sectors can fetch improved products for agricultural,

D. R. Choudhury · L. Sharma · R. Singh (✉)
Division of Genomic Resources, ICAR-National Bureau of Plant Genetic Resources, New Delhi, India
e-mail: laxmi.sharma@icar.org.in; rakesh.singh2@icar.org.in

K. Tripathi et al. (eds.), *Textbook of Plant Genetic Resources*,
https://doi.org/10.1007/978-981-96-5024-8_11

commercial, forest crops and agro-industry. In addition, their utilization in systematic research like evolutionary biology, cytogenetics, biochemical, physiological, phylogenetic, ecological, and molecular studies has the potential to generate novel information that can be further utilized in crop improvement programs. Moreover, the conservation and utilization of PGR are important for improvements in agricultural and forestry production for yield, quality improvement, and resilience to biotic and abiotic stresses. Biotechnological advancements in the past decades and the ongoing evolution of genomic tools present an exciting prospect for continued enhancement and utilization of PGR. The initial concept of genetic material has opened up avenues for modern genetic tools. In this chapter, the basics of the structure and function of DNA have been discussed followed by the conceptualization of genetic markers, their significance, and the development of modern genetic tools and their applications.

11.2 The Genetic Material

11.2.1 Structure of DNA

Nucleic acids, the macromolecules comprising nucleotide polymers, contain an individual's genetic information. Nucleic acids were initially isolated from pus cells in 1869 by Friedrich Miescher. In the 1880s, he discovered the presence of nitrogenous bases (purines and pyrimidines) in nucleic acids. These are mainly of two types: deoxyribonucleic acids (DNA) and ribonucleic acids (RNA). DNA is a long-chain polymer of deoxyribonucleotides, whereas RNAs are poly-ribonucleotides. Each nucleotide is composed of a sugar-phosphate moiety (ribose or deoxyribose sugar attached to a phosphate molecule) and a nitrogenous base. Phosphate groups confer DNA a negative charge that allows polymerization of several deoxyribonucleotides. Four different types of nitrogen bases are present in DNA, categorized into two groups: purines (nine-membered double rings with nitrogen at positions 1, 3, 7, and 9) and pyrimidines (six-membered rings with nitrogen at positions 1 and 3). DNA has two purines, adenine (A) and guanine (G), and two pyrimidines, cytosine (C) and thymine (T) (Fig. 11.1). The prokaryotic DNA is present in the nucleoid and plasmids having a circular structure, while eukaryotic DNA is highly organized into chromatin in the nucleus.

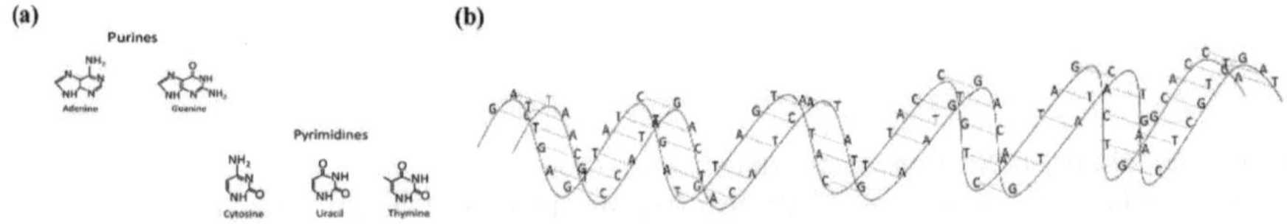

Fig. 11.1 (**a**) Nitrogenous bases found in DNA/RNA and (**b**) double helical structure of DNA

The first accurate and most acceptable model of DNA was proposed by Watson and Crick in 1953. They proposed that

- DNA comprises a double helical structure with two chains docking sugar phosphate on the exterior and nitrogen bases towards the interior of the double helix.
- The nitrogen bases of the two chains form complementary pairs through hydrogen bonds (A-T, C-G).
- Two chains are antiparallel, with one oriented $5' \rightarrow 3'$ and the other $3' \rightarrow 5'$. The helically twisted chains resemble a rope ladder with rigid steps arranged in a spiral, with each turn containing 10 nucleotides.

Chargaff's Rules, proposed in 1950, outline observations on DNA bases and components:

(i) Purine and pyrimidine base pairs are present in equal amounts, meaning adenine + guanine = thymine + cytosine ([A + G] = [T + C]).
(ii) The molar amount of adenine equals the molar amount of thymine, and the molar concentration of guanine is equivalent to that of cytosine ([A] = [T]; [G] = [C]).
(iii) Deoxyribose sugar and phosphate occur in equimolar proportions.
(iv) A–T base pairs are rarely equal to C–G base pairs.
(v) The ratio of [A + T] / [G + C] varies with species, with higher ratios in advanced organisms and lower ratios in primitive ones.

The complementary nature of the two DNA chains arises from specific base pairing. For example, the sequence on one chain, such as AAGCTCAG, has a complementary sequence of TTCGAGTC on the other chain. Further, the twisting of the duplex results in the formation of two types of alternate grooves: major (22 Å) and minor (12 Å). One spiral turn encompasses approximately 10 nucleotides on each DNA strand, covering a distance of about 3.4 nm (34 Å). This arrangement ensures that adjacent nucleotides or their bases are separated by a space of about 0.34 nm (3.4 Å).

Three distinct DNA structures, A-form, B-form, and Z-form, are found in nature. The B-form DNA, the most prevalent one, is a right-handed double helix. The A-form DNA also features a right-handed double helix like the B-DNA form, but it is more condensed. On the other hand, Z-form DNA adopts a left-handed double helix.

RNA is composed of a chain of monomers known as ribonucleotides, each comprising the 5-carbon sugar ribose, a phosphate group, and a nitrogenous base. The phosphate binds to the $5'$ carbon of the ribose, while the nitrogenous base attaches to the $1'$ carbon. The four bases found in RNA are adenine, guanine, cytosine, and uracil. Unlike DNA, RNA is more susceptible to breakdown. The distinctive properties of RNA make it well-suited as a genetic messenger in protein synthesis. The concept of this genetic messenger, mRNA, was originally proposed by François Jacob and Jacques Monod.

11.2.2 Functions of DNA

- Genetic information: DNA as the genetic material carries all hereditary information, typically encoded in its nitrogen bases arrangement.
- Replication: DNA possesses a unique ability to produce identical copies (autocatalytic function), essential for transferring genetic information across generations.
- Chromosomes: DNA is housed within chromosomes, ensuring the equitable distribution of DNA during cell division.
- Recombination: Meiotic processes, particularly crossing over, provide new combinations of genes known as recombination.
- Mutations: Changes in the sequence of nitrogen bases, such as additions, deletions, or incorrect replications, lead to mutations. Mutations are the origin of variations and contribute to evolution.
- Transcription: DNA is transcribed to RNAs representing a heterocatalytic activity of DNA.
- Cellular metabolism: DNA controls cellular metabolic reactions with the assistance of specific RNAs, facilitating the synthesis of particular proteins, enzymes, and hormones.
- Differentiation and development: Due to the distinct functioning of specific DNA regions or genes, various parts of organisms undergo differentiation in terms of shape, size, and functions as well as development.
- DNA Fingerprinting: Using unique hypervariable microsatellite DNA sequences in each individual, identification and relationship deciphering are possible.

Some important functions of the DNA are explained below:

11.2.2.1 DNA Replication

Prokaryotic DNA Replication: Within a typical bacterial cell, which contains significantly fewer base pairs of DNA (ranging from one to four million), the process of DNA replication is categorized into four main stages—initiation, unwinding, primer synthesis, and elongation DNA Replication Steps.

Initiation and Unwinding

- Initiator proteins bind to the replication origin (oriC) and unwind the DNA double helix into two single strands.
- Proteins like DNA helicases, single-strand-binding proteins, and topoisomerases assist in the unwinding process.
- Replication forks move in opposite directions along the single-stranded DNA.

Primer Synthesis

- RNA polymerase enzyme primase synthesizes short RNA primers to provide a 3′-OH group for DNA polymerase to initiate DNA synthesis.
- These RNA primers are later replaced with DNA nucleotides during the elongation step.

Elongation

- DNA polymerase adds nucleotides in the 5′-to-3′ direction to the new DNA strand, using the primers as a starting point.
- Different DNA polymerases in bacteria, such as DNA polymerase III, have specific roles in the elongation process.

Directional Replication

- DNA polymerase can only add nucleotides to the 3′ end, leading to continuous replication on the leading strand.
- The opposite strand, the lagging strand, undergoes discontinuous replication in short segments called Okazaki fragments.
- Okazaki fragments are typically 1000–2000 nucleotides long in bacteria and 100–200 nucleotides long in eukaryotes.

11.2.2.2 Gene Expression

Gene expression in prokaryotes: In prokaryotes, such as bacteria, gene expression occurs in the cytoplasm through transcription and translation. In this process, DNA is directly involved in the synthesis of mRNA with the help of transcription enzymes. RNA polymerase binds to the promoter region on the DNA, initiating the synthesis of a complementary mRNA strand. Prokaryotic mRNA lacks significant processing and is ready for translation as soon as transcription is complete. The coding of mRNA to protein takes place through translation. Ribosomes in the cytoplasm bind to the mRNA, and the process of translation begins. The ribosome reads the mRNA codons and transfers RNA (tRNA) molecules to bring in the corresponding amino acids. This resultant polypeptide chain folds into a functional protein.

Operons: Prokaryotic genes are frequently organized into operons, where multiple genes are transcribed as a single mRNA molecule. The operon structure allows coordinated gene expression regulation in response to environmental conditions. Prokaryotic gene expression is primarily regulated at the transcriptional level. Regulatory proteins, such as repressors and activators, control the access of RNA polymerase to the promoter.

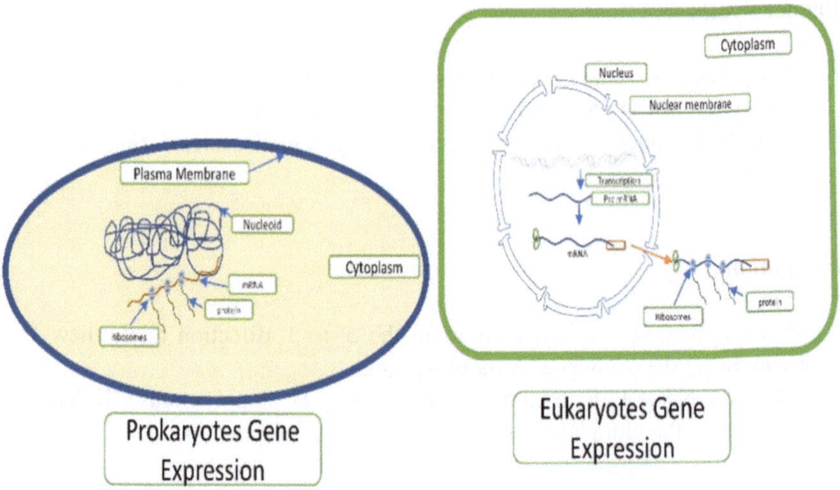

Fig. 11.2 Regulation of gene expression in prokaryotes and eukaryotes

Gene Expression in Eukaryotes: In eukaryotes, gene expression is a more complex process due to the presence of membrane-bound organelles, including the nucleus. The main steps (Fig. 11.2) involved are:

- <u>Transcription</u>: This occurs in the nucleus, where RNA polymerase transcribes the DNA into pre-mRNA. Unlike in prokaryotes, eukaryotic mRNA undergoes significant processing, including the addition of a 5′ cap, a poly-A tail at the 3′ end, and the removal of introns through RNA splicing. The mature mRNA is then transported to the cytoplasm.
- <u>RNA Processing</u>: Eukaryotic pre-mRNA undergoes modifications before it becomes mature mRNA. These modifications include capping, poly-adenylation, and splicing. Introns, noncoding regions, are removed, and exons, coding regions, are joined together.
- <u>Translation:</u> Translation takes place in the cytoplasm, where ribosomes bind to mature mRNA. The mRNA sequence is read in sets of three nucleotides (codons), and tRNA molecules found in the corresponding amino acids. This process allows the synthesis of a polypeptide chain, which folds into a functional protein.
- <u>Regulation:</u> Eukaryotic gene expression is regulated at multiple levels, including transcriptional, post-transcriptional, translational, and post-translational regulation. Transcription factors, chromatin modification, and RNA interference play crucial roles in controlling gene expression against various signals and developmental stages.

11.2.3 Genome Organization

The term "genome" denotes the entire linear DNA sequence encompassing all hereditary material present in the organism. Despite the vast information encoded in the linear DNA, many functional aspects of the genome are influenced by its three-dimensional (3D) structure.

Organization of DNA in Eukaryotes Within the eukaryotic system, chromosomes exhibit a highly organized structure where DNA is packed with DNA-binding proteins, first into chromatin and subsequently into chromosomes. During interphase, the chromosome adopts an uncoiled form known as chromatin. As the cell progresses into metaphase, the chromatin undergoes coiling and condensation, transforming back into visible metaphase chromosomes that can segregate into daughter cells. This condensation process represents a remarkable length contraction of approximately 10,000 times for each chromatin fiber.

Nucleosomes and Chromatin Structure

In eukaryotes, DNA is associated with numerous DNA-binding proteins throughout all phases of the cell cycle, collectively forming chromosomal DNA. These associated proteins are categorized as histones and nonhistone proteins. **Histones** are positively charged proteins rich in lysine and arginine, and interact with the negatively charged DNA. **Nonhistone proteins** are less positively charged and play functional roles in the replication, transcription, and translation of genes rather than chromatin structure. Typically, eukaryotic chromosomes consist of repetitive units with specialized structures known as nucleosomes, where DNA

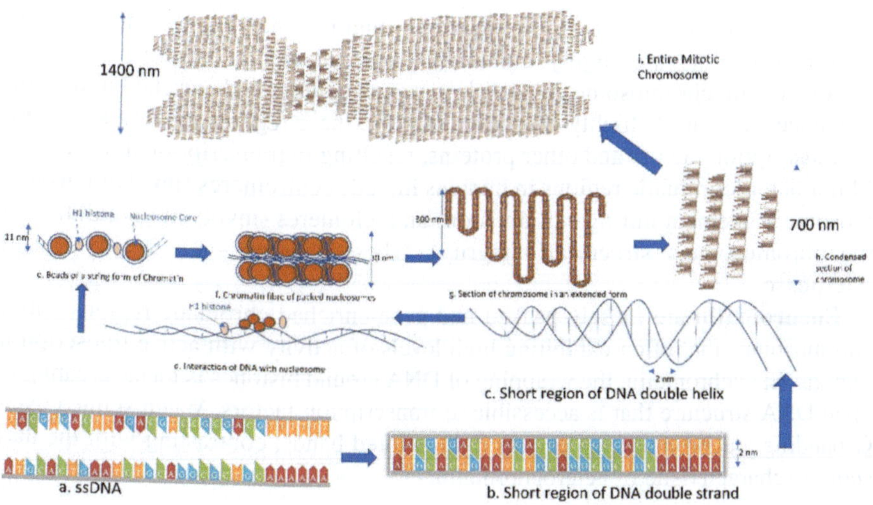

Fig. 11.3 Different levels of organization of nucleosomes and chromatin structure

is wrapped around histones, forming a linear array of chromatin structures (Fig. 11.3).

Primary Level of Organization Initially a 147-bp length DNA strand of 2 nm diameter winds around an octamer of histones. This histone octamer comprises two molecules of H2A, H2B, H3, and H4 each forming a left-handed superhelix structure to create nucleosomes. The DNA completes approximately 1.7 turns per nucleosome, and each nucleosome exhibits an ellipsoidal shape, measuring 11 nm at its longest point. This initial packing level effectively reduces the DNA helix length to about one-third of its original size.

Secondary Level of Organization In this phase, nucleosomes undergo additional packing to generate a solenoid structure of 30 nm diameter. Multiple nucleosomes coil around and stack upon each other, resulting in a sixfold increase in the compaction of DNA. This structure is distinctive of an uncoiled chromatin fiber during the interphase of the cell cycle. Histone H1 assumes a significant role in establishing this second level of packing.

Third Level of Organization In transition to mitotic chromosome, further compaction occurs, leading to metaphase chromosome formation. The 30-nm structures undergo additional folding into a series of looped domains, further condensing the chromatin fiber into a structure with 300 nm diameter resulting in chromosome arms (700 nm condensed chromosome section) which constitute a chromatid, one of the longitudinal subunits of the mitotic phase chromosome.

A chromosome contains a heterochromatic and euchromatic region. Heterochromatins are highly condensed and genetically inactive chromatic regions within chromosomes having lower gene content and enriched noncoding sequences. Due to its highly compact structure, these regions are less accessible to transcription factors and other proteins, resulting in transcriptional repression. Major heterochromatic regions in humans include centromeres (involved in chromosome movement during cell division) and telomeres (involved in maintaining the chromosome's structural integrity). These regions stain deeply during interphase.

Euchromatin are loosely packed and gene-enriched chromatic regions within chromosomes that often-exhibiting high levels of activity with active transcription regions. In euchromatin, the wrapping of DNA around histones is loose, creating an open DNA structure that is accessible to transcription factors. When stained using G-banding, euchromatin appears as light-colored bands, contrasting with the dark staining characteristic of heterochromatin.

11.3 DNA-Level Polymorphisms

Genomic variability at the level of DNA can be in various forms like single nucleotide polymorphisms (SNPs), transposable elements (e.g., Alu repeats), variable numbers of tandem repeats, structural alterations, and copy number variations. This variability can arise in both the nucleus and mitochondria. Two primary sources contribute to these variations: (1) mutations, which may result spontaneously or through external agents such as radiation, and (2) recombination. Once formed, DNA level polymorphisms can be heritable, enabling their tracking from parent to offspring.

Genome can be categorized based on known functional properties, with coding and noncoding regions. Coding regions have DNA sequences that primarily determine the amino acid sequences of encoded proteins. In contrast, noncoding DNA, often with nonidentified functions, may exist as single copy or multiple copies known as repetitive DNA. Regions of DNA devoid of protein-coding functions tend to exhibit more polymorphisms. Recent advances in genome understanding have shifted focus from discovering protein-coding genes to exploring repeat families and variations in copy number that influence gene dosage, playing a crucial role in genome structure, diversity, and evolution.

11.3.1 Types of Polymorphisms

- Single Nucleotide Polymorphisms (SNPs)—A single base change represents "high-density natural sequence variations." SNPs predominantly arise from errors like substitutions, insertions, and deletions. While SNPs may occur intragenic sequences or intergenic sequences, they are often concentrated in noncoding regions, with mostly unknown direct impacts on an individual's phenotype. However, the consequences of SNPs can vary depending on their location, leading to diverse outcomes at the phenotypic level.
- Insertion/Deletion Polymorphisms—This type of DNA variation involves the insertion or deletion of a specific nucleotide sequence, ranging from one to several hundred base pairs. Indels are distributed widely across the genome, with some considering single base pair changes as SNPs or repeat insertion/deletion as indels.
- Polymorphic Repetitive Sequences—DNA repeats are categorized as interspersed repeats or tandem repeats, dispersed throughout the genome, including retro (pseudo) genes and transposons. Tandem repeats, also known as variable number tandem repeats (≥2 bp in length), involve two copies to many thousand copies. Centromeres and telomeres are predominantly composed of tandem repeats. Tandem repeats, occurring in a head-to-tail orientation, can be further classified into macro-satellites, minisatellites, and microsatellites based on the size of each repeat unit. Macro-satellites having sequence repeats longer than 100 bp are the largest of the tandem DNA repeats, found on one or multiple

chromosomes. Minisatellites have moderate length sequences, typically between 10 and 100 bp, typically less than 50 bp. Microsatellites, also known as short tandem repeats (STRs), consist of repeat units below 10 bp.
- Structural and Copy Number Variations—Structural and copy number variations (CNVs) represent another frequent source of genome variability. The term CNVs include previously introduced terms like large-scale copy number variants (LCVs), copy number polymorphisms (CNPs), and intermediate-sized variants (ISVs). A few terms currently used include structural variations, which involve genomic alterations (e.g., inversions) with segments of DNA > 1 kb, copy number polymorphisms, representing duplication or deletion events involving >1 kb of DNA, and intermediate-sized structural variants, which are ~8–40 kb in size and can refer to CNVs or balanced structural rearrangements (e.g., inversions).

11.4 Development of Modern Genomic Tools

The efficient conservation, characterization, and utilization of PGRs have necessitated the interventions of genomic and bioinformatics tools as a prerequisite of crop improvement programs. The availability of tools like next-generation sequencing (NGS), high–throughput sequencing (HTS), high–throughput phenotyping (HTP), high-throughput genotyping platforms etc. has presented landmarks in the efficient utilization of PGRs. Genomic tools are widely used.

- to characterize the genetic diversity of PGR collections,
- to identify accessions with valuable traits,
- to trace the evolutionary history of PGR,
- to understand the origin and distribution of genetic diversity,
- to identify genes associated with desirable traits to accelerate crop breeding,
- to develop genomic selection methods,
- to select individuals with high genetic potential for desired traits,
- identify PGRs that are resilient to pests, diseases, and climate change.

However, challenges associated with genomic tools for PGR management are:

(a) Cost: Genomic tools can be expensive, limiting their accessibility to genebanks.
(b) Data interpretation: Genomic data can be complex to interpret and requires specialized skills.
(c) Lack of standards: There are currently no international standards for the use of genomic tools in PGR management, which creates difficulty in the comparison of data from different studies.

Despite these challenges, the use of genomic tools in PGR management is likely to gain importance in the future due to the reduction in the cost of genomic tools and the increase in the availability of genomic data over time. The genomics era has led to the development of a variety of technologies, such as sequencing and re-sequencing platforms, high-throughput genotyping platforms, SNP arrays, and genome editing tools. Important tools for PGR management including molecular markers, genetic maps, genome-wide association studies, association mapping, selection of candidate genes, SNP arrays, and NGS are discussed in the following sections.

11.4.1 Molecular Markers and Genetic Maps

Among the various available techniques, DNA-based techniques are more reliable and widely used in crop improvement programs. This technology is neither influenced by environmental changes nor by developmental stages and can be used to analyze small samples of plant material that makes them ideal for studying genetic diversity. Molecular markers are the DNA sequences to identify specific genes or genetic variants. Various DNA-based markers, including restriction fragment length polymorphisms (RFLPs), random amplified polymorphic DNA (RAPD), amplified fragment length polymorphisms (AFLPs), inter-simple sequence repeats (ISSRs), diversity array technology (DArT), simple sequence repeats (SSRs), and single nucleotide polymorphisms (SNPs), are available. DNA-based marker technology has been systematically used to study genetic diversity in a wide range of crops, including rice, wheat, mung bean, soybean, potato, carrot, cotton, mulberry, legume crops and many other crops.

Poczai and co-workers in 2013 broadly classified DNA markers into five categories:

- Arbitrarily amplified DNA markers (AAD), such as amplified fragment length polymorphism (AFLP), inter-simple sequence repeats (ISSR), and random amplified polymorphic DNA (RAPD).
- Conserved DNA-based markers (CDM), such as conserved DNA-derived polymorphism (CDDP), P450-based analogue (PBA) markers, and tubulin-based polymorphism (TBP).
- Transposable element-based markers (TEM), such as inter-retrotransposon amplified polymorphism (IRAP), retrotransposon-microsatellite amplified polymorphism (REMAP), inter-primer binding site (IPBS) amplification, and retrotransposon-based sequence-specific amplification polymorphism (SSAP).
- Resistance-gene based markers (RGM), such as resistance-gene analog polymorphism (RGAP).
- RNA-based markers such as EST and cDNA-based markers and targeted fingerprinting markers (TFM), such as direct amplification of length polymorphisms (DALP), promoter anchored amplified polymorphism (PAAP), sequence related amplified polymorphism (SRAP), target region amplification polymorphism (TRAP), conserved region amplification polymorphism (CoRAP), and start codon targeted (SCoT) polymorphism.

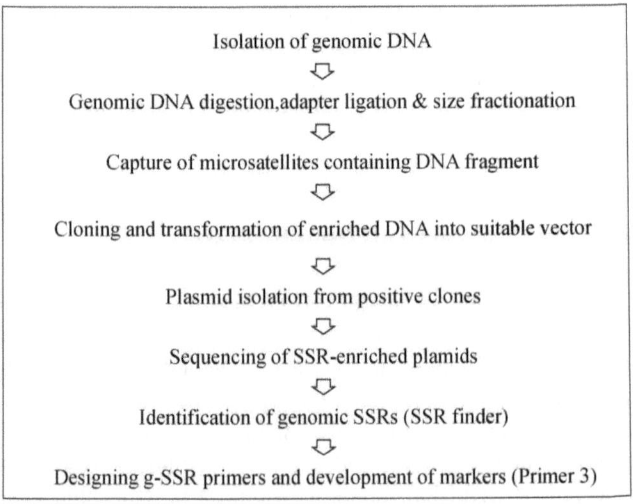

Fig. 11.4 Flowchart describing the development of genomic SSR markers

DNA marker systems have various advantages and disadvantages, and they are differentially used in various crop breeding programs. While AAD markers are easy to use, can scan large populations, less reliable, CDM markers are more reliable, but difficult to use. Again, TEM markers are used to detect transposable elements associated with important traits, RGM markers are used to identify genes that confer resistance to pests and diseases and TFM markers are useful for identifying specific genes or regions of DNA.

Generation of molecular markers: Generating genomic resources can significantly improve the use of PGR. The choice of DNA marker system depends on the specific needs of the crop being studied. Markers like genomic SSRs (simple sequence repeats), ESTs (expressed sequence tags), transcription factors, and small RNAs have been successfully generated through technologies like transcriptome and whole genome sequencing. Over the years, the development of molecular markers as modern tools of molecular genetic approaches has preferred the widespread use of codominant markers. Previously used markers like RAPD (randomly amplified polymorphic DNA), AFLP (amplified fragment length polymorphism), ISSR (inter-simple sequence repeats), SCoT (start codon targeted polymorphism), etc. have low reproducibility and low consistency. Thus, SSRs have been the most widely used markers over the past two decades as they are co-dominant, multiallelic, and highly informative genetic markers that are reproducible and transferable among species. Microsatellite enriched library, whole genome sequencing, and transcriptome profiling have led to the generation of genomic resources in plants like giloe, kalmegh, guava, and bunium. The overall strategy for the development and utilization of microsatellites is summarized in Fig. 11.4.

11.4.1.1 Uses of Molecular Markers

Applications of molecular markers are varied, ranging from sex determination to the management of genebanks. Practically, all DNA marker applications fall under the umbrella definition of "molecular characterization." Various types of molecular characterization, applications utilize a similar set of techniques, but their objectives differ.

Molecular Basis of DNA Fingerprinting

DNA fingerprints arise from differences in nucleotide sequences between individuals. Due to the difficulty in entire genome comparison, genomic fragments are generated, and cultivars are discriminated based on differences in the number, size, and sequence of specific DNA fragments. Various genomic fragments, including tandemly repeated sequences such as mini- and microsatellites, are utilized as markers in DNA fingerprinting. Known gene sequences are used as unique sequence markers, although coding sequences are often not polymorphic enough for routine fingerprinting.

Techniques in Plant DNA Fingerprinting

Fingerprinting techniques aim to use molecular markers that are polymorphic enough to distinguish between cultivars but not excessively polymorphic to reveal differences within individuals of a cultivar. Techniques vary based on how DNA fragments are generated and assayed. Fragment markers provide high-resolution qualitative information about sequence variation. Fragments are generated through restriction digestion or polymerase chain reaction, separated by agarose or polyacrylamide gel electrophoresis, and different detection techniques, such as staining with ethidium bromide or silver. Alternatively, specific target fragments are detected through hybridization with probes carrying radioactive, colorimetric, fluorescent, or chemiluminescent labels. The evaluation of single or low-copy number fragments with varying lengths is termed restriction fragment length polymorphism (RFLP). Multilocus fingerprinting, employed for analyzing repetitive DNA, probes target fragments using known repeat sequences. While single-locus RFLP markers are co-dominant, genetically defined, and highly reproducible, multilocus markers are dominant and lack genetic definition. Multilocus restriction fragment profiling in plant genomes is not suitable for phylogenetic analysis but serves as a robust identification protocol.

PCR-based marker techniques are categorized based on sequence, primer characteristics, and amplified product size range. Designed primers, especially those flanking microsatellites (SSRs), yield locus specific polymorphic amplicons suitable for phylogenetic studies. Randomly amplified polymorphic DNA analysis (RAPD) employs short arbitrary sequences as primers, producing variable fragment patterns but are not locus specific therefore are not much reliable.

Study of Genetic Diversity

Higher genetic diversity in crops contributes to an enhanced capacity for adaptation to varying climatic conditions, while a lack of diversity diminishes genetic variability. A more diverse breeding population is crucial for fostering genetic gain through selection, leading to improved yields, enhanced biotic and abiotic resistance stresses, and the development of other desirable traits. Properly coordinated genebanks can be explored to identify genetic variations within their accessions, aiding in advanced trait discovery and selection of optimal parental combinations for the development of lineages with maximal genetic variability. Knowledge of genetic diversity and population structure forms the foundation for building core sets that adequately represent variations within the entire collection. This process streamlines the collection, generates standardized data, and calculates the potential loss of genetic diversity during conservation and management. Results from molecular marker-based diversity studies which include parameters like gene diversity, heterozygosity, allele number, and polymorphic information content, should be approached with caution, particularly in germplasm conservation initiatives, given the significant role of adaptability in the evolutionary process and the survival of individuals within a population. Various statistical software like Powermarker, PowerCore, STRUCTURE, Genalex, and Popgene have been used for the analysis of sample sets within the populations. These methods depend on parameters such as species' genetic diversity, collection size, grouping, and data type (phenotypic or molecular).

Gap Analysis

Gap analysis is a conservation assessment method that guides the prioritization of biodiversity components for conservation efforts by analyzing deficiencies or "gaps" in the conservation of these elements. In practical terms, for example, the gap analysis of landraces involves comparing the range of diversity maintained by farmers (equivalent to the natural diversity pattern in wild plant species) with the diversity already adequately represented through current on-farm conservation actions (in-situ gap analysis). Additionally, it includes samples of that diversity present in genebank collections (ex-situ gap analysis). While the technique has been utilized to identify gaps

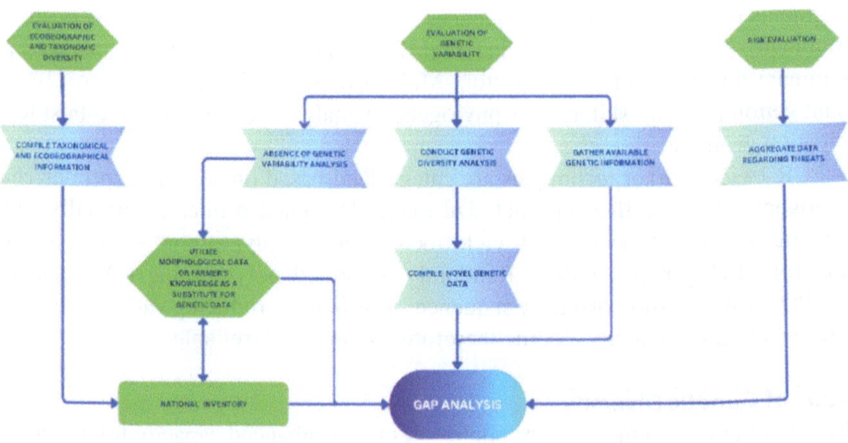

Fig. 11.5 Overview of the gap analysis

in networks of protected habitats for in-situ conservation of genetic resources, such as crop wild relatives (CWR), it is essential to note that environmental gap analysis predominantly emphasizes in-situ conservation alone. In contrast, for the conservation of plant genetic resources for food and agriculture (PGRFA), both in-situ and ex-situ conservation are equally important and complementary techniques. A systematic methodology for genetic gap analysis has been developed, specifically aiming to identify gaps within a crop gene pool and individual species, as illustrated in the case of African *Vigna* wild relatives and landraces. This study evaluated the effectiveness of current in-situ and ex-situ conservation actions, identifying gaps and informing the development of a conservation plan for the crop gene pool. More recently, a gap analysis methodology based on geographic information system (GIS) tools has been tailored to identify crop gene pools. Ecogeographic, taxonomic, and farmers' knowledge of landraces, as well as assessments of threats and genetic diversity, provide crucial information for identifying gaps in landraces conservation. The outcome of an in-situ or ex-situ landraces gap analysis, for example, is a list of landraces populations that require active on-farm or ex-situ conservation. Conducting in-situ and ex-situ gap analyses involves examining various levels, contingent on the available information. Geographic information system (GIS)-based predictive characterization is a valuable tool for identifying populations likely to possess desirable traits, such as resistance to insect pests. The focused identification of germplasm strategy (FIGS) is a predictive characterization technique that can be applied in this context. Figure 11.5 gives an overview of the gap analysis.

Cultivar Identification

For cultivar identification, genotyping with genetic markers like SSR is done using the PCR method. Gel electrophoresis is done to visualize the bands of PCR analysis. Bands are scored based on "0" for absence and "1" for presence to generate a score sheet of that particular marker with the sample. These are the fingerprints of that particular variety or germplasm and these are then matched for similarity index with any unidentified cultivars. A phylogenetic tree is generated to see the varieties forming a close cluster based on the similarity index. Finally, the phylogenetic tree along with other diversity parameters will help in identifying a particular cultivar.

Quantitative Trait Locus (QTL) Mapping

Traits that are influenced by several genes and affected by environmental factors are complex. Quantitative trait locus (QTL) analysis uses statistical tools for the identification of genetic loci that control complex traits using phenotypic and genotypic data. The primary purpose of QTL mapping is the identification of chromosomal regions associated with variation in a quantitative trait. This is accomplished using molecular markers to track the inheritance of traits and identify genomic regions that are co-inherited along with the trait. Moreover, QTL mapping establishes the groundwork for marker-assisted selection (MAS), which accelerates the breeding process, provided that the positions and effects of QTLs are accurately estimated. QTLs that consistently manifest across various environmental conditions are

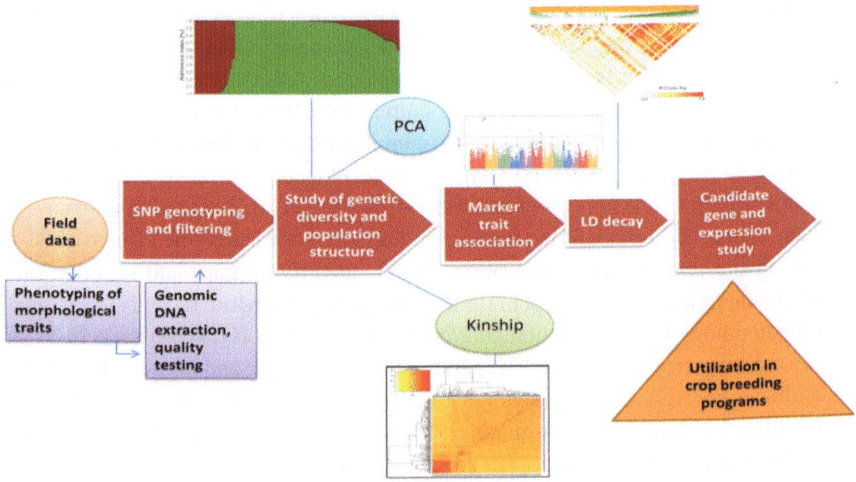

Fig. 11.6 A generalized procedure of genome-wide association study

considered reliable candidates for MAS. The following steps are involved in QTL mapping:

1. Selection of target trait
2. Identification of parents differing in the trait of interest, development of appropriate mapping population, and parental polymorphism survey using markers
3. Screening the population for target trait (phenotyping)
4. Genotyping of mapping population and development of linkage maps
5. Identification of major QTLs controlling the trait
6. Validation of major QTLs across populations
7. Utilization of QTLs in breeding program

There are only a few examples where molecular markers and genetic maps have been used to improve crop breeding. As our understanding of the genetics of crops continues to grow, these tools will become even more desirable in the years to come.

11.4.2 Genome-Wide Association Studies

Genome-wide association studies (GWAS) involve genotyping a diverse set of individuals with numerous genetic markers and comparing their genotypes to corresponding phenotypes to identify associations between genetic markers and desirable traits. This methodology has discovered markers and genes linked to specific traits across various species. Diversity subsets of any crop species from genebank collections serve as valuable resources for conducting GWAS for desirable traits. Figure 11.6 illustrates a generalized procedure for GWAS. The initial step involves selecting appropriate study populations, which should be large and diverse for the target traits. These populations are then genotyped for high density markers such as SNP and phenotyped for the target traits. Subsequent steps include population structure analysis, Kinship studies, linkage disequilibrium mapping, genetic association studies using suitable models, and identifying QTLs or reliable marker-trait associations (MTAs). Further, various annotation studies are followed, to know the functional impact of genetic variants on genes, transcripts, protein sequences, and regulatory regions. The study then assesses gene functions and pathways. Genebank samples often show a strong population structure, which can lead to spurious associations. Therefore, statistical analysis is to summarize the kind of population structure prevailing in the given set of panels before association mapping. Association studies are useful in constructing haplotypes, which further help in allele mining and elucidating the molecular basis of important traits. Additionally, SNP-trait associations mediated by gene expression and transcriptome-wide association studies (TWAS) are performed to predict functional and molecular phenotypes. Thus, GWAS has emerged as a powerful tool for quantitative trait locus (QTL) mapping in plants, allowing access to a broad range of genetic resources for marker-trait associations without limitations on marker availability. This approach has been successfully applied to various traits in crops such as wheat, rice, barley, alfalfa, and others.

Another approach, the candidate gene approach is based on the principle that genes with known functions for traits of interest residing in major loci have been used to map genes/QTL of target traits. The candidate gene (CG) approach comprises a sequence of three chronological phases. Initially, CGs were based on molecular and physiological investigations, which are referred to as functional CGs. They may be derived from linkage data associated with the target locus, with all other closely linked genes considered as positional CGs. The subsequent phase involves the revelation of molecular polymorphism. First, it allows the localization of the CGs on a genetic linkage map, enabling the exploration of genetic linkage between the CG markers and the loci being characterized. Second, it facilitates the calculation of statistical correlations between CG polymorphisms and phenotypic variations within a collection. These two strategies are fundamentally similar and can be executed either concurrently or sequentially. In the third phase, if map co-segregation and/or statistical correlations have been identified, supplementary experiments are conducted to validate the actual involvement of the CG in understanding the variation of the trait. However, the candidate gene approach has a few limitations. First, the candidate gene approach may slip unknown genes that are not involved in the expression of the trait. Second, the candidate gene approach is expensive and time-consuming. Third, the candidate gene approach lacks background knowledge about the biological processes relevant to the trait of interest. This is because the candidate gene approach needs a good understanding of biological processes involved in trait expression to identify the linked genes.

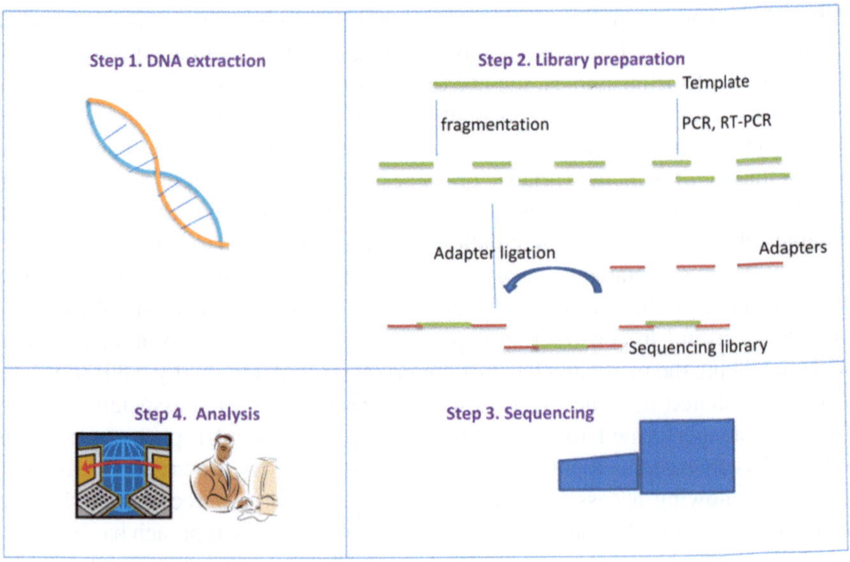

Fig. 11.7 Diagrammatic representation of NGS workflow

11.4.3 Single Nucleotide Polymorphism (SNP) Arrays

SNP arrays are a type of microarray used to genotype thousands to millions of SNPs across the genome. This makes them a valuable high-throughput genotyping tool for the large-scale germplasm collections. SNP arrays can be customized to target specific crop species or it can be commercially available arrays designed for broader applications. These single-base genomic variants can be generated using next-generation sequencing (NGS) technologies, such as Illumina sequencing, which provide a comprehensive view of the genome by generating large amounts of sequence data. Whole genome sequencing (WGS) and RNA sequencing (RNA-seq) are most commonly used in PGR management to identify genetic variations, gene expression patterns, and functional genomics studies. These genomic tools, in addition to advances in bioinformatics, provide valuable understandings of genetic diversity, population structure, and functional genomics of PGR.

11.4.4 Next-Generation Sequencing (NGS)

NGS is a massively parallel sequencing method that runs millions of fragments to be sequenced simultaneously. In addition to whole genome sequencing, NGS, coupled with robust computational pipelines, has innovative and swift methods for transcriptome sequencing, the discovery of molecular markers, gene expression studies, and targeted re-sequencing to pinpoint significant genes in plants. This technique has fastened the sequencing of entire genomes, reducing the timeline from years to mere weeks which further led to the development of various

NGS-based mapping techniques in plant research, substantially reducing the time required for identifying mutations. Figure 11.7 shows the schematic workflow of next-generation sequencing. NGS-based approaches include GBS (Genotyping by sequencing), whole genome re-sequencing (WGR), and RNA-seq (whole exome sequencing approach).

11.4.4.1 Genotyping-by-Sequencing (GBS)

Genotyping-by-sequencing (GBS) is a powerful, high-throughput genotyping technique that relies on restriction enzymes to reduce genome complexity. It facilitates genetic diversity assessment, population structure analysis, and marker-assisted selection for a large collection of germplasms The GBS process begins with the preparation of a DNA library. Genomic DNA is isolated from the samples, such as plant tissue, and digested using restriction enzymes, which cut the DNA at specific recognition sites. In the absence of a reference genome, GBS uses at least 15 restriction enzymes for sequencing the genomes. The choice of restriction enzymes depends on research objectives and the desired level of genomic coverage. After digestion, adapter sequences are ligated to the DNA fragments of each sample. Adapters containing the necessary sequencing priming sites are ligated to the ends of the DNA fragments to facilitate subsequent amplification and sequencing steps. PCR is performed to amplify the DNA fragments within the library. The amplification step ensures that there is an adequate quantity of DNA for sequencing. The amplified library is subjected to high-throughput DNA sequencing using NGS platforms, such as Illumina. The sequencing generates millions of short DNA reads from the library. This is followed by bioinformatics analysis whereby the resulting sequence data are processed and analyzed using bioinformatics tools. The reads are aligned to a reference genome or de novo assembled to generate a consensus sequence. Single nucleotide polymorphisms (SNPs) and other genetic variants are then identified by comparing the sequenced reads to the reference or among individuals in the population. GBS has become a popular method for genotyping due to its cost-effectiveness, scalability, and ability to generate genome-wide SNP data.

11.4.4.2 Whole Genome Resequencing

The whole genome resequencing (WGR) approach has substantially gained importance with the availability of reference genome sequences for numerous commercial crops. In this approach, a new individual's genome is sequenced and compared with the reference genome to identify various forms of polymorphisms, including SNPs, and structural variants such as insertions-deletions (InDels) and gene conversions.

11.4.4.3 RNA-Sequencing Approach

The RNA-Seq approach, also known as whole exome sequencing, involves the direct sequencing of cDNA derived from the total transcripts, making it a valuable tool for comprehensive profiling of gene expression related to QTLs. This approach concentrates on the protein-coding regions of the genome, encompassing

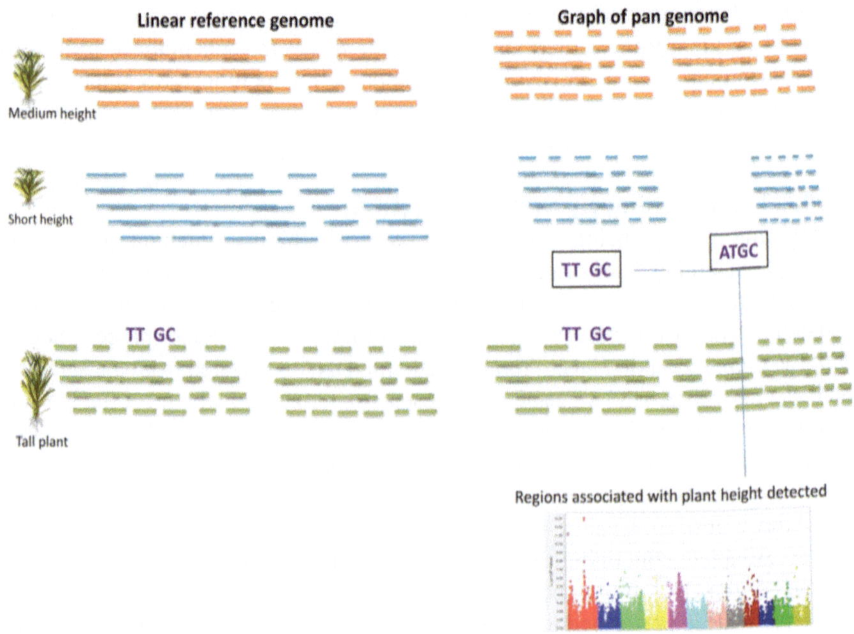

Fig. 11.8 Schematic representation of how pan-genome assembly of a tall plant, identified alleles associated with plant height

approximately 1–2% of the genome. NGS-assisted expression profiling in mutants, compared to their parent plants, has the potential to find candidate genes associated with the desired phenotype. In sorghum, for example, comparative RNA-Seq analysis between parents unveiled 108 differentially expressed genes linked to plant hormone metabolism, glycolysis, and nitrogen metabolism. These differentially expressed genes were found near QTLs for multiple agronomic traits under normal and low-nitrogen conditions. RNA-Seq analysis and QTL analysis can jointly contribute to the identification of the target gene as found in maize and *Glycine max*. Thus, their integrations can enable the exploration of complex trait expressions.

11.4.4.4 Pan Genomes

The pan-genome represents the entire set of genes within a species, consisting of a core genome containing sequences shared between all individuals of the species. Remarkable progress in crop genomics has been observed during recent years, driven by enhancements in sequencing technology, assembly techniques, and computational capabilities. The exploration of structural variations within species and the comprehensive understanding of the pan-genome has unveiled substantial genomic diversity among individuals within a species, revolutionizing the field of crop genomics and its applications. Pan-genomes reveal the extent of genetic variation within a species which is crucial for understanding its natural diversity and response to biotic and abiotic stresses. Pan-genomes often reveal genes that are unique or rare within a species. These genes may be absent in the reference genome but could be essential for specific ecological niches or under different environmental conditions. Pan-genomes can shed light on the functional differences among gene variants. Researchers can investigate how specific gene variations influence protein function, gene regulation, and biological pathways. This is vital for understanding the molecular mechanisms underlying phenotypic variation. Pan-genomes also aid in the study of species evolution. By comparing the genomes of multiple individuals within a species, researchers can trace the evolutionary history and divergence patterns, helping to reconstruct the evolutionary relationships among populations and subspecies. The schematic representation of a pangenome assembly for plant height is shown in Fig. 11.8. In endangered species, studying the pan-genome can offer insights into the genetic diversity and health of populations. This information is essential for conservation efforts, including breeding programs to increase genetic diversity and resilience.

Overall, pan-genomes are powerful tools that expand our understanding of genetic diversity, evolution, and the functional implications of genetic variations within and among species.

11.5 Tools and Techniques for Genetic Manipulation

Genetic manipulation involves the generation of recombinant molecules which requires precise cutting and joining of both vector and DNA through enzymes. These manipulations, performed in vitro, serve as the basis not only for gene cloning but also for exploring DNA biochemistry, gene structure, and gene expression control. Purified enzymes play a crucial role in nearly all DNA modification techniques, as they are involved in essential cellular processes like DNA replication, transcription, DNA breakdown, repair, and recombination. Various techniques are employed in attaining genetic modification, including gene cloning, recombinant DNA technology, and gene editing tools like CRISPR-Cas9. These techniques allow to precisely insert, delete, or modify specific genes within an organism's DNA. Genetic manipulations are mainly popular through gene cloning and genome editing techniques followed by their stable expression.

11.5.1 Transformation

Methodologies for Developing Transgenic Plants.

Genetic transformation refers to the directed transfer of genes from one individual to another and its stable integration. Various gene delivery methods for transferring target DNA to higher plants are available through physical and biological methods. Here we discuss some popular methods of plant transformation.

(i) **Natural Transformation:** Some bacteria can take up foreign DNA from their surroundings naturally. This process is common in certain species of bacteria, such as *Bacillus subtilis*.

(ii) **Chemical Transformation:** This involves treating bacterial cells with calcium chloride or other chemicals to make the cell membrane more permeable to DNA, allowing the uptake of plasmids or other DNA fragments.
(iii) **Electroporation:** Cells are briefly exposed to an electric field, which increases the permeability of the cell membrane. This allows DNA to enter the cells more easily. Electroporation is widely used for various cell types, including bacteria, yeast, and mammalian cells.
(iv) **Microinjection:** A glass micropipette is used to inject DNA directly into the nucleus or cytoplasm of a cell. This method is often employed for transferring DNA into animal cells, including embryos.
(v) **Gene Gun or Biolistics:** DNA-coated microscopic particles (e.g., gold or tungsten) are shot into target cells using a gene gun. This physical method is commonly used for plant cells and some animal cells.
(vi) **Viral Transduction:** Viruses can be engineered to carry specific DNA sequences and then used to infect target cells. The virus injects its genetic material into the host cell, facilitating the transfer of the desired DNA.
(vii) **Lipofection:** Liposomes or lipid-based carriers are used to deliver DNA into cells. The liposomes merge with the cell membrane, releasing the DNA into the cell. Lipofection is commonly used for transfecting mammalian cells.
(viii) **Polyethylene Glycol (PEG) Transfection:** Cells are mixed with DNA and PEG, which induce the formation of temporary pores in the cell membrane, allowing DNA entry. PEG transfection is commonly used for plant cells and mammalian cells.
(ix) **Protoplast Fusion:** Protoplasts (plant cells with the cell wall removed) from different sources are fused, enabling the transfer of DNA between them. This technique is often used in plant genetic engineering.
(x) **Biological transformation method:** The choice of method depends on factors such as the type of cell, the purpose of the DNA transfer, and the specific requirements of the experiment or application. Bacterial DNA uptake efficiency in transformation depends on species, with *Bacillus* and *Streptococcus* being easily transformed, while others like *E. coli* require treatments for enhanced competence.

Agrobacterium-mediated transformation is the most widely used method for genetically modifying plants. The key points are as follows:

Agrobacterium tumefaciens is a soil bacterium that can transfer part of its genetic material (T-DNA) into plant cells, altering their gene expression. During infection, Agrobacterium recognizes and attaches to plant cells through signaling molecules released from wounded plant tissues. The T-DNA from the Agrobacterium is then integrated into the plant's genome, allowing the introduction of desired genes. Agrobacterium-mediated transformation is preferred over other methods like particle bombardment as it results in more stable, single-copy transgene integration.

Principle: The tumor-inducing Ti plasmid (*A. tumefaciens*) and Ri plasmid (*A. rhizogenes*) contain the virulence genes (vir genes) that are activated and are essential for the transfer of the T-DNA region to the host plant cell nucleus. The

T-DNA has left and right border segments critical for recognition. The proteins encoded by *vir* genes facilitate the transfer of the T-DNA complex, containing genes for the biosynthesis of cytokinins and auxins. These genes lead to uncontrolled cell proliferation, causing tumor formation. The T-DNA also encodes enzymes for opine synthesis. The transfer of T-DNA to the plant cell nucleus is believed to be random, potentially occurring in transcriptionally active regions. Regulatory elements in the T-DNA of *A. tumefaciens* enable transcription using the plant eukaryotic transcription system. This method, based on the natural genetic engineering ability of *Agrobacterium*, is a well-established technique for developing transgenic plants. In *A. tumefaciens* infections, segments of a prokaryotic genome (specifically, the T-DNA region of the Ti plasmid) are integrated into the DNA of a eukaryotic organism, a unique natural property of this bacterium.

(xi) ***In planta* or floral dip method:** Various plant transformation methods have emerged with advanced technologies, often tailored to specific crops. Model plants like *A. thaliana* can be easily transformed using techniques like vacuum in-filtration or floral dip. The foreign DNA in the vector is transferred by dipping the floral buds in Agrobacterium suspension containing the desired DNA. It has an advantage over the tissue culture or regeneration method as the foreign DNA is directly incorporated into the seeds and now the seeds are designated as T_0. Plants generated from these seeds are T_1 generation plants.

(xii) **Co-culture method for regeneration:** The agrobacterium suspension containing the foreign DNA segments is co-cultured with the callus of the plants so that the foreign DNA gets incorporated into the callus. The plants regenerated from these calli contain the foreign DNA and the seeds from these plants will be designated as T_0 seeds.

11.5.2 Identification of Recombinants After Transformation

The selection of recombinant DNA molecules is a critical step in genetic engineering for the identification and isolation of cells or organisms that are successfully taken up and expressed as the foreign DNA of interest. Various methods are employed to ensure the selection of cells containing the desired recombinant DNA:

- **Antibiotic Resistance Selection:** Plasmids used in genetic engineering often carry antibiotic-resistance genes. After transformation, cells that have successfully taken up the plasmid and expressed the antibiotic resistance gene survive in the presence of the corresponding antibiotic, while nontransformed cells die. This allows the selection of transformed cells.
- **Reporter Genes:** Reporter genes, such as green fluorescent protein (GFP) or beta-galactosidase, are often linked to the gene of interest in the recombinant DNA. Expression of the reporter gene indicates successful uptake and expression of the foreign DNA. Detection methods, such as fluorescence or enzymatic assays, are then used to identify positive transformants.

- **Auxotrophic Selection:** In some cases, recombinant DNA may include genes for biosynthetic pathways crucial for cell growth. Host cells lacking these pathways (auxotrophic mutants) can grow only if they successfully take up the recombinant DNA carrying the missing genes.
- **Phenotypic Selection:** Recombinant DNA may confer a specific phenotype, such as resistance to a toxin or the ability to utilize a particular substrate. The selection is based on the phenotypic change observed in the transformed cells.
- **Hybridization Techniques:** DNA probes can be designed to specifically bind to the recombinant DNA sequence. This allows for the identification of cells containing the recombinant DNA through hybridization techniques like Southern blotting.
- **PCR Screening:** Polymerase chain reaction (PCR) can be employed to amplify specific regions of the recombinant DNA. This allows for the quick screening of transformed cells to confirm the presence of the desired gene.
- **DNA Sequence:** This selection method depends on the specific goals of the experiment and the characteristics of the recombinant DNA being introduced. DNA sequencing ensures the presence of the desired genetic modification and expression in the host organism or cell line.

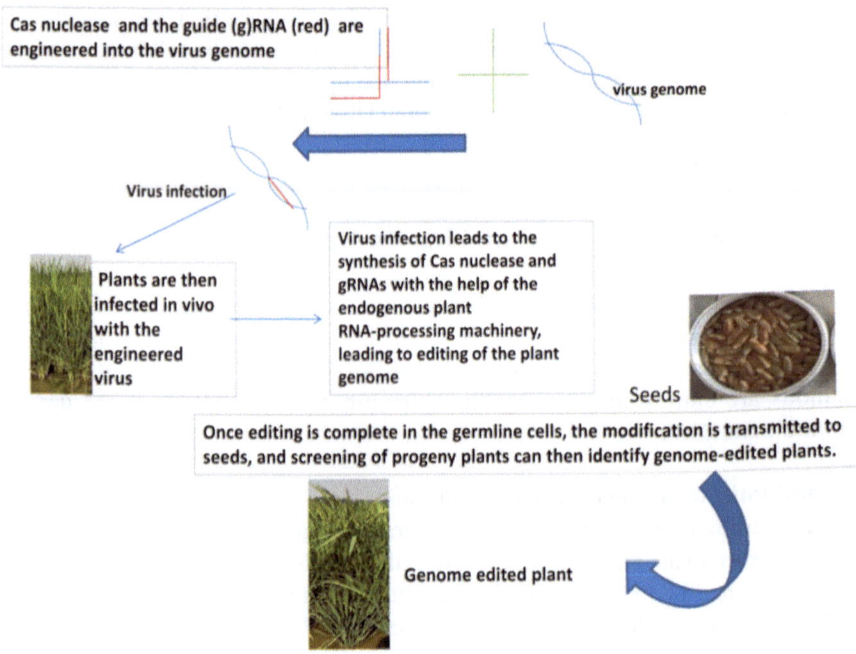

Fig. 11.9 Steps involved in genome editing

11.5.3 Transgenics

Transgenics is a field of biotechnology that involves the genetic modification of organisms by the introduction of genetic material from one species into another. The term "transgenic" is derived from "transgene," which refers to a gene that has been transferred between organisms. This process allows the creation of organisms with novel traits or characteristics that may not naturally occur. This can include traits such as resistance to pests, improved nutritional content, or enhanced tolerance to environmental stress and others. Genetically modified (GM) crops, for example, may be engineered to resist herbicides or pests, leading to increased crop yields and reduced reliance on chemical inputs. Although the use of transgenics raises ethical and environmental concerns, different countries have established regulatory frameworks to assess and monitor the release of transgenic organisms into the environment or their use in agriculture. Regulatory bodies aim to ensure the safety of transgenic products for both human health and the environment.

11.5.4 Genome Editing

Recent advancements in genomics have opened up the possibility of editing genomes, offering important tools for crop improvement programs. Over the past two decades, a range of genome editing technologies have emerged to complement traditional genetic engineering methods. These technologies include antisense-RNA, RNA interference (RNAi), virus-induced gene silencing (VIGS), oligonucleotide-directed mutagenesis (ODM), zinc finger nucleases (ZFN), transcription activator-like effector nucleases (TALENs), and clustered regularly interspaced short palindromic repeats/Cas9 (CRISPR/Cas9). These genome editing tools have the potential to expedite pre-breeding programs by introducing beneficial knockout mutations, such as identifying genes for disease resistance or suppressing unwanted traits associated with desired ones in wild species. Except for RNAi, most genome editing tools operate by inserting, removing, or replacing specific genome regions using specialized nucleases often referred to as "molecular scissors."

GE approaches can be employed to modify genes that harbor defined quantitative trait nucleotides (QTNs) responsible for substantial phenotypic changes. Furthermore, GE tools can diversify the allele pool by creating targeted variations that are valuable for genomic selection. The steps involved in genome editing are explained in Fig. 11.9. It is poised to greatly expand the allelic diversity available for genomic selection, facilitate the correlation of phenotypes with genes through mutant development, especially in primitive landraces and wild species, and chart a new course for pre-breeding programs. Figure 11.9 shows genome editing in plants by viral delivery of CRISPR/Cas9.

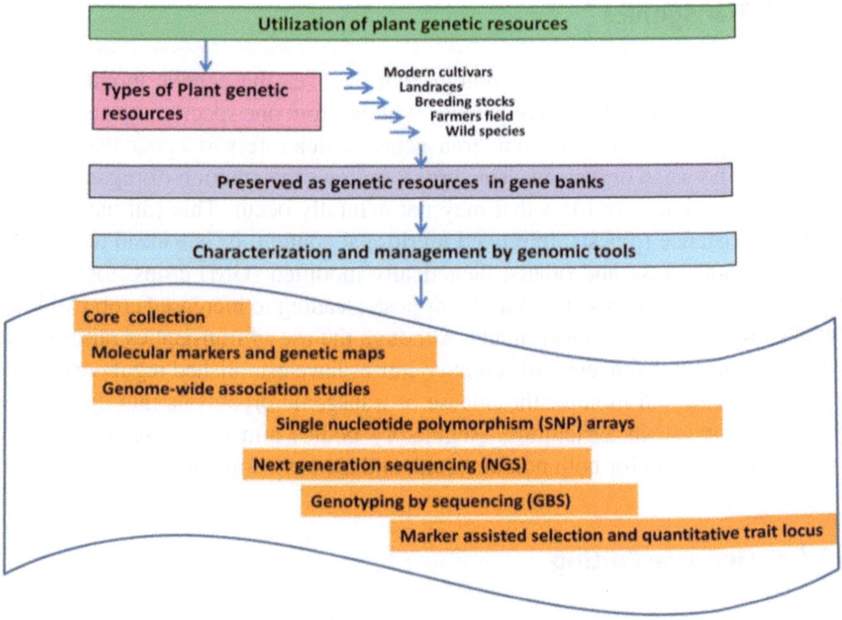

Fig. 11.10 Overview of all genomic tools used for conservation and utilization of plant genetic resources

11.6 Contribution of Bioinformatics to the Advancement of Plant Genetic Resources

"Bioinformatics" is an indispensable tool that plays a substantial role in the development and utilization of genomic tools. The generation of huge biological data gets its meaning through comprehensive analysis and visualization for retrieval of biological information. For PGR management and utilization, bioinformatics tools are deployed in various ways, such as in the development of the core collection, identification of genes, genomic characterization, and the creation and management of databases.

For example, SNP data sets for the entire genome collection of any crop hold great potential for enhancing ex-situ conservation. Due to the large number of SNPs, it may be difficult to assemble collections that maximize diversity for any crop species. Here, bioinformatics programs like "GenoCore" can identify optimized core collections from large genotypic data sets. An important issue while managing ex-situ collections is the integrity of the conserved accessions and duplicates in the genebank, in vitro, or in the field. Furthermore, the use of an association panel with an over-abundance of redundant germplasms in genome-wide association analysis can introduce biases, leading to spurious marker-trait associations. Consequently, a web-based tool known as the germplasm duplicate identification and removal tool (G-DIRT) was developed to identify any duplicates. This tool enables the identification and removal of duplicate germplasms based on single-nucleotide polymorphism (SNP) genotyping information, coupled with the preprocessing of genotypic data.

Again, while analyzing haplotype block structure, the uncertainty surrounding the genome's linked regions or haplotype blocks can impact diversification and optimization. Therefore, determining haplotype block structure typically requires large samples and complex methodologies. In such a situation, bioinformatics aids in simulating basic haplotype block structures through programs like Haploview.

Further, commonly used plant phylogenetic analyses require simple and easy-to-use bioinformatics tools like Powermarker, MEGA, PHYLIP, R, etc. The development of such tools over time required extensive exploration. Some of the notable software in Packages of R, Adegenet, and poppr are utilized to perform minimum spanning networks, principal component plots, and phylogenetic trees. GATK and Heap are tools that detect SNPs, with a specific focus on GWAS. Bioinformatics tools play a crucial role in database development and management. In this context, well-aligned binary data storage, Relational Database Management System (RDBMS), customized data import, and user interfaces must be combined into an all-in-one back-end. Besides these backend services, efficient data access and sharing is a further interesting task for scientific data engineering. For proper utilization of PGRs, the germplasm database system should have information related to quality, accessibility, and presentation to provide the correct format and summarized information the way users want it. The usefulness of germplasm is magnified by putting databases on the web. Some examples of databases include the United States Department of Agriculture (USDA) www.ars-grin.gov, the Center for Genetic Resources. The Netherlands (CGN) www.cgngenis.wur.nl, Amaranth Genomic Resource Database (AGRDB) www.nbpgr.ernet.in:8080/AmaranthGRD/, and database (ApTransDB) www.nbpgr.ernet.in:8080/Andrographis/ of EST-SSR markers and transcription factors for *A. paniculata*. Figure 11.10 shows a few genomic tools used for the conservation and utilization of plant genetic resources.

11.7 Utilization of Genomic Tools for Crop Improvement Program

Genetic resources for food and agriculture are important as they offer a wide range of various traits that can enhance crop quality, resistance to pests and diseases, adaptability to challenging environments, and increased food production. So far, we have discussed the genomic tools required for PGR management. In this section, we describe the application of these genomic tools for crop improvement using PGRs. Genomic tools provide in-depth understanding of plant genomes and enable more precise and efficient breeding strategies. The utilization of plant genetic resources through marker-assisted selection (MAS), genomic selection, and genome editing techniques plays a critical role in crop improvement and agricultural sustainability. Here's an overview of how these methods are applied:

11.7.1 Marker-Assisted Selection

Marker-assisted selection (MAS) is a breeding technique that uses molecular markers to select varieties/cultivars with desirable traits. The use of MAS accelerates the breeding process by reducing the number of generations required to select a desired trait. For example, in the development of submergence-tolerant rice varieties, the Sub1A gene, responsible for submergence tolerance, was identified using molecular markers. MAS was then used to introduce the Sub1A gene into high-yielding rice varieties. MAS has also been used to improve resistance to diseases such as rust in wheat, bacterial blight in rice, and quality traits in maize.

11.7.2 Gene Pyramiding

Gene pyramiding strategically combines desirable genes from various parent plants into one genotype, aiming to enhance specific traits by integrating multiple beneficial genes. It primarily refines superior cultivars by addressing specific deficiencies and targeting traits with identified beneficial genes. However, historical challenges in identifying valuable gene sources and tracking their presence or absence have limited the incorporation of genes into elite cultivars. The advancement of modern molecular and genomics technology has not only expedited the identification of beneficial genes but also broadened the pool of useful gene sources.

Objectives of gene pyramiding include:

- Improving trait performance through the combination of two or more complementary genes,
- Addressing deficiencies by incorporating genes from external sources,
- Enhancing durability of trait expression.

Types of gene pyramiding:

Conventional technique: Serial gene pyramiding, where genes are sequentially integrated into the same plant using methods such as pedigree breeding, backcross breeding, and recurrent selection.

Molecular techniques: These include marker-assisted gene pyramiding and marker-assisted backcrossing. Molecular marker-assisted selection (MAS) speeds up plant trait selection by pinpointing genome regions linked to desirable genes. Marker assisted backcrossing (MABC) relies on marker alleles linked to desired genes, enabling faster outcomes than traditional methods. MABC facilitates gene pyramiding, enhancing crop resilience, and success depends on factors like marker proximity and genetic background. Gene pyramiding aims to enhance qualitative traits such as disease and insect resistance, verified through individual-level phenotyping, with reliable phenotypic performance indicating genotype. Factors like gene inheritance models and linkage/pleiotropy between traits influence its efficacy. Allelic genes cannot coexist, requiring progeny testing for recessive gene evaluation. Co-transfer of closely linked genes may lead to diminished performance, known as linkage drag. Effective gene pyramiding relies on understanding trait genetics and genotype–phenotype correlations.

11.7.3 Allele Mining

Allele mining involves the analysis of naturally occurring allelic variation or suitable alleles within candidate genes that influence important agronomic traits, potentially benefiting crop improvement. This method assists in understanding allele evolution, uncovering new haplotypes, and generating allele-specific molecular markers for marker-assisted selection. Recent advancements in large-scale genome sequencing techniques enable the differentiation of various alleles at the DNA level. Reverse genetics is the identification of DNA-level mutations and subsequent determination of their impact on an individual's morphology. Allele mining involves the discovery of novel alleles in wild germplasm or the selection of superior alleles from natural populations. Given that genetic heterogeneity is essential for genetic advancement in crop varieties, allele mining serves as the initial step in breeding efforts, facilitating the acquisition of genetic diversity. Mutation, as an evolutionary driver, has led to the current allelic diversity in crop species and can be induced through various breeding techniques. Mutations occurring in coding or regulatory genome regions, such as single nucleotide polymorphisms (SNPs) or insertions and deletions (InDels), can significantly impact phenotype by altering amino acid sequences or their functions in encoded proteins. Conversely, mutations in noncoding regions may have subtler effects. True allele mining encompasses the analysis of both coding and noncoding regions of candidate genes. For instance, a mutation in the 5′ splice site of the first intron of the waxy (Wx) gene in rice led to a tenfold increase in the gene's activity, demonstrating the potential impact of mutations beyond coding sequences. It enables the monitoring of allele evolution, the identification of novel haplotypes, and the development of allele-specific markers for utilization in marker-assisted selection (MAS). This capability grants breeders direct access to critical genes associated with traits such as resistance to biotic stress, tolerance to abiotic stress, enhanced nutrient utilization efficiency, increased yield, and improved quality. Furthermore, it sheds light on the molecular mechanisms underlying new phenotypic variations and identifies nucleotide sequence alterations linked to beneficial alleles.

11.7.3.1 Approaches for Allele Mining

The primary methods for detecting sequence polymorphism in a specific gene include targeting induced local lesions in genomes (TILLING), a modified version known as Eco-TILLING, and sequence-based allele mining.

TILLING

It offers a valuable nontransgenic reverse genetic approach for studying gene function. This strategy enables rapid screening for induced lesions in a gene of interest or identification of mutations in genes with known sequences within a population of plant mutants. TILLING utilizes heteroduplex analysis to directly detect allelic variation, particularly point mutations, resulting from induced mutations in target genes. It serves as a potent tool for functional genomics when knockout methods are not feasible.

Prerequisites for TILLING include chemical mutagenic agents, targeted gene sequence data, two differently labelled 5′ end dyes (dye 700 & 800), specialized nucleases (CEL1, S1, and mung bean nuclease), and a Li-Cor genotype (Li-Cor USA).

Eco-TILLING

TILLING has been expanded to include eco-TILLING, proposed by Comai et al. which identifies polymorphisms in naturally occurring populations without chemical mutagenesis. Eco-TILLING harnesses natural genetic diversity without inducing mutations, aiming to identify SNPs and haplotypes in naturally occurring populations. This technique utilizes a single-strand-specific nuclease to enzymatically degrade DNA heteroduplexes, followed by detection using Li-Cor genotyping. Eco-tilling, without mutagenesis, efficiently identifies natural variation within populations or germplasm. It detects DNA polymorphism using SNPs, InDels, haplotypes, and SSR variations.

11.7.3.2 Sequencing-Based Allele Mining

It involves PCR amplification of gene alleles in different genotypes, followed by DNA sequencing to detect nucleotide variation. Steps include sample collection, DNA extraction, primer design, PCR amplification, and allelic variation identification. It efficiently identifies different alleles among cultivars, such as increasing the source of the harmful blast disease and rice blast R gene. Analyzing individuals for haplotype structure and recognizing mutation impacts on gene structure is beneficial for genetic association studies. Sequences are examined for point mutations, SNPs, and InDels to create haplotypes. It requires less complex equipment and fewer laborious stages than TILLING and Eco-Tilling. Successful allele mining relies on genome and gene sequence availability and the genetic foundation of the crop species.

11.7.4 Genomic Selection (GS)

Genomic selection is a versatile breeding approach to predict the breeding values of individuals using genomic information. Introduced by Lande and Thompson in 2000 genomic selection is a form of marker-assisted breeding (MAB). It ensures that every quantitative trait locus (QTL) is in linkage disequilibrium (LD) with at least one genetic marker, facilitating the efficient selection of desired traits. Advances in genome sequencing and genotyping techniques, particularly the identification of numerous single nucleotide polymorphisms (SNPs), have made genomic selection more practical and cost-effective. This can be achieved by combining GS with high-throughput genotyping and phenotyping platforms, and with speed breeding. These advancements enhance the speed and accuracy of the GS-assisted breeding process, leading to better genetic gains per unit of time and expense. Genomic selection allows for rapid crop improvement without any need for in-depth analysis of individual loci. It uses a phenotyped and genotyped training population (TP) to predict the genomic estimated breeding values (GEBVs) of specific lines. Subsequently, a breeding population (BP) can be established from selected individuals and bred over several generations without additional time-consuming phenotyping. In rice, genomic selection (GS) has been used to identify the lines tolerant to blast disease. In maize, GS has proven effective in selecting lines from natural populations tolerant to *Stenocarpella maydis*, which causes ear rot (dos Santos et al. 2016), and from biparental populations for higher yield under severe *Striga* infestation.

Advantages of Genomic Selection
GS has many advantages over traditional breeding methods.

1. It can be used for the selection of complex traits, such as grain yield and tolerance to abiotic and biotic stress, that are difficult only through phenotypic selection.
2. GS can shorten the breeding cycle by allowing early selection of individuals with desirable traits in the breeding process.
3. GS allows the selection of traits that are not easily measured, such as resistance to diseases and pests. GS has been used successfully in a variety of crops, including maize and rice. GS is a promising new breeding technology to improve crop yields and quality. As the cost of genotyping decreases, GS is likely to become more widely adopted by breeders.

11.8 Summary

Genomic tools have profoundly advanced the conservation and management of plant genetic resources (PGRs). This chapter explores the diverse applications of these tools in various aspects of PGR stewardship. Molecular markers are employed to fingerprint and classify PGR accessions, offering more precise information than traditional phenotypic evaluations for the characterization and utilization of PGRs. Marker-assisted selection aids in identifying rare alleles or desirable traits, thereby facilitating the targeted use of PGRs in breeding programs. Additionally, molecular markers help monitor changes in the genetic structure of PGR collections during regeneration cycles, ensuring the maintenance of their genetic integrity. The creation of core collections, which encapsulate the majority of genetic diversity, represents an effective application of genomic tools in PGR management. Advanced genomic technologies, such as next-generation sequencing, have reduced the high costs and technical barriers traditionally associated with the use, management, and conservation of PGRs. Bioinformatics has become an essential component of genomic research, enabling the development of new genomic tools for PGR management and their application in breeding programs. Overall, the integration of genomic tools has significantly enhanced the conservation, characterization, and utilization of PGRs, which is crucial for addressing food security and sustainability in agriculture.

Multiple-Choice Questions

1. The sign of virulence of *Agrobacterium tumefaciens* is carried on
   (a) Chromosomal DNA
   (b) Tumor-inducing plasmid DNA
   (c) Both chromosomal and plasmid DNA
   (d) Cryptic plasmid DNA
2. Direct uptake of protoplast DNA can be stimulated by
   (a) Polyethylene glycol (PEG)
   (b) Decanal
   (c) Luciferin
   (d) All of these
3. The size of the virulent plasmid of *Agrobacterium tumefaciens* is
   (a) 40–80 kb
   (b) 80–120 kb
   (c) 140–235 kb
   (d) >235 kb
4. Molecular markers are used to construct
   (a) Chromosome maps
   (b) Cytogenetic maps
   (c) Physical maps
   (d) All of these
5. RFLP and RAPD have one thing in common except
   (a) RFLP is a faster method than RAPD
   (b) RFLP is more reliable than RAPD
   (c) Species-specific primers are required for RAPD
   (d) Radioactive probes are not required in RAPD
6. Which of the following methods is suitable for cryopreservation?
   (a) Above solid carbon dioxide (at −79 °C)
   (b) Low-temperature deep freezers (at −80 °C)
   (c) In the gas phase of nitrogen (at −150 °C)
   (d) In liquid nitrogen (at −196 °C)
7. Which of the following methods is used for the preservation of germplasm in vitro?
   (a) Cryopreservation (lyophilization)
   (b) Cold storage
   (c) Low pressure and low oxygen storage
   (d) All of them
8. Which of the following is not a strategy for ex-situ preservation of germplasm?
   (a) Seed genebank
   (b) DNA genebank
   (c) Biosphere Reserve
   (d) Botanical garden

9. Genebanks contain
   (a) Seed banks, tissue cultures, and cryopreservation
   (b) Seed banks, tissue cultures, and cryopreservation
   (c) Seed banks and cryopreservation
   (d) Tissue culture and cryopreservation
10. The maximum genetic diversity of crops occurs where agriculture is
   (a) Operation on commercial lines
   (b) Advanced
   (c) Flooded by rain
   (d) Primitive
11. Which will be sought to obtain maximum genetic diversity?
   (a) Latest varieties
   (b) Old varieties
   (c) Wild relatives
   (d) All of these
12. Global rice gene pool is maintained at
   (a) China
   (b) IRRI, Philippines
   (c) Mexico
   (d) None of these
13. The gene pool consists of
   (a) Modern cultivars
   (b) Advanced breeding materials
   (c) Landraces
   (d) All of the above
14. What is the unit of a genetic map?
   (a) Centimeter
   (b) Nanometer
   (c) Angstroms
   (d) Centimorgan
15. What is a genome-wide association study?
   (a) A study that tests whether a disease or trait is hereditary
   (b) A study that tests the association between a disease/trait with every gene in the genome
   (c) Study of the genetic association between all genetic variants in the genome and the disease/trait
   (d) Genetic association studies between the disease/trait and sufficient genetic markers to capture almost all common genetic variation
16. SNP is a kind of
   (a) Frameshift mutation
   (b) Transposition control
   (c) Genetic regulation
   (d) Genetic marker

17. What is DNA sequencing?
    (a) Identification of different cell types in a tissue sample
    (b) Determining the order of nucleotides in a DNA molecule
    (c) Creating synthetic DNA molecules in the laboratory
    (d) Modification of DNA strand structure
18. What advantage does next-generation sequencing (NGS) have over traditional Sanger sequencing?
    (a) NGS is faster and can sequence longer DNA fragments
    (b) NGS uses fewer chemicals for sequencing
    (c) NGS requires less DNA for sequencing
    (d) NGS can sequence only one DNA fragment at a time
19. Which method revolutionized DNA sequencing by allowing multiple DNA fragments to be sequenced simultaneously?
    (a) Polymerase chain reaction (PCR)
    (b) Sanger sequencing
    (c) Next-generation sequencing (NGS)
    (d) Gel electrophoresis

Answers

1. (b)
2. (a)
3. (c)
4. (d)
5. (c)
6. (d)
7. (d)
8. (c)
9. (a)
10. (d)
11. (d)
12. (b)
13. (d)
14. (d)
15. (d)
16. (d)
17. (b)
18. (a)
19. (c)

Suggested Reading

Frankel OH (1984) Genetic perspectives of germplasm conservation. In: Arber WK, Illmensee K, Peacock WJ, Starlinger P, Ehrlich SD, Dagert M, Romac S, Michel B, Levy SB, Goebel W et al (eds) Genetic manipulation: impact on man and society. Cambridge University Press, Cambridge, pp 161–170

Frankel OH, Brown AHD, Burdon JJ (1995) The conservation of plant biodiversity. Cambridge University Press, Cambridge

Gene cloning and DNA Analysis. An introduction, 6th edn. T A Brown.

Heffner EL, Jannink JL, Iwata H, Souza E, Sorrells ME (2011) Genomic selection accuracy for grain quality traits in biparental wheat populations. Crop Sci 51:2597–2606. https://doi.org/10.2135/cropsci2011.05.0253

Holbrook CC, Anderson WF, Pittman RN (1993) Selection of a core collection from the US germplasm collection of peanut. Crop Sci 33:859–861. https://doi.org/10.2135/cropsci1993.0011183X003300040044x

Introduction to plant biotechnology. H. S. Chawla.

Poczai P, Varga I, Laos M, Cseh A, Bell N, Valkonen JP, Hyvönen J (2013) Advances in plant gene targeted and functional markers: a review. Plant Methods 9:6–15. https://doi.org/10.1186/1746-4811-9-6

Principles of Gene Manipulation, 6th edn. Sandy B Primrose, Richard M Twyman and Robert W Old

Xu Y, Crouch JH (2008) Marker-assisted selection in plant-breeding: from publications to practice. Crop Sci 48:391–407. https://doi.org/10.2135/cropsci2007.04.0191

Crop Wild Relatives and their Utilization in Crop Improvement

12

Arun Gupta, Pradeep Kumar, Jyoti Kumari, B. S. Tyagi, Bhaskar C. Patra, Rinky Resma Panda, N. C. Gupta, Vishnu Kumar, Dalamu, M. Vetriverithan, Priya Pardeshi, C. Bhardwaj, Gyanendra Singh, and G. P. Singh

Learning objectives

- Concepts of gene pools, pre-breeding.
- Role of crop wild relatives.
- Pre-breeding in CWRs of cereals, legumes, oilseeds, vegetables.
- Assessment of pre- and post-zygotic barriers in wide hybridization crosses.
- Related topics.

Keywords

Plant genetic resources · Germplasm · PGR · Crop wild relatives · CWR · Utilization · Improvement · Species · Gene pool · Introgression · Embryo rescue

A. Gupta (✉) · P. Kumar · B. S. Tyagi · G. Singh
ICAR-Indian Institute of Wheat and Barley Research, Karnal, Haryana, India
e-mail: arun.gupta@icar.org.in

J. Kumari · R. R. Panda · V. Kumar · P. Pardeshi · G. P. Singh
ICAR-National Bureau of Plant Genetic Resources, New Delhi, India
e-mail: jyoti.kumari@icar.org.in

B. C. Patra
ICAR-National Rice Research Institute, Cuttack, Odisha, India

N. C. Gupta
ICAR-National Institute for Plant Biotechnology, New Delhi, India

Dalamu
ICAR-Central Potato Research Institute, Shimla, Himachal Pradesh, India

M. Vetriverithan
International Crops Research Institute for Semi-Arid Tropics (ICRISAT), Hyderabad, India

C. Bhardwaj
ICAR-Indian Agricultural Research Institute, New Delhi, India

K. Tripathi et al. (eds.), *Textbook of Plant Genetic Resources*,
https://doi.org/10.1007/978-981-96-5024-8_12

12.1 Introduction

Crop wild relatives (CWRs) are the wild species of the same genus or very closely related genera of crop plants but have not been domesticated. It also includes distantly related species within the same or related genera. However, they exist as wild species in natural habitats near their centers of origin. CWRs can be used as donors in the crop improvement programs for their genetic proximity and morphological resemblances with the crop. The wild progenitor and related species and genera are instrumental in providing the genes for:

(i) Resistance to biotic stresses (insects and pests, weeds, nematodes, and pathogens)
(ii) Tolerance to various abiotic stresses (submergence, salinity, heat, cold, and drought)
(iii) Biomass and vigor potential in forage crops
(iv) Grain nutrients, high iron, zinc, and protein content in cereal crops
(v) Fiber strength in fiber-yielding crops
(vi) Oil content in oilseed crops
(vii) Dwarfism in cereals and oilseed crops

Plant domestication started about 10,000 years ago and is continuing, but the criteria have changed to trait-based selection. As a result, there is a reduction in diversity at every level due to the selective sweep, in which positive selection on target locus controlling or associated with the domesticated traits results in a reduction in diversity of that particular locus. Thus, during the process of domestication, certain loci in modern crops have been either reduced or lost. Only a subset of genes and alleles, which are purposely selected from the wild progenitor gene pool is present among the crop cultivars. Such reduction in diversity is known as the "founder effect." Loss of genetic variation occurs when a new population is established by a very small number of individuals from a population with a large number of individuals. Inbreeding, genetic drift, and restricted gene flow are the major factors responsible for the reduction in diversity. As a result, genetic variation has been considerably reduced in several crops, for example, soybean, maize, and rice. Thus, many useful alleles present in the CWRs may be lost during the process of domestication and the loss of diversity affects individual fitness and increases susceptibility to diseases.

12.2 Importance of CWRs

Crop improvement has traditionally relied on conventional methods like introduction, selection, and hybridization using cultivated genotypes. However, these modern breeding techniques have led to narrow genetic diversity, contributing to the loss of about 75% of crop genetic diversity in the twentieth century. The importance of crop wild relatives (CWRs) was first recognized by Russian botanist N.I. Vavilov in the 1920s, who identified the geographic centers of crop origin and diversity. Since the 1990s, the use of wild gene resources in plant breeding has notably increased. Global food insecurity has been rising since 2014, with nearly one-third of the world's population (2.37 billion people) lacking adequate food in 2020. About 30% of crop yield improvements since 1945 are due to the use of CWRs in crop improvement. For example, one wild tomato variety increased solid content by 2.4%, worth $250 million annually, and genes from three wild peanut varieties have increased resistance to root-knot nematode, saving around $100 million yearly worldwide.

CWRs offer extensive genetic diversity, providing novel traits for crop improvement. They help mitigate gene erosion from decades of domestication and intense breeding. Overall, CWRs contribute to disease resistance (39%), pest resistance (17%), abiotic stress tolerance (13%), yield increase (10%), quality improvement (11%), and cytoplasmic male sterility and fertility restoration (4%). Thus, CWRs are invaluable genetic resources for developing new, high-yielding, stress-tolerant varieties suitable for changing climatic conditions, ensuring food and nutritional security.

12.3 Concepts of Gene Pool

The Russian geneticist Alexander Sergeevich Serebrovsky gave the concept of "genofond," which was translated in English as "gene pool" by Theodosius Dobzhansky. A gene pool refers to the total of all the genes and their alleles influencing a wide range of characteristics in a population of a single species. It includes both expressed and non-expressed genes. The term gene pool can be used in a similar way as the term germplasm and it represents the reservoir of diversity. Harlan and De Wet (1971) proposed the concept of gene pool based on crossability between cultivated crop species and CWRs/their taxa and classified them into primary, secondary, and tertiary gene pools.

12.3.1 Primary Gene Pool (GP1)

Based on their taxonomic relationship, taxa can be crossed with the crop and yield fertile hybrids with good chromosome pairing and gene transfer through simple hybridization. It includes plants of the same species, subspecies, and wild or weedy relatives. In these species, genetic segregation is approximately normal and gene transfer is generally simple. Further, the primary gene pool is also known as gene pool 1 (GP1), and the members of GP1 are commonly used in the breeding programmes.

12.3.2 Secondary Gene Pool (GP2)

It consists of the species that are distantly related, and gene transfer is possible, but with some level of difficulty. Based on their taxonomic relationship, taxa that cross successfully with the crop plant but yield partially or mostly sterile hybrid with poor chromosome pairing, making gene transfer through hybridization is possible but difficult. There are some reproductive barriers between the members of GP1 and GP2, which lead to anomalies in the hybrid such as the following: (i) the hybrids are partially sterile or weak, (ii) the chromosomes may pair poorly or not at all, (iii) recovery of the desired phenotypes may be difficult in subsequent generations, (iv) the gene pool is available to be used by a plant breeder or a geneticist with due effort.

12.3.3 Tertiary Gene Pool (GP3)

Members of this gene pool when crossed with GP1 produce either lethal or completely sterile hybrids and are termed as the tertiary gene pool. They are more distantly related to the members of the primary gene pool. The transfer of desirable genes from distant wild relatives through intergeneric or interspecific hybridization requires special breeding techniques such as: (i) embryo rescue (or embryo culture, a form of plant organ culture), (ii) induced polyploidy (chromosome doubling), and (iii) bridging crosses (e.g., with members of the secondary gene pool) to transfer genes. Here gene transfer is not possible without the use of special techniques.

12.3.4 Quaternary Gene Pool (GP4)

An additional concept of gene pool, that is, the quaternary gene pool, was proposed by Gepts and Papa (2003), which is also known as Gene Ocean. Genes can be transferred from any organism to any organism through recombinant DNA technology, such as transformation and genomics.

The crop progenitor and their wild relatives, and the primary, secondary, and tertiary gene pools of some important agri-horticultural crops are given in Tables 12.1, 12.2, and 12.3.

Table 12.1 Progenitor and primary, secondary, and tertiary gene pool of cereal and millet crops

Crop (Botanical name)	Progenitor	Primary gene pool	Secondary gene pool	Tertiary gene pool
Wheat (*Triticum aestivum*)	*T. monococcum* (AA), *Aegilops* (*Ae.*) *tauschii* (DD), *T. urartu* (BB), *Ae. spetoides* (BB)	*T. monococcum* ssp *aegilopoides*, *T. urartu*, *T. timopheevi* ssp *armeniacum*, *T. turgidum* ssp *dicoccoides* and *Secale cerale*	*All the species of genus Aegilops*, *Dasypyrum villosum*, *Agropyron desertorum* and *cristatum*	*Several species of the genus Elymus*, *Leymus*, *Thinopyrum* and *Hordeum*
Asian Rice (*Oryza sativa* L.) **African Rice** (*O. glaberrima Steud.*)	*O. rufipogon/O. nivara* (AA) *O. longistaminata/ O. barthii* (AA)	*O. barthii* *O. longistaminata* *O. nivara* *O. rufipogon* *O. meridionalis* *O. glumaepatula*	*O. schlechteri* *O. brachyantha* *O. granulata* *O. meyeriana* *O. officinalis* *O. eichingeri* *O. rhizomatis* *O. australiensis*	*O. longiglumis* *O. ridleyi* *O. coarctata* *O. latifolia* *O. alta* *O. grandiglumis* *O. minuta* *O. punctata*
Barley (*Hordeum vulgare* L.)	*H. vulgare* L. ssp. *spontaneum*	*H. spontaneum* *H. vulgare* L. ssp. *vulgare*	*H. bulbosum*	*H. chilense* *H. muticum* *H. jubatum* *H. procerum* *H. lechleri* *H. marinum* *H. capense*
Maize (*Zea mays* L.)	*Zea mays subsp. parviglumis (Balsus Teosinte)*	*Z. huehuetenangensis* *Z. mexicana* *Z. parviglumis*	*Z. diploperennis* *Z. luxurians* *Z. nicaraguensis* *Z. perennis*	*Trypsacum species* (all)
Finger millet (*Eleusine coracana* L.)	*E. coracana ubsp. africana*	*E. coracana subsp. coracana* *E. africana subsp. africana*	*E. indica* *E. floccifolia* *E. tristachya*	*E. intermedia* *E. jaegeri* *E. kigeziensis* *E. multiflora* *E. semisterilis*
Pearl millet (*Pennisetum glaucum*)	*P. glaucum* ssp *monodii*	*P. glaucum* ssp *monodii* with its two ecotypes- *P. violaceum* and *P. mollissimum* *P. glaucum* ssp. *stenostachyum*	*P. purpureum* *P. squamulatum*	*P. schweinfurthii* *P. ramosum* *P. mezianum*
Sorghum (*Sorghum bicolor* (L.) Moench)	*S. bicolor* subsp. *verticilliflorum*	*S. propinquum* *S. bicolor* subsp. *verticilliflorum* *S. bicolor* subsp. *drummondii*	*S. halepense* *S. almum*	Species of subgenera *Heterosorghum*, *Chaetosorghum*, *Parasorghum* and *Stiposorghum*

Table 12.2 Progenitor and primary, secondary, and tertiary gene pool of pulses and oilseed crops

Crop (Botanical name)	Progenitor	Primary gene pool	Secondary gene pool	Tertiary gene pool
Pea (*Pisum sativum*)	*Pisum sativum* subsp. *elatius*	*P. sativum* *P. elatius* *P. pumilio* *P. abyssinicum*	*P. humile* *P. formosum* *P. fulvum*	*Vavilovia formosa*
Lentil (*Lens culinaris*)	*Lens culinaris* ssp. *orientalis*	*L. culinaris ssp. culinaris* *L. culinaris ssp. orientalis* *L. odemensis*	*L. ervoides* *L. nigricans*	*L. lamottei* *L. tomentosus*
Chickpea (*Cicer arietinum* L.)	*Cicer reticulatum*	*C. reticulatum* *C. echinospermum*	*C. bijugum* *C. judaicum* *C. pinnatifidum*	*C. chorassanicum* *C. yamashitae* *C. cuneatum* *C. canariense* *C. anatolicum* and other species
Pigeon pea (*Cajanus cajan* L.)	*Cajanus cajanifolius*	*Cajanus cajan* *C. cajanifolius*	*C. acutifolius* *C. albicans* *C. confertiflorus* *C. lanceolatus* *C. latisepalous* *C. lineatus* *C. reticulatus* *C. scarabaeoides* *C. sericeus* *C. trinervius*	*C. goensis* *C. heynei* *C. kerstingii* *C. mollis* *C. platycarpus* *C. rugosus* *C. volubilis* and other species
Mung bean (*Vigna radiata*)	*Vigna radiata* var. *sublobata*	*V. radiata* var. *radiat*, *V. radiata* var. *sublobata*, *V. radiata var. setulosa*	*V. mungo* var. *mungo*, *V. mungo* var. *sylvestris*, *V. aconitifolia*, *V. trilobata*	*V. angularis* *V dalzelliana* *V. glabrescens* *V. grandis* *V. umbellata* *V. vexillata*
Urd bean (*Vigna mungo L.*)	*Vigna mungo* var. *sylvestris*	*V. mungo* var. *mungo*, *V. mungo var. sylvestris*	*V. radiata* var. *radiata*, *V. radiata* var. *sublobata*, *V. radiata* var. *setulosa*, *V. aconitifolia*, *V. trilobata*	*V. angularis* *V. dalzelliana* *V. glabrescens* *V. grandis* *V. umbellata* *V. vexillata*
Common bean (*Phaseolus vulgaris*)	*P. vulgaris* subsp. *aborigineus*	*P. vulgaris* Domesticated cultivated species	*P. coccineus* *P. costaricensis* *P. polyanthus*	*P. acutifolius* *P. lunatus* and other species
Cowpea (*Vigna unguiculata* L.)	*V. unguiculata* ssp. *dekindtiana*	Four cultigroups, landraces, and subsp. *tenuis, denkindtiana, stenophylla*	subsp. *pubescence*	*V. vexillata* *V. radiata*

(continued)

Table 12.2 (continued)

Crop (Botanical name)	Progenitor	Primary gene pool	Secondary gene pool	Tertiary gene pool
Cluster bean (*Cyamopsis tetragonoloba*)	*C. senegalensis*	Cultivars, landraces, *C. tetragonoloba*	*C. dentate*, *C. senegalensis* *C. serrata*	None
Grass pea (*Lathyrus sativus*)	*Lathyrus cicera*	*Lathyrus sativus*	*L. chrysanthus* *L. cicera* *L. gorgoni* *L. marmoratus* *L. pseudocicera* *L. amphicarpos* *L. blepharicarpos* *L. chloranthus* *L. hierosolymitanus* *L. hirsutus*	Remaining *Lathyrus* species
Soybean (*Glycine max* L.)	*Glycine soja*	*G. max* *G. soja*	Unknown	26 Wild species
Groundnut (*Arachis hypogaea* L.)	*Arachis monticola*	*A. monticola*	*A. duranensis* *A. stenosperma* *A. cardenasii* *A. diogoi* *A. linearifolia* *A. correntina* *A. helodes* *A. herzogii* *A. hoehnei* *A. kuhlmannii* *A. microsperma* *A. kempff-mercadoi* *A. schininii* *A. simpsonii* *A. villosa* *A. palustris* *A. praecox* *A. decora* *A. vollsii* *A. glandulifera* *A. williamsii* *A. valida* *A. trinitensis* *A. magna* *A. krapovickasii* *A. gregoryi* *A. cruziana* *A. benensis* *A. batizocoi* *A. ipaensis*	*Procumbentes* *Caulorrhizae* *Heteranthae* *Rhizomatosae* *Extranervosae* *Erectoides* *Trierectoides* *Triseminatae* *Heteranthae*

(continued)

Table 12.3 Progenitor and primary, secondary, and tertiary gene pool of vegetable and commercial crops

Crop (Botanical name)	Progenitor	Primary gene pool	Secondary gene pool	Tertiary gene pool
Tomato (*Solanum lycopersicum*)	*S. pimpinellifolium* *S. cerasiferum*	*S. lycopersicon* *S. pimpinellifolium* *S. cheesemanii*	*S. parviflorum* *S. chmielewskii* *S. hirsutum*	*S. peruvianum* *S. chilense*
Brinjal (*Solanum melongena*)	*S. insanum* L.	*S. incanum* *S. insanum*	*S. anguivi* *S. dasyphyllum* *S. pyracanthos* *S. tomentosum*	*S. torvum* *S. sisymbriifolium* *S. elaeagnifolium* *S. khasianum*
Chilli (*Capsicum annuum*)	*C. frutescens*	*C. annuum* *C. frutescens* *C. chinensis*	*C. baccatum* var.*baccatum*, *C. baccatum* var. *pendulum*	*C. pubescence* *C. chacoense* *C. exemium* *C. cardenasii*
Cucumber (*Cucumis sativus*)	*C. sativus* var. *hardwickii*	*C. sativus* var. *sativus*, *C. sativus* var. *hardwickii*	*C. hystrix* (African cucumber)	*C. melo* L. *Cucurbita sp.*
Potato (*Solanum tuberosum*)	*S. maglia* *S. tarijense*	*S. acaule* *S. berthaultii* *S. brevicaule* *S. candolleanum* *S. infundibuliforme* *S. okadae* *S. vernei*	*S. acroglossum* *S. acroscopicum* *S. agrimonifolium* *S. albicans* *S. albornozii* *S. andreanum* *S. boliviense* *S. bombycinum* *S. buesii* *S. chacoense* *S. chilense* *S. chiquidenum* *S. chomatophilum* *S. clarum* *S. colombianum* *S. demissum* *S. iopetalum* *S. longiconicum* *S. stoloniferum* *S. venturii* *S. verrucosum*	*S. anamatophilum* *S. augustii* *S. bulbocastanum* *S. cardiophyllum* *S. humectophilum* *S. hypacrarthrum* *S. immite* *S. lignicaule* *S. malmeanum* *S. minutifoliolum* *S. mochiquense* *S. pinnatisectum* *S. stenophyllidium* *S. tarnii* *S. trifidum* *S. trinitense* *S. wittmackii*

(continued)

Table 12.3 (continued)

Crop (Botanical name)	Progenitor	Primary gene pool	Secondary gene pool	Tertiary gene pool
Cotton (*Gossypium* species)	*G. arboreum (AA)* *G. herbaceum (AA* *G. raimondii (DD)*	*G. tomentosum* *G. darwinii* *G. mustelinum*	*G. arboreum* *G. herbaceum* *G. anomalum* *G. triphyllum* *G. capitis viridis* *G. barbosanum* *G. longicalyx* *G. thurberi* *G. trilobum* *G. davidsonii* *G. klotzschianum* *G. armourianum* *G. harknessii* *G. turberi,* *G. aridum* *G. lobatum* *G. laxum* *G. raimondii* *G. gossypioides* *G. schwendimanii* *G. davidsonii* *G. barbadense* *G. hirsutum*	*G. sturtianum* *G. robinsonii* *G. nandewarense* *G. benadirense* *G. bricchettii* *G. vollesenii* *G. stocksii* *G. somalense* *G. areysianum* *G. incanum* *G. australe* *G. nelsonii* *G. bickii* *G. costulatum* *G. cunninghamii* *G. exiguum* *G. rotundifolium* *G. enthyle* *G. nobile* *G. pilosum* *G. pulchellum* *G londonderriense* *G marchentii* *G populifolium* *G. anapoides*

12.4 Important Crop Wild Relatives in Indian Gene Center

The Indian gene center is unique and serves as the origin for several cultivated agri-horticultural crops. The Indian gene center houses more than 320 wild relatives of crop plants. The northeastern regions of India are particularly notable for their diversity and abundance of wild relatives of agri-horticultural crops. For example, the Tura range of the Garo Hills in Meghalaya is known for wild relatives of *Citrus and Musa,* while Sikkim is extremely rich in orchid diversity. *Zizania latifolia*, a wild relative of rice, is found in Loktak Lake, Manipur. Various wild *Ziziphus* species are reported from Tripura, and wild species of *Musa* and Zingiberaceae are found in Mizoram. Table 12.4 lists some of the important wild relatives of crop plants in the Indian gene center.

Table 12.4 Notable wild relatives reported from the Indian gene center

Important wild relatives reported from Indian Gene Centre (Source; Pandey *et al.*, 2005)	
Oryza nivara, *O. rufipogon*, *O. coarctata*	*Dioscorea tomentosa*, *D. wallichii*
Vigna radiata var. *sublobata*, *V. mungo* var. *sylvestris*, *V. vexillata*, *V. hainiana*, *V. bourneae*, *V. khandalensis*, *Cajanus cajanifolia*, *C. scarabaeoides*, *Cicer microphyllum*	*Citrus assamensis*, *C. indica*, *C. ichangensis*, *Musa balbisiana*, *Ensete glauca*, *E. superbum*, *Mangifera andamanica*, *M. sylvatica*, *Prunus jacquemontii*, *P. pashia*
Sesamum mulayanum, *S. prostratum*, *S. laciniatum*, *Brassica quadrivalvis*, *Carthamus oxyacanthus*	*Piper bantamense*, *P. beddomei*, *P. hapnium*, *P. nigrum*, *Allium carolinianum*, *A. humile*, *A. tuberosum*, *A. wallichii*
Corchorus depressus, *C. pseudo-olitorius*, *C. tridens*, *C. trilocularis*, *C. urticifolius*	*Erianthus munja*, *E. arundinaceus*, *Saccharum spontaneum*
Abelmoschus tuberculatus, *A. manihot*, *Cucumis callosus*, *C. sativus* var. *hardwickii*, *C. prophetarum*, *Momordica dioica*, *M. denudata*, *M. balsamina*	

12.5 Use of Crop Wild Relatives in Crop Improvement

Crop wild relatives (CWRs) play an important role in maintaining food security. CWRs can be exploited as a valuable source of alleles related to adaptive traits that provide resistance to different abiotic and biotic stresses, and tolerance to climate change, and improve yield potential and nutrition. The CWRs were mainly used for the improvement of food crops namely wheat, rice, barley, tomato, soybean, and mustard. The CWRs along with their use in trait improvement are summarized crop group-wise under the following sub-heads and the potential sources, genes, and traits identified in CWRs in important field crops (cereals, pulses, fiber, and oilseed) are given in Tables 12.5, 12.6 and 12.7.

12.5.1 Cereals

Cereals are staple foods of billions of people across the globe and their production is continuously affected by climate change. Among cereals, wheat, rice, maize, barley, sorghum, pearl millet, and oats are important crops. Pre-breeding can provide an opportunity to regain the genetic diversity lost during domestication and monoculture.

About 95% of global wheat production comes from hexaploid bread wheat (*Triticum aestivum*, AABBDD with $2n = 6x = 42$; genome size--~17 GB), while the rest includes tetraploid durum wheat (*Triticum turgidum*, AABB) and other less economically important species. Undomesticated wild species, crop wild relatives,

Table 12.5 Wild species identified as sources of resistance for abiotic/biotic stresses in cereal and millet crops

Crop	Wild species	Traits/Genes in parenthesis
Wheat	Wild emmer, *Triticum turgidum* ssp. *dicoccoides*	Drought tolerance, Grain protein content, Spike productivity, *Fusarium* head blight resistance, Fe and Zn content, End-use qualities, Leaf rust (***Lr53, Lr64***), Stripe rust (***Yr15, Yr35, Yr36***), Powdery mildew (***Pm3k, Pm16, Pm26, Pm30, Pm31, Pm36, Pm41, Pm42***), Wheat blast resistance genes (***Rmg***)
	Ae. tauschii	Salinity, Drought and freezing stress tolerance, Higher thermostability of the photosynthetic mechanism, Nematode resistance, Powdery mildew (***Pm2, Pm19, Pm34, Pm35, Pm58***), Hessian fly (***H22, H23***), Green bug (***Gb3***), Russian wheat aphid (**Dn3**), Wheat curl mite (**Cmc4**), Leaf rust (***Lr21, Lr22a, Lr32, Lr39, Lr41, Lr40, Lr42, Lr43***), Stripe rust (***Yr28***), Stem rust (***Sr33, Sr45, Sr46***)
	Ae. comosa	Salt resistance, Stripe rust (***Yr8***), Stem rust (***Sr34***)
	Ae. umbellulata	Cold resistance, Powdery mildew (***Pm57***), Leaf rust (***Lr9, Lr76***), Stripe rust (***Yr70***), Powdery mildew (***Pm33***)
	Ae. sharonensis	Drought resistance, Leaf rust (***Lr56***), Stripe rust (***Yr38***)
	Ae. longissima	Drought resistance, Powdery mildew (***Pm132***)
	Ae. crassa	Physiological performance
	Ae. cylindrica	Drought, Cold and Salt resistance
	Ae. speltoides	Transpiration efficiency, Water use efficiency and drought tolerance, Tolerance to high temperature, Nematode resistance, Powdery mildew (***Pm12, Pm32, Pm53***), Green bug (***Gb5***), Stem rust (***Sr32, Sr47***), Leaf rust (***Lr28, Lr35, Lr36, Lr37, Lr47, Lr51, Lr66***), Stem rust (***Sr32, Sr39, Sr47***)
	Ae. neglecta	Salinity tolerance, Cold resistance, Drought tolerance, Stripe rust resistance (***Yr42***), Leaf rust (***Lr62***)
	Triticum monococcum	Russian wheat aphid resistance; Heat, Cold and Salt tolerance, Stem rust (***Sr21, Sr22, Sr35***), Leaf rust (***Lr63***), Powdery mildew (***Pm1b, Pm1c, Pm4d, Pm4a, Pm25***)
	T. turgidum spp. *dicoccum*	Fusarium head blight and Rust resistance, Drought and Salt tolerance
	Ae. geniculata	Heat and Drought tolerance, Dwarf virus resistance, Powdery mildew (***Pm29***), Stem rust (***Sr53***), Leaf rust (***Lr57***), Stripe rust (***Yr40***)
	Ae. ventricosa	Eyespot (***Pch1***), Cyst nematode (***Cre2, Cre5, Cre6***), Root knot nematode (***Rkn3***), Hessian fly (***H27***), Stem rust (***Sr38***), Stripe rust (***Yr17***), Leaf rust (***Lr37***), Wheat blast disease
	Ae. peregrina	Leaf rust (***Lr59***), Cyst nematode (***CreX, CreY***), Root knot nematode (***Rkn2***)
	T. araraticum	Better pasta firmness and bread-making quality, Yield performance with 24% yield advantage
	T. boeoticum	Grain Fe and Zn concentration
	Agropyron elongatum	Drought tolerance, Stem rust (***Sr24***), Leaf rust (***Lr19***)
	Thinopyrum ponticum	Fusarium head blight resistance ***(FHB3, FHB6, FHB7)***, Powdery mildew (***Pm51***), Stripe rust (***Yr69***)

(continued)

Table 12.5 (continued)

Crop	Wild species	Traits/Genes in parenthesis
	Th. intermedium	Powdery mildew (***Pm40***), Stem rust (***Sr44***), Stripe rust (***Yr50***)
	T. timopheevii	Leaf rust (***Lr18, Lr50, Lr52***), Stem rust (***Sr36***), Powdery mildew (***Pm6***)
	Ae. searsii	Drought tolerance, Powdery mildew (***Pm57***)
	T. spelta	Leaf rust (***Lr44, (Yr5***)
	Secale cereale	Leaf rust (***Lr25, (Yr9***), Stem rust (***Sr27, Sr59***), Powdery mildew (***Pm6, Pm56***)
Rice	*Oryza longistaminata*	Drought tolerance, Bacterial blight (***Xa21***), **Blast** (***Pi57***)
	O. rufipogon	Bacterial blight (***Xa23, Xa30***), Brown plant hopper (***Bph27, Bph29, Bph30, Bph36, Bph38,***), Green rice leafhopper (***Grh5***)
	O. nivara	Bacterial blight (***Xa33***), Sheath blight (***qShB6***), Green rice leafhopper/(***Grh4, Bph34***)
	O. rhizomatis	Rice blast (***Pi54rh***)
	O. australiensis	Brown plant hopper (***Bph10, Bph18***)
	O. latifolia, O. officinalis, O. minuta	Brown plant hopper (***Bph12, Bph15, Bph21***)
Barley	*Hordeum bulbosum*	Powdery mildew, Leaf rust and Leaf scald resistance
	H. spontaneum	High fiber content, Beta-glucan, Zn and Fe content
Oats	*Avena barbata*	Powdery mildew and Crown rust resistance, and survival under mesic and xeric conditions
Sorghum	*S. propinquum*	Grain yield, height, and earliness
	S. bicolor subsp. *verticilliflorum*	Grain yield, Drought and High temperature, Parasite resistance
	S. bicolor notho subsp. *drummondii*	Allelopathic properties, Ergot and Nematode resistance
	S. halepense	Resistance to Green bug and Sorghum shoot fly
	S. angustum, S. amplum S. bulbosum	Resistance to sorghum midge
	S. exstans, S. stipoideum S. matarankense	Resistance to shoot fly
	S. leiocladum	Cold tolerance

and landraces are valuable sources of variation for improving wheat to address challenges from climate change and a growing population. *Aegilops* species, a close relative of bread wheat, offers valuable traits for drought, heat, salinity, and disease resistance. Salinity stress, which limits wheat production worldwide, can be mitigated by transferring salt resistance genes from *Festuca* and *Agropyron* species to wheat. Biotic stress resistance in wheat, often controlled by single genes, has been more extensively transferred from wild relatives compared to abiotic stress tolerance. Aegilops species like *Ae. triuncialis, Ae. umbellulata, Ae. kotschyi,* and *Ae.*

Table 12.6 Wild species identified as sources of resistance for abiotic/biotic stresses in oilseed and fiber crops

Brassica	*Brassica tournefortii*	Drought and Heat tolerance, Alternaria and white rust resistance, Cytoplasmic male sterility
	Camelina sativa	Alternaria blight, Mustard aphids, White rust resistance, and drought tolerance
	Sinapis alba	Resistance to Alternaria blight, SSR, High temperature and Drought stress tolerance
	Diplotaxis erucoides	Alternaria blight Resistance
	Brassica fruticulosa	Resistance to Mustard aphids, SSR, Alternaria blight, White rust, Cabbage aphid, Cabbage root fly resistance, and Drought tolerance
	Crambe abyssinica	Resistance to Mustard Aphids and Drought tolerance
	Erucastrum cardaminoides, E. gallicum, Camelina cardaminoides	Resistance to SSR
	Capsella bursa-pastoris	Resistance to SSR, Alternaria blight, Drought tolerance
	Lepidium sativum	Mustard aphids, White rust resistance and Heat tolerance
Sunflower	*Helianthus praecox* accession 1823 and *H. nuttallii* NUT05	Herbicide tolerance (sulfonylurea imazethapyr)
	Wild sunflower (*H. annuus* L).	Sunflower moth resistance
	H. anomalus	Drought resistance, High water use efficiency, Broomrape resistance, Oil concentration, and fatty acid profile
	H. paradoxus	Salt resistance
	H. argophyllus	Drought and Heat resistance
	H. tuberosus	Sclerotinia stalk rot, Phomopsis stem canker, Phoma black stem, and Charcoal rot resistance
	H. maximiliani	*Sclerotinia sclerotiorum* resistance
	H. agrestis, H. debilis subsp. *cucumerifolius* and *H. exilis*	Broomrape resistance
	H. portei, H. atrorubens, and *H. schweinitzii*	Increased α-linolenic acid
	H. deserticola	Oil concentration and fatty acid profile
Linseed	*Linum grandiflorum*	Bud fly and Alternaria blight resistance
Cotton	*G. tomentosum*	Heat, Drought, Salt, Nectariless and Thrips Tolerance
	G. herbaceum, G. darwinii	Drought tolerance
	G. davidsonii, G. aridum, G. klotzschianum	Salt tolerance
	G. thurberi, G. darwinii	Verticillium Wilt resistance
	G. stocksii, G. somalense	Cotton Leaf Curl Virus resistance
	G. barbadense, G. hirsutum, G. raimondii	Jassids tolerance
	G. barbadense	Thrips tolerance
	G. stocksii, G. capitis-viridis	Fiber traits
	G. sturtianum, G. austral, G. bickll	Glandless seed and glanded plant

Table 12.7 Wild species identified as sources of resistance for abiotic/biotic stresses in pulses crops

Crop	Wild species	Traits
Chickpea	*C. reticulatum* *C. echinospermum*	Pod number, Short internodes, Cold tolerance, *Ascochyta* blight, Cyst nematode, Fusarium wilt, Helicoverpa pod borer, and Phytophthora root rot resistance
	C. judaicum	Fusarium wilt, Botrytis grey mould, Ascochyta blight, and Helicoverpa pod borer tolerance
	C. bijugum	Fusarium wilt, Cyst nematode, Phytophthora root rot, Helicoverpa pod borer resistance, and Cold tolerance
	C. pinnatifidum	Botrytis grey mould, Ascochyta blight, Cyst nematode, Phytophthora root rot, and Helicoverpa pod borer tolerance
	C. montbretii	Ascochyta blight resistance
	C. cuneatum	Helicoverpa pod borer tolerance
	C. anatolicum, C. microphyllum, C. montbretii, C. oxydon, C. songaricum	Drought tolerance
Pigeon pea	*Cajanus acutifolius*	Cytoplasmic male sterility, High seed weight, Pod borer, Pod fly, Bruchids, and Phytophthora blight resistance
	C. platycarpus	Phytophthora blight, Pod borer, Bruchids, Fusarium wilt resistance, Salinity tolerance, Cytoplasmic male sterility
	C. scarabaeoides	Pod borer, Sterility mosaic disease, Pod fly, Bruchids, Drought and Water-logging tolerance, Cytoplasmic male sterility, High protein content
	C. sericeus	Cytoplasmic male sterility, High protein content, Pod borer, Sterility mosaic disease, Phytophthora blight resistance, Drought tolerance
	C. albicans	Pod borer, Sterility mosaic disease, High protein content, Salinity tolerance, Drought tolerance
	C. reticulatus	Pod fly and Pod borer resistance
	C. volubilis	Sterility mosaic disease resistance
	C. cajanifolius	High protein content, Cytoplasmic male sterility
	C. volubilis	Cytoplasmic male sterility
	C. lineatus	Drought tolerance, High protein content, Sterility mosaic disease resistance

(continued)

Table 12.7 (continued)

Crop	Wild species	Traits
Lentil	*Lens culinaris* ssp. *culinaris*	Anthracnose resistance, Heat tolerance
	L. culinaris ssp. *orientalis/L. orientalis*	Bruchid weevils, Orobanche resistance, Ascochyta blight, Stemphylium blight, Fusarium wilt, Powdery mildew, Rust resistance, Sitona weevils, Cold, Salinity, Heat, Drought tolerance, Biofortified traits (High Zn, Fe, protein content), Seed weight, Yield attributes
	L. ervoides	Anthracnose, Powdery mildew, Ascochyta blight, Stemphylium blight, Fusarium wilt, Sitona weevil and Rust resistance, Drought tolerance, Orobanche Resistance, Biofortification potential
	L. tomentosus	Fusarium wilt, Powdery mildew resistance, Drought tolerance
	L. lamottei	Anthracnose, Ascochyta blight, Stemphylium blight and Bruchid weevils Resistance
	L. nigricans	Anthracnose, Ascochyta blight, Stemphylium blight, Fusarium wilt, Rust, Powdery mildew, Sitona weevils, Bruchid weevils resistance, Cold, drought, and heat tolerance
	L. montbretti	Ascochyta blight resistance
	L. culinaris ssp. *odemensis/L. odemensis*	Ascochyta blight, Stemphylium blight, Fusarium wilt, Rust, Sitona weevils, Powdery mildew resistance, Orobanche and Drought tolerance
Common bean	*Phaseolus coccineus*	Anthracnose, Root rots, White mold, Bean yellow mosaic virus, and Angular leaf spot resistance
	P. acutifolius	Common blight resistance, Heat and Drought tolerance
	P. angustissimus	Freezing tolerance
	P. filiformis	Salt tolerance
Pea	*Pisum fulvum*	Pea weevil and Ascochyta blight resistance
	P. sativum ssp. Elatius	Fusarium wilt resistance
Soybean	*Glycine soja*	Salt tolerance, *Sclerotinia* stem rot and Soybean cyst nematode resistance
	G. canescens	Powdery mildew
	G. tomentella	Nematode, Leaf rust, Salt resistance
	G. latifolia	Stem rot resistance
	G. argyrea, G. clandestine, G. dolichocarpoa	Soybean cyst nematode and Salt resistance
	G. microphylla, G. tabacina	Salt tolerance resistance

ovata are known to resist yellow and black rust. Two important QTLs (QYrtm.pau-2A identified in *Triticum monococcum* on chromosome 2A, QYrtm. pau-5A identified in *Triticum boeoticum* on chromosome 5A) were identified for stripe rust resistance in wild species of wheat. Using marker-assisted selection, an adult plant resistance (APR) gene for leaf and stripe rust was transferred from *T. monococcum* to bread wheat variety WL711, and another leaf rust gene was transferred to the PBW343 background. Other wild relatives, such as *Triticum dicoccoides*, offer resistance to Fusarium head blight. A spring wheat germplasm line derived from *Ae. tauschii* has recently been made available to breeders and shows resistance to the Hessian fly, a major pest causing significant crop losses in the United States. *Leymus racemosus* chromosome introgression lines in wheat showed increased heat tolerance at seedling and reproductive stages, with one line exhibiting high yield potential.

The genus *Oryza* includes two cultivated and 23 wild rice species. *Oryza* contains 11 different types of genomes including diploid genomes (six) and tetraploid genomes (five). The genome size of *Oryza sativa* ($2n = 24$) is 430 Mb. The genomes of diploids are designated as $n = 12$: AA, BB, CC, EE, FF, and GG, where the AA genome alone consists one-third species of *Oryza* and it forms the primary gene pool of rice. There are four major species complexes in the genus *Oryza*, namely *O. sativa* complex (eight species; AA genome), *O. officinalis* complex (ten species; BB, CC, EE, BBCC, and CCDD genomes), *O. ridleyi* complex (two species; HHJJ genome) and *O. meyeriana* complex (two species, GG genome). The remaining three species are categorized as outliers or unclassified species, namely *O. brachyantha* (FF), *O. schlechteri* (HHKK), and *O. coarctata* (KKLL), and are also considered as most primitive species of the genus *Oryza*. Weedy rice, an introgressed form of wild and cultivated rice (*Oryza sativa* L.), is native to Asia, but was first documented in North Carolina, USA, in 1846. Crop wild relatives (CWRs) have contributed beneficial traits to rice, such as ideal plant architecture and weed suppression. *Oryza rufipogon* provides QTLs for cold and aluminum tolerance, while *Oryza nivara* has genes for strong resistance to grassy stunt virus. Traits like weed competitiveness, drought tolerance, and low input requirements from *O. glaberrima* have been combined with high yield to develop NERICA (New Rice for Africa), which is high yielding, drought and pest resistant, and suited for West Africa. Successful gene transfers from wild to cultivated rice include grassy stunt virus resistance from *O. nivara,* cytoplasmic male sterility from *O. sativa* spp. *spontanea*, bacterial blight resistance (Xa21) from *O. longistaminata*, and rice blast resistance (pi9) from *O. minuta*. An introgression line from wild Dongxiang to cultivated japonica rice showed enhanced salt stress resistance at all stages: germination, establishment, vegetative growth, and seed yield.

In case of barley (*Hordeum vulgare*; $2n = 2x = 14$ and genome size-5.3 Gb), a wild species, *H. spontaneum*, has been shown to improve the response of barley to major biotic and abiotic stresses and has potential to enhance barley yield. *Hordeum spontaneum* has been used for acquiring resistance against many biotic stresses such as leaf stripe, powdery mildew, leaf rust, and leaf scald, and abiotic stresses such as drought and salt tolerance. In *Hordeum bulbosum*, important sources of resistance against powdery mildew, leaf rust, and leaf scald have been identified. Furthermore, higher concentrations of beta-glucan, Zn, and Fe were identified in *H. spontaneum* derived lines. At the International Center for Agricultural Research in the Dry Areas (ICARDA), Syria, a six-row barley cultivar with drought tolerance derived from *H. spontaneum* was released for cultivation in 2004.

In maize (*Zea mays* L.), with a genome size of 2.4 Gb and $2n = 2x = 20$, two important crop wild relatives, namely Teosinte (*Zea mays* ssp. *parviglumis*) and *Tripsacum*, are identified as donors of economically important traits. Introgression in cultivated maize by introducing *Tripsacum* genes for resistance against *Helminthosporium*, *Puccinia*, rootworm, drought, and aluminum stress has been reported. In addition, *Zea mays* ssp. *mexicana* is a donor for downy mildew and Fusarium resistance and *Tripsacum dactyloides for* rootworm resistance. Blight resistant alleles transferred from a wild relative of Mexican maize (*Tripsacum dactyloides* L.) into commercial corn lines is a well-known success story that resolved corn blight in the United States of America. Using genomic in-situ hybridization technology, the chromosomal locations of the *Zea diploperennis*-derived segments in Maize × *Zea diploperennis* crosses have been identified that carried loci resistant to fungal diseases including northern corn leaf blight, southern corn leaf blight, and corn leaf spot diseases.

Pearl millet (*Pennisetum glaucum* L.) with a genome size of 1.76 Gb and $2n = 2x = 14$ is one of the world's hardiest warm-season cereal crops cultivated for food and animal feed in the semi-arid tropics of Asia and Africa. Screening of pearl millet wild species accessions indicated that *P. schweinfurthii* was found to be highly resistant, while *P. violaceum* was susceptible to downy mildew. The screening of many wild species accessions of *P. violaceum*, for blast and rust resistance, resulted in identification of promising sources for use in crop improvement. Some wild species namely *P. glaucum* ssp. *monodii*, *P. purpureum*, and *P. squamulatum* have been used for improving cultivated pearl millet. Two accessions of *P. violaceum*, IP 21544 and IP 21720, originating from Niger and Chad, respectively, have been identified for multiple traits namely heat stress tolerance, high dry fodder yield, and resistance to multiple pathotypes of blast (*Magnaporthe grisea*), and these accessions were used to transfer blast resistance in the cultivated background through advanced backcross method.

12.5.2 Oilseed and Fiber Crops

Oilseeds are major food crops, known for their unique protein and oil-rich characteristics. Major biotic and abiotic stresses are a serious concern for global oilseed production. Use of some important CWRs of oilseed *Brassica* and other minor oilseed crops (Table 12.6) are mentioned below.

In Brassica, wild relatives have been used as genetic resources to develop new cultivars resistant to various stresses. The genus Brassica includes three diploid species: *B. rapa* (AA genome), *B. nigra* (BB genome), and *B. oleracea* (CC genome), and their naturally occurring allotetraploid species: *B. napus* (AACC genome),

B. juncea (AABB genome), and *B. carinata* (BBCC genome). Black mustard (*Brassica nigra*) is a valuable gene source for improving cultivated Brassicas against pathogens. Interspecific hybrids between *B. fruticulosa*, *B. nigra*, and *B. rapa* are used as bridge species for this purpose. Despite the genetic variation in wild Brassicaceae, their use in conventional breeding is limited due to differences in ploidy, linkage drag, and fertilization barriers. Notable wild relatives of Brassica include *Sinapis alba* (White mustard), *B. tournefortii* (African mustard), *B. fruticulosa* (Twiggy turnip), *Camelina sativa* (Gold-of-pleasure), *Diplotaxis species* (various types of wall rocket), *Crambe abyssinica* (Abyssinian mustard), *Erucastrum species* (dog mustard), *Capsella bursa-pastoris* (Shepherd's purse), and *Lepidium sativum* (Garden Cress). Some wild Brassicaceae show resistance to leaf spot/ Alternaria blight, with *Eruca sativa* and *Sinapis alba* being notably resistant. Wild relatives are also used to develop male sterile lines for hybrid production. In *B. juncea*, a CMS (cytoplasmic male sterile) line was developed using *Diplotaxis siifolia* cytoplasm. Another CMS system was created by backcrossing a somatic hybrid of *Moricandia arvensis* and *B. junce*a to *B. juncea*. The Ogura CMS system, developed by crossing *B. napus* with *Raphanus sativus*, is widely used for its complete pollen abortion and high sterility rates. This system is extensively deployed in B. *napus*, *B. juncea*, and *B. oleracea*.

Sunflowers belong to the genus *Helianthus*, native to North America, ranging from central Mexico to southern Canada. The genus includes 52 species and 19 subspecies, with 14 annuals and 38 perennials. The sunflower genome is diploid ($2n = 2x = 34$) with a genome size of 3.6 Gb. Wild annual sunflowers have high resistance to all races of rust (*Puccinia helianthi*), while only two populations (*H. argophyllus* and *H. debilis*) show resistance to all races of powdery mildew (*Erysiphe cichoracearum*). Complete resistance to downy mildew has been found in several annual species (*H. annuus, H. argophyllus, H. debilis, H. petiolaris) and perennial species (H. decapetalus, H. divaricatus, H. eggertii, H. giganteus, H. laetiflorus, H. mollis, H. nuttallii, H. scaberrimus, H. pauciflorus, H. salicifolius, and H. tuberosus*). Annual species *H. debilis* ssp. sylvestris, H. praecox ssp. praecox, and *H. bolanderi*, and 14 perennial species are tolerant to powdery mildew. Wild *H. argophyllus* has been used to improve drought tolerance, and *H. paradoxus* to enhance salinity tolerance. Herbicide resistance to imidazolinone and sulfonylurea, used to control broomrape, was recently introgressed from wild *H. annuus*. CMS from wild *H. annuus* and *H. petiolaris* have been used in commercial sunflower hybrids since 1972, expanding the industry significantly. Today, 100% of US sunflower production and 60–70% of worldwide production use these hybrids.

In linseed (*Linum usitatissimum* L., $2n = 2x = 30$; genome size ~370 Mb), the important wild relative is *Linum tenuifolium*, having potential for lowering linolenic acid in oil to make it edible, *Linum perenne* for drought and cold hardiness, *Linum grandiflorum* for bud fly (*Dasyneura lini*), and *Alternaria* blight resistance.

The cotton genus (*Gossypium*) includes over 50 species, comprising five allotetraploid species and the rest diploid species. The cultivated species of cotton are *Gossypium hirsutum*, *G. barbadense*, *G. herbaceum*, and *G. arboreum*. Additionally, there are about 35 wild species that have desirable traits like good fiber quality and resistance to pests. The world's textile fiber mainly comes from four species: two tetraploid species (*G. barbadense* and *G. hirsutum*) and two diploid species (*G. arboreum* and *G. herbaceum*), from the New and Old Worlds, respectively. Wild cotton species are valuable sources of resistance to biotic stresses (insects and diseases) and abiotic stresses like salinity, cold, drought, and heat. Due to their potential, these wild species (both diploids and tetraploids) are being used in various hybridization programs. Rust resistance genes have been transferred from *G. anomalum* and *G. arboreum* into *G. hirsutum* through interspecific hybridization, and bacterial blight resistance genes from *G. arboreum* into *G. barbadense*. Similarly, cotton leaf curl virus resistance genes from the autotetraploid of the diploid donor species *G. arboreum* have been introgressed into allotetraploid *G. hirsutum*.

12.5.3 Grain Legume

Grain legumes contribute nearly 27% of the world's food production and also act as the single-largest source of vegetable protein (~33%) in the human diet. In pulses, chickpea, pigeon pea, lentil, cowpea, soybean, and groundnut with high nutrition and protein values are the leading legume crops grown in marginal environments in Asia and sub-Saharan Africa. Being the primary source of high-quality dietary protein, it also provides high-quality nutritious fodder for animal consumption. Grain legumes are usually deficient in sulfur-containing amino acids, methionine, cysteine, and tryptophan, but are rich in another essential amino acid, lysine. Changing climatic conditions resulting in high rainfall favor foliar diseases and some soil-borne pathogens such as *Phytophthora, Pythium, Rhizoctonia solani*, and *Sclerotium rolfsii* and low rainfall resulting in decreased moisture leads to the high incidence of *Fusarium* wilt and dry root rot in cool season grain legumes. Wild species are the reservoir of many useful genes/alleles as they have evolved under natural selection to survive climate extremes. The important wild sources of various pulse crops are presented in Table 12.7 and some successful wide hybridizations are mentioned in the following subheads.

Chickpea (*Cicer arietinum* L.) is a true diploid ($2n = 16$) annual food legume species having a genome size of ~738 Mbp. The genus *Cicer* consists of 46 species, of which 10 are annual and 36 are perennial (Harlan and de Wet, 1971). Wild species are well-known for having climate-resilient and productivity-enhancing traits of interest. Chickpea production and productivity are adversely affected by fusarium wilt (*Fusarium oxysporum*), ascochyta blight (*Ascochyta rabiei*), botrytis graymold (BGM, *Botrytis cinerea*), dry root rot (DRR, *Rhizoctonia bataticola*), powdery mildew (*Leveillulataurica*), and pod borer (*Helicoverpa armigera*), and abiotic (drought, heat, cold, and salinity) stresses, and the estimated yield losses by individual pests or diseases vary from 5–10% in temperate and 50–100% in tropical regions. Resistance to diseases such as wilt, foot rot, root rot, and *Botrytis* graymold has been introgressed from *C. reticulatum* and *C. echinospermum* into cultivated chickpea. Wild species such as *C. echinospermum* for *Ascochyta* blight resistance, pod number, and short internode, and *C. reticulatum* and *C. echinospermum* for heat and drought tolerance have been used. The use of *C. reticulatum* accession, ILWC119, in the crossing program developed two cyst nematode resistant chickpea germplasm lines, ILC10765 and ILC10766. Interspecific derivatives possessing a high degree of resistance to diseases such as wilt, root rot, and foot rot, and high yield, have been obtained from *C. arietinum* × *C. reticulatum* crosses. Recently, mapping population (NAM) developed by the interspecific hybridization between cultivated chickpea (commercial variety, Gokce), and wild accessions of *C. reticulatum* and *C. echinospermum* species lead to the identification of QTLs associated with flower initiation and flower color. Furthermore, marker-assisted selection (MAS) helped to introgress leaf miner resistance genes from *C. reticulatum* to cultivated chickpea.

Pigeon pea (*Cajanus cajan* L.) is a sub-tropical and tropical grain legume which originated in the northern region of the Indian sub-continent. The chromosome number of the pigeon pea is $2n = 16$ with a genome size of 833Mbp. Crop wild relatives of pigeon pea are valuable genetic resources for resistance to *Alternaria* blight, *Phytophthora* blight, sterility mosaic virus, podfly and nematodes, and tolerance to salinity, drought, and photoperiod insensitivity. The resistant genes have been transferred from *Cajanus platycarpus* into cultivated pigeon pea for Phytophthora blight. The CMS systems have been developed with cytoplasm derived from wild *Cajanus* species namely *C. sericeus*, *C. scarabaeoides*, *C. volubilis*, *C. cajanifolius*, *C. lineatus* and *C. platycarpus* (Sharma et al. 2013). A high protein line, ICPL 87162 {(30–34% high protein compared to check (23% protein)} was developed from the cross *C. cajan* × *C. scarabaeoides*. Several lines such as HPL2, HPL7, HPL40, and HPL51 with high protein and seed weight have been developed by utilizing wild Cajanus species. The photoperiod-insensitive and extra-early flowering introgression lines were developed by crossing cultivated pigeon pea and its wild relatives, namely *C. platycarpus, C. volubilis, C. acutifolius*, and *C. cajanifolius*. *Cajanus scarabaeoides*, a wild relative of pigeon pea, has good agronomic traits including dwarfness, number of fruiting branches and pods per plant, and resistance to pod borers and phytophthora stem blight.

The genus *Lens* ($2n = 14$), an important source of food, fodder, and dietary protein, is one of the most important members of the family *Fabaceae*. The genome size of cultivated lentil is 4300Mbp. In lentil, approximately 40% of genetic diversity has been lost during the domestication process, which resulted in increased susceptibility to both biotic and abiotic stresses. The wild species namely *L. nigricans* showed drought tolerance and *L. orientalis* has superior alleles for heat and drought tolerance, as well as higher protein and micronutrient concentrations. The *L. odemensis* showed resistance to *Sitona weevil* and *L. ervoides* contains favorable or superior genes or alleles for growth habit, phenology, plant biomass, and seed traits.

Groundnut (*Arachis hypogaea* L.) is an important oilseed crop of tropical and subtropical regions of the world, comprising 48% vegetable oil and 25% protein. Cultivated peanut, an allotetraploid ($2n = 40$), has a genome size of 2540 Mbp. It is native to South America with an "AABB" genome and is likely to have originated from a single hybridization event between two wild diploids *A. duranensis* (A genome) and *A. ipaensis* (B genome) followed by chromosome doubling. A synthetic amphiploid namely TxAG-6 was developed by crossing between an AA-genome donor hybrid (*A. cardenasii* × *A. diogoi*) and a BB genome species (*A. batizocoi*) followed by colchicine treatment of the sterile triploid to produce a fertile hexaploid. This fertile hexaploid is crossable with the cultivated groundnut and produced fertile progenies, which resulted in the release of two cultivars, Coan and NemaTAM, carrying genes for root-knot nematode resistance from *A. cardenasii*. The rust resistance was transferred from *A. cardenasii* to cultivated peanuts and leaf spot resistance genes from wild accessions of *A. cardenasii, A. batizocoi, A. diogoi, A. magna, A. monticola, A. stenosperma*, and *A. valida* transferred by crossing with tetraploid *A. hypogaea*.

12.5.4 Vegetable Crops

The wild relatives of vegetable crops constitute an increasingly important resource for improving vegetable production and a critical resource in ensuring food security. The potential source, genes, and traits identified in CWRs providing resistance/tolerance against different biotic and abiotic stresses in important vegetable and fruit crops are summarized in Table 12.8.

The cultivated tomato (*Solanum lycopersicum*) is one of the most economically important crops, also used as a model crop for vegetables. It is a diploid species with $2n = 24$ and a genome size of approximately 950 Mb. In tomato, the wild-species *L. cheesmani* contains the salt-resistance genes which have been successfully introgressed in the cultivated tomato. *Solanum pennellii* and *S. chilense* are drought tolerant, while *S. cheesmanii* and *S. peruvianum* grow better in salty coastal areas due to their root systems. *Solanum habrochaites* has been an important source of favorable alleles for yield, fruit size, and quality, and also contains genes for resistance to major tomato diseases such as late blight, bacterial canker, gray mold, and early blight. The genome of *S. pimpinellifolium* has been sequenced and contains genes associated with salinity stress.

Potato (*Solanum tuberosum*) is the fourth-highest-produced crop in the world and the most important non-cereal food crop. Wild potatoes contain many resistance genes that can be incorporated into cultivated types through a resistance breeding program. The wild species *Solanum commersonii* was used as a genetic donor of cold tolerance for *Solanum tuberosum*. In the 1900s, resistance genes were transferred from the wild potato species *Solanum demissum* to the potato crops. Late blight resistance genes were introgressed from the wild species namely *S. demissum, S. stoloniferum*, and the cultivated *S. tuberosum* ssp. *andigena* and *S. phureja* into common potato in different parts of the world.

In case of okra (*Abelmomoschus esculentus*), the major disease is yellow vein mosaic disease (YVMD), while shoot and fruit borer, and leaf hopper are major insect pests. The wild relatives such as *A. manihot, A. tuberculatus*, and *A. moschatus* have resistance genes for yellow vein mosaic disease, shoot and fruit borer, and leaf hopper, respectively. The wild species namely *A. moschatus* (IC141055), *A.*

Table 12.8 Wild species identified as sources of resistance for abiotic/biotic stresses in vegetable and some fruit crops

Vegetable crop	Wild species	Traits/genes
Tomato	*Solanum pimpinellifolium L.*	Fusarium wilt/(***I* and *I-2***), Late blight/(***Ph-1*, *Ph-2*, *Ph-3*, *Ph-5***), Early blight/(***EB-5* and *EB-9***), Bacterial wilt/ (***Bwr-6* and *Bwr-12***), Bacterial spot/ (***Rx-1*, *Rx-2*, and *Rx-3***), Bacterial spot/ (***Rx-4***), Gray leaf spot/ (***Sm***), Leaf mold/(**Cf**), Tomato spotted wilt arthotospo virus/(***Sw-1*, *Sw-2*, *Sw-3*, and *Sw-4***), Bacterial speck/(***Pto***), Bacterial canker/(***Cmm* genes**), Tomato leaf curl virus resistance, Fruit number, Fruit weight, High fruit yield,Salt, Heat and Drought tolerance
	S. peruvianum L.	Tomato yellow leaf curl virus/(***ty-5***), Tomato spotted wilt orthotospo virus/(***Sw-5* and *Sw-6***), Root-knot Nematodes/(***Mi***), Tomato mosaic virus/***Tm-2*, *Tm-22* and *Tm-2a***), Verticillium wilt/(***Ve***), Fusarium crown/ (***Frl***) resistance
	S. chilense	High lycopene content, Tomato leaf curl virus resistance
	S. cheesmanii	Salt, Drought and Heat tolerance
	S. pennellii	Fusarium wilt/(***I-3* and *I-7***), Alternaria stem Canker/(***Asc***), Bacterial spot/(***Xv-4***), A wide range of insects/(Acylsugar-related genes), Drought, Heat and Salt stresses tolerance
	S. parviflorum	Salt and Heat tolerance
	S. habrochaites (syn. *Solanum hirsutum*)	Late blight/(***Ph-4***), Leaf mold/(**Cf-4**), Tomato yellow leaf curl virus/(***Ty-2***), Tomato mosaic virus/(***Tm-1***), Powdery mildew/(***Ol-1*, *Ol-3*, *Ol-5***), Bacterial canker/(***Cmm***),Gray mould/(***Rbcq***), Drought, Heat and Salt tolerance, High lycopene content

Potato	*S. demissum*	Salinity and Potato virus X resistance
	S. bulbocastanum	Late blight (***RB*/*Rpi-blb1*, *Rpi-blb2*, *Rpi-blb3***) resistance
	S. stoloniferum	Late blight (***Rpi-sto1*, R genes**) resistance
	S. papita	Late blight (**Rpi-pta1**) resistance
	S. venturii	Late blight (***Rpi-vnt1.1***) resistance
	S. berthaultii, S. microdontum, S. sparsipilum, S. spegazzinii, S. verrucosum	Late blight (**R genes**) resistance
	S. stenotomum, S. commersonii	Bacterial wilt resistance
	S. pinnatisectum	Colorado potato beetle, Potato virus X resistance
	S. acaule	Potato virus X, Frost, Drought and Salinity resistance
	S. etuberosum, S. verrucosum	Potato virus X resistance
	S. stoloniferum	Potato virus Y/(***Ry***) resistance
	S. phureja	Potato virus Y/(***Ry***)**,** Bacterial wilt, Heat resistance
	S. brevidens	Potato virus Y/(***Ry***)**,** Potato virus X resistance
	S. chacoense	Potato virus X, Bacterial wilt, Potato virus Y/(***Ry***)**,** Colorado potato beetle resistance, Salinity resistance
	S. tuberosum ssp. *andigena*	Potato cyst nematode/(***H1gene***)**,** Potato virus Y/(***Ryadg***)**,** *G. pallida*/(***Gpa2***) resistance
	S. vernei, S. spegazzinii	Potato cyst nematode/(***Fb* gene**) resistance
	S. spegazzinii	Potato cyst nematode/(***Gro1***) resistance
	S. vernei, S. spegazzinii, S. gourlayi, S. aplocense, S. linguicaule, S. megistacrolobum	*G. pallida*/(***Gpa2***) resistance
	S. berthaultii, S. capsicibaccatum, S. jamezii, S. trifidum, S. chacoense, S. berthaultii	Colorado potato beetle resistance
	S. commersonii	Frost and Heat resistance
	S. gandarillasii	Drought resistance
	S. juzepczukii	Heat resistance

(continued)

Table 12.8 (continued)

Vegetable crop	Wild species	Traits/genes
Brinjal	*Solanum linnaeanum*	Verticillium wilt, Powdery mildew
	S. torvum	Bacterial wilt resistance
	S. sisymbriifolium, S. indicum	Fruit and Shoot Borer resistance
	S. violaceum	Fruit and Shoot Borer resistance, Male sterility
	S. aethiopicum, S. incanum	Male sterility
	S. viarum	Resistance to shoot and fruit borer
Okra	*A. manihot, A. tuberculatus, A. moschatus, A. caillei*	Yellow vein mosaic disease, Shoot and fruit borer and leaf hopper resistance
Cucumber	*Cucumis hystrix, C. metuliferous*	Gummy stem blight, Downy mildew, Cucumber mosaic virus (CMV-C), Zucchini yellow mosaic virus, Papaya ringspot virus
	C. figarei	Cucumber green mottle mosaic virus, *Fusarium* wilt, Downy mildew Resistance
	C. myriocarpus, C. africanus, C. meeusii, C. ficifolius, C. zeyheri	Cucumber green mottle mosaic virus Resistance
	Cucumis hardiwickii	Resistance to Powdery and Downy mildew

tetraphyllus (IC90476-1), and *A. caillei* under natural epiphytotic conditions and *A. angulosus* (IC203833 and IC470751) under artificial screening conditions are highly resistant accessions to Yellow Vein Mosaic Virus.

12.6 Pre-breeding

Pre-breeding involves identifying desirable traits and genes from unadapted materials (exotics or semi-exotics) that cannot be directly used in breeding programs. This process transfers these traits into elite backgrounds to create new varieties. It requires phenotyping crop wild relatives, identifying key donors with beneficial or novel alleles, and introgressing these alleles into species of interest to generate intermediate material. This step is crucial for linking the valuable traits of CWRs to modern variety development, providing breeders with wild genetic diversity in a more usable form. Pre-breeding, also known as genetic enhancement or developmental breeding, involves transferring beneficial alleles from unused germplasm into elite breeding materials. The term "enhancement" was first used by Jones (1983) to describe the transfer of useful genes from exotic or wild types into agronomically acceptable backgrounds. Rick (1984) first used "pre-breeding" or "developmental breeding," referring to the transfer or introgression of genes or gene combinations from unadapted sources into breeding materials. The goal of pre-breeding is to generate a new base population for breeding programs by using a broader pool of genetic material. Pre-breeding is necessary when there is insufficient variability for the target trait in the available germplasm, or the available variability has already been fully exploited. Various activities, from plant genetic resource management to pre-breeding, are detailed in a flow chart diagram in Fig. 12.1 and some important terminologies are provided in Box 12.1.

12.6.1 Major Applications of Pre-breeding in Crop Improvement

There are four major applications of pre-breeding:

1. Broadening the genetic base and reducing vulnerability.
2. Identifying traits in unadapted species and moving those traits (genes) into material more readily accessed by breeders.
3. Moving genes from wild species into breeding populations when this appears to be the most effective strategy.
4. Identification and transfer of novel genes from unadapted species using genetic transformation techniques.

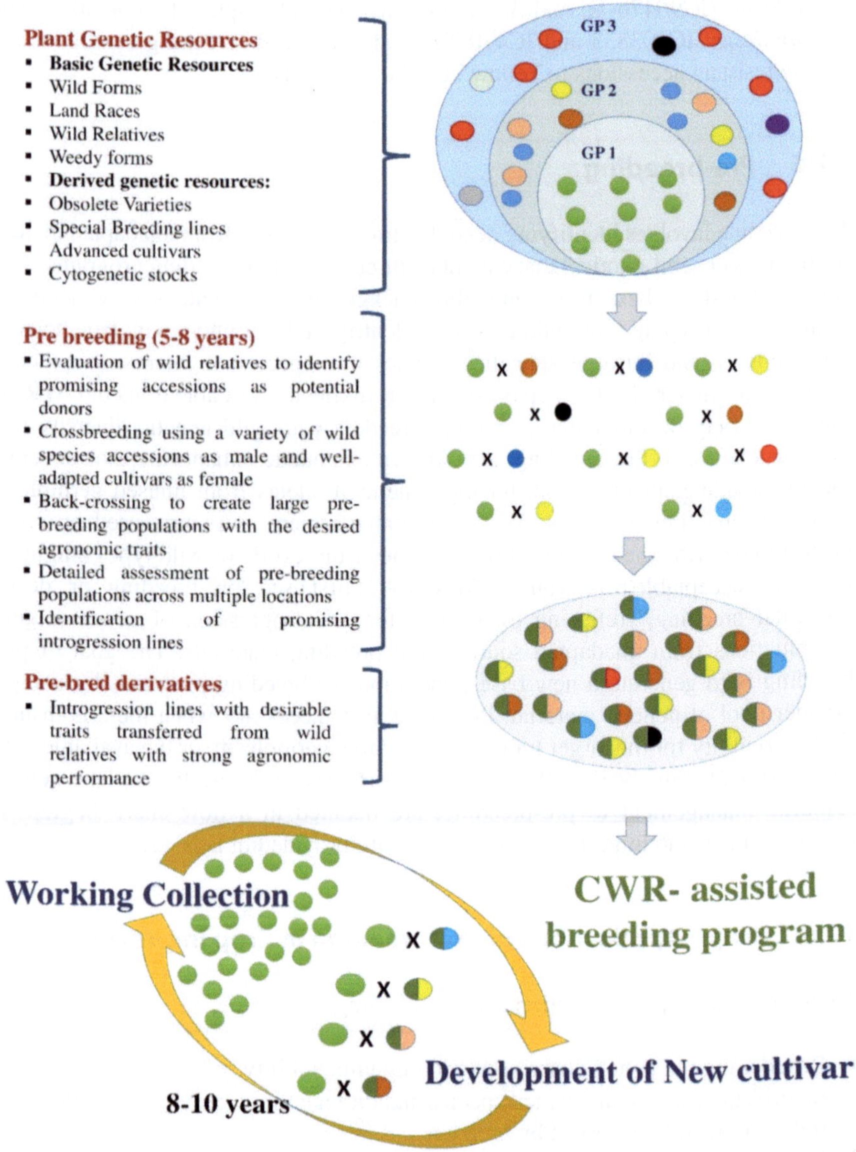

Fig. 12.1 PGR to pre-breeding flow-diagram

Box 12.1 Some Terminologies Related to Pre-breeding

Pre-breeding: Pre-breeding is an essential activity in plant breeding that helps bring new traits and genetic diversity into elite germplasm.

Exotics: According to Hallauer and Miranda Filho (1988), exotics include any germplasm that does not have immediate usefulness without selection for adaptation in a given area. In this sense, exotic germplasms are represented by landraces, traditional populations, inbred lines and CWRS among others.

Semi-exotics: The results of crosses between adapted and exotic materials, where different proportions of introgression are obtained and evaluated, have been denominated as semi-exotic materials. According to Hallauer (1978), the use of semi-exotic populations has been the most common procedure in evaluating exotic germplasms for transferring beneficial alleles into adapted background.

Introgression: Introgression involves the transfer of one or a few gene or gene complexes from the genebank resource material into the breeding material.

Base broadening: Base broadening, also called genetic enhancement, describes the development of new genetically broad, adapted populations with large variations to broaden the genetic diversity under the theory that a broader genetic base may reduce vulnerability of a crop.

Landraces: A landrace of a seed-propagated crop is a dynamic population(s) of a cultivated plant that has a historical origin and a distinct identity, and lacks formal crop improvement. Often it is genetically diverse, locally adapted, has a local name, and is associated with traditional farming systems.

Primary landrace: This is a crop that has developed its unique characteristics through year after year cultivation and selection by farmers under in situ farming and that has never been subjected to formal plant breeding.

Secondary landrace: This is a crop that has been developed in the formal plant breeding sector but is now maintained continuously in situ by farmers' selection and seed saving, and is likely to be genetically distinct from the original bred material.

Traits: Any morphological, physiological, biochemical, or genetic character of a plant including resistance/susceptibility to biotic stresses that can be used to differentiate two genotypes is called a trait.

Selective sweep: A selective sweep is a process through which the frequency of new beneficial mutation increases and becomes fixed in the population, leading to a reduction in genetic variation. In the selective sweep, the positive selection causes the new mutation to get fixed in the population.

12.6.2 Successful Pre-breeding Program

To start a pre-breeding program, understanding characterization, evaluation, genetic diversity, and inter-species relationships is essential. Introgression of traits from crop wild relatives (CWRs) has enhanced the pest and disease resistance; tolerance to drought, heat, and salinity; increased yields; and improved quality in many crops. Pre-breeding using genes from wild/exotic species has been successful in many crops, including rice, wheat, barley, maize, sorghum, pearl millet, soybean, cotton, tomato, groundnut, chickpea, and pigeon pea. Global pre-breeding programs were initiated for maize at CIMMYT and wheat by ICARDA in 1994/1995. Various species from the tribe Triticeae, such as *Aegilops, Agropyron, Dasypyrum, Hordeum, Leymus, Lophopyrum, Psathyrostachys, Secale, and Thinopyrum,* have been successfully crossed with wheat and significantly boosted the wheat production worldwide.

Specifically, crops like rice, tomato, and wheat have well-established pre-breeding programs focused on CWRs, using advanced genomic tools and diverse characterization and evaluation data. For example, marker-assisted selection has been used to transfer resistance genes for leaf rust and stripe rust from *T. monococcum* to bread wheat. In 1990, the *Xa21* gene from wild rice (*Oryza longistaminata)* was transferred for resistance to bacterial blight in IR72, marking the systematic use of the wild rice gene pool. The rice variety IR 36 incorporated resistance to the grassy stunt virus from the wild rice *Oryza nivara*. An introgression line with heat stress tolerance genes from the wild rice *Oryza rufipogon* was identified, significantly increasing rice yields.

High-protein and large-seed breeding lines were developed from wild species of pigeon pea (*Cajanus sericeus, C. albicans, and C. scarabaeoides*). Stable recombinant inbred lines for rust resistance, powdery mildew, and pod number were developed from wild lentil species (*Lens orientalis, L. odomensis, and L. ervoides*). In chickpea, several drought-tolerant varieties were developed using marker-assisted backcrossing by transferring QTLs for root traits from *Cicer arietinum* into JG-11.

Monosomic addition lines from backcrosses of somatic hybrids between *B. oleracea var. botrytis* and *B. nigra* are used for resistance breeding in *B. oleracea*. An introgression line (ICGV-86699) developed from interspecific hybridization and chromosome doubling of diploid species (*A. batizocoi and A. duranensis*) with *A. hypogaea* showed multiple disease resistances. Also, two interspecific derivatives (NRCGCS-296 and NRCGCS-241) were characterized as tolerant to salinity.

Some successful examples of CWR-derived varieties and lines in important crops are presented in Table 12.9.

12.6.2.1 Development of Pre-bred Introgression Lines in Wheat

In case of wheat, a large number of introgression lines have been generated globally, as well as in India, using diploid progenitor parents/tetraploid wild species for different traits such as rust resistance, Karnal bunt resistance, etc. Figure 12.2 describes the steps for developing introgression lines using *Ae. speltoides*.

Table 12.9 Breeding lines/varieties developed using CWRS

Crops	Species	Utilization traits	Population/Varieties
Wheat	*Aegilops uniaristata*	Aluminum tolerance	Wheat substitution lines production
	T. monococcum	Adult plant resistance to leaf rust and stripe rust using marker-assisted selection	WL711
Rice	*Oryza rufipogon*	Blast resistance	Blast resistant gene *Pi33* introgressed in rice var. IR64
	O. longistaminata	Bacterial blight resistance	*Xa21* gene transfer from wild to variety IR72
	O. glaberrima	Drought tolerance, high yield, and Weed competitiveness	NERICA (NEw RIce for AfriCA)
		Grassy stunt virus resistance	IR 36
		Bacterial blight resistance	PB1121 from donor parent PR114-*Xa38*
Barley	*Hordeum spontaneum*	Salt and Dehydration stress, Aluminium and *Fusarium* resistance	*H. spontaneum* × two-rowed malting barley population
	H. marinum	Salt tolerance	Salt tolerant amphiploid production
	H. bulbosum	Powdery mildew, leaf rust and leaf scald Resistance	Advanced back cross population of wild × cultivated
Maize	*Tripsacum dactyloides*	Corn root-worm resistance	Maize ×*Tripsacum* introgression line
Soybean	*Glycine soja*	Salt tolerance	RILs (Cultivated × Wild)
		Sclerotinia stem rot resistance	Germplasm lines
Peanut	*Arachis duranensis* × *A. stenosperma*	Late leaf spot resistance	F_2 population (*A. duranensis* × *A. stenosperma*)
	A. hypogaea × *A. cardenasii*	Cylindrocladium black rot, Tomato spotted wilt, Early leaf spot, and Sclerotinia blight resistance	Lines GP-NC WS 16 and GP-NC WS 17 (SPT 06-07)
	A. Duranensis, A. cardenasii	High germination tolerant index and promptness index for salt stress	Two interspecific derivatives (NRCGCS-296 (J11 × *A. duranensis*) and (NRCGCS-241 (GG 2 × *A. cardenasii*)

(continued)

Table 12.9 (continued)

Crops	Species	Utilization traits	Population/Varieties
Pigeon pea	*Cajansu sericeus* × *C. cajan*	Male sterile plants	BC_3F_1 population
	C. scarabaeoides (ICPW 89) × *C. cajan* (ICPL88039)	Cytolasmic male sterile line	CMS 88039A
	C. cajan × *C. scarabaeoides*	Cytoplasmic male sterile lines	CORG990052A and CORG 990047A
	C. sericeus, C. albicans, C. scarabaeoides	High protein content and seed size	Selected breeding lines (HPL 2, HPL 7, HPL 40, HPL 51)
Chickpea	ICC-4958 accession of *Cicer arietinum*	Root traits	Varieties Pusa-362, Pusa-372 and DCP 92-3 from GJ-11
Tobacco	*Medicago ruthenica*	Genes for salt stress tolerance	Transgenic lines of tobacco
Tomato	*Lycopersicum pennellii*	QTLs for drought stress tolerance	Introgression lines (Wild × Cultivated tomato (cv M82)
Lentil	*Lens ervoides* × *L. culinaris*	Anthracnose resistance	RILs (Wild × Cultivated)
	L. orientalis, L. odomensis and *L. ervoides*	Rust and powdery mildew resistance	RILs population
Sunflower	Wild *H. annuus*	Sulfonylurea herbicide Resistance	SURES-1, SURES-1, RHA-426, and RHA-427
	H. paradoxus-1671 (NM) and *H. paradoxus*-1673 (TX)	Salt stress tolerance	HA 429 and HA 430
	H. argophyllus-T, *H. annuus*-NM, and *H. annuus*-TX TX-16X	Downy mildew resistance	RHA 340, *H. argophyllus*-FL HA, 419/420, HA 428, and HA 458

12.6.2.2 Cultivar Development Using Marker-Assisted Pre-breeding

In the case of chickpea, a landrace "WR 315" possessing a wilt resistance gene was used as a donor parent for introgression of "QTL region" for wilt resistance on LG 2 having QTLs 1, 3, 4, and 5 into the recurrent parent "Pusa 391" using marker-assisted backcrossing (MABC). The MABC-derived variety, Pusa chickpea Manav, was developed by ICAR-IARI, New Delhi and released by the Central Varietal Release Committee (CVRC), for Central India in 2020. Its average 100-seed weight is 19.5 g and it is highly resistant to Fusarium wilt, and moderately resistant to dry root rot, collar rot and stunt. The development of the MAS-derived chickpea cultivar "Pusa Manav" is described in the flow diagram in Fig. 12.3.

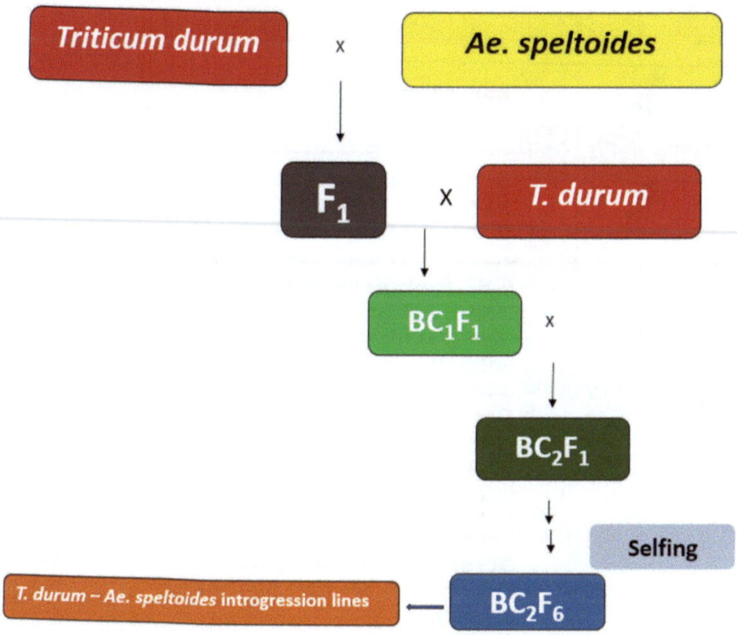

Fig. 12.2 Steps for developing *T. durum—Ae. speltoides* introgression lines

12.6.3 Role of Amphidiploids in Pre-breeding

Amphidiploids are plants created using the methods of wide hybridization and combining the genomes (or specified parts of them) of the parental species involved in the cross. Several antimitotic agents can be used for the chromosome doubling of F_1 hybrids to produce amphidiploids. Colchicine, extracted from the bulbs of the autumn crocus (*Colchicum autumnale*), is widely used. This alkaloid inhibits microtubule polymerization by binding to tubulin, thus inhibiting chromosome segregation during meiosis. Amphidiploids can be used as a bridge to transfer a single chromosome from one species to another species or to create new crops. In addition to their use in bridging the fertility gap, amphidiploids can be valuable tools for basic and applied research in the genomic organization of cultivated species and their relatives.

12.6.3.1 Natural Amphidiploids

The best example of natural amphidiploids is described in the U Triangle of *Brassica* species. The *Brassica* comprises three basic diploid species, viz., *B. rapa* (AA, $2n = 20$), *B. nigra* (BB, $2n = 16$), and *B. oleracea* (CC, $2n = 18$), and three amphidiploids, viz., *B. napus* (AACC, $2n = 38$), *B. juncea* (AABB, $2n = 36$), and *B. carinata* (BBCC, $2n = 34$). As per the theory of U triangle, three allotetraploid species,

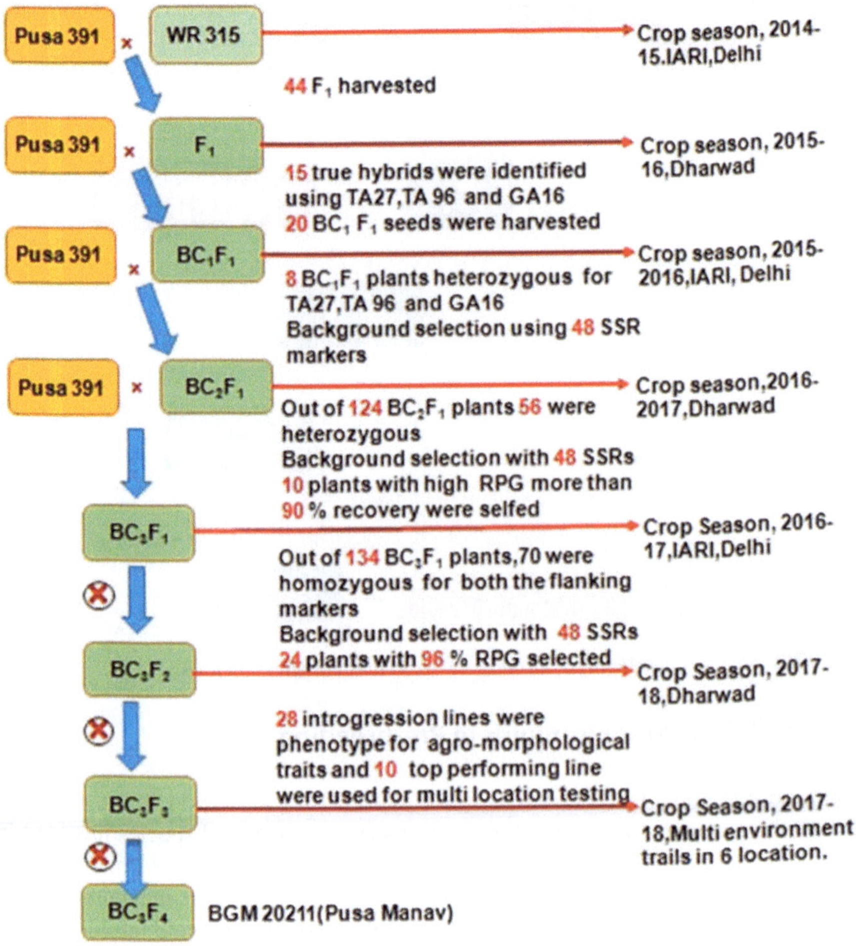

Fig. 12.3 Flow diagram of the development of MAB-derived chickpea cultivar "Pusa Manav"

namely *B. napus, B. juncea*, and *B. carinata*, were derived using natural hybridization of three basic diploid species, *viz*., *B. rapa, B. nigra*, and *B. oleracea*, followed by genome doubling (Fig. 12.4).

12.6.3.2 Artificial Amphidiploids

The generation of artificial amphidiploids can facilitate the transmission of desirable genetic material from wild species to crops. In amphiploid forms, combining both complete parental genomes, *Triticale*, a hybrid of wheat (*Triticum*) and rye (*Secale cereale*, $2n = 2x = 14$, RR), is one of the most successful synthetically developed hybrids. *Triticale* combines the high yield inherited from wheat with the ability to grow in harsher conditions due to the abiotic stress tolerance inherited from rye.

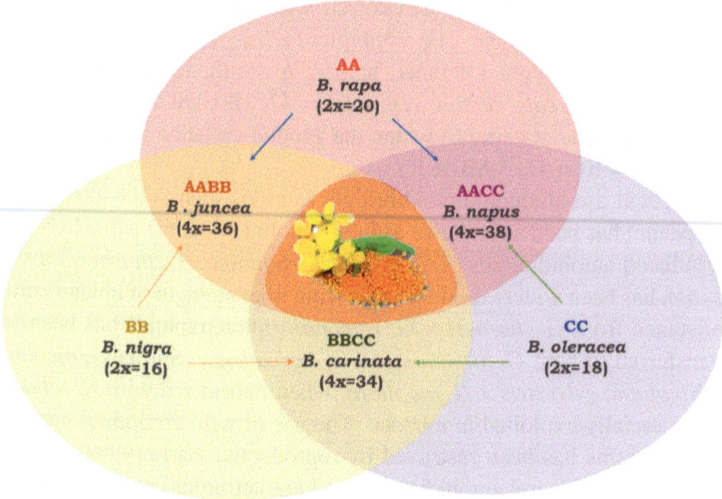

Fig. 12.4 U triangle shows a genetic relationship among six *Brassica* species (Nagaharu U, 1935)

The transfer of leaf rust *(Puccinia triticina)* resistance from *Aegilops umbellulata* to *Triticum aestivum* via an amphidiploid of *T. dicoccoides* × *A. umbellulata* by Sears (1956) is a classic example of an integrated use of amphidiploidy, backcrossing, irradiation, and selection in an interspecific breeding program for the incorporation of a single chromosomal segment. The amphidiploid lines were produced by crossing accessions of wild relatives of wheat (*Amblyopyrum muticum*, *Aegilops speltoides*, *Aegilops umbellulata*, *Aegilops comosa*, *Thinopyrum turcicum*, and *Thinopyrum bessarbicum*) with the bread wheat cultivars, *viz*., Chinese Spring, Paragon, Pavon 76, and Highbury. Several amphidiploid lines, namely *Th. tur* 201 x CS, *Se. ana*142 × HB and *Se. ana*141 × CS, were developed that have the potential for higher biomass and nitrogen use efficiency. The amphidiploid contributes to the production of addition and substitution wheat lines through distant hybridization. The maximum number of created amphidiploids are synthetic hexaploid wheat ($2n = 6x = 42$, AABBDD), which allows for the introduction of various genes from wild species, namely *Aegilops tauschii* ($2n = 2x = 14$, DD). Synthetic lines (*Triticum durum* × *Aegilops tauschii*) and derivatives developed by CIMMYT have been associated with higher leaf photosynthetic rate and grain yield than the recurrent bread wheat parents under optimal conditions. The artificially developed *T. monoccocum/S. cereale* amphiploid was used as a pollen donor for crossing with hexaploid triticale. It was the first successful attempt to transfer the Am genome of *T. monococum* (line Tm16) into a triticale background to increase the leaf rust resistance.

An amphidiploid was obtained by hybridization between *Hordeum chilense* ($2n = 2x = 14$, $H^{ch}H^{ch}$) and *Triticum turgidum* ($2n = 4x = 28$, BBAA), which resulted in the development of ×*Tritordeum martini*. Two cultivars were obtained by backcrossing the amphiploid to wheat in order to change the cytoplasm from barley to wheat. Today tritordeum is commercialized with rheological and baking performances similar to bread wheat. The amphiploid ×*Tritordeum* as compared to wheat also shows better drought tolerance ability. A synthetic amdiphiploid namely *Aegilops kotschyi/Seceale cereale* ($2n = 6x = 42$, UkUkSkSkRR) was used as a donor of wild genetic material to widen the genetic variation in hexaploid triticale "Sekundo" ($2n = 6x = 42$, AABBRR).

The practical use of amphidiploid bridges in crops such as tobacco, potato, cotton, and peanut has been as effective as in wheat. In cotton, the triple hybrid involving the induced amphidiploids of *Gossypium arboreum* × *G. thurberi* crossed with *G. hirsutum* has been widely used for improving fiber strength of upland cotton. The wilt resistance from *G. thurberi* × *G. ramondii* (allotetraploid) has been successfully transferred to both *G. hirsutum* and *G. barbadense.* Similarly, the amphidiploid of *Nicotiana sylvestris* × *N. otophora,* when hybridized with *N. tabacum,* has been commercially exploited in tobacco. The use of wild groundnut germplasm in breeding programs has been restricted by reproductive barriers between wild and cultivated species. Several amphidiploids and autotetraploid groundnuts have been developed using A- and B-genome accessions with high levels of resistance to multiple stresses namely late leaf spot, stem rot, and collar rot diseases. Several amphidiploids, namely An 12 (*Arachisbatizocoi* × *A. kempff-mercadoi*) 4*x*, An 9 (*A. gregoryi* × *A. stenosperma*) 4*x*, and An 8 (*A. magna* × *A. cardenasii*) 4*x*, showed high level of resistance to thrips caused by *Enneothrips flavens*. These amphidiploids introgressed resistance to thrips into cultivated groundnut cultivars. Amphidiploid and autotetraploid groundnut have been developed at ICRISAT, which served as useful genetic resources to transfer useful genes into the cultivated groundnut. The resistance to leaf rust and late leaf spot was successfully introgressed into the cultivated groundnut varieties (ICGV 91114, ICGS 76, ICGV 91278, JL 24, and DH 86) using two synthetic resistance sources, namely, ISATGR 278-18 and ISATGR 5B. Two interspecific derivatives, (NRCGCS-296 (J11 × *A. duranensis*)) and (NRCGCS-241 (GG 2 × *A. cardenasii*)), were characterized as tolerant to salinity. The synthetic amphidiploids from a cross between *A. ipaensis* and *A. duranensis* followed by chromosome doubling produced fertile hybrids when crossed with *A. hypogaea*.

12.7 Wide Hybridization

The improvement in any crop involves crossing among the different varieties or genotypes to create variability followed by selecting the desirable genotype to be released as a variety. Hybridization of the plants/genotypes from two distinct species has been used in plant breeding with the aims of: (1) merging entire genomes of parental species to widen the gene pools of one or the other parental species, (2) merging genomes of parental species to combine their desirable traits (amphiploid form), and (3) introgression of desirable trait(s) from one (frequently wild) species into elite cultivars of another species (introgression forms). Hybridization is the crossing between two genetically dissimilar parents. Therefore, hybridization can be classified into three categories.

12.7.1 Intraspecific Hybridization

When hybridization is carried out between two genetically dissimilar parents of the same species, it is called intraspecific/intervarietal hybridization.

12.7.2 Interspecific Hybridization

When hybridization is carried out between individuals of different species of the same genus, it is called interspecific/intrageneric hybridization. Thomas Fairchild created the first interspecific hybrid between sweet William (*Dianthus barbatus*) and Dianthus carnation (*Dianthus caryophyllus*) species in 1717. When a desirable trait is not found within a crop's species, interspecific hybridization is used. In tobacco, hybridization between two wild diploid species, *Nicotiana sylvestris* ($2n = 24$) and *Nicotiana tomentosa* ($2n = 24$), produces a hybrid. The F_1 hybrid proved to be infertile. When the F_1 plants were given colchicine, they produced a completely viable tetraploid ($2n = 48$) that looked like the *N. tabacum* species. Another hybrid was created between two wild diploid tobacco species, *N. paniculata* ($2n = 24$) and *N. undulata* ($2n = 24$). The F_1 was sterile. Colchicine treatment of F_1 resulted in the generation of viable amphidiploid ($2n = 48$) seeds which were comparable to *N. rustica*.

12.7.3 Intergeneric Hybridization

When hybridization is carried out between individuals from two different genera of the same family, it is called, intergeneric hybridization. Karpechenko created the first intergeneric hybrid between radish (*Raphanus sativus*) and cabbage (*Bassica oleracea*) of the *Cruciferae* family in Russia in 1928 and called it *Raphanobrassica*. The major goal behind the formation of *Raphanobrassica* was to mix the roots of radish with cabbage leaves. However, the new species that resulted had roots similar to cabbage and leaves similar to radish, which was an ineffective combination. The other example of intergeneric hybridization is *Triticale* developed by crossing wheat with rye (Rimpu, 1890).

Both interspecific and intergeneric hybridizations are combinedly known as distant hybridization and such types of crosses are known as distant or wide crosses. These wide crosses allow for the introgression of important alleles of agricultural interest from wild species into cultivated crop species. Wide hybridization is more likely practiced to transfer desirable genes from the secondary and tertiary gene pools to the cultivated species (primary gene pool).

12.8 Limitation to Use CWRs Within Breeding Programs

The ultimate goal of distant hybridization is to transfer useful genes from alien species into cultivated species. Three major factors limit the use of crop wild relatives in plant breeding:

- Poor agronomic performance of CWRs, mostly due to the presence of undesirable traits such as shattering and seed dormancy.
- Crossing barriers between crops and their wild relatives limit the production of fertile offspring. Direct use of CWR is restricted in hybridization programs due to the cross incompatibility, relatedness, and crossability of crop wild relatives. For example, two species of cotton, namely *Gossypium somalense* and G. *longicalyx*, are highly disease-resistant but could not be used directly in hybridization due to their genetic incompatibility and ploidy.
- The transfer of undesirable traits physically linked to the target trait is known as linkage drag. If beneficial wild-derived traits are introgressed into elite material, linkage drags can often negatively impact yield or yield-related traits. For example, in tomatoes, the wild species *Solanum galapagense* is a source of insect-pest resistance, but when it is crossed with cultivated tomatoes, the fruit size and fruit setting reduce significantly. Similarly yield penalty is reported in bread wheat when resistance genes Pch1 (resistance to eyespot) from *Aegilops ventricosa* and Pm16 (resistance to powdery mildew) from *Aegilops speltoides* are introgressed. The rigorous use of CWRs as potential allele donors in conventional breeding programs is hampered due to their different ploidy levels, or certain pre- and post-fertilization barriers. The crossability of cultivars with wild species is a prerequisite for alien gene introgression. However, various sexual barriers result from incompatibility to prevent interspecific or intergeneric hybridization.

12.8.1 Maintenance of CWRs

The crop was domesticated and dispersed from one part of the world to another. During the process of dispersal, new adaptive traits were selected as per the requirement of the environment and the crop was acclimatized in a new environment. For example, wheat was domesticated in the Fertile Crescent and dispersed to New World and South Asian countries. Spring-type wheat was selected in South Asia, while both spring- and winter-type wheat were selected in the New World. However, crop wild relatives were not selected and domesticated by humans. They are still grown in natural environments. The wild relatives are poorly adapted if they are grown outside the natural environment. They need specific requirements for their proper growth and maintenance as detailed in following sections.

12.8.1.1 Seed Germination

Seed germination, a biological process influenced by both internal seed characteristics and external conditions like water, temperature, and light, is typically measured by the percentage of seeds that successfully germinate under optimal conditions within a specific timeframe. This metric is commonly referenced for crops, cultivated trees, and their wild relatives. Maintaining these crop wild relatives (CWRs) involves a comprehensive understanding of various factors influencing seed germination to ensure their conservation and viability. Seed dormancy is a significant barrier to germination, but it can be addressed through various mechanical and chemical treatments. By creating optimal growth conditions that mimic each species' natural habitat, dormancy can be effectively overcome. Here, seed dormancy is the inability of an intact, viable seed to germinate under optimal conditions, serving as an adaptation to ensure germination occurs only during favorable environmental conditions. This seed's inability to sprout can be due to the embryo, endosperm, or seed coat. Dormancy is typically stronger in wild species than in cultivated crop plants.

Other factors that influence seed germination can be temperature, moisture, light, soil or other growing medium, timing, and sowing depth. Optimal temperature ranges vary for different species, and understanding the native habitat conditions can guide temperature settings, with some seeds requiring fluctuating temperatures to mimic natural conditions. Adequate moisture is critical for germination, but overwatering can also lead to fungal growth and seed rot, so maintaining a controlled, moist environment is crucial. Light requirements differ among species, with some seeds needing light to germinate, while others require darkness, necessitating knowledge of the CWR's natural light conditions. The choice of soil or growing medium is crucial for success, with a sterile, well-draining medium preferred to reduce disease risk, and the pH and nutrient content tailored to the specific CWRs. Additionally, the correct timing for sowing seeds can mimic natural cycles and improve germination rates, and sowing depth should be appropriate for the seed size, generally at a depth of 2–3 times their diameter.

12.8.1.2 Special Requirements for Growing CWRs for Flowering and Seed Development

Due to the poor response of wild relatives to new environments, it is important to study the phenological stages of the CWRs. Wild relatives have different phenology than cultivated plants. Traits such as flowering behavior, duration of flowering, anthesis, pollen dehiscence, and grain filling period are considered for effective management and use of CWRs.

Crop wild relatives collected from temperate regions, when grown in tropical regions, do not produce flowers because of higher vernalization requirements. Similarly, the wild relatives of *Vigna* species collected from South India may not be adapted to the conditions of North India.

The wild species of *Cicer* show different vernalization and photoperiod responses. A long photoperiod is required for the transformation from the vegetative to the reproductive phase in wild *Cicer*. Different wild *Cicer* species have different photoperiod requirements. *Cicer reticulatum* require 18 h for efficient regeneration and reducing the vegetative phase, while *C. judaicum* and *C. yamashitae* require photoperiod of 15 h, *C. bijugum* and *C. pinnatifidium* require photoperiod of 15–18 h to reduce the vegetative phase, while *C. chorasanicum* and *C. cuneatum* show weak responses to extended photoperiod treatments. Therefore, flowering in crop wild species is induced by manipulating day length, temperature, chemical spraying, etc. and synchronization of flowering is required for a successful crossing program.

In the case of wheat, chilling/vernalization treatment for 6 weeks at 6–8°C is given to wild species requiring vernalization treatment. Also, an extended photoperiod during the pre-flowering phase is required for 6–8 weeks.

12.9 Reproductive Barriers in Wide Hybridizations

In the production of wide hybrid, parental species must belong to a primary, secondary, or tertiary gene pool. In cultivated tomato, hybrids are easily produced with the closely related species, namely *Lycopericum pimpinellifolium.* In rice, hybrids are easily produced by crossing cultivated species with *Oryza rufipogon.* In food legume crops, several crossability barriers have been reported, the most common being cross incompatibility, embryo abortion at the early growth stage, inviability of F_1 hybrids, and sterility of F_1 hybrids and subsequent progenies. Depending on the time and place of occurrence, barriers in interspecific hybridization can be broadly classified as pre- and post-zygotic barriers and their detailed mechanisms are discussed in the following sections.

12.9.1 Pre-zygotic Barriers

In pre-zygotic barriers, pollination with other species is restricted by their natural habitat (geographic isolation), flower structure (i.e., shape or color), flowering time, or pollen mediator. Pre-fertilization cross incompatibility between parent species arises when pollen grains do not germinate, the pollen tube does not reach the ovary, or the male gametes do not fuse with the female gametes. The lack of crossability in the interspecific crosses due to incompatibility in the fertilization process is called interspecific incompatibility. In self-pollinated and compatible crosses, fertilization takes place between 24 to 72 h after pollination (depending on the environmental conditions, especially the temperature). In normal and interspecific crosses, pollen pistil interaction must pass through certain stages for successful hybridization, which include pollen germination on stigma due to hydration, pollen tube growth and penetration, and finally entry into female gametophyte for fusion to form zygote. At the pre-zygotic level, compared with self-pollination, pollen tubes grew much slower in incompatible crosses and rarely reached the ovules.

In the interspecific crosses, inhibition of pollen tube growth was reported in *Lycopersicon*, *Linum*, *Brassica*, and *Trifolium*. Similarly, failure of pollen germination and penetration of pollen tubes into the stylar tissues were reported in interspecific crosses between *Cicer arietinum* and *C. soongaricum.* In wheat (*Triticum aestivum*), two loci carrying dominant genes, *Kr1* and *Kr2*, located on the long arm of chromosomes 5B and 5A, respectively, affect the crossability in inter-specific crossing between wheat and rye. Dominant alleles of *Kr1* and *Kr2* showed poor crossability caused by inhibition of pollen germination and pollen tube growth before fertilization. In normal hybridization, the pollen tube delivers two sperm nuclei to the embryo sac; one fertilizes the egg cell and the other fertilizes the two polar nuclei to form a triploid endosperm nucleus. In wide crosses, these events are disturbed, as observed in Wheat × Maize crosses, where only embryos were formed without endosperm.

12.9.2 Post-zygotic Barriers

The post-zygotic barriers act at different tissues and stages ranging from early embryo development to the subsequent growth and reproduction of the offspring. In some of the interspecific crosses, hybridization takes place to form zygote, but the development of the zygote is affected at different stages. Besides hybrid embryo lethality, which results in seed abortion, distant hybridizations often come with seedling lethality, morphological abnormalities like hybrid weakness, and death at later developmental stages. Several post-zygotic barriers are briefly discussed below:

12.9.2.1 Chromosome Elimination

After successful pollination and zygote formation, regular mitosis is essential for the proper development of the embryo. Chromosome elimination was observed in interspecific hybridization within a genus like *Nicotiana*, *Hordeum*; *Solanum* and *Elymus*. In barley, the chromosome elimination of one parent (*Hordeum bulbosum*) takes place during the early mitotic divisions in interspecific crosses between *Hordeum vulgare* and *Hordeum bulbosum*, resulting in the production of haploid barley. The "*bulbosum*" method was the first haploid induction method to produce large numbers of barley haploids used in crop breeding programs. Similarly in the intergeneric crosses between *T. aestivum* and *H. bulbosum*, *bulbosum* chromosomes were eliminated producing haploid wheat. Further, chromosome elimination was observed in another interspecific cross between *H. vulgare* and *H. marinum* crosses. In these interspecific crosses, dual chromosome elimination occurs, in which *H. marinum* chromosomes are eliminated in the embryo stage, while *H. vulgare* chromosomes are eliminated in the endosperm stage.

12.9.2.2 Lethal Genes and Hybrid Lethality

Hybrid lethality is a type of post-zygotic reproductive barrier in which hybrids do not mature after fertilization between different species. In this case, the hybrid embryos either die before germination or the hybrid seedlings exhibit lethal symptoms, such as browning, withering, and yellowing after germination. Hybrid lethality has been reported in a wide range of plant species, including wheat, rice, and cotton. The intergeneric hybrids between einkorn wheat and *Aegilops umbellulata* either died in the seedling stage or reached maturity depending on whether the lethal gene l or its non-lethal allele L was present. In einkorn wheat, three alleles for early lethality (Le), late lethality (LL), and viability (I) are found. In crosses between

T. durum and rye, hybrid lethality is characterized by abnormal development of embryo and endosperm, and developed seeds do not germinate.

12.9.2.3 Hybrid Breakdown

When the F_1 hybrid of interspecific crosses is vigorous and fertile but F_2 progeny is weak and sterile, it is known as a hybrid breakdown or hybrid sterility, and is commonly encountered during wide hybridization studies. Hybrid sterility is caused by epistatic interactions between nuclear genes or between the nuclear genome and the maternally derived mitochondrial genome, termed cytoplasmic male sterility (CMS). Inter-specific hybrids between distantly related species are generally sterile or partially sterile. Reasons for hybrid sterility are reduced chromosome pairing during meiosis or disharmonious interaction between combinations of the genes of parental species. In crosses between two *indica* and *japonica,* subspecies of *Oryza sativa,* varying degrees of hybrid sterility and hybrid breakdown were observed. Certain rice varieties produce fertile F_1 hybrids when crossed to either *ssp. indica* or *ssp. japonica* and thus are called widely compatible varieties, whereas sometimes, hybrid breakdown is found in advanced generation progenies from completely fertile F_1 hybrids. Many F_1 sterility genes have been identified, and some of them have already been localized on the rice genetic map. In wheat, *Triticum boeoticum, T. urartu, Ae. squarrosa*, and *Ae. speltoides* exhibited reciprocal differences in hybrid seed morphology, endosperm development, and embryo viability. Genetic experiments involving hybrids of *T. boeoticum, T. urartu*, and *T. monococcum* showed that a factor is present in pollen or male gametes, which shows dosage effect and by interacting with the maternal genome, leads to endosperm abortion. In chickpea, *Cicer echinospermum* is crossable with cultivated chickpeas, but the resulting hybrids are mostly sterile. In tomato, low seed setting in the hybrid fruits and the occurrence of brown wrinkled seeds inside the fruit in an interspecific cross between *Solanum lycopersicum*× *S. peruvianum* suggested post-zygotic barrier.

12.9.2.4 Hybrid Necrosis

Hybrid necrosis occurs in some important genera such as *Triticum, Aegilops, Oryza, Lycopersicon*, and *Gossypium*. Hybrid necrosis is one of the negative results of both intra- and inter-specific hybrids. Necrosis can be temperature dependent and causes low growth, wilting, discoloration, and lethality. Hybrid necrosis, usually lethal or semi-lethal, has been frequently observed in F_1 hybrids between genotypes of common wheat (*T. aestivum* L.) and tetraploid wheat (*T. turgidum* L.). Hybrid necrosis is controlled by the complementary dominant genes *Ne1* and *Ne2*, which are located on chromosome arms 5BL and 2BS, respectively, in wheat. Both *Ne1* and *Ne2* genes are widely distributed among different subspecies, varieties, and commercial cultivars of common wheat. Temperature-dependent hybrid weakness is observed in interspecific hybrids in rice. One high-temperature dependent hybrid weakness is caused by the interaction between *Hwi1* in wild rice, *Oryza rufipogon*, and *Hwi2* in indica rice, *Oryza sativa. Hwi1* and *Hwi2* encode two LRR-RLK genes and putative subtilisin-like protease, respectively, and it has been considered that the activation of the autoimmune response by the interaction between these two genes causes hybrid weakness. In chickpea, cross *C. arietinum* × *C. pinnatifidum* resulted in a poor seed set due to high proportion of hybrid embryo necrosis, albinism, and poor growth to form roots.

12.10 Methods for Overcoming Crossability Barriers in Wide Hybridization

There are different types of crossing barriers which occur frequently in distant hybridization. The knowledge about the processes involved in pollen germination, pollen tube growth, and fertilization enhances the opportunities to manipulate these processes toward the development of viable and fertile interspecific hybrids. The methods to overcome the reproductive barrier depend on the nature of their types and the stage at which these occur.

12.10.1 Methods for Overcoming Pre-zygotic Barriers

The methods to overcome pre-zygotic barriers are as discussed below:

12.10.1.1 Manipulation of Chromosome Number

The crossing between species with different ploidy levels is difficult compared to crossing between species with the same ploidy level. In some crops such as wheat, potato, cotton, tobacco, oats, and *brassicas*, several wild relatives of these crops are diploids or have lower chromosome numbers than the cultivated crop species. Several other crops such as rice, maize, and barley have wild species with higher ploidy levels than the cultivated species. In wide crosses, the chromosome doubling of either of the parent is essential for successful hybridization. Based on the chromosome behaviors of wide hybrids and the resulting chromosome constitutions in their progenies, chromosome manipulation of wide hybrids in crop improvement is classified into three main categories:

Manipulation for Single Chromosome In this technique, the incorporation of a single chromosome or chromosome fragment from a wild species into a cultivated crop is being used to enhance crop genetic diversity. The resulting alien chromosome substitution, addition, or translocation lines help breeders to transfer beneficial traits from wild and/or weedy plants to cultivated crop species. The transfer of beneficial traits from wild plants into crops so far has been bridged by alien chromosome translocation lines.

Genome Manipulation by the Loss of Alien Genome The incorporation of all the alien chromosomes by chromosome doubling produces amphidiploids. Sometimes, the incorporation of partial alien chromosomes will lead to partial amphidiploids. Amphidiploids can be used for the development of a new crop. The man-made crop Triticale is an amphidiploid between wheat and rye. Amphidiploid can be further used as a bridge for the development of alien gene introgression or alien chromosome addition, substitution, and translocation lines.

Alien Addition and Substitution Lines An alien addition line carries one chromosome pair from an alien species in addition to the normal somatic chromosome number of the parent species. When only one chromosome is present, it is referred to as an alien addition monosome. An alien substitution line had one chromosome pair of a different species in place of a chromosome pair of recipient species. Alien addition and substitution lines have been developed in polyploid species like wheat, oats, and cotton as they have buffering capacity to be viable.

Genome Manipulation by the Elimination of Alien Genome The elimination of all alien chromosomes induces crop haploid production. Haploid is very useful in double-haploid breeding of a true-breeding crop like wheat and rice since it can quickly fix genetic recombination and thus enhance breeding efficiency or facilitate genetic analysis.

12.10.1.2 Bridging Species

Transferring genes from wild species to cultivated plants can be challenging or impossible due to compatibility issues, often resulting in infertile hybrids. When two plants cannot be directly crossed, a third plant that is compatible with both can be used as a bridge to facilitate the transfer of genes. Here are some examples:

Wheat and lymegrass: Lymegrass (*Leymus arenarius*) can cross with wheat species (*Triticum aestivum, Triticum carthlicum*) and is used as a bridge to transfer stress tolerance and disease resistance traits.

Tritordeums: These are hybrids created by crossing wild barley (*Hordeum chilense*) with various wheat species. Tritordeums have resistance to several diseases and pests, and can tolerate drought and salt stress. They are used to transfer genes from wild barley to durum wheat.

Tobacco: *Nicotiana sylvestris* is used to transfer nematode resistance from *N. repananda to N. tabacum.*

Lentils: Genes for resistance to ascochyta blight and anthracnose from *Lens lamottei* and *L. nigricans* are transferred to cultivated lentils (*L. culinaris*) using *L. ervoides* as a bridge, with the help of embryo rescue techniques.

Azuki beans: *Vigna nakashimae* is used as a bridge to transfer bruchid resistance genes from *V. umbellata* to *Vigna angularis* (azuki bean).

Tomatoes: *Lycopersicon chilense* is used to overcome the incompatibility between cultivated tomatoes and *L. peruvianum. L. peruvianum* is crossed with *L. chilense*, and their progeny is then crossed with *L. esculentum.*

Peanuts: Rust resistance from *Arachis batizocoi* (BB) is transferred to *A. hypogaea* (AABB) using an autotetraploid of *A. batizocoi* (BBBB).

Potatoes: Some Mexican Solanum species with desirable traits do not cross directly with *Solanum tuberosum*. Here, *Solanum acaule* is used as a bridge to transfer genes from these incompatible Mexican species into *S. tuberosum.*

Thus, using bridge species helps overcome genetic barriers and introduces beneficial traits from wild species into cultivated crops.

12.10.1.3 Use of Growth Hormones

Different types of growth hormones such as gibberellic acid (GA), naphthalene acetic acid (NAA), and kinetin or 2, 4-D (dimethylamine) help in the successful crossing of cultivated species with wild species. Application of gibberellic acid (75 ppm) 1 or 2 days before or after pollination promotes pollen tube growth and ovary development. Growth of pollen tube, development of embryo, and stigma receptivity to prevent the early abscission of flowers have been promoted by these hormones. These growth hormones can be used for the growth of hybrid embryos of interspecific crosses, where hybrid embryos die due to their small size. The application of growth regulators after pollination is successfully obtained in interspecific crosses of *Phaseolus*, *Cajanus*, and *Cicer*, wheat × rye crosses. In chickpea, growth regulators were applied to pollinated pistils, which prevented initial pod abscission and helped save the aborting hybrid embryos. The application of 2, 4-D at low concentrations (approximately 1.0 mg/l) before or after pollination promoted the retention of *Vigna vexillata* flowers pollinated with cowpea and subsequently, pods formed from the cross-pollination. Similarly, the application of 2, 4-D followed by GA3 before pollination has allowed for intergeneric crosses between *Hordeum vulgare* and species of *Avena*, *Triticum*, and *Lolium*. Application of growth hormones is needed to ensure proper pod and seed set in interspecific crosses involving crossable species.

12.10.1.4 Shortening the Style

In some cases, the differences in the floral morphology affect the hybridization process. In such situations, the style of one species is too long or short for the pollen tubes of the other species to reach the ovary. This type of floral morphology is found in maize and restricts the hybridization between *Zea mays* and *Tripsacum dactyloides*. For effective hybridization, the part of the stigma of maize is cut down when it is crossed as female with *Tripsacum*.

12.10.1.5 Use of Mentor Pollen

When pollen grains of one species do not germinate on the stigmas of another species, pollen grains of incompatible pollen grains are mixed with the killed maternal pollen grains to increase or to obtain the germination of the incompatible pollen grains. When the compatible pollen is killed with ethanol and mixed with incompatible pollen for use in pollination, the proteinaceous recognition factors released from the walls of the killed compatible pollen grains mask the rejection reaction of the recipient stigma, thus allowing the alien pollen grains to germinate. The killed maternal pollen is called recognition or mentor pollen. Recognition pollen has been used for overcoming self-incompatibility in sesame (*Seasmum indicum*×*S. mulayanum*) cross in the Pedaliaceae family.

12.10.1.6 Advanced Backcross-QTL Strategy

The use of wild species in crossing programs for the transfer of genes is limited due to the high tendency of linkage between desirable and undesirable genes. The Advanced Backcross-Quantitative Trait Loci novel breeding strategy (AB-QTL) is used to minimize the negative effect of linkage drag associated with alien gene introgression. The productivity-enhancing genes/QTLs have been introgressed in oats from *Avena sterilis*, in tomato from *Lycopersicon pimpinellifolium* and *L. parviflorum*, in rice from *Oryza rufipogon*, and in chickpea from *Cicer reticulatum*. In the AB-QTL approach, the molecular markers associated with improved background are used to genotype the advanced backcross progenies (BC_2/BC_3), and the progenies having minimum linkage drag are selected based on genome recovery of improved genotypes. In rice, the AB-QTL approach was used successfully for the development of salt-tolerant genotypes by the introgression of salt tolerance genes *Saltol* from genotype 'FL478' into high-yielding varieties 'BT7'.

12.10.2 Techniques for Overcoming Post-zygotic Barriers

The methods used for overcoming the post-zygotic barriers are discussed below:

Reciprocal Crosses The reciprocal cross may be attempted when a unidirectional cross combination fails. Reciprocal differences in wide crosses can be due to chromosomal imbalance in the endosperm, the role of the sperm nucleus in differential endosperm development, or the alteration of endosperm development by pollen through the effects of antipodal cells. Sometimes, the disharmony between the genome and the cytoplasm of different species creates fertilization barriers. Therefore, the reciprocal crossing technique has been found effective in some crops such as tobacco, wheat, chilli, brinjal, and potato. In tobacco, when the crossing is attempted between *Nicotiana debneyi* and *N. tabacum*, F_1 was male sterile whereas the reciprocal cross was fertile. Similarly, the *T. timopheevi* cytoplasm with a nucleus of *T. aestivum* produces a male sterile hybrid, whereas the reciprocal cross between these species produces a fertile hybrid. The crossing between *Vigna mungo* × *Vigna radiata* was also unsuccessful, while reciprocal crossing produced successful viable hybrids. As a general rule, the use of species with higher chromosome numbers as a female parent is more successful than the reciprocal method.

Chromosome Doubling In wide crosses, the transfer of alien genes is restricted due to the sterility of F_1 hybrids caused by the lack of chromosome pairing in F_1 generation, which may be solved by chromosome doubling of either of the parental species before crossing or of the F_1 itself by colchicine treatment. In several crops such as wheat, *Vigna*, and cotton, the allopolyploid species have been developed using the chromosome doubling technique from most of the semi-fertile and completely sterile F_1 hybrids to transfer alien genes using allopolyploids as a bridge species. *Triticale* is the classical example produced through chromosome doubling of wheat × rye hybrids. *Tritordeum* is another example of amphiploid produced by chromosome doubling of F_1s resulting from cross *Hordeum chilense*×*T. turgidum*.

Backcrossing In wide crosses, plants in initial generations are generally inferior with a poor expression of desired traits, thus for the recovery of desired types, the cross populations may be advanced up to F_8/F_9 generations. Although both wild and cultivated species are compatible for crossing, the expression of desirable traits is poor due to delayed segregation. Here the backcrossing approach, also called introgression breeding, is used, which involves interspecific hybridization followed by one or more backcrossing with the recipient parent. This approach is an effective way to transfer agronomically beneficial alleles from wild relatives/species to cultivated varieties of crops. The successful example is 1B/1R translocation in wheat, where the short arm of rye chromosome 1 (1RS) was transferred into bread wheat to develop a 1RS.1BL chromosome. The 1RS.1BL contains several genes for resistance to pathogens and pests of wheat. The grassy stunt virus resistance gene from *O. minuta* was transferred to *O. sativa* through backcrossing. In addition to this, genes for white-backed plant hopper were transferred from *O. officinalis* to *O. sativa* through repeated backcrossing to their sterile F_1s. In pigeon pea, the genome of the wild species, *Cajanu platycarpus* was transferred into cultivated species by backcrossing of F_1 hybrids rescued through embryo culture, followed by in vitro culture of aborting embryos of BC_1 progeny.

Limited Recombination The effective use of wild species depends on the transfer of specific chromosome segments that contain genes of interest. The reduced homology or uneven chromosome pairing and reduced recombination is generally found in the F_1 generation of distant crosses. The problem of lack of recombination can be solved by treating the F_1 ionizing radiations, inducing chromosomal translocations, and by manipulating the chromosome pairing control system. The ionizing radiations was used to induce chromosomal breaks in transferring a gene-controlling resistance to leaf rust from *Ae. umbellulata* to cultivated wheat. In wheat, three genomes in one nucleus without intergenomic recombination is a classic example of a pairing control mechanism, governed by the *Ph1* gene located on the long arm of chromosome 5B. Looking into the importance of the Ph1 locus, a homoeologous pairing promoter gene, Hpp-5 Mg (derived from *Ae. geniculata* with *T. aestivum*), promotes homoeologous recombination, and multiple crossovers between wheat and wild relative chromosomes lead to an enrichment of the genetic base in prebreeding materials. Three methods in wheat had been used to manipulate recombination and achieve homoeologous pairing: (1) crossing alien addition lines with wild species; (2) crossing nullisomic 5B stocks of wheat with alien addition lines; and (3) crossing a ph mutant of wheat with alien addition lines. The stem rust resistance gene has been transferred from *Agropyron elongatum* to common wheat and stripe rust resistance from *Ae. comosa* ssp. *comosa*. The ph1b deletion line has been used to introgress powdery mildew resistance genes from *A. triuncialis* (5 U) to bread wheat (5A) by inducing homoeologous meiotic pairing. In addition to Ph1, a Ph2 locus encodes repair protein MSH7-3D, which plays a crucial role in the genomic and meiotic stabilization of allopolyploidy. Radiation-induced translocations for transferring alien chromosomal segments have also been used in oats for transferring powdery mildew resistance. A wild species, namely *Avena longiglumis*,

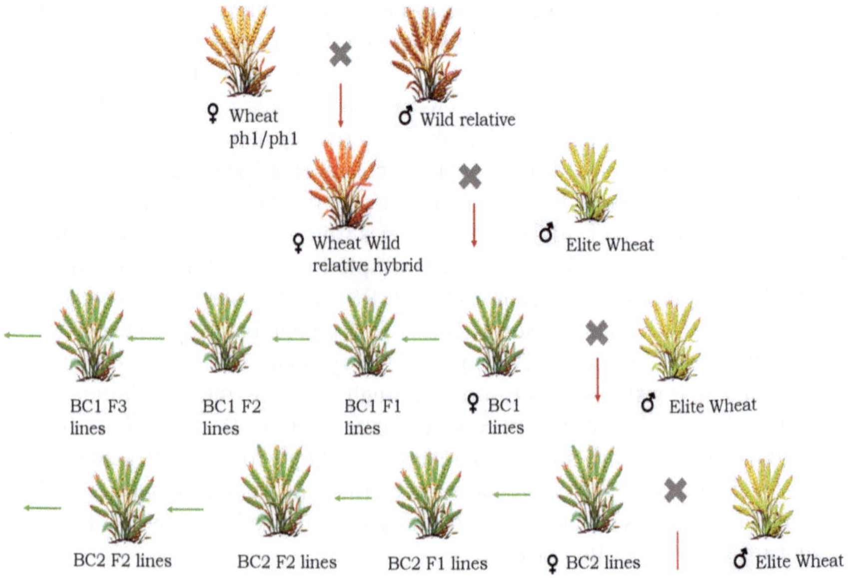

Fig. 12.5 Overview of the approach using ph 1 mutant to introgress relevant DNA fragments from wild species in the wheat genome of elite lines. A primary cross is made between the ph 1 mutant and the wild species to generate a hybrid. This plant is grown until flowering, and homoeologous recombination can occur during meiosis leading to introgressions. This plant is then crossed with elite lines to introduce the alien fragment in their genome. The elite background is recovered either through successive backcrosses (BC; orange arrow) or self-fertilization (BC F; green arrow)

was used to transfer powdery mildew resistance from *A. barbata* to cultivated oats through induced pairing. An overview of the approach using a ph 1 mutant to introgress relevant DNA fragments from wild species in the wheat genome of elite lines is depicted in Fig. 12.5.

12.11 Strategies to Improve the Efficiency of Pre-breeding

12.11.1 Fluorescence Microscopy

Among the strategies to improve the pre-breeding practice, a pollen study of CWRs plays a pivotal role and serves as a foundational aspect in the development of improved plant varieties. By assessing pollen parameters, breeders can strategically select parental lines with desirable traits and highly viable pollen for crossing, thus facilitating successful hybridization. Understanding pollen characteristics also aids in the identification of genetic diversity within populations and the introgression of beneficial traits from wild or unadapted species. Moreover, pollen studies inform breeders about pollination mechanisms, optimal timing for crosses, and compatibility/incompatibility issues, enabling an efficient selection of compatible parents and the implementation of techniques to overcome barriers to successful fertilization. Among the various parameters used to measure pollen performance, pollen viability stands out as one of the quickest and simplest to evaluate. This is because it does not necessitate waiting for seeds to develop or pollen tubes to grow. Extensive research has been conducted on pollen viability, revealing its significant correlation with pre-zygotic success, thus underscoring its importance in estimating reproductive success in plants.

Fluorescence microscopy plays a major role in this regard by advancing our understanding of pollen biology by offering a detailed view of pollen structure and function. This technique allows researchers to visualize intricate features of pollen grains, such as their size, shape, and surface characteristics, providing essential insights into pollen morphology. Moreover, fluorescence microscopy enables the detection of vital stains (acetocarmine, FDA) that distinguish between viable and non-viable pollen grains, facilitating accurate assessments of pollen viability. By tracking pollen germination and pollen tube growth in real time, fluorescence microscopy provides dynamic observations of pollen physiology, revealing crucial information about pollen fertility. Furthermore, this technique supports high-throughput analysis, making it valuable for large-scale studies aimed at elucidating the complex mechanisms underlying pollen development, pollination, pollen pistil interaction, and reproductive success. Overall, it serves as an indispensable tool in pollen biology research, driving advancements in plant breeding.

12.11.2 Pollen Physiology and Storage

One of the significant challenges in using CWRs is the asynchronous flowering patterns between wild and cultivated species. This disparity in flowering times can hinder natural pollen exchange and gene flow between wild and cultivated populations, limiting the introgression of desirable traits into breeding programs. To overcome this challenge, conservation efforts focus on preserving the genetic integrity of CWRs through ex-situ conservation methods like seed banks and pollen storage facilities. Pollen conservation of crop wild relatives (CWRs) is vital for safeguarding genetic diversity and ensuring the resilience of cultivated crops. By maintaining viable pollen samples of CWRs, researchers can facilitate controlled pollination experiments and hybridization programs aimed at introgressing beneficial traits from wild relatives into cultivated crops. This can be made possible by a detailed study of pollen physiology and storage parameters.

The study of pollen physiology encompasses a broad spectrum of research aimed at understanding the intricate processes involved in the formation, composition, function, and storage of pollen grains. Researchers investigate various aspects, including the development of microspores within anthers, the biochemical composition of pollen grains, and the factors influencing their viability and germination. Through meticulous experimentation and analysis, scientists seek to unravel the complex interactions between genetic factors, environmental conditions, and cellular mechanisms that govern pollen physiology. This multidisciplinary approach not only sheds light on the fundamental aspects of plant reproduction but also has practical implications, such as improving crop yield through enhanced pollen viability and storage techniques. Additionally, pollen physiology studies contribute to our understanding of pollen-mediated reproductive barriers, which is essential for ensuring successful fertilization and seed production. Pollen grains are metabolically active structures, and their viability depends on factors such as temperature, humidity, and nutrient availability. Proper storage techniques are crucial for conserving their pollen viability over extended periods, facilitating breeding programs, genetic conservation, and research endeavors. Various methods, including cryopreservation, desiccation, and low-temperature storage, are employed to maintain pollen viability. Cryopreservation, in particular, involves freezing pollen at ultra-low temperatures (−196 °C) to effectively halt metabolic activity and conserve true-to-type genetic material. This method has proven valuable for the long-term conservation of pollen from endangered, economically important plant species and crop wild relatives. Additionally, desiccation techniques involve removing moisture from pollen grains to prevent cellular damage during storage. Optimal storage conditions, such as low temperatures and controlled humidity levels, help prolong pollen viability and ensure its availability for future use in breeding programs or scientific studies. Overall, a comprehensive understanding of pollen physiology and effective storage strategies are essential for maintaining genetic diversity, conserving plant species, and supporting agricultural innovation in the face of environmental challenges.

12.11.3 Advances in In Vitro Techniques

In vitro techniques can be a potential solution for manipulating the desired trait, enriching the genetic base and recovering desirable variation. In vitro techniques using somatic hybridization or embryo rescue are the best options. Several embryo rescue techniques such as embryo culture, ovary culture, ovule culture, anther culture, and protoplast culture protect embryos from successful hybridization and premature abortion. These techniques help overcome the crossing barriers and have greatly increased the opportunity to transfer the alien gene from cross-incompatible species/genera of crop plants.

Embryo Rescue Embryo rescue is one of the successful forms of in vitro culture used to assist in the development of plant embryos that are either inherently weak or might not survive to become viable plants. This technique has been widely used to produce plants from hybridizations in which endosperm failure causes embryo abortion. In embryo rescue procedures, the artificial nutrient medium replaces the endosperm, allowing the embryo to continue development. In this technique, the immature/weak/degenerating embryos could be rescued by providing nutrition under artificial conditions to survive and undergo normal developmental phases to produce a complete plant. Embryo rescue may be referred to as embryo, ovule, or ovary culture, depending on the organ cultured. The embryo culture procedure involves excising plant embryos and placing them on culture media. The first hybrid zygotic embryo was rescued from interspecific hybridization between *Linum perenne*×*Linum austriacum* by Laibach (1925) by excising the embryos early during seed development and cultivating them in vitro. Interspecific incompatibility is the common cause of embryo abortion causing the production of small, degenerated, shrunken seeds following wide hybridization. The embryo rescue technique can be attempted to overcome embryo abortion through various steps: cutting of proembryo within 3–5 days of pollination, culturing of proembryo on media containing agar, collection of plantlets obtained from the embryo, and transplantation of the plantlet in soil.

Protoplast Cultures Somatic hybridization is an important tool for plant breeding and crop improvements through the production of inter-specific and inter-generic hybrids. The technique involves the fusion of protoplasts of two different genomes followed by the selection of desired somatic hybrid cells and regeneration of somatic hybrid cells. The somatic hybridization using protoplast fusion has great ability to overcome the barriers of gene transfer from wild species to cultivated species. The somatic fusion, also known as protoplast fusion, is a sort of genetic manipulation in plants in which two distinct species of plants are fused together to generate a new hybrid plant possessing the features of both species. The regeneration of plants using protoplast cultures is possible in *Pisum*, *Trifolium*, and *Melilotus*. Asymmetric protoplast fusion has been used for *Medicago* improvement. Recently, a suspension-derived protoplast fusion was performed to develop asymmetric somatic hybrids between bread wheat and *Agropyron elongatum*.

12.11.4 Application of Embryo Rescue

Among the post-zygotic barriers, abortion of embryos is the most common problem in distant crosses, which may occur during different stages of embryo development. These embryos can be artificially cultured on nutrient medium and can be grown as mature plants. The embryo rescue technique has been used for producing wide crosses in different genera such as *Hordeum*, *Agropyron*, *Triticum*, *Oryza*, *Solanum*, *Lycopersicon*, *Brassica*, *Gossypium*, *Phaseolus*, *Vigna*, *Cicer*, and *Cucumis*. The embryo rescue technique has been successfully used to produce intergeneric crosses in cereals such as barley × rye, wheat × barley, wheat × rye, oat × maize, wheat × maize, *Hordeum* × *Elymus*, and *Elymus* × *Triticum*. Disease resistance has been successfully introduced from *Triticum timopheevi* to bread wheat, via embryo rescue. Wide crosses were made between *Thinopyrum scirpeum* and hexaploid wheat for transferring salt tolerance into wheat by regenerating F_1 hybrids through embryo culture. Resistance to blast and bacterial blight diseases from wild rice (*Oryza minuta*) has been transferred to cultivated rice (*Oryza sativa*) with the aid of embryo rescue. Similarly, embryo rescue has been used in leguminous crops for producing interspecific hybrids. Hybridization of cultivated lentil with *Lens ervoides* and *L. nigricans* results in pod development that is arrested within 10–16 days after pollination and finally, yields shrivelled, non-viable seeds, but can be rescued by a two-step in vitro method of embryo-ovule rescue to obtain successful distant hybrids. In chickpea, embryo rescue has been successfully used to obtain viable embryos from *Cicer arietinum* × *C. bijugum* and *C. arietinum* × *C. pinnatifidum* crosses.

Embryo rescue techniques such as immature embryo culture were used to develop an interspecific hybrid ACC between *B. napus* "Zhongshuang 9" and *B. oleracea* "6 m08". In interspecific crosses between *B campestris* and *B. oleracea*, the ovaries were excised and cultured on artificial media containing minerals and vitamins, and after a few weeks, the embryos were excised and cultured on the same media. The embryo rescue approach has been used effectively in transferring resistance to tobacco mosaic virus, tomato spotted wilt virus, root-knot nematode, and pathogenic fungi in cultivated species from the wild species *L. peruvianum*. In potato, potato leaf roll virus resistance has successfully been introgressed from *Solanum etuberosum* to *S. tuberosum* through embryo rescue. The successful use of in vitro techniques in interspecific and intergeneric hybridization with examples are presented in Table 12.10.

Haploid and Doubled Haploid Production via Embryo Rescue Haploid plants have a single set of chromosomes (n). When these chromosomes double, the plant becomes a "doubled haploid" (DH). For example, DH wheat plants are created by doubling the chromosome set of a gamete to maintain a balanced hexaploid genome. The DH technology speeds up breeding and enhances genetic gains by instantly creating homozygous lines, avoiding the need for 6–10 generations of inbreeding by selfing or sib-crossing. The DH technology is widely used in crops like Brassicas, wheat, barley, rice, and maize. It involves two main steps: haploid induction and genome doubling. There are two types of haploid induction: in vitro-based (IVB) and in vivo-based. In vitro-based haploid induction relies on the development of immature male or female gametophytic cells, whereas in vivo haploid induction systems are based on inter/intraspecific hybridization and subsequent uniparental chromosome removal. Producing DH plants from male cells is called in vitro androgenesis, while producing them from female cells is called in vitro gynogenesis. Androgenesis is the most effective method but is challenging for some crops like tomatoes, peppers, eggplants, wheat, rye, and barley. For the first time, androgenesis was reported by Guha and Maheshwari in *Datura innoxia* anthers ($2n = 2x = 24$) to develop haploid plants ($n = 12$).

Haploids can also result from pollination between different species or genera. For example, crossing *Hordeum vulgare* (barley) with *H. bulbosum* produces hybrid embryos. However, *H. bulbosum* chromosomes are eliminated during development, resulting in haploid embryos with only *H. vulgare* chromosomes. These embryos must be rescued and grown in vitro before endosperm degeneration.

In wheat and barley, DHs obtained from immature microspores or in vitro ER following interspecific hybridization are important strategies for breeding and generating populations for gene mapping purposes. The culture of embryos obtained by

Table 12.10 In vitro techniques used for crop improvement

Crop	In vitro techniques	Function/Traits
Interspecific hybridization		
Aegilops tauschii × *Triticum aestivum* *Aegilops ovata* × *Triticum aestivum* *Aegilops cylindrica* × *Triticum aestivum* *Aegilops speltoides* × *Triticum aestivum*	Embryo culture	Alien gene introgression
Triticum dicoccum × *Triticum timopheevii*	Immature embryo culture	Creation of genetic variation
Wheat (*T. aestivum* × *Dasypyrum villosum*)	Embryo rescue	Powdery mildew resistance Stem rust resistance
Rice (*Oryza sativa* × *Oryza australiensis*)	Young embryo rescue	Interspecific hybrid
Barley (*Hordeum vulgare* × *H. bulbosum*)	Embryo rescue	Powdery mildew resistance
Lentil (*Lens culinaris* × *L. tomentosus*)	Ovule rescue	Interspecific hybrid
Pea (*Pisum sativum* × *P. fulvum*)	Immature embryo culture	Interspecific hybrid
Chick pea (*Cicer arietinum* × *C. songaricum*) Chick pea (*C. arietinum* × *C. pinnatifidum*)	Embryo rescue via ovules	Interspecific hybrid
Rajma (*Phaseolus vulgaris* × *P. coccineus*)	Immature embryo	Ascochyta blight resistance Bean fly resistance
Rajma (*P. vulgaris* × *P. polyanthus*)	Immature embryo	Bean golden mosaic virus resistance
Soybean (*Glycine max* × *G. tomentella*)	Immature embryo	Cyst nematode resistance
Brassica (*B. kaber* × *B. napus*)	Embryo rescue via ovule	Herbicides tolerance
Brassica (*B. juncea* × *B. kaber*)	Immature embryos culture	Herbicides tolerance
Brassica (*B. juncea* × *B. tournefortii*)	Embryo rescue	Aphids tolerance Drought tolerance
Brassica (*B. rapa* × *B. kaber*)	Immature embryos culture	Herbicides tolerance
Brassica (*B. rapa* × (*B. juncea* × *B. rapa*)	Embryo rescue via ovule	Interspecific hybrid
Brassica (*B. napus* × *B. rapa*)	Embryo rescue	Clubroot-resistant
Brassica (*B. napus* × *B. nigra*)	Embryo rescue	Blackleg disease resistance
Brassica (*B. oleracea* var. botrytis × *B. carinata*)	Embryo rescue	Black rot resistance
Brassica (*B. oleracea* × *B. napus*)	Embryo rescue	Atrazine resistance

(continued)

Table 12.10 (continued)

Crop	In vitro techniques	Function/Traits
Brassica (*B. napus*× *Brassica carinata*)	Embryo rescue	Powdery mildew resistance
Cucumber (*Cucumis melo* × *C. metuliferus*	Embryo rescue	Interspecific hybrid
C. metuliferus × *C. anguria*	Embryo rescue	Interspecific hybrid
C. melo × *C. anguria*	Embryo rescue	Interspecific hybrid
Tomato (*Solanum lycopersicum* × *S. peruvianum*)	Embryo rescue	Tobacco mosaic virus resistance Tomato spotted wilt virus Root-knot nematode resistance
Intergeneric hybridization		
Wheat × rye (*T. aestivum* × *Secale cereale*)	Embryo rescue	Embryo lethality
Wheat × Barley (*T. aestivum* × *H. vulgare*)	Embryo rescue	Intergeneric hybrid
Wheat× rye (*T. aestivum* × *Secale cereale* (addition line [1] ch. 6R)	Embryo rescue	Powdery mildew resistance
Wheat × rye (*T. aestivum* × *Secale cereale* (addition line [2] ch. 4R)	Embryo rescue	Multiple fungal disease resistance
T. aestivum × *Psathyrostachys huashanica*	Embryo culture	Resistance to powdery mildew
Rice (*Oryza sativa* × *Leersia perrieri*)	Embryo rescue	Intergeneric hybrid
Brassica (*Sinapis alba* × *Brassica juncea*)	Somatic hybridization through protoplast fusion	*Alternaria brassicae* and heat stress resistance
Brassica (*Brassica juncea* × *Sinapis alba*)	Somatic hybridization through protoplast fusion	Development of a yellow-seeded stable allohexaploid

hybridization between *T. aestivum* and *Zea mays* ($2n = 2x = 20$) is a widely used method for obtaining haploid and, later, homozygous lines (DH) of wheat. This technique is quite laborious and involves several steps: (a) emasculation of the wheat flower and pollination of the emasculated flower with maize pollen, (b) hormone treatments of the flowers with growth regulators applied auxin to each wheat floret at 1 DAP. (c) embryo rescue, (d) regeneration of the haploid shoot in culture medium (during this stage, maize chromosomes were eliminated) and (e) chromosome doubling. The DH production efficiency mainly depends on the maize and wheat genotypes used in the crossing, the health status of the plants, and their excellent development.

Amphidiploid via Embryo Rescue Hybrids were created by crossing *Ae. ovata* ($2n = 4x = 28$) with *S. cereale* ($2n = 2x = 14$), which resulted in offspring with $2n = 3x = 21$ chromosomes. *Ae. ovata* was used as the female parent, and treated with GA3 to aid embryo development. These hybrids were initially sterile but regained fertility after chromosome doubling with colchicine treatment. The resulting amphidiploids ($2n = 6x = 42$) resembled the hybrid plants but had lower tillering ability, broader leaves, and larger spikes.

12.12 Summary

Crop wild relatives are the wild plant species closely related to crop plants and contain many useful genes for various abiotic and biotic stresses, quality, nutritional, and agronomic traits. However, during the process of domestication, useful alleles have been lost, which is affecting the species' fitness for changing climatic conditions and various stresses. CWRs are critical resources to meet food security needs and the challenges of new production systems especially in response to climate change. The wild relatives of primary, secondary, and tertiary gene pools of important crops are enumerated in this chapter. To use these extremely important CWRs, it is of utmost importance to conserve these crop wild relatives in ex-situ as well as in-situ depending upon biological factors, seed behavior, and type of propagation. However, wild relatives have different phenology than cultivated plants and for effective management and use of CWRs, it is important to study the phenotypic characteristics of the crop plant. The genes associated with resistance/tolerance against various biotic and abiotic stresses, quality traits, and yield potential in various crops have been summarized in this chapter. The transfer of these genes from CWRs is not an easy job and requires phenotyping of crop wild relatives, identifying the key lineage with beneficial or novel alleles and introgressing the donor alleles in species of interest through wide hybridization. However, there are various limitations and barriers to the use of wide hybridization. Special breeding techniques such as embryo rescue technique and induced polyploidy and bridging crosses can be used in hybridization programs for the transfer of desirable genes.

Multiple-Choice Questions

1: Which of the following is used as a bridge for gene transfer from wild to cultivated species of wheat?
  (a) Amphidiploids
  (b) Synthetic hexaploid wheat
  (c) *Triticum durum*
  (d) All of the above

2: Which of the following is an allohexaploid crop?
  (a) *Triticum aestivum*
  (b) *Oryza sativa*
  (c) *Triticum durum*
  (d) None of these

3: The use of crop wild relatives/wild species is limited in plant breeding due to
  (a) Presence of undesirable traits/genes
  (b) Lack of fertile offspring due to crossing barriers
  (c) Presence of linkage drag
  (d) All of the above

4: Quaternary gene pool is also known as…………..
   (a) GP4
   (b) Gene Ocean
   (c) Genes can be transferred among organisms through recombinant DNA technology
   (d) All of the above
5: Which of the following is the progenitor of lentil?
   (a) *Lens culinaris ssp. culinaris*
   (b) *Lens culinaris ssp. orientalis*
   (c) *Lens odemensis*
   (d) *Lens ervoides*
6: Blast resistant gene *Pi33* in rice var. IR64 is introgressed from which species?
   (a) *Oryza rufipogon*
   (b) *Oryza glaberrima*
   (c) *Oryza longistaminata*
   (d) All of the above
7: New rice for Africa (NERICA) developed by crossing between *Oryza sativa* and *Oryza glaberrima* grown under the following conditions.
   (a) Weed competitiveness and high yield
   (b) Drought tolerance
   (c) Resistance to insect and pest
   (d) All of the above
8: Which method can be used to overcome pre-zygotic barriers in wide hybridization?
   (a) Cut-style or graft-on-style techniques
   (b) Application of pollen mixture from several species
   (c) In vitro ovule pollination
   (d) All of the above
9: Embryo abortion is one of the following barriers in wide hybridization.
   (a) Pre-zygotic barrier
   (b) Post-zygotic barrier
   (c) Reproductive barrier
   (d) None of these
10: Origin of urdbean is
   (a) Tropical Africa
   (b) South America
   (c) India
   (d) China
11: Cultivated durum wheat is ……….. with basic chromosome ……………..
   (a) Diploid; x = 12
   (b) Diploid; x = 7
   (c) Tetraploid; x = 7
   (d) Tetraploid; x = 12

12: The ....disease led to epidemic resulting in "Irish famine of Potato" Ireland during 1845–49.
(a) Common scab
(b) Late blight of potato
(c) Early blight
(d) Wart

13: The grassy stunt virus resistance gene was transferred to ***Oryza sativa*** from....
(a) *O. minuta*
(b) *O. nivara*
(c) *O. glaberrima*
(d) All of the above

14: Late blight of potato is ....disease caused by the pathogen................
(a) Fungal; *Pythium aphanidermatum*
(b) Fungal; *Pytophthora infestans*
(c) Bacterial; *Ralstonia solancearum*
(d) Viral; virus

15: The wild species as a source of resistance against late blight is ....
(a) *S. demissum*
(b) *S. phureja*
(c) *S. stoloniferum*
(d) *S. microdontum*
(e) All of the above

16: Loktak Lake in Manipur is famous for wild relatives of........
(a) Wheat
(b) Rice
(c) Maize
(d) Pearl millet

17: Aegilops species is a wild relative of................
(a) Wheat
(b) Rice
(c) Maize
(d) Pearl millet

18: Teosinte is wild relative of................
(a) Wheat
(b) Rice
(c) Maize
(d) Pearl millet

19: Common name of dicoccum wheat................
(a) Spelta
(b) Emmer
(c) Einkorn
(d) Polish

20: The resistant source of bacterial blight gene Xa21 in rice is................
(a) *O. minuta*
(b) *O. longistaminata*
(c) *O. nivara*
(d) None

Answers

1. (d)
2. (a)
3. (d)
4. (a)
5. (b)
6. (a)
7. (d)
8. (d)
9. (b)
10. (c)
11. (c)
12. (b)
13. (a)
14. (b)
15. (a)
16. (b)
17. (a)
18. (c)
19. (b)
20. (b)

Suggested Reading

Bohra et al (2022) Reap the crop wild relatives for breeding future crops. Trends Biotechnol 40(4):412–431

Dempewolf et al (2017) Past and future use of wild relatives in crop breeding. Crop Sci 57:1070–1082

Engels JMM (1990) Conservation strategies: a holistic approach. Paper presented at the International Okar Genetic Resource Workshop at NBPGR, New Delhi

Gupta et al (2018) Uses and importance of wide hybridization in crop improvement. In: Bharadwaj DN (ed) Advance molecular plant breeding: meeting the challenge of food security. Apple Academic Press, Toranto, pp 107–137

Hajjar R, Hodgkin T (2007) The use of wild relatives in crop improvement: a survey of developments over the last 20 years. Euphytica 156(1):1–13

Kapazoglou et al (2023) Crop wild relatives: a valuable source of tolerance to various abiotic stresses. Plants 12:2–20

Kaur et al. Role of crop wild relatives in crop improvement under changing climatic conditions

Kumar et al (2013) Distant hybridization and alien gene introgression. In: Pratap A, Kumar J (eds) Biology and breeding of food legumes. CAB International, pp 81–110

Kumar et al (2022) A walk towards wild grasses to unlock the clandestine of gene pools for wheat improvement. A review. Plant Stress 3:1–12

Kumar J, Pratap A (2013) Alien gene transfer: challenges and opportunities. In: Pratap A, Kumar J (eds) Innovations, methods and risk assessment, vol 1. Springer Science and Business Media, New York

Liu et al (2014) Distant hybridization: a tool for interspecific manipulation of chromosomes. In: Pratap A, Kumar J (eds) Alien gene transfer in crop plants, vol 1. Springer, New York, NY, pp 25–42

Pandey et al (2005) Wild relatives of crop plants in India: collection and conservation. National Bureau of Plant Genetic Resources, New Delhi, p 73

Ren et al (2017) Novel technologies in doubled haploid line development. Plant Biotechnol J 15:1361–1370

Rogo et al (2023) Embryo rescue in plant breeding. Plants 12:2–60

Singh et al (2022) Unlocking the hidden variation from wild repository for accelerating genetic gain in legumes. Front Plant Sci 13:1–24

National and International Genebanks: Conservation Status of Plant Genetic Resources

13

Sherry Rachel Jacob, Anju Mahendru Singh, and Veena Gupta

Learning objectives

- Concept of genebank.
- Global status of genebanks.
- Global information databases for genebanks.
- Future challenges.

Keywords

Plant genetic resources · Germplasm · PGR · Genebanks · Conservation · Genebank database platforms

13.1 Introduction

The term 'Genebank' refers to an organisational unit that has the objective of managing plant genetic resource (PGR) collections for their conservation and to facilitate their use. They are institutionalised ex situ sites where PGR is conserved in the form of seed, plant tissue, pollen or DNA sequences and are often complemented by associated field genebanks. The efforts towards PGR conservation began in the past century when breeders, farmers and researchers made conscious efforts to collect and safeguard unique genetic resources of potential value. The first significant effort in this direction was that of Nikolai Vavilov, the Russian botanist, in 1920s, which led to his classic work on centres of origin. The first genebank in the world was also

S. R. Jacob (✉) · A. Mahendru Singh · V. Gupta
Division of Germplasm Conservation, ICAR-National Bureau of Plant Genetic Resources, New Delhi, India

K. Tripathi et al. (eds.), *Textbook of Plant Genetic Resources*,
https://doi.org/10.1007/978-981-96-5024-8_13

established in Leningrad, Russia, in 1920. The work on collection and conservation was further taken ahead by Jack Harlan, the American botanist. He highlighted the case of loss of diversity in barley and proposed the need for more focused efforts in collection and conservation. Subsequently, through the efforts of Sir Otto Frankel, an Australian genetic conservationist, the first conference on 'Exploration, utilisation and conservation of plant genetic resources' was convened in 1967, under the aegis of the Food and Agriculture Organization (FAO) and the International Biological Programme. This event was a major milestone in the history of PGR conservation since it led to the charting of an international plan for PGR conservation (Frankel and Bennett 1970). In the same year, Jack Harlan and J.M.J deWet jointly started the Crop Evolution Laboratory at the University of Illinois, where they worked on and published the concept of 'genepools' and their relevance for systematic utilisation of genetic resources (Harlan and deWet 1971). They also expanded the Vavilovian concept of centre of origin to include space, time and variation thereby giving rise to crop-specific patterns of evolution. His classic work 'Our Vanishing Genetic Resources' was published in 1975 in *Science* wherein the importance of genebanks and ex situ conservation was well highlighted. All these background efforts laid the foundation for the establishment of International Plant Genetic Resources Institute (IPGRI) in 1974, by the 'Consultative Group on International Agricultural Research' (CGIAR), to promote an international network of genetic resources centres to further collection, conservation, documentation, evaluation and use of germplasm.

Today, the global system for conservation of PGR is a pan-continental network integrating more than 2000 national and international genebanks, out of which 130 genebanks hold more than 10,000 accessions and 8 have more than 100,000 accessions. The first and second reports on the *State of the World's Plant Genetic Resources for Food and Agriculture* provide a comprehensive and authoritative account of PGR status at the global level (FAO 1998, 2010). The PGR Genesys platform, which hosts compiled data on accessions conserved in ex situ genebanks, reports a total of more than four million accessions maintained by over 450 institutes across the globe (https://www.croptrust.org/). The largest genebanks, in the order of size of their ex situ germplasm collections, are: (i) National Centre for Genetic Resources Preservation (NCGRP) in the USA, (ii) National Bureau of Plant Genetic Resources (NBPGR) in India, (iii) Institute of Crop Germplasm Resources, Chinese Academy of Agricultural Sciences (ICGR-CAAS) in China, (iv) and N.I. Vavilov All Russian Scientific Research Institute of Plant Industry (VIR) in the Russian Federation. As per the information available on multiple databases, around 45% of germplasm is held by only seven countries. Estimates suggest that more than 70% of the genetic diversity of some 200–300 crops is already conserved in genebanks (Khoury et al. 2010). In addition, there are over 2500 botanic gardens maintaining samples of some 80,000 plant species (FAO 2010). The National Genebank of India (NGB) currently holds a base collection of 462,606 accessions in 2049 species. These collections have been conserved for more than 35 years and cater to all major crop improvement programmes of the National Agricultural Research System (NARS).

There are several examples of classical breakthroughs achieved using novel alleles discovered from conserved genetic resources. Jack Harlan, in 1948, chanced upon a wheat accession in Turkey, which he described as 'miserable looking, tall, thin stemmed, lodges badly, is susceptible to leaf rust and has poor baking qualities'. Eventually, when wheat stripe rust became serious in the USA, this accession became the major source of resistance to 4 races of stripe rust as well as 35 races of common bunt, 10 races of dwarf bunt and also good tolerance to flag smut (Harlan et al. 1975). Similarly, in paddy, the source of resistance to grassy stunt virus is a single accession of *Oryza nivara* that was collected by explorers from the northern plains of India and conserved in International Rice Research Institute (IRRI) Genebank (Plucknett et al. 1987), which today forms a component of the genetic background of several modern rice varieties. It has been estimated that there is a benefit of around 115 billion dollars annually for the world economy, from using wild materials from genebanks to develop environmentally resilient crops.

13.2 Major Seed Genebanks

Today more than 2000 genebanks of national, regional and international status conserve more than 7.4 million samples of crop diversity. Some of the major genebanks holding more than one lakh accessions are described below.

13.2.1 USDA Genebank

The genebank system that functions under the aegis of the Agricultural Research Service of the United States Department of Agriculture (USDA-ARS) is called as the National Plant Germplasm System (NPGS). It is a network of more than 20 genebanks, together maintaining over 617,000 accessions of plant germplasm (more than 16,000 plant species) (Fig. 13.1). NPGS does extensive germplasm distribution and data shows an average distribution of over 200,000 samples yearly for various stakeholders across the globe. The base collection of the NPGS is maintained at the National Laboratory for Genetic Resources Preservation (NLGRP), Fort Collins, Colorado, where accessions are maintained at −18 °C as a long-term PGR backup facility. Approximately 82% of the NPGS seed accessions are currently stored at the NLGRP (www.npgsweb.ars-grin.gov/gringlobal).

13.2.2 Leibniz Institute of Plant Genetics and Crop Plant Research (IPK)

The IPK Genebank or Federal Ex-situ Genebank is located in Germany. The history of IPK dates back to 1943, when it was founded as the 'Kaiser Wilhelm Institute for Crop Plant Research', near Vienna. It was relocated to Gatersleben in 1945 and was reconstituted as IPK in 1992. The Federal Ex-situ Genebank for

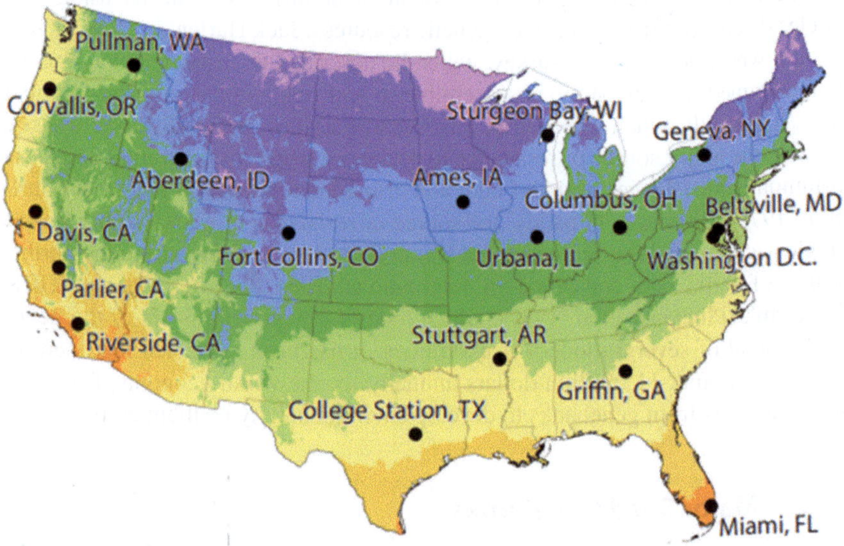

Fig. 13.1 Map of NPGS sites (Source: www.ars.usda.gov)

Agricultural and Horticultural crops was institutionalised in its present form, in 2003. Its germplasm holding currently comprises 151,348 accessions belonging to 2912 species. The collection is distributed across three locations—the main site is at Gatersleben; oil and fodder crops at Malchow; and potato collection at Groß Lüsewitz. Security duplicates of more than 54,000 accessions (36% of the total collection) have been deposited by IPK at the Svalbard Global Seed Vault (SGSV). The genebank has its own Genebank Information System (GBIS), which allows query-based retrieval of information about germplasm availability. The genebank has strong collaboration with the European Co-operative Programme for Plant Genetic Resources (ECPGR) and currently, is responsible for the management of its European Search Catalogue for Plant Genetic Resources (EURISCO) web catalogue (*www.ipk-gatersleben*).

13.2.3 The Australian Grains Genebank (AGG)

Established in 2014 at the Grains Innovation Precinct Horsham, AGG is the national bio-digital resource centre for crop genetic resources and conserves over 190,000 accessions of seed covering 1250 species. The seeds are maintained in long-term storage facility at −20 °C, within airtight aluminium foil packets. It functions via a strategic partnership between the Grains Research and Development Corporation (GRDC) and the State of Victoria. All AGG accessions are backed up in the National Safety Duplication Facility and the Svalbard Global Seed Vault (www.agriculture.vic.gov.au).

13.2.4 National Agriculture and Food Research Organization (NARO), Japan

In Japan, seed preservation facilities were initially constructed in 1966 at the National Institute of Agricultural Science located at Hiratsuka. The current Japanese central genebank facility is known as the National Agriculture and Food Research Organization (NARO) and is located in Tsukuba, Ibaraki, Japan. This central facility conserves plants, animals, microorganisms and DNA materials related to agriculture. In addition to this central bank, plant genetic resources are also conserved at Japan International Research Center for Agricultural Sciences and National Livestock Breeding Center. The genebank has a total of 224,000 accessions, which includes a core collection of Japanese rice landraces, maize landraces, Adzuki bean, wheat and soybean. At NARO, the active collection is maintained at −1 °C and 30% relative humidity (RH). For long-term storage, decompressed and hermetically sealed tin cans are used and they are maintained at −18 °C (www.naro.go.jp).

13.2.5 N.I. Vavilov All Russian Scientific Research Institute of Plant Industry (VIR)

VIR was established in 1894 as the Bureau of Applied Botany. It has expanded its activities since its inception and is currently Russia's national centre for conservation, research and utilisation of PGR. The long-term storage facility is located at St. Petersburg, where a base collection of seed genebank is maintained within vacuum-sealed aluminium pouches, at −10 °C. The centre has 11 main branches and experimental stations distributed in the diverse climatic zones of the country. Its ex situ collection is one of the oldest and the fourth-largest global collections in the world. The current information available in the public domain shows a collection size of 346,666 accessions (as of 2017) representing 2169 species. The ex situ collection includes 29,611 accessions maintained in the field genebank (vegetatively propagated perennial crops), 750 accessions maintained as in vitro collection and 1824 accessions in its cryo-preservation collection. VIR has a safety duplicate genebank at Kuban (Dzybenko 2018; www.vir.nw.ru).

The PGR Genesys database has information on 200,717 accessions held at VIR, which includes 45,297 landraces, primarily of wheat, barley, maize, oats and soybean. The genebank also holds 4711 accessions of Indian origin, as depicted in the database.

13.2.6 Germplasm Resources Unit (GRU), John Innes Centre, Norwich, UK

The GRU conserves significant gene pool diversity in its ex situ collection. The major crops are wheat, pea, barley and oats. GRU has a unique historic/legacy collection of 27,000 accessions, which includes the A.E. Watkins wheat landrace collection of 1920s and the John Innes Pisum Global Collection of 1960s. In 2012, GRU was upgraded as a Biotechnology and Biological Sciences Research Council (BBSRC) National Capability facility and since then it functions as an international crop genebank. India has also benefited from this facility through the repatriation of more than 600 unique heritage wheat landraces (www.jic.ac.uk).

The PGR Genesys database shows information on 27,712 accessions of GRU, which include 10,251 accessions of barley, 9846 accessions of wheat, 3360 accessions of pea and 2109 accessions of oats. The collection has a total of 5266 landraces and 1589 accessions are of Indian origin.

13.2.7 Royal Botanic Garden Conservation Repository

The Royal Botanic Garden (RBG) conservation repository is the largest of its kind repository having a holistic panorama of conservation strategies. It comprises of Royal Botanic Garden at Kew and Millennium Seed Bank (MSB) at Wakehurst Place, Sussex. Established in 1840, RBG has around 27,000 taxa as living collections while the herbarium possesses more than 8.5 million plants and fungal specimens. Kew's Millennium Seed Bank (MSB) at Wakehurst Place is the largest wild plant seed bank in the world containing about 13% of the world's wild plant species, holding some 2.25 billion seeds representing over 39,000 diverse species from 189 countries (https://www.kew.org/). The seed genebank is well supported by research labs, glass houses and a botanical garden for the multiplication of seeds and their conservation. The seeds after pest-free checkup are conserved in sealed double-layered glass bottles and are conserved in an underground module maintained at −18 °C. The Royal Botanic Garden, Kew, has the world's largest living plant collections representing plants of all climatic zones. The botanical garden maintains more than 2700 different taxa in glass houses under controlled conditions. The other facilities are: alpine house, spread over 4880 m^2 (temp. <20 °C); the Nash repository for aroid plants; the Palm house; and Princess of Wales Repository having ten computer-controlled micro-climate zones to maintain different orchids, water lilies, carnivorous plants and bromeliads. The Kew Herbarium is one of the largest repositories having more than seven million specimens representing the globe (Fig. 13.2). The herbarium is widely used for taxonomic studies by researchers, students and botanists. The world's largest underground mycological repository comprises 1.25 million dried specimens of fungi, making it the most diverse Fungarium.

13.2.8 National Agrobiodiversity Center (RDA Genebank), Republic of Korea

The National Agrobiodiversity Center under the Rural Development Administration (RDA) in Jeonju, Republic of Korea, is the major national genebank of Korea. The facility was established as low-temperature seed storage facility in 1975 and today it conserves 2.7 million accessions from 1599 species. About 75% are food crops. It

Fig. 13.2 RBG, Kew herbarium and the storage cabinets with herbarium specimens

has a medium-term, long-term and cryo-conservation facilities (https://genebank.rda.go.kr/).

13.2.9 National Genebank of India (NGB)

The conservation activities within India are carried out through an extensive network, which is collectively called the 'National Agricultural Research System (NARS)'. It includes crop-based institutes of the Indian Council of Agricultural Research (ICAR) and State Agricultural Universities. Within this system, 49 institutes have been officially designated as the National Active Germplasm Sites (NAGS). The nodal organisation for the coordination of PGR activities by all these institutes is ICAR-National Bureau of Plant Genetic Resources (NBPGR), located at Pusa Campus, New Delhi. It houses the National Genebank of India, which holds the second-largest ex situ germplasm collection in the world, next only to the

USDA-NPGS collection. The NGB was established in 1986 with the funding support of the US Agency for International Development (USAID) through an Indo-USAID PGR project that led to the establishment of four Long Term Storage (LTS) modules. It was further expanded in 1996 and upgraded/refurbished in 2022. The NGB facility currently has 12 long-term storage modules and 5 medium-term storage modules. Its long-term storage facility conserves over 0.46 million germplasm accessions belonging to more than 2160 species, as the base collection. In LTS, seeds are maintained at −18 °C, within vacuum sealed tri-layered aluminium pouches. The medium-term storage modules are maintained at 4 °C and 30% RH. Within the NAGS there are additional Medium Term Storage (MTS) facilities and field genebanks, where crop-specific active collections of selected accessions are maintained. The Indian National Genebank also hosts the multi-crop In Vitro Genebank for vegetatively propagated crops and a Cryo-Genebank for the conservation of recalcitrant species. Germplasm of more than 50 species are maintained in tissue culture repository and 485 species are conserved in cryo genebank (www.nbpgr.ernet.in).

13.2.10 The National Crop Genebank of China

The Institute of Crop Science, Chinese Academy of Agricultural Sciences harbours the National Crop Genebank of China (NCGC), which was built in 1986 to conserve crop germplasm of the Chinese region. It is built across an area of 3200 m^2 with a storage capacity of four million accessions. The accessions are conserved at −18 °C under 50% relative humidity. At present NCGC conserves more than 392,000 accessions belonging to 180 crops along with their wild relatives. Above 300 accessions covering 35 vegetative crops are conserved in In-vitro Genebank and 100 accessions belonging to 13 crop species are also conserved in the cryo-conservation facility. For safety duplication of this germplasm, all accessions are also conserved in genebank in Qinghai province (https://ics.caas.cn).

13.2.11 Genebanks of Consultative Group on International Agricultural Research (CGIAR) Centres

The Consultative Group on International Agricultural Research (CGIAR) are primary centres of PGR management and 11 of these Consultative Group (CG) centres have established their genebank facilities that cater to their breeding programmes and also are major sources of global germplasm exchange (Fig. 13.3). These genebanks together conserve 736,210 accessions and over the past 10 years, CGIAR genebanks have distributed more than an estimated million samples to various indenters from across the world (Table 13.1).

The majority of the material conserved by CG centres has been mandatorily declared as available under the multilateral system of the International Treaty on Plant Genetic Resources for Food and Agriculture (ITPGRFA). They are available

Fig. 13.3 Location map of genebanks of CG centres (Source: www.genebanks.org)

to stakeholders through the Standard Material Transfer Agreement (SMTA), as outlined in the Treaty.

13.2.12 Svalbard Global Seed Vault

The 'Svalbard Global Seed Vault (SGSV)' is a state-of-the-art facility that was established in 2008 as a safety duplicate site for germplasm collections. The establishment of the SGSV serves as the ultimate safety net for seed samples from the world's most important collections (Khoury et al. 2010). This is required for preparedness in unforeseen and extreme situations such as nuclear or civil war or any other catastrophic failure. The best-cited global example of its utility is the case of ICARDA, which could safely duplicate 80% of its collection before the genebank was closed down in 2014 because of war. ICARDA used the Svalbard samples to rebuild its collections in their new campuses in Morocco and Lebanon. It is also recommended as per the International Genebank Standards of the FAO, that each genebank should have provision for the conservation of safety duplicates of potentially unique germplasm at a distant place far from the location of the genebank.

The facility functions under the overall responsibility of the Norwegian Government. The Nordic Genetic Resource Centre (NordGen) oversees daily operations and the Crop Trust provides partial funding for its management. There is also an international advisory panel comprising depositors and stakeholders, for general management of the facility. It is built inside a mountain in a man-made tunnel on the frozen Norwegian island of Spitsbergen, 1307 km from the North Pole. The area's permafrost keeps the vault below the freezing point of water and the seeds are protected by 1-m thick walls of steel-reinforced concrete. The natural temperature is

Table 13.1 Ex-situ conservation status of CG Genebanks

S. no	CG Genebank	Location	Major mandate crop/s	No. of conserved accessions
1	Africa rice	Côte d'Ivoire	*Oryza glaberrima*	20,000
2	Bioversity	Belgium	Banana and plantain	1500
3	International Center for Tropical Agriculture-Centro Internacional de Agricultura Tropical (CIAT)	Colombia	Beans, tropical forages and cassava	67,600
4	International Maize and Wheat Improvement Center - Centro Internacional de Mejoramiento de Maíz y Trigo (CIMMYT)	Mexico City	Wheat and maize	181,000
5	The International Potato Center-Centro Internacional de la Papa (CIP)	Peru	Potato, sweet potato and Andean root and tuber crops	
6	International Center for Research in Agroforestry (ICRAF)	Kenya	Agro-forestry tree species used for fruit, timber, medicines and other products	180 tree species are held as seed accessions and 43 species are conserved in field collections at 38 sites in 15 countries
7	The International Center for Agricultural Research in the Dry Areas (ICARDA)	Morocco and Lebanon	Barley, chickpea, grasspea, lentil and wheat and their crop wild relatives (CWRs)	145,000
8	International Crops Research Institute for the Semi Arid Tropics (ICRISAT)	India	Chickpea, groundnut, various millets, sorghum and pigeonpea	125,000
9	International Institute of Tropical Agriculture (IITA)	Nigeria	Banana, cassava, yam, maize, soybean, Bambara groundnut, cowpea, African yam and bean	Seeds (23,000 accessions) and clonal collections (9000)
10	International Livestock Research Institute (ILRI)	Ethiopia	Tropical forage species	19,000
11	IRRI	The Philippines	Rice and its CWRs	128,000

Source: www.genebanks.org

about −3 to −4 °C, and −18 °C temperature is maintained with the help of power backup. It can store 4.5 million accessions, and by 2023, over 1,267,000 seed samples of 6143 species, submitted by 102 genebanks, have been duplicated there (www.seedvault.no).

13.3 Genebank Database Platforms

The global genebank system is ably supported by multiple database platforms that facilitate the streamlining of germplasm exchange and documentation by various stakeholders across the globe. Databases that deal with germplasm information are of various categories. GRIN-Global is a relational database for genebanks whereas Genesys, EURISCO and FAO-WIEWS (World Information and Early Warning System on Plant Genetic Resources for Food and Agriculture) provide information on genebank inventories in various countries. Global Biodiversity Information Facility (GBIF) is a portal that documents and makes available, information on species occurrence. All these databases are constantly upgraded to meet the emerging requirements of stakeholders. APIs (Application Programming Interfaces) and DOIs (Digital Object Identifiers) are recent interventions through which database interoperability has been made possible. APIs are mechanisms that enable two software components to communicate with each other. An example of this is breeding API (BrAPI), which specifically involves plant breeding data. It acts as a standard interface for plant phenotype and genotype databases to serve their information to crop breeding applications. DOIs have been instituted by ITPGRFA under the Global Information System (GLIS) as a means to bring uniformity in information available in various genebanks. It provides a single unique identity for each accession with which cross-referencing of accessions can be done in multiple databases or documents.

The major databases involved in the information systems management of plant genetic resources are described below.

13.3.1 PGR-Genesys

Genesys is the most widely used online database portal on plant genetic resources and is the most extensive compilation of genebank inventories across the world. It was developed as a collaborative project between Bioversity International, the Crop Trust and the Secretariat of ITPGRFA. Currently, it hosts information on 4,345,574 accession records (around half of all seed samples conserved in genebanks worldwide), which are held in over 450 institutes/organisations, as submitted by 53 data providers. It has accession-wise information on passport data, core collections, characterisation and evaluation data and various crop descriptive lists. As depicted in the biological section status of the database, 21% of the records are of landraces and crop wild relatives (CWRs) constitute 11% of the total reported collection. The site of safety duplication of donor institutes, availability under multilateral system and safety duplication status at Svalbard are the other relevant information available on this platform. Though the national genebank inventory of India is not ported in Genesys, the provenance section of the database has information on more than 131,000 Indian-origin accessions held in various genebanks across the world. Genesys also integrates information on GRIN-Global and WIEWS databases. It also forms a core component of GLIS (Box 13.2). The database can be accessed at www.genesys-pgr.org.

13.3.2 Germplasm Resource Information Network: Global Version (GRIN-Global)

GRIN-Global is a relational database platform that has been developed as a genebank information management tool, through the joint effort by the Global Crop Diversity Trust (GCDT) (details given in Box 13.1), Bioversity International and the Agricultural Research Service of the USDA. It is an upscaled version of GRIN software that was developed earlier by the NPGS of the USA. GRIN global uses open-source software that has all the components for genebank inventory management. It has many fields between data tables and there is interoperability between genebank and genomic databases due to the alignment of fields with BrAPI (Breeding API). The database also includes images of the documented germplasm accessions. The GRIN–NPGS database (the GRIN-Global database used by ARS-NPGS) displays records of 621,123 germplasm accessions maintained as active collections at its 30 PGR centres across the USA. Since 2011, GRIN-Global has been available to all genebanks free of cost and instead of developing individual systems, genebanks can modify this software to meet their specific requirements and implement it in their respective genebanks. The database can be accessed at www.grin-global.org.

Box 13.1 The Crop Trust (Formerly The Global Crop Diversity Trust)
The Crop Trust was founded in 2004 by the Food and Agriculture Organization of the United Nations (FAO) and Bioversity International on behalf of the CGIAR. Its functions are:

- Providing long-term funding support for genebanks.
- Providing financial and technical support to conservation and food security projects.
- Collaborating for the development of Information Systems like GRIN-Global and Genesys.
- Compilation and documentation of Global Conservation Strategies.
- Collaborating for management of the safety duplication site—Svalbard Global Seed Vault.

13.3.3 The European Search Catalogue for Plant Genetic Resources

The European Search Catalogue for Plant Genetic Resources, EURISCO is a web catalogue that has been hosted under the aegis of the Secretariat of the European Co-operative Programme for Plant Genetic Resources (ECPGR) in collaboration with the national focal points of its 43 member countries. The catalogue brings together information on their ex situ conservation status and includes both passport and characterisation data. During the initial years of its inception (2003–2014), EURISCO was the responsibility of Bioversity International, Rome, Italy. Subsequently, it has been maintained by the Leibniz Institute of Plant Genetics and Crop Plant Research (IPK), Gatersleben, Germany. EURISCO currently depicts a total of 2,092,387 accessions from 413 institutes that comprise 45,128 species. The largest proportion of the collection is represented by *Arabidopsis* (33%) followed by wheat, barley and maize. Around 20% of the EURISCO accessions are available for exchange under the multilateral system of the Plant Treaty. The database can be accessed at: www.eurisco.ipk-gatersleben.de

13.3.4 Global Biodiversity Information Facility (GBIF)

GBIF is an open access platform representing museums, collections, herbaria, biodiversity conservation groups etc. The secretariat of GBIF is situated in Copenhagen, Denmark. GBIF was launched after a 1999 recommendation of the Biodiversity Informatics subgroup of the Organisation for Economic Cooperation and Development (OECD), for the establishment of a 'Global Biodiversity Information Facility' that would make biodiversity information more globally accessible.

The participants of this inter-governmental initiative sign a GBIF Memorandum of Understanding whereby they agree to the open access norms of sharing biodiversity data. A common data standard, like the Darwin Core, forms the basis for the huge record of species available in this database. The Darwin Core Standard (DwC) is a flexible community-developed standard framework for the compilation of biodiversity data. This was originally developed by the Biodiversity Information Standards Community, a non-profit organisation that was historically known as Taxonomic Databases Working Group (TDWG). This standard enables the use of common terminologies by a wide range of data users and providers and also simplifies the process of publishing biodiversity datasets. The database can be accessed at: www.gbif.org

13.3.5 World Information and Early Warning System on Plant Genetic Resources for Food and Agriculture (WIEWS)

It is an FAO database that hosts information on conservation status in various countries and it comes under the purview of the Commission on Genetic Resources for Food and Agriculture (CGRFA) of the FAO (details given in Box 13.3). Currently, the database has information on 5.9 million accessions from over 7000 genera conserved in over 870 genebanks from 115 countries and also 17 international/regional centres. WIEWS is the information system from which FAO derives data for the preparation of its country reports and global conservation/utilisation status reports. Member countries send annual reports to the nodal point at FAO and every 5 years, extensive reports are uploaded by the countries directly on the WIEWS portal. Through WIEWS, FAO monitors the implementation of (a) the plant component of indicator 2.5.1 of the Sustainable Development Goal (SDG) 2, Zero Hunger. The UN Sustainable Development Goal 2 (SDG 2) recognises that genebanks are critical to the achievement of global food security; Target 2.5 calls for the maintenance of genetic diversity of seeds through soundly managed and diversified seed and plant banks at national, regional and international levels. Through WIEWS, FAO also monitors the (b) Second Global Plan of Action for PGRFA (adopted in 2011) where it monitors its 18 priority activities that are grouped in four main areas, viz. (1) In situ conservation and management, (2) Ex situ conservation, (3) Sustainable Use, and (4) Building sustainable institutional and human capacities. The database can be accessed at: www.fao.org/wiews

Box 13.2 Global Information System (GLIS) for Plant Genetic Resources for Food and Agriculture

It is an integration of all global data on conservation, management and utilisation of PGR. It was conceived and instituted by the FAO under Article 17 of the Plant Treaty, which states that 'the contracting parties shall co-operate to develop and strengthen a global information system to facilitate the exchange of information, based on existing information systems'. GLIS has collated information that is sourced from Genesys, SADC (Southern African Development Commission) PGR centre, DataCite (a global non-profit organisation that provides DOIs for research data), EURISCO, GRIN-Global, WIEWS, SBDM (Statistics, Bioinformatics, Data Management) of ICRISAT and FAO Information Technology Services Division. It is continuously expanding its list of partners to include the maximum possible datasets about PGR.

Box 13.3 Commission on Genetic Resources for Food and Agriculture (CGRFA)

The CGRFA is a permanent inter-governmental body that works under the aegis of FAO and specifically addresses the components of biological diversity for food and agriculture. The commission coordinates global and cross-sectoral assessments of genetic resources that are done in a participatory manner, based on status/implementation reports submitted by member countries. Subsequent to these global assessments, the Commission identifies gaps and challenges that are then collated to conceive a Global Plan of Action that governments are encouraged to adopt and integrate in their policy decisions. The implementation reports also serve as a monitoring tool for member countries with regard to various policy instruments that are required to ensure the conservation and sustainable use of all genetic resources. The outcome of the implementation is reviewed and compiled through a global assessment once in every ten years. In India, information on each component of the genetic resources sector (plant, animal, insect, fish, microbes) is provided by the respective National Bureau. The membership of CGRFA is open to all FAO members and currently, 179 countries and the European Union are the listed members.

13.4 Challenges Faced by Genebanks

Extensive efforts for germplasm conservation have been carried out during the last five decades due to which huge quantities of germplasm have been acquired and conserved in almost all national and international genebanks. This abundance of ex situ collection as well as the skewed nature of the collection has led to several challenges in its management.

1. **Uneven representation of genetic diversity in ex situ collections:** It has been reported that about 50% of global ex situ germplasm consists of only 10 crop species. Rice, wheat and barley alone cover 28% of total ex situ conserved germplasm in the world (Kilian et al. 2012). Similarly, CWRs and landraces, which are the most unique sources for novel alleles, are hugely under-represented. Landraces constitute only one-third of the global collections, whereas CWRs account for only 15%. Data shows that the major components in all genebanks are breeding lines of agriculturally important crops. Minor crops and underutilised species do not have any significant presence.
2. **Redundancy in ex situ collections:** Multiple exploration missions are being undertaken under different programmes and in many cases lack of expertise with regard to multiple crops causes errors of over-estimation of diversity (Agrobiodiversity [PGR] 2000). In many cases, the maintenance of crops under open pollination leads to variations in phenotype, which cannot be attributed to inherent genetic variation. This skewed estimation made at the stage of collection, subsequently leads to the accumulation of genetically redundant accessions in ex situ repositories. Thus, over several years, a large number of collections have been made from each geographical location, without accounting for their genetic similarities. It has been estimated that out of 2.5 million accessions belonging to 37 different crops, which were surveyed from different genebanks for genetic distinctness, only 35% of the stored accessions were reported as unique (Plucknett et al. 1987). Such redundant collections become a huge liability for genebanks and add to the conservation cost, without contributing to diversity. For genetic and economic reasons, the identification of redundant accessions

has been included as a component of Good Genebank Practices (IPGRI 2002), for enhancing genebank efficiency. One of the primary benefits of rationalisation is that it will facilitate the effective development of representative reference sets in genebanks. The IPGRI Genebank management guidelines recommend optimisation of such collection through bulking (Engels and Visser 2003) and several global genebanks have attempted this in selected crops. Genebank curators must examine and analyse each accession in the context of the following question: 'Are accessions sufficiently different to justify managing them as distinct entities?' Only those that are significantly distinct should be taken ahead for conservation (IPGRI 2002).

3. **Breeding bottlenecks in utilisation:** Modern crop breeding has made consistent genetic improvements in commercial grain yield. However, the entire crop improvement programme, especially in cross-pollinated crops, relies on a limited number of elite lines, thus resulting in a very narrow genetic base. The lack of genetic diversity makes the crop genetically vulnerable to the adversities of climate change and to the several newly emerging insect-pests and diseases. Landraces traditionally cultivated by the farmers are regarded as the 'gold mine' for the novel traits. However, very few landraces have been utilised in the breeding programme. The low utilisation has been linked to the time-consuming pre-breeding work involved in characterisation, evaluation, trait identification and inbred development from these highly heterozygous and heterogeneous populations. Breeders have to develop inbred lines by repeated selfing for six to seven generations, which is time-consuming. Further, intense inbreeding depression during inbred development from landraces leads to the loss of a large proportion of lines. The genetic load due to the recessive lethal alleles creates a major bottleneck in retrieving the desirable output. It should be noted that the hybrid development programme of our country has not yielded a single hybrid possessing traits that have been sourced from our indigenous landraces. The genebanks should work towards the development of trait-validated inbred lines and their subsequent high-throughput genotyping to ensure effective utilisation of these resources. The set of inbred lines can serve as a complementary conservation unit, which would functionally represent its parental landrace, for specific traits. Crop breeders can directly utilise these resources in their crop improvement programmes without the burden of large-scale evaluations that would have extended for several cropping seasons. This has been attempted in the case of lettuce landraces, as reported by Hamilton et al. (2002).
4. **Loss of genetic integrity in ex situ conserved germplasm:** The seed quantity required for ex situ conservation in genebanks has been standardised as per the probability of sampling all potentially useful alleles that may exist even at very low allelic frequencies. While fixing the sample size, due consideration has been given to the fact that germplasm populations do not generally have a genetic architecture in steady-state Hardy-Weinberg frequencies. However, during the plant exploration missions, since each germplasm collection is done from diverse geographical locations, it is not possible to capture the original phenotypic variance in its entirety to ensure ecological continuum. Our limited understanding of

complex traits, many of which are polygenic and highly influenced by environment, further compromises the desirable allelic frequency. There are also certain traits, for example, root architecture, which are highly relevant for climate resilience but cannot be considered during collection missions, due to practical reasons. It should be noted that it is impossible to anticipate the many possible traits that each accession might be utilised for, in future. In addition to all these causes of allelic loss at the point of acquisition, ex situ conservation has a prominent disadvantage that it puts an abrupt end to the natural process of genetic variation that prevailed through gene flow among landraces and also the local adaptation that occurs through continued cultivation.

Subsequent to conservation in genebanks, another cause for major loss of genetic integrity is the regeneration procedure, which leads to genetic drift and inbreeding.

13.5 Future Strategies for Efficient Genebanking

The recent developments in the field of PGR management offer great scope for effectively addressing the various challenges faced by genebanks. Though core development was a promising approach for the management and utilisation of huge genebank collections, even four decades after the introduction of this concept there is no suggestive evidence of this approach being the theoretically ideal solution for enhancing genebank utilisation. A major concern is the possibility of losing rare and adaptive alleles that may be crucial for the expression of functional traits. Meanwhile, parallel developments in the field of genomics have taken ahead the germplasm utilisation strategy in multiple dimensions. One example is the study conducted by the Chinese Academy of Agricultural Sciences, BGI-Shenzhen and International Rice Research Institute, where they sequenced over 3000 rice genomes (3 K-RG) (Wang et al. 2018) from accessions conserved in ex situ genebank and the study could successfully establish a core-gene, which is the gene family that is present in all rice accessions (a genome parallel to the germplasm core and mini-core concept).

Another promising development is the successful combination of Focused Identification of Germplasm Strategy (FIGS) and core set strategies, wherein the discovery of rare traits was ensured through the use of modern phenotyping and high throughput genotyping approaches and these adaptive traits were confirmed to be preserved within the core set (Haupt and Schmid 2020). Advances in machine learning (ML) and synthesis of ML algorithms that can enable the capturing of a wide array of variations during phenotyping are the futuristic tools that can add precision to such techniques.

Currently, high throughput genotyping technologies are now being extended for de novo genome assembly of several CWRs, which will facilitate the application of advanced genomic tools that can carry out genetic dissection of their complex traits and thereby ensure their effective utilisation. Genome editing is another biotechnological tool that holds great potential in enhancing utilisation of the genebank CWRs, for crop improvement.

13.6 Summary

Genebanks have the long-term mission of conserving plant genetic resources as an agricultural legacy for future food, nutritional and health security. Genebanks also cater to the requirements of crop improvement programmes by contributing novel genes for the development of resilient cultivars. The first genebank was established in 1894 by N.I. Vavilov and currently, there are more than 2000 genebanks spread globally. NPGS under USDA-ARS is the largest genebank with a network of 20 genebanks in different states of the USA. Other prominent global genebanks are the IPK (Germany), CGIAR genebanks functioning in different countries, AGG (Victoria, Australia), NARO (Japan), RBG (Kew, UK), RDA (Korea), CAAS (China) and NGB (India). The SGSV (Norway) is a unique facility established in 2008 to comply with the various international treaties for the safe conservation of germplasm at a duplicate site. One hundred and two genebanks of the world have contributed their accessions for safe duplicate conservation at SGSV. The global genebank system is ably supported by multiple database platforms like GRIN-Global, Genesys, EURISCO and FAO-WIEWS. However, the genebanks are also facing several challenges, especially from the abundance of ex situ collections and redundancy issues, which is creating a major bottleneck for its effective utilisation. However, successful attempts have been recently made by researchers, to develop protocols and procedures using developments in the field of genomics, to address these pertinent issues.

Questions

1. Who gave the concept of gene pool?
2. Expand the given acronyms:
   (a) CGIAR
   (b) CIMMYT
   (c) ITPGRFA
   (d) GCDT
   (e) GRIN-Global
   (f) WIEWS
3. Give the full name of the listed genebanks:
   (a) SGSV, Norway
   (b) NARO, Japan
   (c) JIC, Norwich
   (d) IPK Germany
   (e) NGB, India
4. Which of the following is not an example of ex situ conservation?
   (a) Herbal Garden
   (b) Gene Sanctuary
   (c) Seed Genebank
   (d) Cryo-Genebank

5. Define the term 'genebank'.
6. What are the major challenges faced during utilisation of genebank material?
7. What is the major mandate of Crop Trust?

Suggested Reading

Dzyubenko NI (2018) Vavilov's Collection of Worldwide Crop Genetic Resources in the twenty-first Century. Biopreserv Biobank 16(5):377–383. https://doi.org/10.1089/bio.2018.0045

Engels JMM, Visser L (eds) (2003) A guide to effective management of germplasm collections. IPGRI, Rome

FAO (1998) The state of the world's plant genetic resources for food and agriculture. Food and Agriculture Organization of the United Nations, Rome

FAO (2010) The second report on the state of the world's plant genetic resources for food and agriculture. Food and Agriculture Organization of the United Nations, Rome

Frankel OH, Bennett E (eds) (1970) Genetic resources in plants-their exploration and conservation. Blackwell, Oxford

Hamilton NRS, Engels JM, Van Hintum TJ (2002) Accession management: combining or splitting accessions as a tool to improve germplasm management efficiency (No. 5). Bioversity International

Harlan JR, Frankel OH, Hawkes JG (1975) Seed crops. Practical problems in exploration. In: Frankel OH, Hawkes JG (eds) Crop genetic resources for today and tomorrow. Cambridge University Press, Cambridge, pp 111–115

Haupt M, Schmid K (2020) Combining focused identification of germplasm and core collection strategies to identify genebank accessions for central European soybean breeding. Plant Cell Environ 43:1421–1436

IPGRI (2002) In: Engels JMM, Ramanatha Rao V, Brown AHD, Jackson MT (eds) Managing plant genetic diversity, Rome

Khoury CB, Laliberte, Guarino L (2010) Trends in ex-situ conservation of plant genetic resources: a review of global crop and regional conservation strategies. Genetic Resour Crop Evol 57:625–639

Plucknett DL, Smith NJH, Williams JT, Anishetty NM (1987) Genebanks and the world's food. Princeton University Press, Princeton

Wang W, Mauleon R, Hu Z et al (2018) Genomic variation in 3010 diverse accessions of Asian cultivated rice. Nature 557:43–49

Remote Sensing and Geographic Information System (GIS) in PGR Management: Concepts and Applications

14

Debasish Chakraborty, Prasenjit Ray,
Roomesh Kumar Jena, Dinesh Prasad Semwal, Bappa Das,
Pooja Rathore, Swayam Vid, and Debashis Chakraborty

Learning objectives

- Remote sensing and its role in plant genetic resource (PGR) management
- Geographic information system (GIS)
- Role of GIS in PGR management

Keywords

Plant genetic resources · Germplasm · PGR · Geographic information systems · Remote sensing · Grid mapping · Diversity analysis

The original version of the chapter has been revised. A correction to this chapter can be found at https://doi.org/10.1007/978-981-96-5024-8_18

D. Chakraborty
ICAR-Research Complex for NEH Region, Umiam, Meghalaya, India

P. Ray (✉)
Division of Soil Science and Agricultural Chemistry, ICAR-IARI, New Delhi, India

R. K. Jena
ICAR-Indian Institute of Water Management, Bhubaneswar, Odisha, India

D. P. Semwal
ICAR- National Bureau of Plant Genetic Resources, New Delhi, India

B. Das
ICAR-Central Coastal Agricultural Research Institute, Old Goa, Goa, India

P. Rathore · S. Vid
ICAR-IARI, New Delhi, India

D. Chakraborty
CIMMYT, New Delhi, India

K. Tripathi et al. (eds.), *Textbook of Plant Genetic Resources*,
https://doi.org/10.1007/978-981-96-5024-8_14

14.1 Introduction

The effective management of plant genetic resources (PGRs) stands as a linchpin in the quest for global food security, agricultural sustainability, and ecological resilience. To navigate the vast and intricate landscapes where these resources reside, modern technology has gifted us with invaluable tools: remote sensing (RS), global positioning system (GPS) and geographic information system (GIS). These technological marvels have transformed the landscape of PGR management, offering unprecedented insights into the exploration, conservation, and utilization of these vital resources. Remote sensing, with its ability to capture detailed imagery of vegetation distribution and health, serves as our eyes in the sky. Through spectral signatures and vegetation indices, it unveils hidden pockets of genetic diversity, guiding our exploration efforts with unparalleled precision. By leveraging this spatial data, we can orchestrate targeted collection missions, ensuring a comprehensive representation of genetic diversity within our genebanks.

Meanwhile, GIS emerges as a powerful set of tools for conservation endeavors owing to voluminous spatial data on habitat suitability, threats like deforestation, and protected areas. Through habitat suitability models that integrate GIS data with climate and soil variables, we can predict species ranges and identify vulnerable populations, enabling proactive conservation strategies to safeguard PGRs for posterity. But GIS does not merely stop at conservation—it serves as a nexus for knowledge exchange and collaboration. By bridging the gap between data and knowledge, GIS facilitates the establishment of comprehensive online databases and interactive platforms. These repositories house a treasure trove of PGR information, including passport data, genetic characteristics, and geographical distribution, fostering collaboration across borders and research institutions.

The utilization of PGRs for sustainable agriculture demands integration between genotype and environment, a challenge addressed by GIS-generated bioclimatic maps. By integrating climate models and soil data, these maps predict growth potential for different species under diverse ecological conditions. Using this knowledge, researchers can select PGRs tailored to specific landscapes, optimizing crop yields while minimizing environmental impact and enhancing resilience to climate change. In the rich tapestry of India's agricultural heritage, GIS plays a crucial role in analyzing gaps in ex situ collections and charting future conservation strategies. Through gap analysis studies, facilitated by GIS tools, researchers identify unexplored diversity-rich areas and prioritize collection efforts to mitigate the risk of genetic erosion. Through diversity distribution mapping, which integrates remote sensing, global positioning system (GPS), and GIS, gap areas for future explorations are identified ensuring the preservation of India's agricultural diversity.

In essence, remote sensing and GIS stand as beacons of hope in the realm of PGR management. They empower us to navigate the complexities of our natural world, safeguarding genetic diversity and promoting sustainable agriculture on a global scale. With each technological advancement and collaborative effort, we may achieve a sustainable future where food security, environmental resilience, and the vibrant tapestry of plant life are intertwined in harmony.

14.2 Remote Sensing

Remote sensing is a powerful scientific technique of gathering information about an object or area without making physical contact with the object. It involves the detection and interpretation of information about an object or phenomenon from a distance, typically using sensors onboard satellites, aircraft, drones, handheld devices, etc. It plays a crucial role in various fields such as environmental monitoring, agriculture, biodiversity, plant genetic resources mapping, disaster management, and natural resource exploration.

14.2.1 Principles of Remote Sensing

In remote sensing, sunlight is the source of energy and electromagnetic radiation (EMR) is the medium of interaction with the Earth's objects. The sun's energy reaches the Earth and some parts of them are absorbed, reflected, transmitted, or emitted by the object/target, and reflected energy is recorded by the satellite sensor and stored in digital form for further image analysis. Remote sensing represents a technology for synoptic acquisition of spatial data and the extraction of specific/selected area information.

Detection and discrimination of Earth objects or surface area means detecting and recording radiant energy reflected or emitted by objects. Every object reflects different amount of energy in different wavelength range (bands) of the electromagnetic spectrum, incident upon it.

14.2.1.1 Electromagnetic Energy

Remote sensing relies on the principle that objects emit, reflect, or transmit electromagnetic energy/radiation (EMR) across the electromagnetic spectrum. This energy interacts with the Earth's surface and atmosphere, providing valuable information for analysis.

14.2.1.2 Energy Matter Interaction

Remote sensing sensors measure the reflected, absorbed, or emitted EMR from the target object. Understanding these interactions, governed by principles like absorption spectroscopy and scattering theory, is crucial for interpreting the collected data.

14.2.1.3 Energy Detection

Sensors capture and measure the energy reflected or emitted by the Earth's surface. Different sensors are designed to detect specific wavelengths or bands of the electromagnetic spectrum, ranging from visible light to microwaves and beyond.

14.2.1.4 Data Processing and Analysis

Collected remote sensing data goes through various processing steps, including geometric correction, radiometric correction, and atmospheric correction, to adjust for distortions and ensure accurate representation of the target object. Extracting meaningful information from the processed data involves specialized analysis techniques like image classification, spectral signature analysis, and change detection.

14.2.2 Resolution

Remote sensing sensor captures information at different resolutions that are crucial for their applications.

14.2.2.1 Spatial Resolution

Spatial resolution refers to the smallest object that can be distinguished, measured in length (meters or centimeters) or ground sampling distance (GSD) by satellite sensors. High-spatial-resolution images offer detailed views, while low-resolution images cover larger areas.

14.2.2.2 Spectral Resolution

Spectral resolution refers to the range of EMR wavelengths a sensor can detect. Multispectral sensors capture data in a few selected bands, while hyperspectral sensors gather data across a continuous spectrum, enabling detailed spectral analysis.

14.2.2.3 Temporal Resolution

Temporal resolution refers to the frequency of data acquisition, measured in days, weeks, or months. Frequent revisit times allow for monitoring dynamic processes, while infrequent observations provide long-term trends.

14.2.2.4 Radiometric Resolution

The information content in an image is determined by the number of digital levels (quantization levels) used to express the data collected by the satellite sensors. The smallest change in intensity level that can be detected by a sensing system is known as radiometric resolution.

14.2.3 Electromagnetic Energy Bands Used for Remote Sensing

The electromagnetic spectrum, energy waves ranging from radio waves to gamma rays, forms the lifeblood of remote sensing. Each region within this spectrum offers unique insights into the Earth's surface and atmosphere, leading to a diverse world of remote sensing applications.

14.2.3.1 Visible and Near-Infrared (VNIR)

This familiar range, most of it perceptible to our eyes (visible, i.e., 0.4–0.7 μm), reveals the beauty of landscapes, differentiates vegetation types, and highlights human-made structures. VNIR wavelength ranges from 0.4 to 1.0 μm. VNIR sensors are commonly used in satellite and aerial imagery, providing high-resolution details essential for various applications. This region is mainly used for vegetation studies, land cover mapping, urban planning, aerial photography, color analysis, etc.

14.2.3.2 Shortwave Infrared (SWIR)

SWIR penetrates beyond the visible spectrum, revealing information hidden from our eyes. SWIR wavelength ranges from 1.0 to 2.5 μm. It allows us to identify different rock types, assess soil moisture content, and detect subtle changes in vegetation health, critical for mineral exploration, agriculture, and drought monitoring. The predominant use is in mineral exploration, soil moisture estimation, vegetation stress detection, water resource management, etc.

14.2.3.3 Middle Infrared (MIR)

The MIR region delves into the realm of thermal energy. MIR ranges from 2.5 to 5 μm. It allows us to measure surface temperature variations, detect active volcanoes, identify fires, and even monitor the concentration of greenhouse gases like methane and carbon dioxide, crucial for climate change studies.

14.2.3.4 Longwave Infrared (LWIR)

LWIR penetrates clouds and fog, providing vision in darkness. LWIR ranges from 8 to 14 μm. It facilitates night-time imaging for security purposes, allows us to study cloud properties and track weather patterns, and even enables thermal remote sensing of oceans and polar regions. It has applications in night-time imaging, cloud cover assessment, weather forecasting, security and surveillance, etc.

14.2.3.5 Microwave

Microwaves offer all-weather capabilities, penetrating clouds and foliage. The wavelength ranges from 1 mm to 1 m. Radar sensors generate highly detailed images, useful for mapping topography, monitoring ocean currents, estimating forest biomass, and measuring soil moisture content. It is used in radar imaging, sea surface topography, biomass mapping, moisture estimation, etc.

14.2.3.6 Radio Wave

Radio waves have the longest wavelengths and can penetrate deep. Its wavelength ranges from 1 m to 1 km. They enable precise measurement of surface elevation, map sea ice extent, explore geological structures, and even probe the atmosphere to retrieve information about temperature, pressure, and humidity. It is applied in radar altimetry, sea ice mapping, subsurface exploration, atmospheric sounding, etc.

14.2.4 Remote Sensing Based on Energy Source

Based on the source of electromagnetic energy, remote sensing can be categorized into two key types: active and passive remote sensing.

14.2.4.1 Active Remote Sensing

These systems emit the energy, typically in the form of electromagnetic waves, and then measure the reflected or backscattered energy from the target object.

Types of Active Sensors

Radar: Sends out radio waves and analyzes the reflection to create detailed images of terrain, sea surface, vegetation or any object, etc. Ideal for all-weather applications and penetrating vegetation cover.

LiDAR (Light Detection and Ranging): Emits laser pulses and measures the reflected time to accurately determine precise distances and elevations. Perfect for creating three-dimensional (3D) terrain models and studying landforms.

Sonar: Uses sound waves, particularly underwater, to map the seafloor or detect submerged objects. Valuable for oceanographic studies and shipwreck exploration.

Advantages

- Works in any weather and lighting conditions: Not reliant on natural sunlight, giving flexibility for day/night and even cloud-covered situations.
- High spatial resolution: Sensor outputs allow for detailed and precise measurements of features and distances.
- Direct control over signal properties: Ability to adjust wavelength and power based on specific needs and target characteristics.

Disadvantages

- Higher cost and complexity: Sensors and systems may be expensive and require specialized expertise to operate and interpret data.
- Limited penetration depth: Certain wavelengths may not penetrate deeply into vegetation or ground, restricting information access.

14.2.4.2 Passive Remote Sensing

Passive remote sensing relies on existing naturally occurring energy like sunlight to measure the interaction with the target object.

Types of Passive Sensors

Multispectral and Hyperspectral Sensors: Capture reflected sunlight in specific wavelength bands, revealing information about vegetation health, mineral composition, and other physical properties. Useful for environmental monitoring and resource exploration.

Thermal Sensors: Detect infrared radiation emitted by objects, providing temperature maps and helping identify hot spots (volcanoes, fires) or assess thermal properties of materials.

Radiometers: Measure the total incoming or outgoing radiation within a specific wavelength range, offering insights into atmospheric properties and energy fluxes.

Advantages

- Lower cost and simpler systems: Often more affordable and easier to operate compared to active sensors.
- No environmental impact: Relying on natural energy ensures minimal ecological disturbance.
- Provides complementary information: Can reveal features invisible to active sensors, offering a broader picture of the target.

Disadvantages

- Reliant on external light sources: Limited to daylight hours and can be affected by cloud cover or atmospheric conditions.
- Lower spatial resolution: Data may be less detailed compared to active sensors, with limitations in identifying small features.
- Indirect interpretation of signal: Analyzing the interaction with natural energy often requires advanced modeling and data processing techniques.

14.2.5 Sensor Platforms for Remote Sensing

The platform from which sensors collect data significantly impacts the type of information we can get and the applications we can tackle. There are three main categories of remote sensing platforms:

14.2.5.1 Ground-Based Platforms

Advantages

- High spatial resolution, allowing for detailed inspections of specific areas.
- Flexible deployment and targeting, enabling close-up study of specific features.
- Lower cost compared to air and space-based systems.

Limitations

- Limited coverage area, restricting studies to localized areas.
- Weather and accessibility can hinder deployment and data collection.
- May not provide a synoptic view of larger landscapes.

Applications: Precision agriculture, soil analysis, archaeological surveys, environmental monitoring of specific sites.

14.2.5.2 Airborne Platforms

Advantages

- Larger coverage area compared to ground-based platforms.
- Higher spatial resolution than satellites, offering detailed observation of features.
- More flexible deployment and targeting than satellites, allowing for quick response to events.

Limitations

- Higher cost compared to ground-based systems and can be weather-dependent.
- Flight altitude limitations impact spatial resolution compared to satellites.
- Operational range is limited, requiring refueling from base locations.

Applications: Aerial photography, land cover mapping, disaster assessment, forest monitoring, wildlife tracking.

14.2.5.3 Spaceborne Platforms

Advantages

- Global coverage, providing near-real-time data for vast areas.
- Repeatable observations, enabling monitoring of long-term changes and trends.
- Ability to capture synoptic views and track large-scale phenomena.

Limitations

- Lower spatial resolution compared to ground and air-based platforms, limiting detail in observations.
- Long revisit times for specific locations, potentially missing rapid changes.
- High cost and complexity of satellite development and launch.

Applications: Weather forecasting, climate change monitoring, global land cover mapping, oceanographic studies, resource exploration.

14.2.6 Pre-Processing and Image Enhancement in Remote Sensing

Before the stunning visuals and insightful data hidden within remote sensing images can be fully extracted, they undergo a critical metamorphosis: pre-processing and image enhancement. These crucial steps remove unwanted noise, sharpen features, and boost information extraction, paving the way for accurate analysis and interpretation.

14.2.6.1 Pre-Processing

Geometric Correction: Rectifies distortions due to sensor viewing angle, terrain elevation, and Earth's curvature. Ensures accurate spatial alignment between different images and real-world locations.

Radiometric Correction: Adjusts pixel values for variations in sensor sensitivity, illumination, and atmospheric effects. Guarantees consistent brightness and reflectance values across an image and between different dates.

Atmospheric Correction: Compensates for the blurring and absorption of light by the atmosphere, crucial for analyzing scenes with haze, clouds, or smoke. Unveils true surface reflectance for accurate quantitative analysis.

14.2.6.2 Image Enhancement

Contrast Stretching: Expands the range of pixel values, making subtle features more visible. Ideal for highlighting boundaries, vegetation changes, and landforms.

Edge Detection: Identifies and enhances sharp transitions between different features, aiding in object recognition and delineation. Useful for mapping linear features like roads, rivers, and geological structures.

Filtering: Removes unwanted noise and artifacts, such as sensor noise, striping, and atmospheric disturbances. Improves image clarity and facilitates accurate information extraction.

Color Manipulation/Density Slicing: Emphasizes specific color bands or indices to reveal hidden information. For example, highlighting vegetation health through normalized difference vegetation index (NDVI) or identifying mineral deposits using spectral signatures.

Advantages

- Improved visual appearance and interpretability.
- More accurate quantitative analysis and data extraction.
- Enables comparison of images collected at different times and with different sensors.
- Facilitates specific applications focused on particular features or information.

Limitations

- Over-processing can introduce artifacts and distort information.
- Requires knowledge of the appropriate technique and parameters for specific data and applications.
- Some techniques may not be effective for all types of noise or image characteristics.

14.2.7 Types of Remote Sensing Imageries

In the case of remote-sensing imagery, the use of color is very important. But not all colors are created equal. Different techniques manipulate and combine spectral data to unveil unique perspectives on our planet.

14.2.7.1 Panchromatic Images (PAN)

This technique combines information from all visible wavelengths (approximately 0.4–0.7 μm) into a single grayscale image.

Advantages: High spatial resolution, often exceeding that of color images from the same sensor. Simple and efficient data storage and transmission.

Limitations: Lacks spectral information, limiting identification of different materials and features. Requires additional processing for colorization or spectral analysis.

Applications: Base layers for geographic information system (GIS), terrain analysis, change detection, aerial photography, and resource exploration.

14.2.7.2 True Color Composites (TCC)

This approach combines red, green, and blue bands from the visible spectrum, mimicking the colors we perceive with our eyes.

Advantages: Familiar and intuitive for human interpretation. Useful for land cover mapping, urban planning, vegetation studies, and visual reconnaissance.

Limitations: Limited spectral information compared to other techniques. May not differentiate features with similar colors in the visible spectrum.

Applications: Aerial photography, visual assessment of landscapes, public outreach and education, basic land cover mapping, disaster assessment.

14.2.7.3 False Color Composites (FCC)

In this method wavelength bands from different regions of the electromagnetic spectrum are combined, often including invisible wavelengths like near-infrared (NIR). The most common FCC is where NIR is treated as a red band, red is treated as a green band, and green is treated as a blue band, and in the resulting image, the vegetations look red.

Advantages: Highlights specific features based on their unique spectral signatures. Enables identification of vegetation health, mineral composition, soil moisture, and water quality.

Limitations: Unfamiliar color combinations require training for interpretation. May not be visually intuitive for beginners.

Applications: Precision agriculture, mineral exploration, environmental monitoring, water resource management, forest health assessment, geological studies.

14.2.8 Energy–Matter Interaction and Spectral Signatures

The world of remote sensing revolves around the intricate interaction between energy and matter. Each object on Earth interacts with electromagnetic radiation (EMR) uniquely, leaving behind a fingerprint called its spectral signature. By understanding these interactions and interpreting the resulting signatures, we can unlock a wealth of information about our planet's diverse landscapes.

14.2.8.1 Energy–Matter Interaction

EMR interacts with objects in various ways:

- Reflection: EMR bounces off the surface, returning toward the sensor. This forms the basis for most remote sensing techniques.
- Absorption: EMR is absorbed by the object, converting it into heat or other forms of energy.
- Transmission: EMR passes through the object, revealing information about its internal structure.

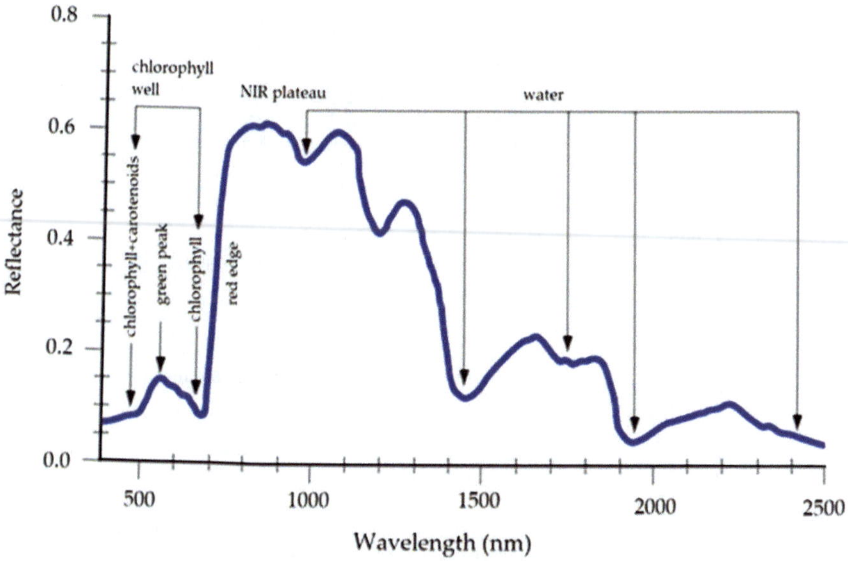

Fig. 14.1 Spectral signature of vegetation. (Source: Moroni et al. 2019)

14.2.8.2 Spectral Signatures of Different Objects

Different objects interact differently with the EMR, creating their unique spectral response, which is called spectral signatures. Each material has specific spectral signatures due to variations in their chemical composition and physical properties. These signatures play a key role in identifying and characterizing objects in remote sensing imagery.

Vegetation

The spectral signature of vegetation is mainly determined by the pigmentation, physiological structure, and leaf moisture content (Fig. 14.1). In the visible spectrum, the pigmentation affects the response with a high absorption in blue and red regions, leading to the green appearance. High reflectance in the NIR region is due to leaf structure, while in the SWIR region, the varying reflectance is based on water content.

Soil

The spectral signature of soil is mainly determined by the moisture content, particle size distribution, organic matter content, iron oxide content, soil structure, and soil mineralogy (Fig. 14.2). In the visible spectrum, the broad reflectance is influenced by overall color and mineral composition affects specific absorption bands, while in the SWIR region, strong absorption is due to the presence of water molecules.

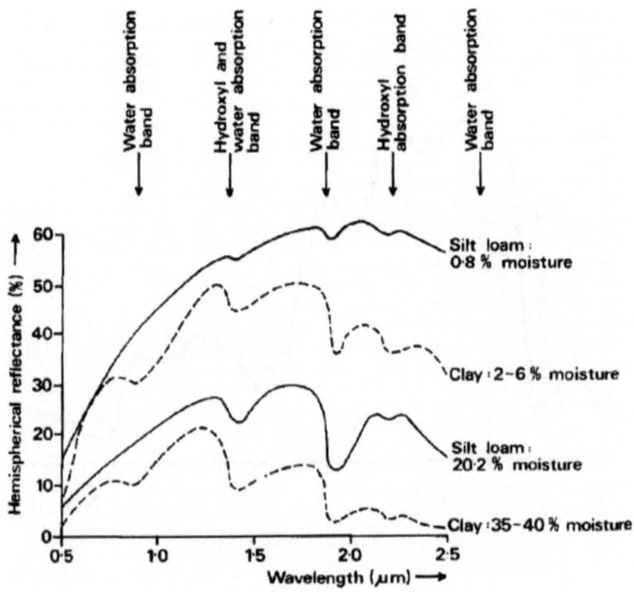

Fig. 14.2 Spectral signature of soils. (Source: Bowers and Hanks 1965)

Water

The spectral signature of water is mainly characterized by the depth of water, materials within the water, and surface roughness. In the visible wavelengths of electromagnetic radiation little light is absorbed, a small amount is reflected, and the majority is transmitted (Fig. 14.3). Water absorbs near-infrared and middle-infrared wavelengths strongly, leaving little radiation to be either reflected or transmitted.

14.2.9 Spectral Indices

The world of remote sensing is brimming with valuable tools, and spectral indices stand out as powerful instruments for extracting precise information from imagery. These mathematical formulas combine reflectance values from different spectral bands, highlighting specific features and unlocking hidden insights. Following are some of the spectral indices, focusing on some prominent examples and their applications (Table 14.1).

Advantages

- Quantitative analysis: Unlike visual interpretation, spectral indices offer numerical values for specific features, facilitating accurate measurements and comparisons.
- Enhanced features: Indices can amplify subtle differences in reflectance that the human eye might miss, highlighting specific features like vegetation stress or water content.

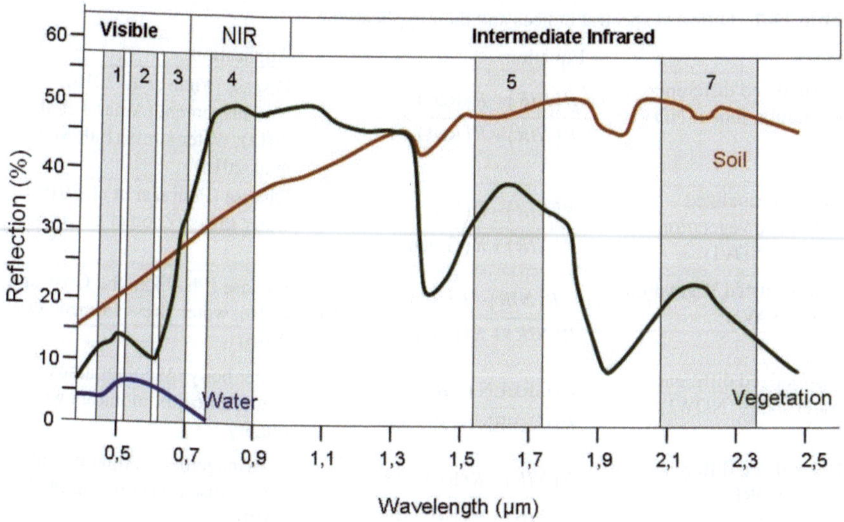

Fig. 14.3 Spectral signature of water, soil, and vegetation. (Source: https://seos-project.eu/remotesensing/remotesensing-c01-p06.html)

- Standardization: Many indices are widely used and have well-established interpretations, enabling comparison of data across different studies and sensors.
- Reduced sensitivity to external factors: Indices can minimize the influence of illumination and atmospheric effects, providing more consistent results.

Limitations

- Oversimplification: Spectral indices represent complex interactions between light and matter with a single value, potentially masking important details.
- Sensor dependence: The effectiveness of an index can vary depending on the specific spectral bands available on the sensor used.
- Calibration needed: Some indices require calibration for specific applications and vegetation types to interpret results accurately.
- Context matters: Interpretation of indices needs to consider the broader context of the landscape and other environmental factors.

14.3 Geographic Information System (GIS)

The geographic information system (GIS) is a tool to capture, store, query, analyze, and display geospatial data. GIS can be defined as software, which is capable of processing and presenting geographic information as maps. It is imperative to mention that there is no universally acceptable definition of GIS, and it is usually considered a decision-making tool for geospatial data and information. The ability of GIS to analyze geospatial data revolves around the descriptive attributes associated with the data. GIS allows the transformation of data through spatial operation and

Table 14.1 Different spectral indices and their applications

Index	Equation	Applications
Normalized difference vegetation index (NDVI)	$\frac{R(NIR)-R(\text{Red})}{R(NIR)+R(\text{Red})}$	Disease (Yuan et al. 2016); N-management (Amaral et al. 2015); Water stress (Ballester et al. 2018)
Green normalized difference vegetation index (GNDVI)	$\frac{R(NIR)-R(\text{Green})}{R(NIR)+R(\text{Green})}$	Disease (Pourazar et al. 2019); water stress (Zhou et al. 2018)
Soil Adjusted Vegetation index (SAVI)	$\frac{R(NIR)-R(\text{Red})}{R(NIR)+R(\text{Red})+L}(1+L)$	Disease (Phadikar and Goswami 2016); water stress (Marino et al. 2015)
Normalized difference water index (NDWI)	$\frac{R(\text{GREEN})-R(NIR)}{R(\text{GREEN})+R(NIR)}$	Water body mapping; assesses wetland health; monitors water quality
Normalized difference red edge (NDRE)	$\frac{R(NIR)-R(\text{Red Edge})}{R(NIR)+R(\text{Red Edge})}$	N-management (Amaral et al. 2015); disease (Pourazar et al. 2019)
Normalized pigment chlorophyll ratio index (NPCI)	$\frac{R680-R430}{R680+R430}$	Water stress (Klem et al. 2018)
Photochemical reflectance index (PRI)	$\frac{R531-R570}{R531+R570}$	Disease (Abdulridha et al. 2020); water stress (Ballester et al. 2018)
Water balance index	$\frac{R1500-R531}{R1500+R531}$	Water stress (Rapaport et al. 2017)

overlaying of different data layers for further integration and analysis to serve as a decision-making tool. To combine different types of information and attributes, it must be ensured that each layer is associated with the same geographic coordinate system (GCS). This way, each layer occupies the same space and can be superimposed on top of each other. It is also important to know that any layer in the GIS comprises the fundamental features, namely, points, lines, and polygons.

Points: A point is suitable for representing information where it is necessary to show the location of a feature, but its physical shape is not essential (e.g., the location of a botanical garden in a city).

Lines: A line can be used to represent linear features, such as rivers or roads.

Polygons: Polygons are shapes with multiple sides that represent a solid area. To accurately depict an area, we must draw a polygon, ensuring that all attributes associated with the record such as land use/land cover types are included within the boundary.

A geographic information system requires a few essential components to work with geospatial data. These are computer hardware, GIS software, data, skilled people or users, and methods or procedures. The GIS software is required to store, analyze, and display geospatial data and information as maps. Many GIS software are available for the users, and the use of such software is application specific. The commonly used software are ArcGIS, MapInfo GIS, and Geomatica. There are many open-source software available, namely, QGIS, DIVA-GIS, and SAGA-GIS. In view of the high-cost involvement in the licensed GIS software, the open-source software is gaining popularity and is equally effective in processing the geospatial data. GIS data can be grouped into geographic (spatial) and attribute (non-spatial) data. Spatial data has location information represented by coordinates, while attribute data includes descriptive information about a location. Important data sources include maps, remote sensing imagery, data from global positioning system (GPS), and other databases. To manage spatial data, most organizations use database management system (DBMS). People associated with GIS can be classified as viewers, general users, and GIS specialists. The latter category is responsible for maintaining the geographic database and providing technical support to the other two classes of users. The method or procedure is another important component, and the output of any GIS operation depends on the successful use of appropriate methods or procedures. The ability of GIS to perform spatial analysis depends on the techniques or methodology applied by the GIS specialist.

The use of GIS as a decision-making tool is well documented in diversified agricultural fields. These studies justify the fact that GIS is commonly viewed as a tool, which is primarily based on the application approach to GIS. However, the other two approaches to understand GIS include the developer approach and the technology approach. The developer approach emphasizes the innovation in the GIS depending on the user demand, and the technology approach to GIS involves the diffusion of GIS technology.

14.3.1 Map: Scale, Coordinate System, and Projection

Maps and mapping are integral components of any activities involving GIS. A map is considered as a representation of the features of interest of any place and its accurate geographic location, usually on a flat surface. In general, maps are of three types, namely, reference map, thematic map, and interactive map. The reference map represents the accurate geographic location and reality, whereas the thematic map provides information on the spatial distribution of the theme of interest. The interactive maps are dynamic and involve user interaction for a particular information. The map elements and features can easily be changed in interactive maps. The "Survey of India Topo-sheet" is an example of a reference map, whereas the spatial distribution map of deciduous forests in India is an example of a thematic map. Google maps of any road network or locality are examples of interactive maps, where the map scale can easily be changed by zooming in or zooming out the map on an electronic device.

The scale of a map can be described as the ratio of the distance on the map to the corresponding distance on the ground. Such a ratio is known as Representative Fraction (RF). A map with RF of 1:50,000 represents 1 unit on the map that corresponds to 50,000 units on the ground. The units could be in centimeters or any other units. It is important to keep in mind that GIS technology enables the analysis of different layers at different scales. However, it is crucial to use data that is at the appropriate scale to avoid drawing false conclusions in GIS analysis.

A coordinate system is a reference framework fixed onto an area's surface to indicate the position of its features. Geographic coordinates measure the positions of objects on the Earth's spherical surface. Although latitude and longitude can locate precise positions on the Earth's surface, they are not uniform units of measurement. One degree of longitude does not equal one degree of latitude, except along the equator. To overcome measurement difficulties, data are often transformed from three-dimensional geographic coordinates to two-dimensional projected coordinates. A geographic coordinate system (GCS) is a reference system that helps identify locations on the curved surface of the Earth. These locations are measured in angular units from the center of the Earth relative to two planes—the equator and the prime meridian. The prime meridian crosses Greenwich, England. Therefore, a location is defined by two values—a latitudinal value and a longitudinal value in GCS. Because of the spherical shape of the Earth, it is highly essential to define the orientation of latitudinal and longitudinal lines on the spheroidal Earth. A datum is useful in defining the orientation and origin of these lines concerning the center of the spheroidal Earth. A datum can be local (e.g., NAD83) or global (e.g., WGS84). A projected coordinate system (PCS) is a reference system used to identify locations and measure features on a flat (map) surface (Fig. 14.4). It consists of intersecting lines that form a grid at right angles. Projected coordinate systems are based on Cartesian coordinates and include an origin, an x-axis, a y-axis, and a linear unit of measure. To convert from a GCS to a PCS, mathematical transformations are required, which is known as map projection.

Map projection is defined as a system of transformation of the spherical (curved) surface ("The Globe") onto a plane (flat) ("The Map") surface. When we try to

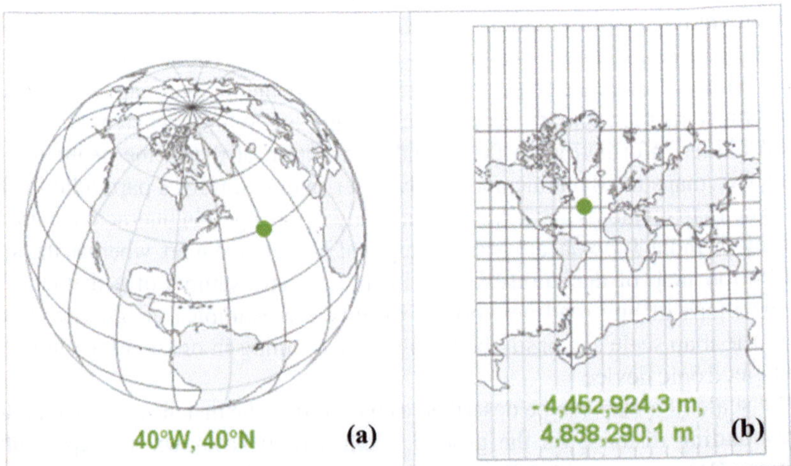

Fig. 14.4 Geographic Coordinate System (**a**) and Projected Coordinate System (**b**) (note the units of measurements). (Source: ESRI)

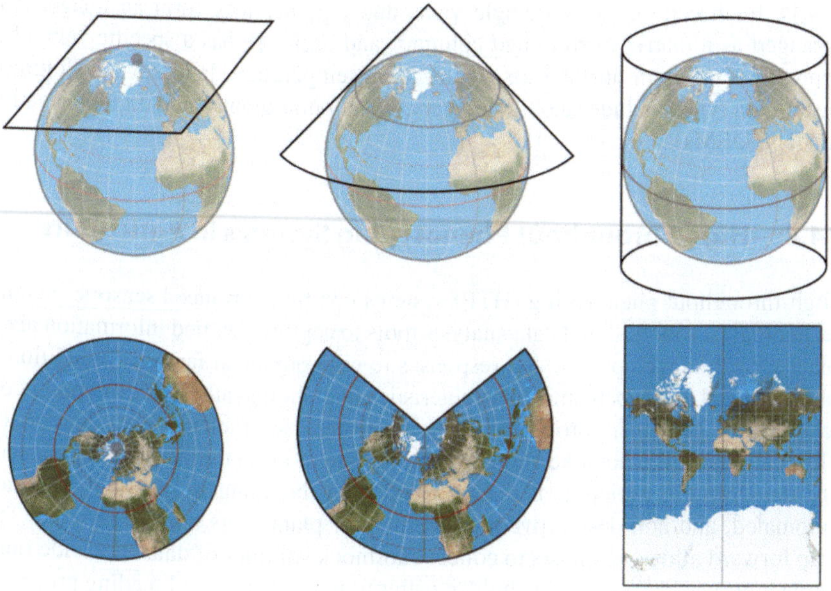

Fig. 14.5 Map projection surfaces

represent the round Earth on a flat surface, it causes distortions in some of its spatial properties, like (1) Distance (equidistant projection), for example, Azimuthal Equidistant Projection; (2) Area (equal area or homolographic projection), for example, Albers Equal Area Conic projection; (3) Shape (conformal or orthomorphic), for example, Mercator projections; and (4) Direction (Azimuthal projection), for example, Lamber Equal Area Azimuthal. The map projection is classified based on which of the above attributes it preserves. No projection can preserve all these properties at once. Therefore, every flat map has distortions to some extent. There are three major groups of map projection: Planar or Azimuthal (projection onto a plane), Cylindrical (projection onto a cylindrical surface, and then flattened into a plane), and Conic (projection onto a cone, and then flattened to a plane) (Fig. 14.5).

14.3.2 Data Models in GIS

The vector data model and raster data model represent the geographic data in the GIS platform. The vector data models represent the geographic information using points, lines, and polygons. The geographic features are represented by the vector data the same way the paper maps do. In the vector data model, the points are recorded as a single coordinate, while the lines are recorded as the series of x and y coordinates. Polygons are also represented by a series of x and y coordinates with an area. To define spatial relationships in vector data model, topological concepts are used. The raster data model represents spatial information as a grid of cells or pixels. Each cell contains a single value and together they form an image. It is arranged as a matrix of rows and columns, and each cell has a specific value that represents a certain attribute (e.g., elevation, temperature, land cover). Common raster data types include satellite imagery, aerial photography, and digital elevation models (DEMs).

14.4 High-Throughput Phenotyping Systems in Agriculture

High-throughput phenotyping (HTP) systems combine advanced sensors, imaging technologies, robotics, and data analysis tools to capture detailed information about plant growth, development, and responses to environmental factors. In addition to physiological and biochemical characteristics like photosynthesis, chlorophyll content, sugars, proteins, nutrient content, and water use efficiency, they may assess physical characteristics like shoot height, leaf area, and grain yield. HTP systems are transforming agricultural research and plant breeding by enabling accurate, automated, and non-destructive measurement of plant traits on a large scale. This leap forward allows scientists to collect enormous volumes of data on a wide range of plant properties, uncovering hidden patterns and accelerating breeding programs.

14.4.1 Platforms of High-Throughput Phenotyping Systems

The HTP can be performed from three platforms: ground-based, airborne, and spaceborne platforms. Ground-based platforms often use cameras, scanners, and other sensors mounted on movable rigs or drones to monitor plants in controlled environments like greenhouses or field plots. Drones using multispectral or hyperspectral sensors provide high-resolution airborne photographs of large fields, providing insights into crop health and yield variability. Remote sensing from satellites and airplanes offers large-scale data collection that provides essential information for precision agriculture management.

14.4.2 Quantitative Imaging and Image Processing Techniques Used in HTP

Quantitative imaging and image processing techniques have emerged as powerful tools in the field of automated germplasm characterization, revolutionizing the way plant genetic resources are evaluated and utilized in agricultural research and breeding programs. Traditional methods of superior germplasm identification and characterization often rely on manual phenotypic assessments, which can be time-consuming, subjective, and error-prone. Quantitative imaging and image processing techniques offer a powerful and objective alternative for automated germplasm characterization, opening doors to high-throughput analysis and deeper insights. These techniques leverage advanced imaging technologies, such as hyperspectral and multispectral imaging, to capture detailed information about the morphological, physiological, and biochemical characteristics of plant germplasm at various developmental stages.

By employing a spectrum of wavelengths, hyperspectral imaging allows for the acquisition of valuable data beyond the visible spectrum, enabling the detection of subtle variations in plant biochemical composition and physiological status in response to stresses. Multispectral imaging, on the other hand, captures information in discrete bands, offering a balance between spectral detail and data simplicity. These imaging modalities are complemented by sophisticated image processing algorithms that analyze and extract relevant features from the captured images. Image segmentation techniques like thresholding and edge detection are applied on hyperspectral and multispectral images to automatically segment images into meaningful regions (e.g., leaf area, root length, grain number). Various quantitative features like shape descriptors (area, perimeter, aspect ratio), texture analysis (graininess, smoothness), and color descriptors (brightness, hue, saturation) are extracted from segmented regions. The application of machine learning and computer vision techniques facilitates the automation of germplasm characterization, enabling the rapid and accurate assessment of diverse traits, including plant architecture, leaf morphology, and disease resistance. Machine learning algorithms like support vector machines (SVMs) and random forests classify germplasm based on extracted features, associating them with desired traits like disease resistance or yield potential. Advance imaging techniques like X-ray microtomography and LiDAR provide detailed 3D reconstructions of plant organs that can reveal internal structures and hidden morphological features of germplasms.

Deep learning techniques, particularly convolutional neural networks (CNNs), have been instrumental in automating the analysis of HTP data by extracting complex features directly from raw images, bypassing the need for manual feature extraction. CNNs can process images of plants, cells, or tissues, extracting relevant features and patterns without the need for explicit feature engineering. This allows researchers to analyze phenotypic traits at an unprecedented scale and speed. CNNs trained on labeled datasets where images are associated with known traits can automatically predict trait values in new unlabeled images. At the same time, CNNs can also find hidden patterns and clusters within unlabeled data, enabling the discovery of novel phenotypic variations and potentially new traits of interest.

14.4.2.1 Advantages of High-Throughput Phenotyping (HTP) Systems

One of the primary advantages of HTP through quantitative imaging in germplasm characterization is the non-destructive nature of data acquisition. Traditional methods often involve destructive sampling, limiting the ability to study the same plant over time or to assess traits dynamically. With quantitative imaging, researchers can monitor and analyze plant traits throughout the entire growth cycle without causing harm to the plants. This longitudinal approach enhances the accuracy of trait measurements, providing insights into the temporal dynamics of trait expression and the plant's response to environmental conditions. Furthermore, the high-throughput capabilities of quantitative imaging systems enable the simultaneous characterization of large germplasm collections, expediting the evaluation process and facilitating the identification of superior genotypes with desirable traits.

In addition to their non-destructive nature and high-throughput capabilities, quantitative imaging techniques contribute to the generation of accurate multidimensional datasets of various traits, minimizing subjective bias and improving data quality leading to robust datasets for analysis. These datasets capture intricate details of plant phenotypes, offering a comprehensive view of the genetic diversity present in germplasm collections. Advanced image processing algorithms play a pivotal role in handling and analyzing these complex datasets. Techniques such as feature extraction, segmentation, and pattern recognition allow for the identification of subtle variations in traits that may not be apparent through manual inspection. The integration of artificial intelligence and deep learning models further enhances the accuracy of trait prediction and classification, enabling the discovery of novel associations between genotypic and phenotypic variations. Automatic image capturing and analysis pipelines in HTP significantly lower labor costs and time required for data collection compared to traditional manual methods. HTP data helps identify superior genotypes with desirable traits, accelerating breeding cycles and improving crop varieties. HTP information can guide site-specific management practices, optimizing resource allocation and maximizing yields.

Quantitative imaging has also proven invaluable in assessing traits related to abiotic and biotic stress responses. For instance, hyperspectral remote sensing can detect stress-induced changes in pigments, and biophysical and biochemical composition, providing early indicators of biotic and abiotic stresses in plants. This information is crucial for identifying germplasm with resilience to environmental challenges, facilitating the development of climate-resilient crop varieties. Moreover, the automated detection of disease symptoms and pest infestations through imaging allows for timely interventions in breeding programs and crop management strategies. Thus, the field of quantitative imaging and image processing is revolutionizing HTP, offering a powerful tool for plant breeders and scientists and their integration into mainstream agricultural research holds tremendous potential for accelerating the development of resilient and high-yielding crop varieties to address global food security challenges.

14.4.3 Challenges of HTP Systems

- Data analysis and interpretation: The vast amount of data generated by HTP systems requires robust storage, advanced computational tools, and expertise for efficient analysis and meaningful interpretation.
- Standardization and integration: Diverse imaging platforms of HTP systems and data formats hinder cross-platform comparisons and limit data sharing. Standardization efforts are crucial to facilitate data integration and knowledge sharing.

- Integration with plant models and genomics: Connecting HTP data with plant models and genetic information promises a deeper understanding of plant biology and tailoring breeding strategies.
- Continuous algorithm development: Advancing image processing and machine learning algorithms will further enhance accuracy, enable analysis of more complex traits, and unlock new possibilities for germplasm characterization.

14.5 Geographic Information System Applications in PGR Studies

The geographic information system also enhances the efficiency of PGR management by generating new information through spatial analysis of exploration, evaluation, characterization, and locality data. Based on the distribution of crop and herbarium specimens, GIS tools have been used in species distribution and mapping of the diversity index of *Piper* spp. (Parthasarathy et al. 2006), and diversity assessment of pod characteristics in black gram. This tool is also used in diversity distribution, species richness, and grid mapping in *Brassica* spp. collected from different states of the country. Semwal et al. (2018) have also used GIS tools to map the trait-specific germplasm and geographic distribution of wheat and pigeonpea in India, respectively.

14.6 Remote Sensing Applications in PGR studies

Remote sensing satellite data combined with ground-truthing provides valuable information on different crops/species by using temporal (time interval) digital data. RS can collect real-time data of the same area of the Earth's surface at different periods, which is one of the most important features in satellite technology. Spectral features (reflectance value of a particular object/crop) may change over time and these changes can be detected by collecting and comparing multi-date satellite data using the ratio vegetation index (RVI) and normalized difference vegetation index (NDVI). Since spectral characteristics (reflectance value) of different crops/species vary at the same period of observations, these could be used to locate their availability and acreage estimation. During the growing season, most of the crops (rice, wheat, maize, barley) and wild relatives of crops are in a frequent state of change and the capability to monitor those subtle changes by remote sensing depends on how frequently the satellite data is analyzed and used.

14.7 Geospatial Data and Acquisition

14.7.1 Species Observation Data

Data about species observations can be gathered from a variety of sources, including scientific publications, field surveys, citizen science programs, and museum collections. Public resources like the European Variation Archive (EVA), as well as academic institutions and research organizations such as Indian Council of Agricultural research (ICAR)-National Bureau of Plant Genetic Resources (NBPGR), GRIN-Global Project GenBank, and Plant GARDEN are frequently used to access genetic markers, genotypic and phenotypic data, and genebank collections. Species occurrence data is fundamental in understanding the distribution, abundance, and ecology of organisms. Species occurrence datasets of species can be downloaded or accessed from one of the many Internet resources such as the Global Biodiversity Information Facility (GBIF) having more than 1.5 million species data. The other data resources include Atlas of Living Australia (ALA) with occurrence records for more than 100,000 species, NatureServe Explorer with information about >70,000 species, and the Catalogue of Life.

14.7.2 Remote Sensing Data

Satellites are frequently utilized for extensive worldwide surveillance, although drones and airplanes can give higher spatial resolution data at the local level for mapping and managing natural resources. Some of the key satellite missions and data sources for remote sensing (RS) are: Landsat (L1–L9), Sentinel (Sentinel-1, Sentinel-2, Sentinel-3), MODIS (Moderate Resolution Imaging Spectroradiometer), ASTER (Advanced Spaceborne Thermal Emission and Reflection Radiometer), and IRS (Indian Remote Sensing). Landsat satellites have provided high-resolution imagery of the Earth's surface since the 1970s, whereas Copernicus's Sentinel is fairly a newer program offering radar and multispectral imagery. MODIS sensors onboard NASA's (National Aeronautical Space Agency's) Terra and Aqua satellites which provide global coverage with moderate spatial resolution but high temporal resolution. The data from these satellites are used in feature extraction, and categorization of PGR facilitated by customized programs and algorithms. Satellite imagery is available from many sources, including the United States Geological Survey's (USGS's) Earth Explorer; NASA's Earth Observing System Data, Alaska Satellite Facility, and Information System (EOSDIS); European Space Agency's Earth Online; Indian Geo Platform of ISRO-Bhuvan; and private suppliers like Planet Labs and DigitalGlobe. Various GIS applications, from in-situ conservation to germplasm collection and documentation, are discussed in Table 14.2.

Table 14.2 Examples of GIS software applied in different aspects of the conservation of plant genetic resources

Software	Distribution	Website	Observations	Example of use
Arc-GIS/Arc info	Commercial	http://www.esri.com	Specialized in vectorial format	Evaluation of eco-geographical representativeness in ex situ collections (Parra-Quijano et al. 2007)
DIVA-GIS	Free	http://www.diva-gis.org	It provides specific tools for specialists in plant genetic resources	Analysis of the geographic distribution of wild potato species (Hijmans and Spooner 2001)
GRASS GIS	Free	http://grass.osgeo.org	It can be used with R software environment	Evaluation of climate change on the distribution of tree species
gvSIG	Free	http://www.gvsig.org/web	Spatial data infrastructure	Evaluation of on-farm conservation
IDRISI	Commercial	http://www.clarklabs.org	Specialized in raster format	Gap analysis in *Medicago* crop wild relatives
Marxan	Free	http://www.uq.edu.au/marxan	Specific to aid systematic reserve design on conservation planning; it is not a GIS, but it requires complementary GIS tools	Selection of areas for in situ conservation
R	Free	http://www.r--project.org	Statistical model that can perform geospatial analysis (e.g., Raster, sp, or dismo)	Focused identification of germplasm strategy (Bari et al. 2012)

Source: Parra-Quijano et al. (2012)

14.7.3 Environmental Data

Environmental data describes the conditions of the locations where species occur or are distributed. The most common types of environmental variables are climate regimes such as temperature, precipitation, land use, land cover, altitude data, and soil type. Altitude affects temperature, and therefore indirectly affects species ecology and distributions. Online data portals such as WorldClim and Chelsa provide a large collection of global climate layers of past, current, and future climate. The global soil database that describes the mineral nutrients regime is available at the Food and Agriculture Organization (FAO). Climate data, soil maps, and topographic data are also available at agencies such as India Meteorological Organization's Climate Data Service Portal, The National Oceanic and Atmospheric Administration

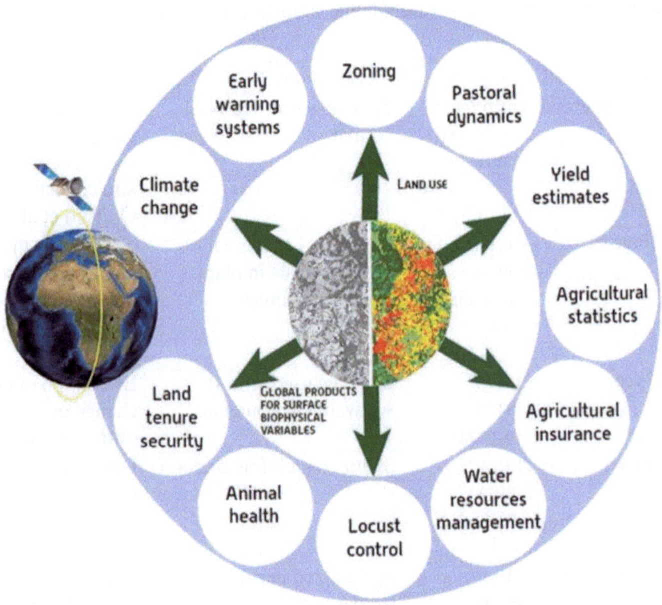

Fig. 14.6 The main fields of agricultural applications using EO data or EO-derived products (vegetation indices, land use maps, biophysical variables, etc.). (Source: Tonneau et al. 2019)

(NOAA), and the United States Geological Survey (USGS). The main fields of agricultural applications using earth observation (EO) data or EO-derived products (vegetation indices, land use maps, biophysical variables, etc.) are mentioned in Fig. 14.6.

14.8 Methodology for PGR Data Analysis

14.8.1 Mapping and Inventorying Plant Genetic Resources

GIS allows visualization and a better understanding of spatial patterns of genetic diversity as a key input to optimize the conservation and use of plant genetic resources. GIS is used for many tasks related to PGR administration and conservation, including geo-referencing germplasm in genebanks, developing plans for expeditions and collections, and identifying conservation sites. RS and GIS tools are also helpful in planning field surveys of PGR by supporting sampling design by stratifying the study area into different land cover classes, ensuring representative data collection and optimization of resources. Temporal monitoring aligns field data collection with satellite overpasses, enabling the tracking of changes over time. Researchers provide ground truth data on biophysical parameters through field surveys, which complement and further validate satellite observations.

To create a more contextualized conservation strategy, researchers can accurately map the spatial distributions of species and correlate them with pertinent socioeconomic, ecological, and physical data. This can help PGR managers, farmers, and conservationists create a pest monitoring map that will help them effectively address the effects of pest infestation in wild species or on landraces. The mapping of data acquired via the use of GIS makes it easier to incorporate the genetic findings into the development of conservation strategies and the implementation of conservation actions (Jarvis et al. 2010). Sabran (2016) explored how GIS technology aids in identifying and cataloguing the distribution of plant species, genotypes, and traits.

In a study by Pacicco et al. (2018), the first methodological approach to identify agrobiodiversity hotspots, termed MAPAs (Most Appropriate Areas), was established using geographic information system (GIS) in the Italian landscape. The research compared two prioritization strategies: the Restrictive Strategy and the Additive Strategy. The additive strategy combines standardized index values to establish a general threshold, while the restrictive strategy employs distinct criteria for each biodiversity component. The study emphasizes the importance of conserving agricultural genetic resources and highlights the need for enhanced data availability and international collaboration to effectively preserve agrobiodiversity. Methods to develop proxies of genetic differentiation to identify key conservation areas can also be developed using geospatial data. The method accounts for evolutionary processes, including historical and environmental drivers of genetic diversity, as well as threat processes, such as species-specific tolerance to human-modified habitats and extinction risk status.

14.8.2 Utilizing GIS for Germplasm Conservation and Characterization

14.8.2.1 Species Distribution Modeling (SDM)

SDM can play a crucial role in plant genetic resource (PGR) mapping by providing insights into the spatial distribution, habitat suitability, and environmental drivers of plant species' occurrences. By integrating SDM with PGR mapping efforts, researchers can identify regions with high genetic diversity, rare or endemic species, medicinal species, and unique genetic traits that are valuable for crop improvement and breeding programs. SDM can upscale ex situ conservation strategies by identifying suitable sites for establishing genebanks, botanical gardens, and seed banks. Additionally, SDM can guide in situ conservation efforts by identifying priority areas for habitat restoration, protected area designation, and sustainable land management practices to conserve plant genetic resources in their natural habitats. Overall, the integration of species distribution modeling with plant genetic resources mapping enhances our understanding of the spatial distribution, conservation status, and ecological requirements of plant species, contributing to more effective and informed conservation and management strategies for plant genetic resources.

14.8.2.2 Multi-Criteria Decision Analysis (MCDA)

MCDA integrates multiple criteria, including biodiversity value, ecosystem services, and vulnerability assessments, to identify priority areas for conservation. This approach allows decision-makers to weigh the importance of different factors and prioritize actions accordingly. Prioritizing areas based on ecosystem services can ensure the protection of critical resources. GIS allows researchers to combine genetic, environmental, and geographical data to evaluate the distribution, diversity, and structure of germplasm collections. GIS and RS tools help to visualize genetic diversity hotspots, identify possible conservation areas, and analyze population connections. Furthermore, GIS enables the creation of geographic models for predicting germplasm distribution patterns, directing germplasm collection techniques, and determining conservation priorities. GIS can also play a pivotal role in managing and analyzing eco-geographical data to enhance understanding of species adaptations. The study highlighted challenges such as limited resources and a prevailing focus on curator-centric data rather than breeder needs.

14.8.3 Geospatial Techniques for Plant Genetic Resource Management

14.8.3.1 Remote Sensing for Habitat Monitoring

RS techniques, such as satellite imagery and aerial photography, provide valuable data for monitoring habitat conditions and changes over time using approaches such as change analysis. RS data can be used to assess land cover changes, vegetation dynamics, and habitat fragmentation using temporal RS data and GIS software, which are essential for understanding the status and trends of PGR habitats.

14.8.3.2 Genetic Mapping and Marker-Assisted Selection (MAS)

Geospatial techniques are used in genetic mapping and MAS to identify and characterize genetic markers associated with desirable traits in plant populations. Spatial analysis of genetic data helps identify genetic loci linked to specific traits, assess genetic diversity within and among populations, and develop breeding strategies for crop improvement and conservation. The efficiency of tools like DIVA-GIS and FIGS (Focused Identification of Germplasm Strategy) in facilitating germplasm selection for breeding purposes has already been established. The main component of mapping plant genetic resources using geospatial technologies are given in Fig. 14.7.

14.8.3.3 Spatial Decision Support Systems (SDSS)

SDSS integrates geospatial data, modeling tools, and decision support algorithms to facilitate decision-making processes related to PGR management. SDSS helps stakeholders prioritize conservation actions, allocate resources, and design conservation strategies based on spatially explicit information about genetic diversity, threats, and conservation priorities. Geospatial techniques offer powerful tools for managing plant genetic resources by providing spatially explicit information and

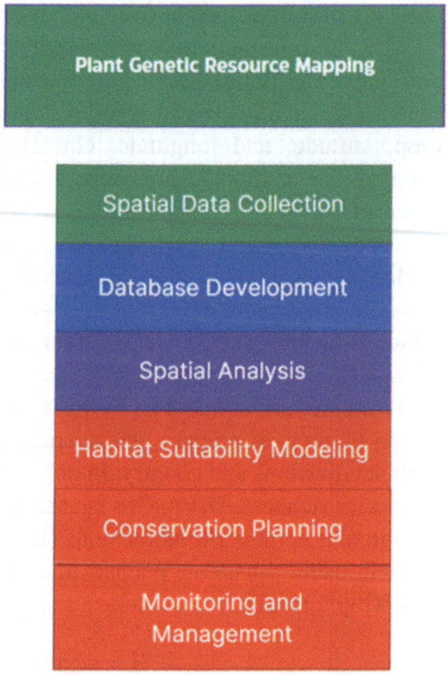

Fig. 14.7 The main component of mapping plant genetic resources using geospatial technologies

decision support for conservation, breeding, and sustainable use initiatives. Overall, by integrating geospatial data and analysis tools into PGR management workflows, researchers, policymakers, and practitioners can make informed decisions to safeguard genetic diversity and ensure the resilience and adaptability of plant populations in the face of environmental change and emerging threats.

14.8.3.4 Germplasm Material and Scrutiny of Passport Data

There are about 472,557 accessions in the national genebank of the Bureau while total germplasm including the accessions under cryobank, in-vitro bank, and field genebank is quite high (conserved IC number: 6.49 lakhs). The majority of the germplasm collection data have locality information (village/block/tehsil/district/state) and geo-coordinates (latitude and longitude) recorded by GPS. However, germplasm collected before 1998 does not have collection site coordinates, which now can be taken from secondary sources (Gazetteer, Google map, Survey of India map, etc.) used in geo-referencing and diversity distribution mapping. Before mapping and analysis, the passport data require detailed scrutiny and checking concerning the geo-coordinates of places/sites of collection (village/block/district) using GIS tools. Several new states and districts have been formed either by merging some parts or by carving out from large states/districts. Hence, there is a need to reassign their correct geographic identity.

Data Format Required in GIS

Geographic information system requires the PGR data in digital format to accept and support attribute data for further analysis. This is usually done by creating a text file with four parameters in different columns: a unique identifier, name of the plant/crop, latitude, and longitude. Global positioning system-acquired coordinates should be in decimal degrees only. In many passport databases geo-coordinates are given in degrees (°), minutes (′), and seconds (″), together with an indication of hemisphere (north or south, and east or west). In any GIS system, geo-coordinates in decimal degrees are needed for linking germplasm collection sites for use in diversity mapping. Conversion of degrees, minutes, and seconds (latitude and longitude values) in decimal degrees can be calculated by using the geo-calculator of DIVA-GIS.

This software consists of WGS84 geographic projection system and geodetic datum systems to prepare geo-referenced maps of desired crops and regions and their subdivisions and country. In order to know spatial distribution and assessment of species richness, DIVA-GIS is useful for point data into grid analysis using simple-circular neighborhood methods (Hijmans et al. 2001) and diversity can be calculated using Shannon-Wiener diversity index (Magurran 1988) through the following formula:

$$H = -\Sigma\left[\left(ni/N\right)\ln\left(ni/N\right)\right]$$

H = Shannon diversity index
ni = individual of a species
N = total individuals of all species
ln = natural log

Habitat suitability model viz. "*EcoCrop*" is implemented in DIVA-GIS (Hijmans et al. 2007) and has been interfaced with climate data for temperature and precipitation suitability.

14.8.3.5 Diversity Analysis Based on Grid Mapping

DIVA-GIS tools have been widely used for eco-geographic mapping of collection sites, and the diversity distribution of crops and wild species. GIS-based grid mapping technique is used to know the diversity-rich areas, variability assessment, and occurrence of trait-specific germplasm of selected crops in different parts of the world (Hijmans et al. 2001; Guarino et al. 2002). A grid of definite size (1° × 1°) is assigned on the map to the points representing collected germplasm. The grid size depends on the size of the geographical area. At the country level (large geographical area) a grid of 1° × 1° (111.1234 × 111.12 km) size, while for states/subdivisions (smaller area) the grid of 0.1° × 0.1° to 0.9° × 0.9° size (Semwal et al. 2013; Laxmisha et al. 2022) may be used depending upon the area and diversity.

An exploration database having more than 2000 species of different agri-horticultural species including their crop wild relatives (CWRs) with locality information and geo-coordinates can easily import data into a GIS environment for geo-referencing and diversity mapping.

14.8.3.6 Mapping of Diversity and "BioClim" Modeling

Variability mapping of germplasm of wild *Vigna* species was geo-referenced for diversity mapping. Modern tools such as geographic information system (GIS), global positioning system (GPS), crop models (CMs), and species distribution models (SDMs) were used to pinpoint diversity distribution mapping and predict suitable sites for future collection of the crops and their wild relatives using data on collected germplasm and different climatic variables based on other crop models (Semwal et al. 2016; IPCC 2014; Hijmans et al. 2005). The creation of the WorldClim dataset is a considerably advanced method used for global environmental characterization as it provides high-resolution (i.e., nearly 1 km) climatic variables derived from historical records from several weather stations across the globe (http://www.worldclim.org). In the BioClim model, climatic variables are derived from the monthly precipitation and temperature values to generate more biologically suitable variables for different crops. Considering optimal conditions and limits to adaptation, a set of 19 bioclimatic variables can be derived (Busby 1991). Prediction maps were drawn based on collection data and climatic variables.

BioClim is implemented in DIVA-GIS interfaced with climate data for temperature suitability or temperature-derived data: BIO1 = annual mean temperature to BIO11 = mean temperature of the coldest quarter; rainfall-derived data: BIO12 = annual precipitation to BIO19 = precipitation of coldest quarter (DIVA-GIS 7.5). BioClim model was applied to the occurrence of species distribution using the current climate data (1950–2000) derived from WorldClim (climate grids with a spatial resolution of 2.5 km^2) as baseline data.

A Case Study in Foxtail Millet (*Setaria italica*)

Geo-Referencing of Collected Germplasm Data

In this study, foxtail millet data was analyzed for geo-referencing and diversity distribution mapping. A total of 5772 accessions have so far been collected from various parts of the country. Among these, 3821 accessions had passport data with locality details of collection sites (state, district, and village). Analysis of passport data shows that the majority of foxtail millet germplasm accessions have been collected from Tamil Nadu (681) followed by Andhra Pradesh (667), Karnataka (352), Maharashtra (337), Uttar Pradesh (247), Madhya Pradesh (243), and Himachal Pradesh (248) (Fig. 14.8).

Figure 14.8 depicts high diversity-rich areas from where foxtail germplasm was collected as diverse spots/regions of the country. Red-colored grid with high diversity values (1.447–2.000) is present mostly in North-Western, Eastern, and North Eastern Hill (NEH) regions of the country, which are the actual areas of the minor millets cultivation in India.

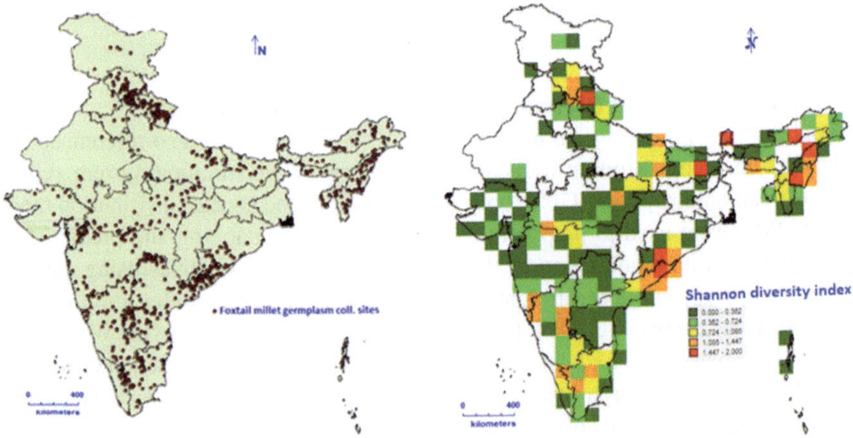

Fig. 14.8 Geo-referenced map (**a**) and GIS-based grid map (**b**) of foxtail millet (*Setaria italica*) collected from different states of the country

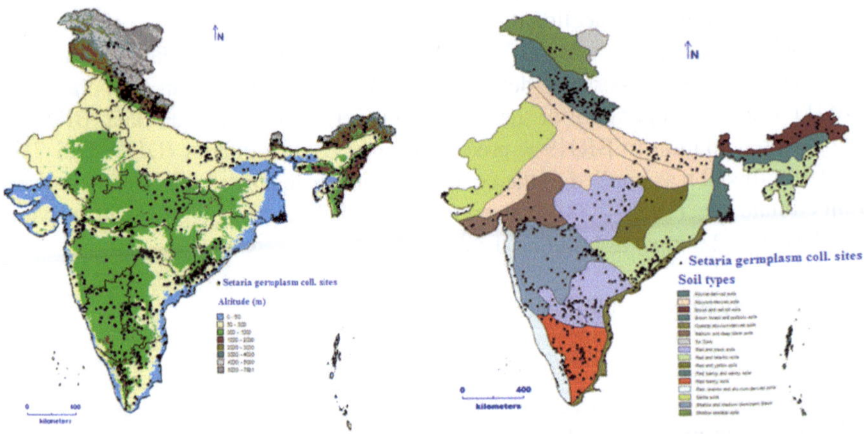

Fig. 14.9 Geo-referenced map of foxtail millets collected from different states of the country along with different map layers (soil types and topographic data)

Mapping of Diversity Collected in Foxtail Millet

To know the eco-geographic diversity distribution of minor millets (12,016 accessions) through mapping the sites, passport data of collected germplasm were shortlisted to find out the gaps in exploration and germplasm collection. Mapping of diversity collected indicates some of the diversity-rich areas, viz., parts of northwestern Himalaya (Jammu & Kashmir, Himachal Pradesh, and Uttarakhand), western parts of Central India (Madhya Pradesh, Maharashtra, and Gujarat), and peninsular region (Andhra Pradesh, Karnataka, and Tamil Nadu) (Fig. 14.9). GIS maps show that foxtail germplasm is distributed in different agroecological zones across the country. The majority of germplasm collections are from good rainfall

(<1500 mm), shown in orange and light green color. Maximum germplasm collected from mean annual temperature (15–28 °C) areas is shown in light brown color. Foxtail millet germplasm was collected from hot semi-arid to hot sub-humid agroecological zones and between 200 and 1500 m altitudinal ranges. Diversity was collected from medium to deep black, lateritic, loamy, red loamy, and yellow soil regions of the country (Fig. 14.9).

Habitat Prediction Mapping Using EcoCrop/BioClim Models

Wild rice: Crop wild relatives (CWRs) having desirable traits/genes, which are generally not available in crops, are increasingly recognized as a valuable resource for the crop improvement program of the country. Different wild rice species have provided many valuable traits/genes such as disease and pest resistance, high yield, abiotic stress tolerance, and cytoplasmic male sterility. Exploration and survey of wild rice species form a priority, notably after the success achieved in the utilization of annual species *Oryza nivara*; its germplasm collection from eastern Uttar Pradesh served as a novel source of resistance to *Rice grassy stunt virus* and *Rice tungro virus*.

Semwal et al. (2016) analyzed wild rice germplasm data from the eastern region of India and generated a suitability map for wild rice species in the current context based on "EcoCrop" model (Fig. 14.10a). Five categories were recognized based on the current climate data: "very high suitable area" (67.9%) and "high suitable area" (18%) have been shown through light green and orange color, respectively, along with "moderately suitable area" (9.7%, dark green), "low suitable area" (pink, 1.6%), and "unsuitable area" (3.1%, brown) in the Gangetic Plains and eastern region of India (Fig. 14.10a). The model predicted that most of the areas of Gangetic Plains form either suitable or very highly conducive habitats for wild rice. Generally, "not suitable areas" were located in mountain areas with slopes >30% and elevations above 1500 m mean sea level (MSL). Also, collected germplasm as well as literature inputs matched well with the habitat suitability map generated using current climate data (Fig. 14.10).

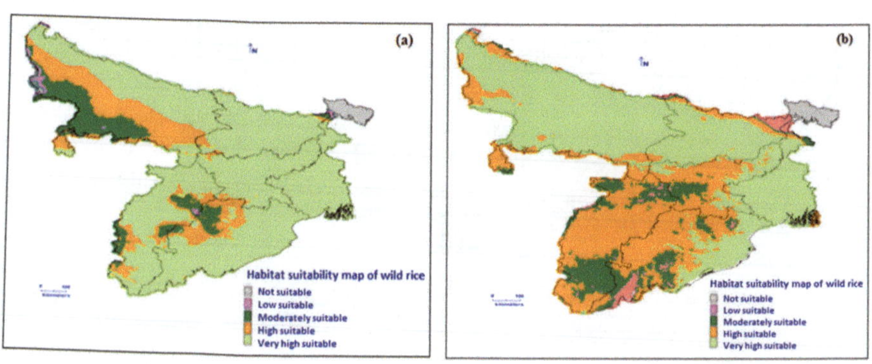

Fig. 14.10 (**a**) Wild rice habitat suitability map (1950–2000); (**b**) habitat prediction map for 2050 in Gangetic Plains and eastern region of India. (Source: Semwal et al. 2016)

Makhana (*Euryale ferox*): An aquatic cash crop, "makhana" (*Euryale ferox* Salisb.), also known as foxnut, is a monotypic genus in the family Nymphaeaceae. Habitats of the foxnut are mainly wetlands ranging from lakes, ponds, and puddles to shallow water bodies. Foxnut was studied for predicted habitat suitability through "BioClim" model for the years 1950–2000 and projected climate data for 2050 and 2070 for the Indian region based on primary data from surveys and exploration for germplasm collection.

For predicting the suitable habitat of foxnuts, data on occurrence were interpolated with climatic variables. The predicted habitat suitability map (1950–2000) for the current climate showed in present conditions highly suitable area (Fig. 14.11a) for "makhana" cultivation spread to four states (Bihar, Jharkhand, Uttar Pradesh, and West Bengal) of eastern India. Maximum suitable areas were observed for Bihar (65.0%) followed by Uttar Pradesh (12.3%), Jharkhand (12.2%), and West

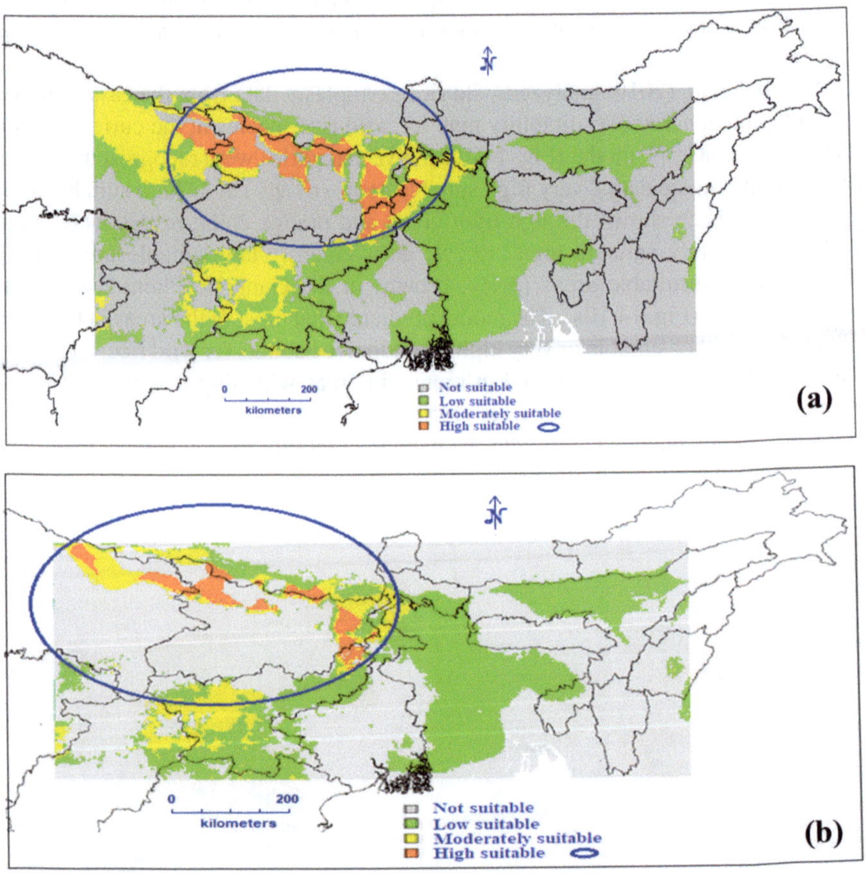

Fig. 14.11 (**a**) Foxnut habitat suitability map prepared from occurrence data of species and current climatic conditions (1950–2000); (**b**) future predicted climatic conditions (2050). (Source: Semwal et al. 2021)

Bengal (10.5%). Habitat suitability areas were mainly eastern and northern districts, *viz.*, East Champaran, Gopalganj, Darbhanga, Sitamarhi, Madhubani, Araria, Katihar, and Purnea in Bihar; Godda and Sahibganj districts in Jharkhand; Kushinagar and Maharajganj districts in Uttar Pradesh; and Malda and Uttar Dinajpur districts in West Bengal (Fig. 14.11a).

In Fig. 14.11b, it is depicted that the suitability areas have shifted from areas mentioned in Fig. 14.11a to other surrounding areas/districts. The predicted map shows that highly suitable areas (Uttar Dinajpur district of West Bengal) were replaced by moderate to low suitable sites while Gopalganj, Madhubani, and Sitamarhi districts of Bihar were replaced completely as unsuitable areas in the future. In Uttar Pradesh, new districts, viz., Balrampur and Shravasti, are presently falling under the highly suitable area for "makhana" distribution range due to the rise of 0.7 °C in temperature.

14.9 Challenges and Future Directions

14.9.1 Limitations of Current Geospatial Analysis Techniques

One of the main limitations of current geospatial approaches is the spatial resolution of remote sensing platforms, which often struggle to accurately characterize small-scale genetic diversity and distribution patterns of plant populations. The variability in the availability and quality of geospatial data can introduce inconsistencies and uncertainties in PGR mapping results, hindering the reliability and accuracy of spatial analyses. Integrating spatial and temporal scales of genetic data may not align with those of environmental variables, leading to scale mismatches and potential biases in spatial analysis results. Spatial biases and inaccuracies in PGR mapping can result from biases in sampling designs and data collection procedures, particularly in locations that are under-sampled or lack sufficient documentation. Furthermore, uncertainties associated with future climate projections can have an impact on the confidence of habitat suitability models and PGR dispersion forecasts in the face of changing environmental conditions. These issues highlight the need for robust analytical techniques and data interoperability standards to ensure accurate and meaningful integration of multi-source data in PGR mapping.

14.9.2 Emerging Trends and Technologies

More precise and complex mapping of vegetation traits and environmental variables relevant to PGR mapping is made possible by developments in remote sensing technologies like hyperspectral imaging and LiDAR (Light Detection and Ranging). Furthermore, the amalgamation of artificial intelligence and machine learning techniques with geospatial analysis has novel prospects for automated feature extraction, identification of species, and predictive modeling in the context of PGR mapping. Another developing approach is to utilize citizen science and crowdsourced data collection platforms to increase involvement and engagement in PGR mapping projects. Furthermore, open-access data repositories and the use of interoperable data standards encourage data sharing and collaboration among academics and stakeholders, boosting the reproducibility and scalability of geospatial analysis in PGR mapping. Collectively, these developing trends and technologies have considerable potential for improving our understanding of PGR distribution, variety, and conservation, allowing stakeholders to make educated decisions about sustainable agricultural growth and biodiversity preservation.

14.9.3 Recommendations for Future Research and Application

It is essential to make recommendations for future research and the application of geospatial approaches for mapping PGR to overcome the current challenges. For characterizing PGR distribution and habitat suitability, satellite images and aerial photography must have higher spatial and spectral resolution. The automatic feature extraction, species identification, and predictive modeling in PGR mapping may be facilitated by the incorporation of cutting-edge machine learning and artificial intelligence techniques with geospatial analysis. Encouraging cooperation and reproducibility in geospatial analysis workflows requires efforts to support data interoperability, standardization, and sharing through open-access repositories and compatible data standards. In order to provide researchers, practitioners, and policymakers with the necessary skills and understanding to use geospatial approaches for PGR mapping and management, it is critical to construct capacity-building and training programs.

The advanced GIS tools in conjunction with the plant genetic resource (PGR) database, collected herbarium specimens, and high-resolution remote sensing technology could serve as a potential information treasure house to the scientific community in general and plant genetic resource scientists in particular. In PGR studies, geospatial technology could be of great use in the management of plant genetic resources particularly in integrated remote sensing and geospatial technology can be used in both operational and real gap analysis, planning, and execution of future exploration programs at the national level. Germplasm passport data information, germplasm reporter data, remote sensing data spectral signature, and climate analog tools can be used in diversity distribution mapping and prediction of diversity-rich areas for different crops and their species.

Field remote sensing data could help in delineating CWR species using a Spectroradiometer to record the spectral signature of targeted crops at points/sites in the field or farm. Drone remote sensing and hyperspectral remote sensing can be used in distinguishing and identification of crop wild relatives of cultivated plants. Drone technology, and remote sensing integrated with a GIS system could be used in mapping and monitoring field crops to know the moisture level, growth conditions, and maturity of crops at a temporal scale. High spatial resolution (0.40 m) satellite data can be used in mapping disease symptoms in crops and their wild relatives at a fine scale.

14.10 Summary

The application of geospatial analysis in mapping plant genetic resources holds immense significance in understanding and conserving biodiversity. Through the integration of various geospatial techniques, such as SDM and spatiotemporal analysis, researchers can gain valuable insights into the distribution, dynamics, and threats facing plant species. Despite the challenges posed by limitations in current methodologies, the emergence of new technologies and trends presents promising opportunities for future research and application. It is crucial to continue exploring and refining geospatial analysis techniques to address the complex conservation and management needs of plant genetic resources.

Multiple-Choice Questions

1. GIS stands for ________.
   (a) Geographic internal system
   (b) Geographic information system
   (c) Global information system
   (d) None of the above
2. GIS tools allow the user to perform which of the following task?
   (a) Create searches
   (b) Store data
   (c) Edit data
   (d) All of the above
3. GIS represents a location in ______ dimensional coordinates.
   (a) 2
   (b) 3
   (c) 5
   (d) All of the above
4. GIS was coined by ________.
   (a) Roger Tomlinson
   (b) Roger James
   (c) Richard
   (d) None of the above
5. A process where a survey plan or hard copy is a map and is represented in digital medium using geo-referencing capabilities and CAD tool is called ______ technique.
   (a) Modulation
   (b) Digitization
   (c) Demodulation
   (d) None of the above
6. National Genebank data is used for--------------
   (a) Gap analysis
   (b) Trait specific mapping
   (c) Geo-mapping
   (d) None of the above

7. GIS deals with which kind of data?
   (a) Numeric data
   (b) Binary data
   (c) Spatial data
   (d) Complex data
8. What is *Metadata*?
   (a) It is "data about data"
   (b) It is "meteorological data"
   (c) It is "oceanic data"
   (d) It is "contour data"
9. Interpolation is made possible by a principle called
   (a) Spatial auto-correction
   (b) Spatial autocorrelation
   (c) Thematic autocorrelation
   (d) Thematic auto-correction
10. A GIS package cannot be called full-fledged if the following capabilities are missing.
   (a) Data capture and preparation
   (b) Data storage
   (c) Data analysis
   (d) All of the above

Answers

1. (b)
2. (d)
3. (b)
4. (a)
5. (b)
6. (a)
7. (c)
8. (a)
9. (b)
10. (d)

Suggested Reading

Abdulridha J, Ampatzidis Y, Kakarla SC, Roberts P (2020) Detection of target spot and bacterial spot diseases in tomato using UAV-based and benchtop-based hyperspectral imaging techniques. Precis Agric 21:955–978

Amaral LR, Molin JP, Portz G, Finazzi FB, Cortinove L (2015) Comparison of crop canopy reflectance sensors used to identify sugarcane biomass and nitrogen status. Precis Agric 16:15–28

Arora RK (1991) Plant diversity in Indian Gene Centre. In: Paroda RS, Arora RK (eds) Plant genetic resources conservation and management: concept and approaches. IBPGR Regional Office for South and South East Asia, New Delhi, pp 25–54

Ballester C, Zarco-Tejada PJ, Nicolás E, Alarcón JJ, Fereres E, Intrigliolo DS, Gonzalez-Dugo VJPA (2018) Evaluating the performance of xanthophyll, chlorophyll and structure-sensitive spectral indices to detect water stress in five fruit tree species. Precis Agric 19:178–193

Bari A, Street K, Mackay M, Endresen DTF, De Pauw E, Amri A (2012) Focused identification of germplasm strategy (FIGS) detects wheat stem rust resistance linked to environmental variables. Genet Resour Crop Evol 59:1465–1481

Bolstad P (2016) GIS fundamentals: a first text on geographic information systems. Eider Press, Minnesota

Bowers SA, Hanks RJ (1965) Reflection of radiant energy from soils. Soil Sci 114:477–485

Burrough PA (1986) Principles of geographical information systems for land resources assessment. Oxford University Press, New York

Busby JR (1991) BIOCLIM – a bioclimate analysis and prediction system. Plant Prot Q 6:8–9

Campbell JB (2011) Introduction to remote sensing. Guilford Press

Das B, Sahoo RN, Pargal S, Krishna G, Verma R, Tiwari R, Viswanathan C, Sehgal VK, Gupta VK (2015) Spectral based non-invasive estimation of plant chlorophyll content. J Agric Phys 15:88–102

Fick SE, Hijmans RJ (2017) WorldClim 2: new 1-km spatial resolution climate surfaces for global land areas. Int J Climatol 37(12):4302–4315

Flemons P, Guralnick R, Krieger J, Ranipeta A, Neufeld D (2007) A web-based GIS tool for exploring the world's biodiversity: the Global Biodiversity Information Facility Mapping and Analysis Portal Application (GBIF-MAPA). Eco Inform 2(1):49–60

Gascon F, Cadau E, Colin O, Hoersch B, Isola C, Fernández BL, Martimort P (2014) Copernicus Sentinel-2 mission: products, algorithms and Cal/Val. In: Earth observing systems XIX, vol 9218. SPIE, pp 455–463

Guarino L, Jarvis AR, Hijmans RJ, Maxted N (2002) Geographic information systems (GIS) and the conservation and use of plant genetic resources. In managing plant genetic diversity. Proceedings of an international conference, Kuala Lumpur, Malaysia, 12-16 June 2000 (pp. 387–404). Wallingford UK: CABI publishing

Hijmans RJ, Spooner DM (2001) Geographic distribution of wild potato species. Am J Bot 88(11):2101–2112

Hijmans RJ, Guarino L, Cruz M, Rojas E (2001) Computer tools for spatial analysis of plant genetic resources data: 1. DIVA-GIS. Pl Genet Resour News Lett 127:15–19

Hijmans RJ, Cameron SE, Parra JL, Jones PG, Jarvis A (2005) Very high resolution interpolated climate surfaces for global land areas. Int J Climatol 25:1965–1978

Hijmans RJ, Gavrilenko T, Stephenson S, Bamberg J, Salas A, Spooner DM (2007) Geographical and environmental range expansion through polyploidy in wild potatoes (*Solanum* section Petota). Global Eco Biogeo 16(4):485–495

Hintum Van T, Menting F, Van Strien E (2011) Quality indicators for passport data in *ex-situ* genebanks. Plant Genet Resour: Charact Utiliz 9(3):478–485

IPCC (2014) In: Core Writing Team, Pachauri RK, Meyer LA (eds) Contribution of Working group I II and III to the Fifth Assessment Report of the Intergovernmental Panel on Climate Change, IPCC, Geneva, Switzerland

Jarvis A, Touval JL, Schmitz MC, Sotomayor L, Hyman GG (2010) Assessment of threats to ecosystems in South America. J Nat Conserv 18(3):180–188

Jena RK, Bandyopadhyay S, Pradhan UK, Moharana PC, Kumar N, Sharma GK, Roy PD, Ghosh D, Ray P, Padua S, Ramachandran S (2022) Geospatial modelling for delineation of crop management zones using local terrain attributes and soil properties. Remote Sens 14(9):2101

Klem K, Záhora J, Zemek F, Trunda P, Tůma I, Novotná K, Hodaňová P, Rapantová B, Hanuš J, Vavříková J, Holub P (2018) Interactive effects of water deficit and nitrogen nutrition on winter wheat. Remote sensing methods for their detection. Agric Water Manag 210:171–184

Laxmisha KM, Semwal DP, Gupta V, Katral A, Bisht IS, Mehta PS, Arya M, Bhardwaj R, Bhatt KC (2022) Nutritional profiling and GIS-based grid mapping of Job's tears (*Coix lacryma*-jobi L.) germplasm. Appl Food Res 2(2):1–10. https://doi.org/10.1016/j.afres.2022.100169

Lillesand TM, Kiefer RW, Chipman J (2014) Remote sensing and image interpretation. Wiley

Magurran AE (1988) Ecological diversity and its measurement. Princeton University Press, Princeton, New Jersey, USA, p 215

Marino S, Cocozza C, Tognetti R, Alvino A (2015) Use of proximal sensing and vegetation indexes to detect the inefficient spatial allocation of drip irrigation in a spot area of tomato field crop. Precis Agric 16:613–629

Pacicco L, Bodesmo M, Torricelli R, Negri V (2018) A methodological approach to identify agro-biodiversity hotspots for priority in situ conservation of plant genetic resources. PLoS One 13(6):e0197709

Parra-Quijano, M., Draper, D., Torres, E. and Iriondo, J.M., 2007. Ecogeographical representativeness in crop wild relative *ex situ* collections.

Parra-Quijano M, Iriondo JM, Torres E (2012) Applications of ecogeography and geographic information systems in conservation and utilization of plant genetic resources. Span J Agric Res 10(2):419–429

Parthasarathy U, Saji KV, Jayarajan K, Parthasarathy VA (2006) Biodiversity of piper in South India – application of GIS and cluster analysis. Curr Sci 91:652–658

Phadikar S, Goswami J (2016) Vegetation indices based segmentation for automatic classification of brown spot and blast diseases of rice. In 2016 3rd International Conference on Recent Advances in Information Technology (RAIT) (pp 284–289). IEEE

Pourazar H, Samadzadegan F, Dadrass Javan F (2019) Aerial multispectral imagery for plant disease detection: Radiometric calibration necessity assessment. Eur J Remote Sens 52(sup3):17–31

Priyanka V, Kumar R, Dhaliwal I, Kaushik P (2021) Germplasm conservation: Instrumental in agricultural biodiversity—a review. Sustainability 13(12):6743

Semwal DP, Bhandari DC, Bhatt KC, Singh R (2013) Diversity distribution pattern in collected Germplasm of Rapeseed-Mustard using GIS in India. Indian J Plant Genetic Resour 26(01):76–81

Semwal DP, Pradheep K, Ahlawat SP (2016) Wild rice (*Oryza* spp.) germplasm collections from Gangetic Plains and Eastern region of India: Diversity mapping and habitat prediction using Ecocrop model. Vegetos 29:96–100

Semwal DP, Ahlawat SP, Pradheep K (2018) Pigeonpea (*Cajanus cajan* (L.) Millsp.) and its *wild* spp. germplasm collection status, diversity distribution and trait-specific germplasm mapping using GIS tools in India. Legume Res 41(5):656–662

Semwal DP, Pandey A, Gore PG, Ahlawat SP, Yadav SK, Kumar A (2021) Habitat prediction mapping using BioClim model for prioritizing germplasm collection and conservation of an aquatic cash crop 'makhana' (*Euryale ferox* Salisb.) in India. Genet. Resour Crop Evol 68:3445–3456

Ramírez-Villegas J, Khoury C, Jarvis A, Debouck DG, Guarino L (2010) A gap analysis methodology for collecting crop genepools: a case study with Phaseolus beans. PLoS One 5(10):e13497

Rapaport T, Hochberg U, Cochavi A, Karnieli A, Rachmilevitch S (2017) The potential of the spectral 'water balance index'(WABI) for crop irrigation scheduling. New Phytol 216(3):741–757

Ray P, Chattaraj S, Bandyopadhyay S, Jena RK, Singh SK, Ray SK (2021) Shifting cultivation, soil degradation, and agricultural land-use planning in the northeastern hill region of India using geo-spatial techniques. Land Degrad Dev 32(14):3870–3892

Sabran M (2016) Plant genetic resources management in Indonesia: conservation, uses, and policy. Pre-Breeding and Gene Discovery for Food and Renewable Energy. Security:3

Tonneau JP, Bégué A, Leroux L, Augusseau X, Faure JF, Mertens B, Pinet C, Tomasini L (2019) Geospatial information for African agriculture: a key investment for agricultural policies. Perspective (English edition) 51:1–4

Vaughan DA, Morishima H, Kadowaki K (2003) Diversity in the *Oryza* genus. Curr Open Plant Biol 6:139–146

Yuan L, Pu R, Zhang J, Wang J, Yang H (2016) Using high spatial resolution satellite imagery for mapping powdery mildew at a regional scale. Precis Agric 17:332–348

Zhou J, Khot LR, Boydston RA, Miklas PN, Porter L (2018) Low altitude remote sensing technologies for crop stress monitoring: a case study on spatial and temporal monitoring of irrigated pinto bean. Precis Agric 19:555–569

Plant Genetic Resources, Intellectual Property Rights, Law and Policy

15

S. Elumalai and Pratibha Brahmi

Learning objectives

- Policy issues important to access and utilization of plant genetic resources
- Major international agreements and their objectives, namely CBD, ITPGRFA, Nagoya Protocol, and TRIIPS, related to biological diversity management
- Categories of intellectual property and legal provisions for their protection
- Basics of patent searches

Keywords
Germplasm · PGRFA · Policy · Law · IPR · CBD · BDA

15.1 Introduction

Plant genetic resources (PGRs) comprise crop plants and their wild/weedy-related species of actual or potential use. PGRs have been developing since time immemorial, and have several values attached to them including intrinsic, ecological, environmental, social, economic, scientific, cultural, spiritual, aesthetic, and others. The development of improved types used today and those that would be cultivated in future is based on the effective utilization of PGRs. There is a continuous search for newer resources and also rediscovering the forgotten food crops to meet the future

S. Elumalai
Tamil Nadu Ambedkar Law University, Chennai, Tamil Nadu, India

P. Brahmi (✉)
Germplasm Exchange and Policy Unit, ICAR-National Bureau of Plant Genetic Resources, New Delhi, India

K. Tripathi et al. (eds.), *Textbook of Plant Genetic Resources*,
https://doi.org/10.1007/978-981-96-5024-8_15

demands that arise with the emergence of climate change, new diseases, and enhanced demand for food and nutritional security. PGRs are exchanged and searched continuously for specific traits to improve crops in terms of yield and nutritional value. All countries are, therefore, interdependent for sources of PGRs, since many crops cultivated in a country have not originated there. Even biodiversity-rich regions depend for more than 30% of their food production on crops originating from other countries and this interdependence plays a very important role in the international collection and exchange of germplasm. Every nation is concerned with the acquisition of diverse and superior germplasm for conservation and utilization. This chapter highlights the salient features of international and national agreements/convention/acts pertaining to policies governing access to PGRs, and intellectual property rights (IPRs) related to PGRs, including the way forward for a more inclusive, sound, viable, and sustainable intellectual property (IP) regime that takes care of the interest of the conservers of PGRs including the Indigenous Traditional Knowledge (ITK) holders, tribes and farmers relating to the PGRs, and their informal and formal innovations carried out by scientists including breeders.

15.2 International Legal Instruments Related to Access and Use of Genetic Resources

As agriculture progressed, a number of species that were not native were introduced from different parts of the world. Thus, exchange of PGRs offers enormous opportunities for sustainable agriculture as there is a continuous need of PGRs for utilization in various crop improvement programs. For sustainable agriculture, and food and nutritional security, the planners and policymakers at the national and international levels decided to strengthen the activities related to PGR management. Three international negotiations impacted this rise to prominence. These are the Convention on Biological Diversity (CBD), the International Treaty on Plant Genetic Resources for Food and Agriculture (ITPGRFA), and the Agreement on Trade-Related Aspects of Intellectual Property Rights (TRIPS) as part of the Agreement establishing the World Trade Organization (WTO). These negotiations, and the rapid advancements in the fields of molecular biology, biotechnology, and bioinformatics, have led to the emergence of new legal, political, and technological regimes that regulate access to PGR. The issues of ownership and utilization of germplasm and benefits therefrom, and related discussions, resulted in a paradigm shift from PGR being a "common heritage of mankind" to "sovereign rights of nations." The current barriers to the free flow of genetic resources (GRs), increasing application of intellectual property rights, and the vast potential of biological wealth yet to be tapped through bioprospecting and genetic engineering have placed greater demands on nations to adjust to the changing scenario of PGR management.

The free availability of germplasm, especially from ex situ collection of Consultative Group on International Agricultural Research (CGIAR) centers, helped most countries to strengthen plant breeding programs. India also benefited by importing germplasm, varieties, and new crop species from other countries, which have improved agricultural productivity and led to diversification of crop species. Things changed fast and the paradigm shift from the free flow of genetic resources to a restricted exchange was introduced when CBD came into force. Developing countries began to argue that common heritage implied the existence of common responsibilities, an important point given concern over the loss of diversity in farmers' fields and the absence of any global system, or mechanism, for preventing genetic erosion. The concept behind this was that profits accrued from developing commercial products such as new varieties or new compounds should be shared with the provider of the genetic material/resources used to develop the product, and thus various issues related to access to germplasm came into the forefront. The exchange scenario changed fast and for regulating access to PGRs and ensuring equitable sharing of benefits, policies and programs are now well placed and are regulated under different Acts and Treaties. These international agreements and conventions are described briefly and followed by the extant national legislation with regard to the management of PGRs. A summary of these is also provided in Tables 15.1 and 15.2 as a reference.

15.2.1 Convention on Biological Diversity (CBD)

The extinction of species is a natural phenomenon that is largely attributed to human activity. This, coupled with heightened awareness of development inequalities between countries or communities that have abundant biological resources and those wanting to conserve or harvest the same resources, led to the development of a multilateral agreement, namely, the Convention on Biological Diversity (CBD). The idea of a global convention for conserving biodiversity emerged from the recommendation of the International Union for Conservation of Nature and Natural Resorces (IUCN) General Assembly in the mid-1980s. The United Nations Environment Program (UNEP) established a working group on biodiversity. Although formal negotiations did not begin until November 1990, the Convention was concluded only 18 months later in May 1992. Issues covered in the final draft went far beyond original expectations. While the in situ conservation of resources is a pivotal aspect of the Convention, many other far-reaching issues were also included. These were sustainable use, incentive measures, IPRs, technology transfer, access to genetic resources (GRs), and financial mechanisms to assist countries in implementing the Convention.

As the negotiations started and proceeded, developing nations brought the issues relating to the burden of conservation of biodiversity and the utilization of biological resources into the scope of negotiations. The developing nations argued that they had already exploited biodiversity sourced from developing countries. Conservation of biodiversity by the developing countries entailed costs, which the developed nations had to share as a *quid pro quo*. This led to the incorporation of provisions in the Convention for setting up a financial mechanism and adequate generation of predictable additional and new resources by the developed nations to help developing nations to conserve biodiversity.

Table 15.1 International and National legal instruments related to access and use of genetic resources

International legal instruments *(Relevant Indian legislations)*	Scope/applicability	Year of adoption/status
Convention on Biological Diversity (CBD) *(Biological Diversity Act, 2002)*	• Covers all biodiversity-related issues • Provides general principles for access and benefit sharing (ABS)	• Legally binding for countries that have ratified • Ratifying countries must adopt appropriate legislation/regulations and/or bring existing ones into harmony with the Convention
CBD/International Protocol on Biosafety (Cartagena Protocol) *(Guidelines for Safe Movement of Living Modified Organisms [LMOs] by Department of Biotechnology (Govt. of India) (DBT), 2002, under the Environment Protection Act [EPA], 1986)*	• Provides international regulatory framework for trade and protection of environment from adverse effects of products of modern biotechnology	• Adopted in January 2000
International Regime on Access and Benefit Sharing (*Nagoya Protocol on ABS*)	• Internationally recognized framework for fair and equitable sharing of benefits arising from utilization of genetic resources through designated check points and ensuring compliance with mutually agreed terms • Binding for all parties	• Adopted on October 29, 2010 • Entered into force on October 12, 2014
International Treaty on Plant Genetic Resources for Food and Agriculture (ITPGRFA)	• Covers all PGRFA (does not regulate non-food and non-agricultural uses) and addresses diverse topics, including conservation, use, international cooperation, technical assistance and farmer's rights • Establishes multilateral system (MLS) for selected crops (over 35 food crops and 29 forages species) • Sets rules for access and benefit sharing for these materials, both ex situ and in situ, while respecting property rights	• Adopted by Food and Agriculture Organization (FAO) Conference in 2001 • Entry into force after the 40th country ratified in November 2004 • Legally binding for all countries that ratify and not applicable to those that do not • Countries that ratify will be required to bring national laws and regulations into conformity with the Treaty • CGIAR centers have signed agreements with the Treaty's Governing Body in order to adhere to the Treaty formally, and put all the material available with them under the MLS

(continvued)

Table 15.1 (continued)

International legal instruments *(Relevant Indian legislations)*	Scope/applicability	Year of adoption/status
WTO-TRIPS Agreement *(Protection of Plant Varieties and Farmer Rights Act, 2001; Patent (Amendments) Act, 2005; Geographical Indications (GIs) of Goods (Registration and Protection Act, 1999)*	• Member States must comply with minimum standards of protection of IP • Must ensure protection of microorganisms, non-biological and microbiological processes and plant varieties that meet protection criteria	• Entered into force in January 1995 after the Marrakesh Ministerial Meeting in April 1994 • Legally binding on all WTO Members • Has possibility for trade sanctions for those found not in compliance
International Plant Protection Convention *Plant Quarantine (Regulation of Import into India) Order, 2003, and its amendments issued under the Destructive Insects and Pests (DIP) Act, 1914; Seed Bill, 2010*	• An instrument and forum for harmonizing phytosanitary measures used in regulating international exchange of plants and planting material	• Established in 1952 (last revised in 1997)
Union for Protection of Plant Varieties (UPOV)	• UPOV aims to maximize plant breeding efforts by providing a model for securing protection under UPOV for plant breeders' rights for plant varieties	• Four versions; only 1991 is still open for joining • Legally binding on Parties

The Convention was signed by 150 countries at the United Nations Conference on Environment and Development in 1992 in Rio de Janeiro (the Earth summit) and entered into force on December 29, 1993. As of 2023, the Convention has 196 Parties, which includes 195 states and the European Union, with the notable exception of the USA.

The Convention is a historic international treaty, in that it not only reflects the commitment of the global community to the conservation and sustainable use of biodiversity but also visualizes the sharing of benefits arising out of the utilization of GRs with the countries of origin. The three principal objectives of the Convention are given in Box 15.1.

Box 15.1 Objectives of CBD

The Convention of Biological Diversity (CBD) entered into force on December 29, 1993. It has three main objectives:

1. The conservation of biological diversity
2. The sustainable use of components of biological diversity
3. The fair and equitable sharing of the benefits arising out of the utilization of genetic resources

Table 15.2 National legal instruments related to access and use of genetic resources

National legislations (Act/guidelines)	Objectives	Scope/applicability
Protection of Plant Varieties and Farmer Rights (PPVFR) Act, 2001) (Effective from January 2005)	• To provide for an effective system of protection of plant varieties • To protect rights of farmers and plant breeders • To ensure availability of high-quality seeds/improved varieties to farmers	• New varieties to be protected (registered) under the Act should be novel, distinct, stable, and uniform (NDUS), and shall be subjected to tests by the PPVFR Authority • Application for new varieties should be accompanied with complete passport data of all the genetic resources used in the development of new material • Proof of lawful acquisition of all materials used in development of new varieties to be furnished
The Biological Diversity Act (Effective from April 2004) The Biological Diversity (Amendment) Act, 2023	• Conservation and sustainable use of biological diversity • Regulation of access to biological resources • Secure sharing of benefits out of use of Indian biological resources/use of traditional knowledge	• The access to Indian biological resources by non-Indians shall be subject to the approval of National Biodiversity Authority (These conditions do not apply to transfer of genetic resources under any collaborative project approved by the Government of India, and also transfer of results under approved projects)
Patent (Amendments) Act, 2005 (Effective from January 2005)	• New amendments include patenting on newly developed (genetically modified) microorganisms and microbiological processes that conform to the criteria of novelty, inventive step, and utility	• Source of biological resources used in the development of patents is to be acknowledged • All biological resources used in development of patents to be deposited in the designated repositories before sealing of patents
Geographical Indications of Goods (Registration and Protection) Act, 1999	• Applicable to goods such as agricultural goods, natural goods or manufactured goods as originating and manufactured in a territory of a country where the given quality/reputation is essentially attributable to its geographic region (Landraces of crop specific to certain agro-climatic regions of the country with attributable quality and reputation)	• Agricultural products granted GI, e.g., Darjeeling tea, Malabar cardamom, Hill banana of Karnataka, Malihabadi mango

The objective of equitable sharing of benefits has to be further seen in the reaffirmation of sovereign rights of states over their biodiversity, which entitles a country to equity and farmers in sharing of benefits arising out of the utilization of their genetic resources.

The important point to note about the Convention is that facilitation of access by developing countries is linked with equitable sharing of benefits, thus making it a two-way process. Technology flows are also founded on the principle of equity with the developing countries providing resources and traditional knowledge (TK), and developed countries sharing and transferring technologies including biotechnology and providing financial resources to help developing countries meet their commitments and realize benefits.

The rights of the Parties to the CBD inter alia include:

- Sovereign rights to exploit their own resources
- Access to GRs subject to national legislation
  - Mutually agreed terms (MAT)
  - Prior informed consent (PIC)
- Developing countries' right to access/transfer of technology
  - Fair and most favorable terms
  - Concessional and preferential terms where mutually agreed
  - Use of financial mechanism for the purposes
- Right to participate in biotechnological research to the provider of GRs

The obligations and commitments of all the parties include

- Conservation and sustainable use by developing national strategies, plans, or programs.
- Rehabilitate and restore degraded ecosystems.
- Protect threatened species and their populations through legislation or regulation.
- Ensure safety in use and release of living modified organisms (LMOs).
- Legislative, administrative, and policy measures for technology transfer issues for sharing in an equitable way the results of research and development and benefits arising from the commercial and other utilization of genetic resources with the party providing such resources.
- Cooperate to ensure that IPRs are supportive of the objectives of the convention.
- Other miscellaneous obligations, for example, incentives, research training, public education and awareness.

15.2.2 The Biological Diversity Act (BDA), 2002 and Biological Diversity Rules (BDR), 2004

Being a signatory to the Convention on Biological Diversity (CBD), Government of India enacted the Biological Diversity Act (BDA), 2002 and also notified the Biological Diversity Rules (BDR), 2004 and established National Biodiversity Authority (NBA) in 2003. As per the BDA, 2002, no person from outside India or a corporate body, association, or organization incorporated or registered in India having non-Indian participation in its share capital or management can access any biological resources or knowledge associated, for research or for commercial utilization, or for bio-prospecting and bio-utilization, without proper approval of the NBA. No person can apply for any IPR in or outside India for any invention based on any research or information on a biological resource obtained from India without obtaining approval from the NBA. The entities requiring to take the approval of the NBA are namely a person who is not a citizen of India; a citizen of India who is a non-resident as defined in clause (30) of Section 2 of the Income Tax Act, 1961; and a corporate body, association, or organization (a) not incorporated or registered in India, or (b) incorporated or registered in India under any law for the time being in force, which is controlled by a foreigner within the meaning of Clause (27) of Section 2 of the Companies Act, 2013 as defined in Section 3(2) of BDA, 2002.

The definition of biological resources in the Act includes plants, animals, and microorganisms or parts thereof, and their genetic material and byproducts (excluding value-added products) with actual or potential use or value, but does not include human genetic material. At the national level, access to biological resources occurring in India is regulated by BDA, 2002 and BDR, 2004. Collaborative research projects involving the transfer or exchange of biological resources or information between Government-sponsored institutions of India, and such institutions in other countries, are however exempted, if they conform to the policy guidelines approved by the Central Government. Under this provision of exemption, the National Bureau of Plant Genetic Resources (NBPGR) facilitates the export of PGRs, which are covered under Collaborative Research Projects through established procedures.

Recently the Biological Diversity Act 2002 has been amended known as The Biological Diversity Bill 2021, and the major amendments and concerns are listed in Box 15.2.

Box 15.2 Key Provisions of the Biological Diversity (Amendment) Bill, 2021

- In December 2021, the Biological Diversity (Amendment) Bill, 2021 was introduced in Lok Sabha.
- **The Bill seeks to amend the Biological Diversity Act, 2002 to:**
  - Encourage the Indian system of medicine and cultivation of wild medicinal plants
  - Facilitate fast-tracking of processes for research, patent application, and transfer of research results; decriminalize offences; and encourage foreign investment in the sector
- **Access to biological resources and associated knowledge:**
  - The Biological Diversity Act requires prior approval or intimation to the regulatory authority based on the origin of the entity for obtaining biological resources occurring in India or associated knowledge.
  - The Bill amends the classification of entities, list of activities requiring intimation, and adds exemptions.
- **Approval for Intellectual Property Rights (IPR):**
  - The Act specifies that approval of National Biodiversity Authority (NBA) is required before: (a) applying for IPR involving biological resources obtained from India, or (b) sealing of patent.

- The Bill provides that approval will be required before the grant of IPR instead of before the application itself.
- **Benefit Sharing:**
  - Under the Act, NBA is required to determine terms of benefit sharing while granting approvals for various activities.
  - Benefit sharing refers to requiring applicants to share monetary and non-monetary benefits with benefit claimers and local people.
  - Benefit claimers are conservers of biodiversity, or creators or holders of associated traditional knowledge.
  - The Act makes benefit sharing provisions applicable to research, commercial utilization, as well as bio-survey and bio-utilization for certain entities.
  - The Bill removes its applicability from research, and bio-survey and bio-utilization.
- **Offences and Penalties:**
  - The Bill decriminalizes the offences and makes offences punishable with a penalty between 1 lakh rupees and Rs 50 lakh.

Criticism of the Biological Diversity (Amendment) Bill, 2021

- **Reduced role of Local Bodies:**
  - The benefit claimers and local people will not be directly involved in setting the terms and conditions.
- **Removal of Prior Informed Consent:**
  - There is no provision for a mechanism for obtaining prior informed consent of the local and indigenous communities. This may be in contrast with the framework under Nagoya Protocol.
- **Change of Adjudicating Authority:**
  - The Bill changes the adjudicating authority from a Judge to a government official.
  - The penalty decisions will be based on an inquiry instead of a judgement after arguments in an open court.

15.2.2.1 Implementation of the Act

National Biodiversity Authority (NBA) was established in the year 2003 for implementing the Act at the national level. States have established State Biodiversity Boards (SBBs) as provided for in Section 22 of the Act, and Biodiversity Management Committees (BMCs) at the local level as provided for in Section 41 have also been established. Similarly, the Union Territories (UTs) are in the process of establishing the UT Biodiversity Council.

The Act regulates the activities of access to biological resources and/or associated knowledge for research, commercial utilization, bio-survey, and bio-utilization (Section 3); Transfer of results of research (Section 4); Applying for intellectual property rights in or outside India, based on any research or information on biological resources (Section 6); and transfer of accessed biological resources (Section 20). The NBA is implementing the provisions of the BDA and BDR thereof.

15.2.3 Nagoya Protocol on Access and Benefit Sharing

India has ratified the Nagoya Protocol (NP), which is an international agreement that provides a transparent legal framework for the effective implementation of one of the three objectives of the CBD: the fair and equitable sharing of benefits arising out of the utilization of genetic resources. The Protocol aims for appropriate access to genetic resources and appropriate transfer of relevant technologies, taking into account all rights over those resources and to technologies, and by appropriate funding. NP entered into force on October 12, 2014. The regulations for access to PGR under the provisions of BDA, 2002 are in harmony with the Convention and the Nagoya Protocol (NP). The regulatory mechanism for access to biological resources thus ensures compliance with the Convention and NP.

NP provides a strong base for legal certainty and transparency for both providers and users of genetic resources. As a non-monetary benefit, NP provides for the provision of transfer of knowledge and technology that make use of genetic resources, including biotechnology, or that are relevant to the conservation and sustainable utilization of biological diversity to the provider of PGR under fair and most favorable terms, including on concessional and preferential terms where agreed. It recognizes the importance of promoting equity and fairness in negotiation between providers and users.

These are based on bilateral agreements guided by the principles of PIC-MAT. Each party needs to put in place domestic laws and regulations to define the conditions of access and equitable sharing of benefits arising out of the utilization of genetic resources. The Protocol calls for designated check points in each country to provide access and to ensure benefit sharing. The Protocol also recognizes the importance of other international agreements that are already in place to ensure benefits on utilization of genetic resources that are not contrary to the objectives of CBD. Ministry of Environment, Forests, and Climate Change (MoEFCC) notification (Guidelines on Access to Biological Resources and Associated Knowledge and Benefits Sharing Regulations, 2014) has put in place the guidelines for benefit-sharing rates depending on the type of biological resources and category of utilization. The Nagoya Protocol thus creates incentives for conservation and sustainable utilization of genetic resources and therefore enhances the contribution of biodiversity to development and human well-being.

15.2.4 International Treaty on Plant Genetic Resources for Food and Agriculture (ITPGRFA)

CBD does not categorize the nature of biological resources and does not differentiate the treatment of plant genetic resources for food and agriculture (PGRFA). The special nature of PGRFA and the interdependence of countries, hence need to be treated differently than biological resources as access to PGRFA is critical for ensuring food security. Legally binding ITPGRFA was negotiated as a direct response to CBD in 2001, and came into force in 2004, to facilitate access to PGRFA in harmony with CBD, through an efficient mutually agreed multilateral system (MLS) of access and benefit sharing. Access is only for research, breeding, and training and not for chemical, pharmaceutical, non-food/feed, or industrial use. No intellectual property rights can be claimed on the form received from MLS that limits the facilitated access to PGRFA/genetic parts or components. To ensure that the germplasm is used only legally, the Standard Material Transfer Agreement (SMTA) binds terms and conditions on the recipient to use it. The treaty works through intergovernmental agreements as required by CBD. Two key components of the treaty are: (a) each party should facilitate germplasm exchange for a list of crops covered in Annexure I (includes 35 food crops and 29 forages), which are important for food security and for which countries are interdependent; and (b) it provides a system to enforce an equitable share of benefits, through a central fund for obligatory payments to the country of origin.

FAO conference established the FAO Commission on PGR in 1983 (first intergovernmental forum) to deal with agricultural biological diversity, which formulated an international undertaking (IU) for PGR management and use. IU became the Commission on GRFA (CGRFA) in 1995 and broadened its scope to cover all components of biological diversity. Thus, CGRFA in harmony with CBD negotiated the revision of IU on PGRFA from November 1994 and continued till November 2001. The revised text of IUPGR was submitted to the 31st Session of FAO Conference that adopted it as the International Treaty on Plant Genetic Resources for Food and Agriculture (ITPGRFA) on November 3, 2001. Legally binding ITPGRFA was thus negotiated as a direct response to CBD in 2001, and came into force in 2004 to facilitate access to PGRFA in harmony with CBD, through an efficient mutually agreed system of access and benefit sharing. Access here is only for research, breeding, and training and not for chemical, pharmaceutical, or non-food/feed industrial use. No intellectual property rights can be claimed on PGRFA in the form received from the multilateral system that limits the facilitated access to PGRFA/genetic parts or components. The Treaty aims to guarantee the future availability of the diversity of PGR for Food and Agriculture. India has signed and ratified the ITPGRFA in 2003.

The Treaty is historic, and for the first time the enormous contribution made by the farmers to the development and conservation of crop diversity was recognized in the Treaty. The objectives of the Treaty are given in Box 15.3.

The text of the Treaty is organized in various Articles in ten Parts.

Box 15.3 Objectives of ITPGRFA

The Treaty entered into force on June 29, 2004. The three objectives of the Treaty are:

1. Conservation of plant genetic resources for food and agriculture (PGRFA)
2. Sustainable use of PGRFA
3. The sharing of the benefits arising from the use of PGRFA in a fair and equitable way

Article 1: Overall objectives and directions of the Treaty.
Article 2: Provides the use of terms and definitions.
Article 3: Scope of the treaty.
Articles 4–8: Conservation and sustainable utilization of plant genetic resources for food and agriculture. General provisions contained in Part II.
Article 9: Farmer's rights (FRs) are contained in Part III.
Articles 10–13: Multilateral system (MLS) of access and benefit sharing contained in Part IV.
Article 18: Financial provisions contained in Part V.
Articles 19–35: Supporting components contained in Part VI.

Different committees of the Treaty are represented in Box 15.4. The key area of work of the Treaty is represented by six different themes, namely benefit-sharing fund, global information system, sustainable use, farmer's rights, MLS, and compliance. Articles 4–8 of the Treaty indicate policies and measures for the national government about the conservation and sustainable use of PGRFA. Sustainable utilization is not defined in the Treaty; however, it implies making use of crop diversity to meet the food security needs of present generations, without compromising its availability, as the basis of food security for future generations. General provisions of the Treaty apply to all PGRFA, not just Annexure I crops (Annexure 1 crops are available for exchange under MLS). Article 9 of the Treaty provides internationally agreed common ground for FRs, and provides measures for contracting parties to take at the national level for the protection and promotion of FRs. The Treaty provides protection of traditional knowledge, rights to participate in the sharing of benefits arising out of the use of PGRFA, and the right to participate in decision-making related to PGRFA. The Treaty promotes in situ conservation through on-farm management of crops and ex situ conservation by establishing a network of genebanks. In situ farm management allows crops to continuously evolve and adapt to the changing environmental conditions. Ex situ conservation facilitates enhanced access and utilization, in addition to serving as a safety backup. The Treaty established a unique mechanism of the MLS as contained in Articles 10–13 of the Treaty. It covers 64 crops listed in Annexure I of the Treaty (Article 11).

Box 15.4 Different Committees of ITPGRFA

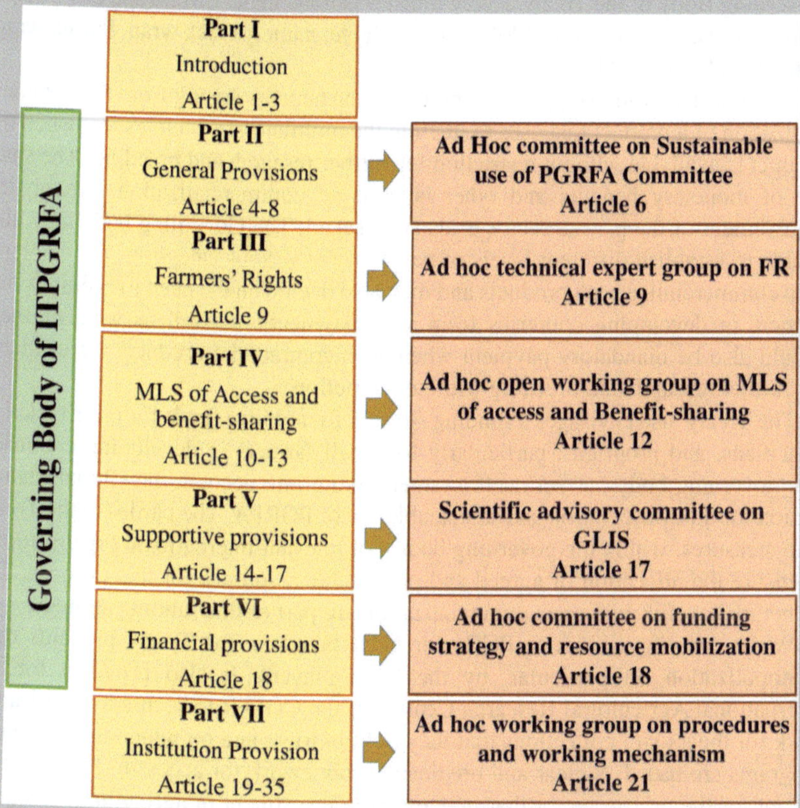

The Treaty being legally binding has the international commitment to the improvement of the world's key food and feed crops. Its centerpiece is a multilateral system of facilitated access and benefit sharing that directly supports the work of breeders and farmers everywhere. Each ratifying government agreed to ensure the conformity of its laws, regulations, and procedures with its obligations under the Treaty. The Governments of the countries that ratified the Treaty form its governing body. Currently, 151 countries are the contracting parties including India who have ratified the Treaty.

The Treaty provides for facilitated access to a specified list of PGRFA, balanced by benefit sharing in the areas of information exchange, technology transfer, capacity building, and commercial benefit sharing. Presently, the multilateral system applies to a list of over 64 plant genera, including 35 food crops and 29 forage plants, agreed on the basis of interdependence and food security and are referred to as Annexure 1 crops. The list may be revised or expanded based on the criteria of food security and interdependence. The conditions for access and benefit sharing are set out in a "Standard Material Transfer Agreement" (SMTA), adopted by the Governing Body of the Treaty. Every transfer is done under SMTA and can be concluded online. Four forms of SMTA are available, namely Click wrap, Shrink wrap, Signed, and Easy SMTA.

An important point is equitable sharing of the benefits arising from the commercialization of a product that uses PGR from the multilateral system except when the product is available without restriction for further research and breeding. The sharing of monetary benefits and other benefits of commercialization through the involvement of the private sector is also envisaged. The Governing body may also decide to establish different levels of payment for various categories of recipients who commercialize such products and may also decide on the need to exempt small farmers in developing countries from such payments, as well as whether there should also be mandatory payment when such commercial products are available for further research and breeding without restriction.

The Treaty also envisages a funding strategy to mobilize funds for priority activities, plans, and programs, particularly for small farmers in developing countries. The governing body has set itself a target, taking into account the Global Plan of Action for the conservation and sustainable use of PGRFA. The parties to the Treaty take measures, within the governing bodies of relevant international organizations, to ensure the allocation of agreed and predictable resources for this purpose. The monetary benefits paid on commercialization are part of this funding strategy.

While ensuring access to PGRs for plant breeders, the Treaty prevents their monopolization, in particular, by the large players. It also provides for the International Agricultural Research Centers of the CGIAR to secure a legal framework for the ex situ collections that they hold in trust, and on which their research programs are based. A clear and predictable framework for access to PGRs is also set out for the private sector that may promote investment in agricultural research. The Treaty thus recognizes the enormous contribution that farmers and farming communities have made and continue to make to the conservation and development of PGRs, and puts the responsibility for realizing farmers' rights on national governments. This could be done through protecting relevant traditional knowledge, and the right to take part equitably in developing the mechanism for benefit sharing and in decision-making regarding the conservation and sustainable use of PGRs. The Treaty necessitates the contracting parties to include all PGRFA (of the agreed list of crops) that are under their management and control and in the public domain. Further, it requests countries to take appropriate measures to encourage natural and legal persons within their jurisdiction who hold such PGRFA to include them in the multilateral system. The countries are required to provide facilitated access to PGRs. Access to PGRFA needs to be accorded expeditiously, without the need to track individual accessions and free of charge, or, if a fee is charged, it shall not exceed the minimal cost involved. The available passport data of the germplasm provided shall also be made available to the recipients.

The obligations defined above, therefore, entail an appropriate national network to strengthen a single window system of exchange. Genetic resources of all the crop genera specified in the list of crops need to be maintained at respective crop institutes/designated sites for expeditious supply. Since countries including India have started exercising sovereign rights over biological resources as established under CBD, by enacting various legislations, the new rules potentially create an environment to reduce biopiracy-patents on natural product inventions, unless all sources of raw material and public knowledge about the species are fully disclosed. Also, if a biological material is removed illegally from a source country and profits are drawn from it, the source country could recover some of the profits, based on a theory of misappropriation. It is, therefore, probable that any germplasm that has not been obtained through legitimate means would not be accepted subsequently. To ensure that access and benefit-sharing mechanisms are met, there is an urgent need to have a watchdog system for documentation, continuous monitoring, and analyses of germplasm use and flow. As a single window system of exchange in India, ICAR-NBPGR (Indian Council of Agricultural Research–National Bureau of Plant Genetic Resources) is regulating access of germplasm to other countries as per the current legal framework. It is continually serving the researchers/breeders/other stakeholders by making available the diverse PGRFA. The purpose of access, however, is solely for utilization and conservation for research, breeding, and training.

15.2.4.1 Implementation of ITPGRFA

In India, the implementation of the ITPGRFA is through the National Focal Point (NFP) at the Department of Agriculture Cooperation and Farmer's Welfare (DACFW) in close coordination with ICAR-NBPGR. ICAR-NBPGR coordinates and promotes the conservation and sustainable use of PGRFA through an integrated approach. Over 0.4 million accessions of agricultural and horticultural crops are conserved in National Genebank, which includes diversity of seeds of cultivated plants and related wild species.

To promote access, and to ensure fair and equitable sharing of benefits arising from the utilization of genetic resources and associated traditional knowledge, gazette notifications are already in place for the facilitated exchange of Annexure 1 crops (MoEFCC Notification S.O. 3232 (E) dated December 17, 2014). Regulation for facilitated access to PGRFA under the Multilateral System for Annexure 1 crops is notified and a list of crops is available at DACFW website PGR Portal. India has notified more than 26,500 accessions to the MLS of the Treaty.

PGRFA collected before 1993 and held in "trust" by International Agricultural Research Centres (IARCs) is also made available in accordance with the provisions of SMTA after the agreements between IARCs and FAO. All requests for import of PGRFA listed in Annexure I to the Treaty should be addressed to NFP by the Contracting Parties.

15.2.5 World Trade Organization-Trade Related Intellectual Property Rights (WTO-TRIPS)

GATT (General Agreement on Tariff and Trade) Uruguay Round led to the establishment of WTO in 1995. WTO has half a dozen intergovernmental agreements with direct impact on agriculture, namely Applications of Sanitary and Phytosanitary Measures (SPS) and Union Internationale pour la Protection des obtentions Végétale (UPOV; International Union for the Protection of New Varieties of Plants; Conventions in 1968, 1972, 1979, 1991). UPOV aims to maximize plant breeding efforts by providing a model for securing protection for plant breeders' rights for plant varieties. Version 1991 is still open for joining. It is legally binding on Parties but India is not a member of UPOV.

One of the most comprehensive agreements on intellectual property (IP) is the WTO Agreement on Trade-Related Aspects of Intellectual Property Rights (TRIPS) that entered into force in January 1995 after the Marrakesh Ministerial Meeting in April 1994. It plays a central role in facilitating trade in knowledge and creativity and is legally binding on all WTO Members. WTO-TRIPS led to competition and secure intellectual creations using IPR tools. The agreement asks for minimum requirements for the protection of intellectual property rights, including patents, copyrights, trademarks (TMs), industrial designs (IDs), trade secrets, and integrated circuits. India being a member country is required to offer protection under TRIPS. An important article of the agreement related to PGR is Article 27.3 of TRIPS, which provides for exclusion of plants and animals (other than microorganisms) and essentially biological processes for the production of plants or animals other than non-biological and microbiological processes. However, the member countries are required to protect varieties either by patents or by an effective *sui generis* system or by any combination thereof. In this chapter, detailed information on the *sui generis* system developed in India and also followed by IPR Protection under various forms is described.

To comply with the mandate of Article 27.3(b) of TRIPS, India chose the option of *sui generis* system and adopted the Protection of Plant Varieties and Farmer Rights Act (PPVFRA), 2001—the genesis of the Indian legislation. The existing Indian Patent Act, 1970 excluded agriculture and horticultural methods of production from patentability. The *sui generis* system for the protection of plant varieties was developed integrating the rights of breeders, farmers, and village communities, and taking care of the concerns for equitable sharing of benefits. It offers flexibility with regard to protected genera/species and level and period of protection when compared to other similar legislations existing or being formulated in different countries. The Act covers all categories of plants, except microorganisms. The genera and species of the varieties for protection are notified through a gazette after the appropriate rules and by-laws are framed for the enforcement of the Act. The salient features of the Act are described here.

15.2.6 Protection of Plant Varieties and Farmers Rights Act (PPVFRA)

The objectives of the Act are: (a) to provide for the establishment of an effective system for the protection of plant varieties; (b) to provide for the rights of farmers and plant breeders; (c) to stimulate investment in research and development and to facilitate the growth of the seed industry; and (d) to ensure the availability of high-quality seeds and planting materials of improved varieties to farmers.

The Act has 11 chapters and is divided into 97 clauses. The first chapter has the title, and the definitions used in the context of the Act. The last chapter is on miscellaneous clauses. The other nine chapters deal with PPVFR authority, registration of plant varieties, duration and effect of registration and benefit sharing, surrender and revocation of certificate, farmer's rights, compulsory license, plant varieties protection appellate tribunal, finance, accounts, audit, infringement, offences and penalties, etc.

15.2.6.1 PPVFR Authority

The Authority has a chairperson and 15 members as representatives of concerned ministries and departments, seed industry, farmers' organizations, tribal communities, State-level women's organizations, etc.

15.2.6.2 Eligibility Criteria

For a variety to be eligible for registration, it must conform to the criteria of novelty, distinctiveness, uniformity, and stability (NDUS), as described below. For the Act, a new variety shall be deemed to be:

(a) *Novel*, if, at the date of filing of the application for registration for protection, the propagating or harvested material of such a variety has not been sold or otherwise disposed of by or with the consent of its breeder or his successor for the purpose of exploitation of such variety:

- In India, earlier than 1 year.
- Outside India, in the case of trees or vines earlier than 6 years, or, in any other case, earlier than 4 years, before the date of filing such applications.

  – Provided that a trial of a new variety that has not been sold or otherwise disposed of shall not affect the right to protection.

  – Provided further that on the date of filing the application for registration, the propagating or harvested material of such variety has become a matter of common knowledge other than through the aforesaid manner and shall not affect the criteria of novelty for such variety.

(b) *Distinct*, if it is clearly distinguishable by at least one essential characteristic from any other variety whose existence is a matter of common knowledge in any country at the time of filing of the application.

(c) *Uniform*, if subject to the variation that may be expected from the particular features of its propagation, it is sufficiently uniform in its essential characteristics.
(d) *Stable*, if its essential characteristics remain unchanged after repeated propagation or, in the case of a particular cycle of propagation, at the end of each such cycle.

15.2.6.3 Application Form

Every application for registration will have to be accompanied by the following information:

(a) Denomination assigned to such variety by the applicant.
(b) An affidavit sworn by the applicant that such variety does not contain any gene or gene sequence involving terminator technology.
(c) The application should be in such form as may be specified by regulations.
(d) A complete passport data of the parental lines from which the variety has been derived along with the geographical location in India from where the genetic material has been taken and all such information relating to the contribution, if any, of any farmer, village community, institution, or organization in breeding, evolving, or developing the variety.
(e) A statement containing a brief description of the variety, bringing out its characteristics of novelty, distinctiveness, uniformity, and stability as required for registration.
(f) Such fees as may be prescribed.
(g) Contain a declaration that the genetic material or parental material acquired for breeding, evolving, or developing the variety has been lawfully acquired.
(h) Such other particulars as may be prescribed.

The conditions stated above (a–h) shall not apply in respect of an application for registration of farmers' varieties.

15.2.6.4 Period of Protection

The certificate of registration issued under Section 24 or Subsection 98 of Section 23 shall be valid for 9 years in the case of trees and vines and 6 years in the case of other crops, and may be reviewed and renewed for the remaining period on payment of such fees as may be fixed by the rules made on this behalf subject to the conditions that the total period of validity shall not exceed:

1. In the case of trees and vines, 18 years from the date of registration of the variety.
2. In the case of extant varieties, 15 years from the date of the notification of that variety by the Central Government under Section 5 of the Seed Act, 1996.
3. In the other case, 15 years from the date of registration of the variety.

15.2.6.5 Payment of Annual Fee

The Authority may, with the prior approval of the Central Government, by notification in the Official Gazette, impose a fee to be paid annually, by every breeder of a variety, agent, and licensee thereof registered under this Act determined on the basis of benefit or royalty gained by such breeder, agent, or licensee, as the case may be, in respect of the variety, for the retention of their registration under this Act.

15.2.6.6 Breeders' Rights

The certificate of registration for a variety issued under this Act shall confer an exclusive right on the breeder or his successor or his agent or licensee, to produce, sell, market, distribute, import, or export the variety.

15.2.6.7 Researchers' Rights

The researchers have been provided access to protected varieties for Bonafide research purposes. This Section states, "Nothing contained in this Act shall prevent (a) the use of any variety registered under this Act by any person using such variety for conducting experiments or research, and (b) the use of a variety by any person as an initial source of a variety to create other varieties provided that the authorization of the breeder of a registered variety is required where the repeated use of such variety as a parental line is necessary for commercial production of such other newly developed variety."

15.2.6.8 Farmers' Rights

The farmers' rights of the Act define the privilege of farmers and their right to protect varieties developed or conserved by them (Chap. 6). Farmers can save, use, sow, resow, exchange, share, and sell farm produce of a protected variety except sale under a commercial marketing arrangement (branded seeds). Further, the farmers have also been protected against innocent infringement when, at the time of infringement, a farmer is not aware of the existence of breeder rights.

A farmer who is engaged in the conservation of genetic resources of landraces and wild relatives of economic plants and their improvement through selection and preservation, shall be entitled in the prescribed manner for recognition and reward from the Gene Fund, provided the material so selected and preserved has been used as donor of genes in varieties registrable under the Act.

The expected performance of a variety is to be disclosed to the farmers at the time of sale of seed/propagating material. A farmer or a group of farmers or an organization of farmers can claim compensation according to the Act, if a variety or the propagating material fails to give the expected performance under given conditions, as claimed by the breeder of the variety.

15.2.6.9 Communities Rights

The rights of the communities as defined, provide for compensation for the contribution of communities in the evolution of new varieties in quantum to be determined by the PPVFR Authority.

15.2.6.10 Registration of Essentially Derived Varieties

The breeder of the essentially derived variety shall have the same rights as the plant breeder of other new varieties, which include production, selling, marketing, and distribution, including export and import of the variety. The other eligibility criteria for an award of registration area are the same as for new variety registration under the Act.

15.2.6.11 Compulsory License

The authority can grant a compulsory license in case of any complaints about the availability of the seeds of any registered variety to public at a reasonable price. The license can be granted to any person interested in taking up such activities after the expiry of a period of 3 years from the date of issue of the certificate of registration to undertake production, distribution, and sale of the seed or other propagating material of the variety.

15.2.6.12 Benefit Sharing

Sharing of benefits accruing to a breeder from a variety developed from indigenously derived plant genetic resources has also been provided. The Authority may invite claims of benefit sharing of any variety registered under the Act, and shall determine the quantum of such award after ascertaining the extent and nature of the benefit claim, after providing an opportunity to be heard, to both the plant breeder and the claimer.

15.2.6.13 National Gene Fund

The National Gene Fund to be constituted under the Act shall be credited there to:

(a) The benefit sharing from the breeder.
(b) The annual fee payable to the Authority by way of royalties.
(c) By the compensation provided to the communities.
(d) Contribution from any national and international organization and other sources.

The fund will be applied for disbursing shares to benefit claimers, either individuals or organization, and for compensation to village communities. The fund will also be used for supporting conservation and sustainable use of genetic resources, including in situ and ex situ collection and for strengthening the capabilities of the Panchayat in carrying out such conservation and sustainable use.

15.3 Intellectual Property Rights (IPRs)

15.3.1 History and Introduction

Though there is no official record of the origin of IP, it is believed that a rudimentary form of IP was being practiced around 500 Before the Common Era (BCE) in Sybaris, a state of Greece. The natives of Sybaris were granted a year's protection for using their intellect to create any new improvement in luxury. A practical and pragmatic approach to IP governance started taking shape in medieval Europe. In 1623, Britain passed an Intellectual Property Legislation that entitled guilds (associations of artisans or merchants) to create innovations and bring them to market for trade purposes. However, this legislation brought a lot of resentment among the public, and thus was replaced by the "Statute of Monopolies," which gave the rights to the original creator/inventor for 14 years. Another legislation, "Statute of Anne," was passed by the British parliament in 1710. This legislation aimed at strengthening copyrights by providing rights to the authors for recreation and distribution of their work. The work could also be renewed for another 14 years. By the end of the eighteenth century and the beginning of the nineteenth century, almost every country started laying down IP legislation to protect their novel inventions and creations.

Anything of value today is being protected not by mere physical property but in the form of IPRs. IPRs and PGR is a very contentious topic involving complex relationships of various stakeholders for the simple reason that the PGRs are not evenly distributed across the globe. Largely, the developing countries are very rich in biological resources that include PGRs. On the other hand, the investment technology and intellectual property (IP) over PGRs is owned and controlled by the multinational companies (MNCs) belonging to the developed countries; thus, there is always a contention between these stakeholders.

IP is created by human intellect (mind) in the fields of arts, literature, science, trade, etc. Since IP is a novel creation of the mind, it is intangible (i.e., invisible and indivisible) in nature and differs from tangible property, such as land, houses, gold, and cars with which we are quite familiar. IPRs on the other hand are the rights given to the inventor (of IP) in conformity with the laws. These rights are given to the creator/inventor in exchange for revealing the process of creation/invention in the public domain. The inventor is conferred with the special rights to use, sell, and distribute the invention while restricting others from using the invention without his prior permission.

Broadly, IP comprises two branches, that is, Copyrights and Related Rights and Industrial Property Rights. Copyrights and Related Rights "refer to the creative expressions in the fields of literature and art, such as books, publications, architecture, music, wood/stone carvings, pictures, sculptures, films, and computer-based software/databases." Industrial Property Rights "refer to the Patents, Trademarks, Trade Services, Industrial Designs, and Geographical Indications." The economic and social development of a society is largely dependent on creativity. The protection provided by the IPR to the creators/innovators is an act of providing incentives for encouraging them to create more and motivating others to create new and novel things.

However, if IPR is practiced rigidly, it may have a negative impact on the progress of society. For example, compliance with the Trade-Related Aspects of Intellectual Property Rights (TRIPS) agreement has affected the farming community as they are unable to store seeds for the next crop. MNCs regulate the price of seeds, which is generally beyond the reach of a majority of the farmers. To circumvent the negative impact of IPR, certain laws, exceptions, and limitations associated with IPR have been enacted to maintain a balance between the interests of the creators/inventors and the community. For example, farmers' rights under the Protection of Plant Varieties and Farmers Rights (PVP&FR) Act, 2001 entitles them to many privileges. For example, Rights on seeds provides rights to the farmers to save seeds, use seeds, and share, exchange, or sell seeds to other farmers. Right to protect the farmers from infringement and other legal accusations levied upon them due to legal ignorance in using other plant varieties, as detailed above, is also provided. The use of copyrighted material for education and religious ceremonies is exempted from the operation of the rights granted in the Copyright Act. Similarly, a patent can be revoked in favor of compulsory licensing by the government during an emergency or a natural calamity. In addition, if an invention/creation is not in the interest of society, it is not registered by the government for grant of any rights associated with IP. For example, cloning of human embryos is banned for IP protection, and so is the creation of super microbial pathogens, which can play havoc with human lives. In addition, India is enriched with massive biodiversity and genetic resources and their use is embodied in what is referred to as Traditional Knowledge (TK). However, the use of such knowledge and resources is not limited to local contexts as many innovations relate to and draw on them. Therefore, the main issue of concern is to protect TK and genetic resources, which are rapidly coming under the governance of sometimes conflicting IPR policies. Different forms of IPRs are described as follows and the validity of different categories of IP Protection is summarized in Table 15.3.

Table 15.3 Validity of different categories of IP protection

Sl. no.	IPR	Maximum protection	Renewal	Act/Rule
1	Patent	20 years	[a]Every year (Mandatory)	The Patents Act, 1970 (Amended in 2005)
2	Trademark	Life long	After 10 years	The Trade Marks Act, 1999 (Amended in 2010)
3	Design	15 years	After 10 years for next 5 years	The Design Act, 2000, and Designs (Amendment) Rules, 2014
4	Copyright	60 years	Not required	The Copyright Act, 1957 (Amended in 2012)
5	Geographical Indication (GI)	Life long	After 10 years	The Geographical Indications of Goods (Registration and Protection) Act, 1999
6	Plant variety	15/18 years	6 years	PPVFRA, 2001

[a]Patents will cease and be transferred to the public domain if patents are not renewed within 6 months of the expiry of the concerned year by paying the renewal fee

15.3.2 Patents

The history of the Indian patent system dates back to the pre-Independence era of British rule. The first patent-related legislation in India was Act VI of 1856, adapted from the British Patent Law of 1852. The objective of this legislation was to encourage the inventions of new and useful manufacturers. The rights conferred to the inventor were termed as "Exclusive Privileges." In 1859, certain amendments were made to the Act, such as:

- Grant of exclusive privileges to useful inventions.
- Increase of priority time from 6 months to 12 months.
- Exclusion of importers from the definition of the inventor.

The world's first patent was granted in 1790 to Samuel Hopkins in the USA for the "making of pot ash and pearl ash by a new apparatus and process." In India, the first patent (known as "Exclusive Privileges" at that time) was awarded in 1856 to a civil engineer, George Alfred DePenning from Calcutta, for his invention, "An Efficient Punkah Pulling Machine."

A few years later, it was felt that Designs could also pass the criteria of the invention and thus should be included in the Patent Act. The new Act was rechristened as The Patterns and Designs Protection Act under Act XIII of 1872. This Act was further amended in 1883 (XVI of 1883) to include the provision of protection for Novelty in the invention. At the beginning of the twentieth century, all the earlier Acts related to inventions and designs were done away with the introduction of The Indian Patents and Designs Act, 1911 (Act II of 1911). As per this Act, the governance of patents was placed under the management of the Controller of Patents. In the next three decades, many amendments were introduced for reciprocal arrangements with other countries for securing priority dates. These amendments dealt with:

- Use of invention by the government
- Patent of Addition
- Enhancing the term of the patent from 14 years to 16 years
- Filing of Provisional Application and submission of Complete Application within 9 months from the date of filing the application

After India got Independence in 1947, many patent experts felt the need to review the Indian Patents and Designs Act, 1911, keeping the national interest (economic and political) in mind. A dedicated committee, chaired by a renowned Justice Bakshi Tek Chand (retired Judge of Lahore High Court), was constituted in 1949 to review the advantages of the patent system. The committee submitted a plethora of recommendations, including:

- Misuse of patent rights needs to be prevented.
- There must be a clear indication in the Act that food, medicine, and surgical and curative devices should be made available to the masses at the cheapest rate by giving reasonable compensation to the owner of the patent.

- Amendments in Sections 22, 23, and 23A of the Patent and Design Act, 1911 on the lines of the UK Patent Act.

These recommendations were introduced in the Act XXXII of 1950. Two years later, another amendment (Act LXX of 1952) was made to provide compulsory licensing of patents related to food, drugs, and chemicals killing insects and microbes. Based on these amendments, a Bill was presented in the parliament in 1953 but was rejected. In 1957, the central government constituted yet another powerful committee under the chairmanship of Justice N. Rajagopala Ayyangar to seek inputs for further strengthening the Indian Patent Law. The committee submitted its report to the government in 1959. It comprised two segments addressing (a) General aspects of the patent laws, and (b) Bill rejected back in 1953. The revised patent legislation was submitted to the Lok Sabha in 1965. After many hiccups, clarifications, and modifications, the Patents Act, 1970 (http://www.ipindia.nic.in/writereaddata/Portal/IPOAct/1_31_1_patent-act-1970-11 march2015.pdf) was introduced, superseding all the previous laws related to the patents.

However, the Indian Patents and Designs Act of 1911 remained applicable for designs only till 1994. With the signing of TRIPS Agreement and after a transition period of 10 years (1995–2005), to make domestic laws compatible with the international treaty, in 1999, The Patents (Amendment) Act, 1999 was introduced providing for the filing of applications for "Product Patents" in the areas of drugs, pharmaceuticals, and agrochemicals (earlier, only processes were protected under the Patent Act). The new Patent Act also included provisions for the grant of Exclusive Market Rights (EMRs) for the distribution and sale of pharma products on fulfilment of certain conditions. The second amendment to the 1970 Act was made through the Patents (Amendment) Act, 2002 (Act 38 of 2002). This Act introduced new Patent Rules, 2003, thus replacing the earlier Patents Rules, 1972. The major amendments were:

- The protection term of 20 years for all inventions from the date of filing
- Scope of non-patentable inventions including Traditional Knowledge expanded
- Disclosure of source and geographical origin of biological material made compulsory
- Provisions concerning convention countries simplified
- Establishment of Appellate Board
- Compulsory license provisions strengthened
- Simplification of procedures
- Harmonization with Patent Cooperation Treaty (PCT) provisions

With the rapidly changing scenario of IPR at a global level, a need was felt to further amend the Patent Act, 1970. The highlights of the Patents (Amendments) Act, 2005, were:

- Product patent for inventions in all fields of technology
- New forms of known substances are excluded to prevent ever greening of the patent
- Rationalization of the opposition procedure

- Introduction of pre-grant opposition by representation
- Introduction of post-grant opposition
- Compulsory license for export purposes
- Compulsory license for manufacture
- Extension of grace period from 6 months to 12 months for filing a patent, if published in government exhibition

Patent searches are an important part of any project proposal to enable researchers to assess the worthiness of their research outcomes as patentable or not and also to seize any such opportunity well in time. A patent search, also known as a patentability search, is a search of the existing things closest to the claimed invention through published works, patents, and other publicly accessible records (also known as "prior art"). The first step in the patent process is the patent search. Finding out how unique an invention is from what has already been invented is the purpose of a patent search. Patent search procedures are described in Box 15.5 below.

Box 15.5 Patent Searches

Patent search is essential before planning and execution of a research or before filing a patent application, which could aid in fulfilling the criteria of patentable invention.

Key points to consider for patent search:

1. Understanding the invention and identifying the keywords for conducting patent search
2. Searching for related patent documents (for a prior art) and non-patent literature
3. Searching all publicly accessible information

Patent information can be searched by using:

Keywords (Boolean Operators [AND, OR, or AND NOT] are used to combine or exclude keywords in a search, resulting in more focused and productive results)

The use of Boolean Operators is illustrated by the given theoretical example: if we want to search for "*markers linked to* ***anthracnose*** *resistance in* ***sorghum***," the keywords to search the patents may be depicted as given in Fig. A.

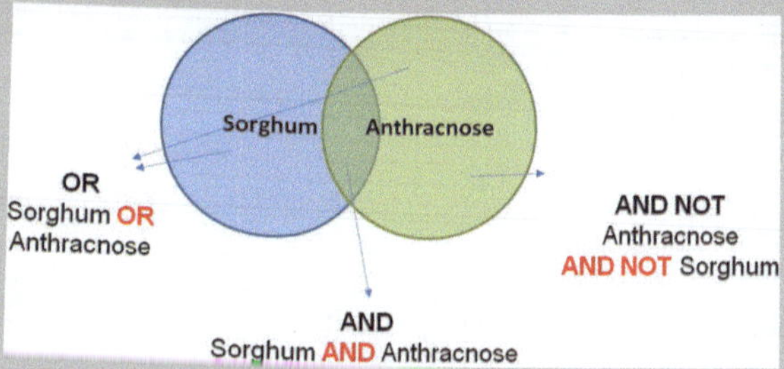

Fig. A Use of Boolean Operators (AND, OR, or AND NOT) for adding keywords in patent search

Dates (e.g., priority date, application date, publication date, grant date)

Patent reference or identification numbers (application number, publication number, patent number)

Names of applicants/assignees or inventors

Webpages/databases for searching the existing patents:

Webpage link	Patent Office/Maintained by	Search tool
https://iprsearch.ipindia.gov.in/publicsearch	Office of the Controller General of Patents, Designs and Trademarks (Ministry of Commerce and Industry), Government of India	Indian Patent Advanced Search System (inPASS)
http://www.uspto.gov/patft/index.html	United States Patent and Trademark Office (USPTO)	
http://ep.espacenet.com	European Patent Office	Espace net
https://patentscope.wipo.int/search/en/search.jsf	World Intellectual Property Organization (WIPO)	Patent scope
https://patents.google.com/	Google	Google Patents

The search of patents using Google Patents by entering different modes of Keywords is shown in Fig. B.

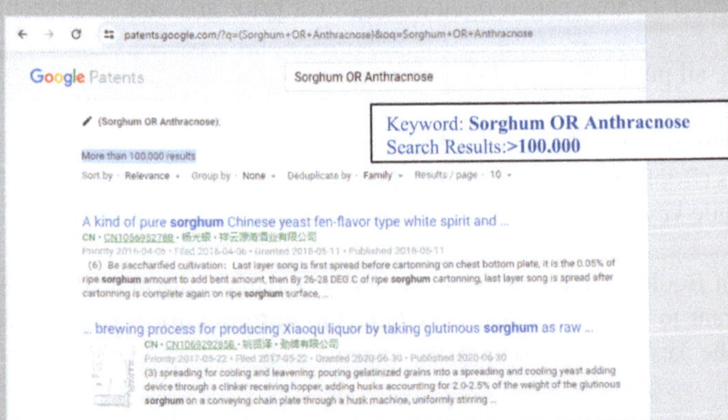

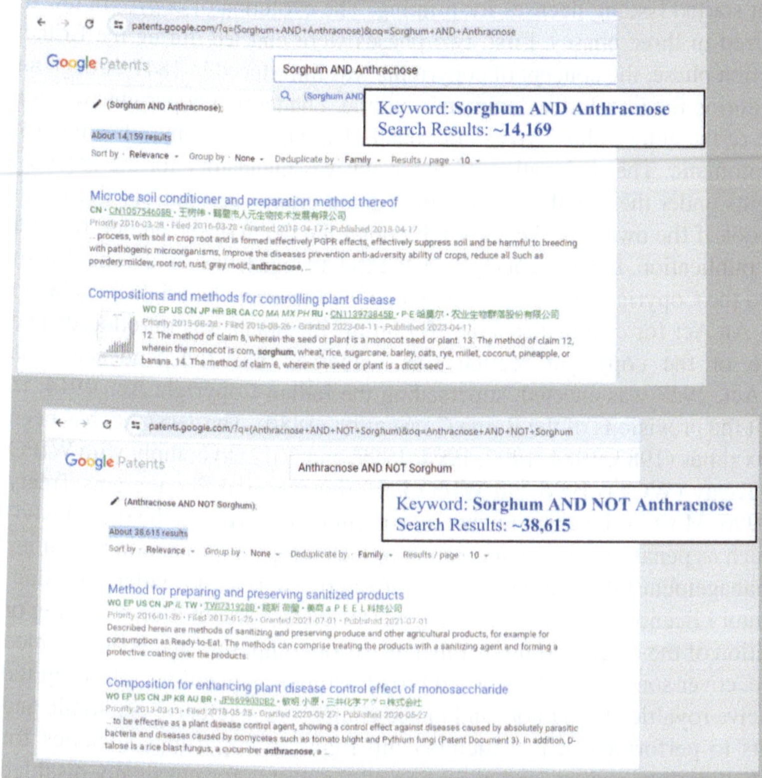

Fig. B An illustration of the search of patents using Google Patents search tool

Once a patent search is performed before deciding on a researchable issue, it can be modified according to the patent search results. For the filing of a patent, the patent search helps in analyzing the patentable criteria of new invention in comparison to the prior art/already existing patents in the area so that claims of the patent can be identified accordingly.

15.3.3 Copyrights and Related Rights

The concept of copyrights started way back in the fifteenth century. However, the actual need for copyright law was felt only after the invention of printers and copiers. Before the invention of printers, writing could be created only once. It was highly laborious and the risk of errors was involved in the manual process of copying by a scribe. During the fifteenth and sixteenth centuries, printing was invented and widely established in Europe. Copies of the Bible were the first to be printed.

The government had allowed the printing of the documents without any restrictions, but this led to the spreading of a lot of governmental information. Subsequently, the government started issuing licenses for printing. The evolution of copyright law in India occurred in three phases. First, two phases were enacted during the British Raj. In the first phase, the concept of copyrights was introduced in 1847 through an enactment during the East India Company's regime. The term of copyrights was for the lifetime of the author plus 7 years after death. Unlike today, copyrights in work were not automatic. The registration of copyright was mandatory for the enforcement of rights under the Act. The government could grant a compulsory license to publish a book if the owner of the copyright, upon the death of the author, refused to allow its publication. In the second phase, the Indian legislature, under the British Raj, enacted the Copyright Act of 1914 based on the Imperial Copyright Act (1911) of the UK. An Act for criminal sanction for an infringement was introduced. The third phase of the copyright regime was witnessed post-Independence. The Copyright Act, 1957 was enacted, superseding the Indian Copyright Act, 1914, in order to suit the provisions of the Berne Convention (1886). The 1957 Act has been amended six times (1983, 1984, 1992, 1994, 1999, and 2012) to comply with WIPO Copyright Treaty (WCT), 1996 and WIPO Performances and Phonograms Treaty (WPPT), 1996. Most of the amendments in copyright laws were in the digital environment, such as penalties for circumvention of technological protection measures; rights of management information; liability of Internet service provider; introduction of statutory licenses for the cover versions (the cover version is re-recording or re-composition of the original song by other artists or composers and is also termed as a remake, cover song, revival, etc.) and broadcasting organizations; ensuring the right to receive royalties for authors and music composers; exclusive economic and moral rights to performers; equal membership rights in copyrights societies for authors and other right owners; and exception of copyrights for physically disabled to access any works. India is an active member of nearly all significant international Conventions/Treaties related to Copyright Law, e.g., the Berne Convention as modified in Paris in 1971, the Universal Copyright Convention (1951), the Rome Convention (1961), WCT, WPPT, and TRIPS, 1995.

15.3.4 Trademarks

The first statutory law related to Trademarks (TM) in India was the Trade Marks Act, 1940, which was carved out from the Trade Marks Act, 1938 of the UK. It was followed by the incorporation of provisions of TM stated in the Indian Penal Code, Criminal Procedure Code, and the Sea Customs Act. Later on, Trade Marks Act, 1940 was rechristened as Trade and Merchandise Marks Act, 1958. Nearly four decades later, this Act was repealed by the Trade Marks Act, 1999. The need for this occurred to comply with the provisions of the TRIPS. It is the current governing law related to registered TM. Written as "trade mark" or "trademark," it is a type of IP consisting of a recognizable sign, design, or expression that identifies products or services and distinguishes them from others. The owner can be an individual, business organization, or any legal entity. It can be located on a package, as a label, a voucher, or on the product itself. It is used to identify services called service marks.

15.3.5 Trade Secrets

Although India has no specific Trade Secrets Laws, Indian courts have upheld Trade Secrets protection under various statutes, including contract law, Copyright law, the principles of equity, and the common law action of breach of confidence (which in effect amounts to a breach of contractual obligation).

15.3.6 Semiconductor Integrated

In the twenty-first century, Information Technology (IT) has revolutionized the economic and societal growth of the world economy. The rapid and tremendous scientific advancements in the field of IT resulted in the creation of a new class of IP called the Layout-Design of the Semiconductor Integrated Circuits. Various organizations, including WTO and TRIPS Agreement, laid down rules and regulations regarding the protection of Semiconductor Integrated Circuits and Layout Designs (SICLD). India being a member passed an Act called the SICLD Act, 2000. This Act is TRIPS compliant and fulfils the conditions of the TRIPS agreement concerning the protection of SICLD.

15.3.7 Industrial Designs

A design is a creation of the human mind, which is appealing to the eyes and attracts the attention of the viewers. The need to protect Industrial Designs (IDs) was recognized in the eighteenth century and the Indian legislation enacted the Patterns and Designs Act in 1872 for the first time. The Act was enacted to protect the rights over the creation of designs and novel patterns by the inventors. The Act was replaced by the British Patents and Designs Act in 1907, which later became the basis for the Indian Patents and Designs Act, 1911. In 1970, a separate Act was enacted for the patent, that is, the Patent Act, 1970. The Indian Patents and Designs Act, 1911, remained in force for designs only. Finally, in the year 2000, a dedicated Act for the ID was passed, which came into force in 2001.

15.3.8 Geographical Indications

Geographical indications (GIs) have been defined under Article 22(1) of the WTO Agreement on TRIPS and thus India, enacted the Geographical Indications of Goods (Registration and Protection) Act in 1999, which came into force with effect from September 15, 2003. The Act provides for the registration and protection of

GIs relating to goods in India. The Act is administered by the Controller General of Patents, Designs, and Trade Marks, who is the Registrar of GI. The GI Registry is located in Chennai. The GI is generally owned by a community that belongs to that particular location. It can be defined by words or numbers or maybe a combination of both. Goods or packages are used to indicate the place of origin of the goods. In an international transaction, marks of origin include marks of the country of origin.

Registering GI is always beneficial as the owner can prevent others from unauthorized usage or from commercializing the registered product. However, the registration of GI is not mandatory in India; unregistered GIs are protected under passing off cases, but it is always advisable to register the geographical origin as no further proof is required. The persons dealing with categories of goods are covered under the term Producer and three categories can be identified: (1) Agricultural Goods include the production, processing, trading, or dealing; (2) Natural Goods include exploiting, trading, or dealing; and (3) Handicrafts or Industrial goods include making, manufacturing, trading, or dealing. Some GIs cannot be registered in India such as things that are determined to be generic names or indications of goods and are, therefore, not protected or ceased to be protected in their country of origin, or which have fallen into disuse in that country. Things that would otherwise be disentitled to protection in a court include any matter likely to hurt the religious susceptibilities of any class or section of the citizens of India, contrary to any law for the time being in force; scandalous or obscene matter; matter likely to deceive or cause confusion; and things that although literally true as to the territory, region, or locality in which the goods originate, but falsely represent to the persons that the goods originate in another territory, region, or locality.

Any association of persons, producers, organizations, or authority established by or under the law can apply for GI and the applicant must represent the interest of the producers. The application should be in writing in the prescribed form and addressed to the Registrar of GIs along with a prescribed fee.

15.3.8.1 Registration Process of Geographical Indication

- An application shall be made in the prescribed format with the Form GI-1.
- A Registration of GI for any product shall be included in any one class.
- An application to register a GI for a specification of goods included in any class from a convention country.
- A single application is required for the Registration of the GI from any class.

Along with it, the applicant must fulfil the various requirements and criteria defined in the GI rules. The applicant must give the following details/documents:

- A statement as to how the GI serves to designate the goods as originating from the concerned territory in respect of specific quality, reputation, or other characteristics.
- The three certified copies of a class of goods to which the geographical indication relates.
- The geographical map of the territory.

- The particulars of the appearance of the geographical indication words or figurative elements or both.
- A statement containing such particulars of the producers of the concerned goods proposed to be initially registered, including a collective reference to all the producers of the goods in respect of which the application is made.

The statement contained in the application shall also include the following:

- An affidavit as to how the applicant claims to represent the interest of the association of persons or producers or any organization or authority established under any law.
- The standards benchmark for the use of the GI or the industry standard as regards the production, exploitation, making, or manufacture of the goods having specific quality, reputation, or other characteristic of such goods that are essentially attributable to its geographical origin with the detailed description of the human creativity involved, if any, or other characteristic.
- The particulars of the mechanism to ensure that the standards, quality, integrity, and consistency or other special characteristics are maintained by the producers or manufacturers of the goods. Three certified copies of the map of the territory, region, or locality.
- The particulars of special human skill involved or the uniqueness of the geographical environment or other inherent characteristics associated with the GI.
- The full name and address of the association of persons or organization or authority representing the interest of the producers of the concerned goods, as well as particulars of the inspection structure.
- In case of a homonymous indication, the material factors differentiating the application from the registered GI and particulars of protective measures adopted.

15.3.8.2 Duration of Registration

- The Registration of a GI is valid for 10 years. It can be renewed from time to time for further period of 10 years each after submitting the Form and requisite amount of fee. If a registered GI is not renewed, it is liable to be removed from the register.

15.3.8.3 Infringement

- When an unauthorized user uses a GI that indicates or suggests that such goods originate in a geographical area other than the true place of origin of such goods in a manner that misleads the public as to the geographical origin of such goods.
- When the use of GI results in unfair competition including passing off in respect of registered GI.
- When the use of another GI results in false representation to the public that goods originate in a territory in respect of which a registered GI relates.

15.3.8.4 Benefit for Registration of Geographical Indications

- It confers legal protection to GI.

- Prevents unauthorized use of a Registered GI by others.
- It provides legal protection to GI which in turn boosts exports.
- It promotes the economic prosperity of producers of goods produced in a geographical territory.

15.3.9 IP Governance

Since IP is an integral component of human society, every nation has dedicated agencies for laying out the guidelines, and for implementation and enforcement of IP-related matters. In India, many organizations/agencies deal with various aspects of IP. The governance of all categories of IP, except the Plant Variety and Farmer Rights Act, is carried out by the Department for Promotion of Industry and Internal Trade (DPIIT) under the aegis of the Ministry of Commerce and Industry, Government of India. There are a few other dedicated organizations/departments established by the government to promote patent-ecosystem (patent awareness, patent filing, and patent commercialization); for example, Technology Information Forecasting and Assessment Council (TIFAC), National Research Development Corporation (NRDC), and Cell for IPR Promotion and Management (CIPAM). In order to create a hassle-free exchange of IP-related activities among all the nations, it is imperative to have minimum standards of rules and regulations pertaining to all aspects of IP including rights, empowerment, exceptions, etc. To achieve this goal, the United Nations (UN) has established an organization called the World Intellectual Property Organization (WIPO). This agency is at the forefront of imparting knowledge about IP and governs international filing and registration of IP through various Conventions and Treaties like Paris Conventions, Patent Cooperation Treaty (PCT), Rome Convention, and Berne Convention.

15.4 Summary

The chapter discusses the policies, regulations, and guidelines related to the management of PGRs, namely the Convention on Biological Diversity (CBD), the International Treaty on Plant Genetic Resources for Food and Agriculture (ITPGRFA), the Nagoya Protocol, and relevant national compliances in detail. Major provisions related to PGR management are described for students as these policies are important to deal with access to genetic resources for exchange within and outside India for research use while planning research projects. The types of intellectual property and the provisions for their protection are presented. The provisions of the Protection of Plant Variety and Farmers Rights Act, 2001, Geographical Indications Registration and Protection Act, 2009, and Patent Search procedures are also described in detail.

Multiple-Choice Questions

1. The CBD came into force in the year
   (a) 1991
   (b) 1992
   (c) 1993
   (d) 1996
2. The CBD covers conservation of
   (a) All PGR
   (b) All agro-biodiversity
   (c) All above ground biodiversity
   (d) All biological diversity
3. One of the rights of contracting parties to the CBD is
   (a) Sovereign rights over their biological resources
   (b) Deny access to their resources
   (c) Not allow research on transgenic material
   (d) Not allow benefit sharing to local people
4. The Biological Diversity Act, 2002 has been amended in
   (a) 2010
   (b) 2023
   (c) 2014
   (d) 2004
5. The Nagoya Protocol deals with
   (a) Research on Living modified organisms
   (b) Cross-country collection protocols
   (c) Japanese inter-university student exchange
   (d) Internationally accepted access and benefit sharing on the use of biological resources
6. The ITPGRFA is a revised version of
   (a) CBD
   (b) IUPGRFA
   (c) Nagoya Protocol
   (d) BDA, 2002
7. The Annexure I of the ITPGRFA is
   (a) Legal obligations of member countries
   (b) List of crops listed for exchange under the Treaty
   (c) Financial provisions
   (d) List of crops not to be exchanged among countries
8. The National Focal Point for the implementation of the ITPGRFA in India is
   (a) Ministry of Environment, Forests, and Climate Change
   (b) Ministry of Law
   (c) ICAR-NBPGR
   (d) Department of Agriculture Cooperation and Farmers Welfare

9. Intellectual Property is
   (a) A law
   (b) A regulation
   (c) A right
   (d) An intangible novel creation of mind
10. The protection period of a patent in India is
   (a) 15 years
   (b) 20 years
   (c) 18 years
   (d) 60 years
11. "Darjeeling Tea" and "Kangra Tea" are examples of
   (a) Finest export-quality tea
   (b) Tea from West Bengal
   (c) GI-protected tea brands
   (d) Trademark goods
12. The total period of protection for trees and vines permitted under the PPVFR Act, 2001 is
   (a) 15 years
   (b) 6 years
   (c) 9 years
   (d) 18 years
13. Traditional Knowledge Digital Library (TKDL) is a repository of
   (a) Formulations of various systems of Indian medicine
   (b) Tribal knowledge
   (c) Patents on medicinal plants
   (d) Traditional knowledge related to agricultural practices
14. Farmers have ……. rights as breeders under PPVFRA, 2001.
   (a) Lesser
   (b) More
   (c) No
   (d) Equal
15. Eligibility criteria for new varieties in the Indian PPVFR Act are
   (a) DUS
   (b) NDUS
   (c) VCU
   (d) NUS
16. Important crops not covered under the Annexure I of the ITPGRFA are
   (a) Apple and banana
   (b) Groundnut and onion
   (c) Wheat and barley
   (d) Lathyrus and Atriplex
17. World's first patent was granted in the USA in the year
   (a) 1856
   (b) 1920
   (c) 1720
   (d) 1790

18. Indian Patents and Design Act was established in
    (a) 1911
    (b) 1947
    (c) 1952
    (d) 1970
19. The GI Act, 1999 and PPVFR Act, 2001 were established as an obligation to be members of
    (a) CBD
    (b) WTO
    (c) Nagoya Protocol
    (d) WTO-TRIPS
20. The national framework for the development of genetically modified plants and their commercial exploitation is given under the
    (a) Wildlife Protection Act
    (b) Biological Diversity Act
    (c) Patent Act
    (d) Guidelines for Safety of Biotechnology by DBT under the EPA 1986

Answers

1. (c)
2. (d)
3. (a)
4. (b)
5. (d)
6. (b)
7. (b)
8. (d)
9. (d)
10. (b)
11. (c)
12. (d)
13. (a)
14. (d)
15. (b)
16. (b)
17. (d)
18. (a)
19. (d)
20. (d)

Suggested Reading

Arora RK (1981) Plant genetic resources exploration and collection: planning and logistics. In: Mehra KL, Arora RK, Wadhi SR (eds) Plant exploration and collection. NBPGR, New Delhi, pp 46–54. NBPGR Sci. Monogr. 3

De Jong R, Li KY, Bootsma A, Huffman T, Rohloff G, Gameda S (2001) Crop Yield and Variability under Climate Change and Adaptive Crop Management Scenarios. Final Report for Climate Change Action Fund Project A080, Eastern Cereal and Oilseed Research Centre (ECORC). Government of Canada Department of Agriculture and Agri-Food Canada, Ottowa, Ontario, Canada. 49 pp

Dhillon BS, Saxena S (2003) Conservation and access to plant genetic resources. In: Mandal BB, Chaudhury R, Engelmann F, Mal B, Tao KL, Dhillon BS (eds) Conservation biotechnology of plant germplasm. NBPGR, New Delhi/IPGRI, Rome/ FAO, Rome, pp 3–18

Evenson RE, Rosegrant M (2003) The economic consequences of CGI programs. In: Evenson RE, Gollin D (eds) Crop variety improvement and its effect on productivity: the impact of international agricultural research. CAB International, Wallingford, UK

Myers N, Mittermeier R, Mittermeier C et al (2000) Biodiversity hotspots for conservation priorities. *Nature* 403:853–858. https://doi.org/10.1038/35002501

National Seed Policy (2002) Department of Agriculture and Co-operation, Ministry of Agriculture. Govt. of India

Sanjeev Saxena BP, Dhillon BS (2004) The Protection of Plant Varieties and Farmers' Rights Act of India. Curr Sci 86(3):392–397

Saxena S, Dhillon BS (2002) Acritical appraisal of the Protection of Plant Varieties and Farmers' Rights Act 2001, India. NATP Trainers Training Jan 2002, Compilation of Experts lecture notes. NBPGR, New Delhi, p 9

Tewari R, Bhardwaj M (2021) Intellectual property: a primer for academia intellectual property (IP). Manager Press Panjab University, Chandigarh. 263 pp

The Protection of Plant Varieties and Farmers Rights Bill (2001) Seed Association of India. Newsletter 15:1–35

Tyagi V, Singh AK, Chand D, Singh RV, Dhillon BS (2006) Plant introduction in India during pre- and and post-CBD periods – an analysis. Ind J Plant Genet Res 19(6):436–441

Vavilov NI (1926) Centres of origin of cultivated plants. Bull Appl Bot Genet Plantbreed 16(2):248

Zeven AC, de Wet JMJ (1982) Dictionary of cultivated plants and their regions of diversity; excluding ornamentals, forest trees and lower plants. CAPD, Wageningen, Netherlands

Application of Statistics and Artificial Intelligence in Plant Genetic Resources

16

Girish Kumar Jha, Dwijesh Chandra Mishra,
Sudhir Srivastava, and Sneha Murmu

Learning objectives

- Gain insight into the role of statistics and artificial intelligence (AI) in analyzing and leveraging plant genetic resources (PGR) data for plant breeding and genetic enhancement.
- Learn about various statistical techniques used in PGR research, including experimental design, univariate and bivariate analysis, as well as multivariate analysis methods like principal component analysis and cluster analysis.
- Explore the application of machine learning (ML) and deep learning (DL) techniques in PGR research, including genomic selection, genome-wide association studies, and decision support systems for crop breeding and management.

Keywords

Plant genetic resources · Germplasm · PGR · Statistical tools · Experimental designs · Data analysis · Artificial intelligence · Neural network

16.1 Introduction

Plant genetic resources (PGR) encompass the genetic diversity present within plant species, including their wild relatives and cultivated varieties. The utilization of PGR in breeding programs aimed at improving traits such as yield, disease

G. K. Jha (✉) · D. C. Mishra · S. Srivastava · S. Murmu
Division of Agricultural Bioinformatics, ICAR-Indian Agricultural Statistics Research Institute, New Delhi, India
e-mail: girish.jha@icar.org.in; dcmishra.iasri@icar.org.in; Sudhir.iasri@icar.org.in; Snehamurmu.iasri@icar.org.in

K. Tripathi et al. (eds.), *Textbook of Plant Genetic Resources*,
https://doi.org/10.1007/978-981-96-5024-8_16

resistance, and nutritional quality is crucial for ensuring food security, adapting to changing environmental conditions, and addressing global challenges such as climate change and population growth.

Traditionally, plant breeders relied on classical breeding methods, which involved the selection and crossbreeding of plants based on observable traits. However, this approach is time-consuming and often limited by the availability of desirable traits within the breeding population. With advancements in statistical methods and artificial intelligence (AI), plant breeders now have powerful tools at their disposal to harness the full potential of PGR and accelerate the breeding process.

Statistics plays a fundamental role in analyzing and interpreting data obtained from plant breeding experiments and for PGR research as well. Statistical techniques such as analysis of variance (ANOVA), regression analysis, and experimental design are used to assess the significance of experimental results, identify genetic variations, and quantify the heritability of traits. These statistical methods enable breeders to make informed decisions about which genotype to select for further breeding and which traits to prioritize for improvement.

In recent years, AI has emerged as a game-changer in the field of plant breeding. AI techniques, including machine learning (ML) and deep learning (DL), have revolutionized the analysis of large-scale genomic and phenotypic data. ML algorithms can process vast amounts of genetic information and identify patterns, correlations, and predictive models that traditional statistical methods may overlook. DL algorithms, inspired by the structure and function of the human brain, excel at recognizing complex patterns in multidimensional data, making them well-suited for tasks such as image analysis and genomic sequence prediction.

Before we get into the details of statistical methods, it is important to understand the basics of different types of data and their generation processes. Data, the foundation of statistical analysis, primarily originates through two methods: experimental design and sampling. Experimental design involves intentionally setting up controlled experiments to observe and measure specific variables under defined conditions. Experiments are planned and executed to elucidate causal relationships and study the effects of independent variables on dependent variables. On the other hand, sampling involves the collection of data from a subset, or sample, of a larger population. In this case, the population exists in its way and the experimenter does not have any control over the population unlike in the case of experimental design. Through sampling, inferences can be drawn about the characteristics of the entire population while minimizing resource and time constraints. Understanding these fundamental data generation processes lays the groundwork for applying statistical methods effectively in various scientific endeavors, including plant genetic resources research.

There are two broad categories of data type, viz., categorical (qualitative) data and numerical (quantitative) data.

16.1.1 Categorical (Qualitative) Data

Categorical (qualitative) data tells about the features of the data and it categorizes the data into a set of categories. It can be categorized into two or more than two categories:

- **Nominal Data:** It consists of categories or names that cannot be ordered or ranked. Therefore, the meaning would not change if we change the order of its values. Examples include gender (male or female), leaf pubescence (present or absent), and flower color (red, white, purple).
- **Ordinal Data:** Ordinal data consists of categories that can be ordered or ranked. The main limitation of ordinal data is that the differences between the values are not known. Examples of ordinal data include social class (upper, middle, lower), and host plant resistance (from 1 = highly tolerant to 9 = highly susceptible).

16.1.2 Numerical (Quantitative) Data

Numerical (quantitative) data represents the numerical value of the data. It can be further classified into two categories:

- **Discrete Data:** This type of data only uses discrete values, that is, the data can only take on certain values. These types of data have values that can be counted as whole numbers. Examples of the discrete data include the number of heads in 100-coin tosses, the number of pods per plant, and the number of ears per plant.
- **Continuous Data:** Continuous data represents the data in a continuous range. Their values cannot be counted but can be measured. Examples include plant height and seed weight. It can be further categorized into two categories, viz., interval data and ratio data.
  - *Interval Data*: This data type is measured along a scale, in which each point is placed at an equal distance from one another. The problem with interval values data is that they do not have a true zero. Therefore, a lot of descriptive and inferential statistics cannot be applied. One can only perform addition or subtraction operations on these values. However, one cannot multiply or divide the values. Examples of interval data include temperature, and time interval.
  - *Ratio Data*: Ratio data is an ordered data unit that can only be measured and ordered with equidistant items. It is the same as the interval data, except that it does have an absolute zero. Examples of ratio data are plant height and seed weight.

In the following sections, we have discussed some of the basic statistical and AI techniques and their applications in plant genetic resources research.

16.2 Statistical Techniques

16.2.1 Descriptive Statistics

The first step in the analysis of any dataset is to compute descriptive statistics such as mean, median, variance, and skewness. Descriptive statistics helps in describing and understanding the features of a dataset by providing summaries about the samples and about the observations that have been made. Much of the information contained in the data can be obtained by computing certain descriptive statistics. Most common descriptive statistics used are measures of central tendency and measures of dispersion (spread). Common univariate statistical techniques include measures of central tendency (mean, median, mode), and measures of dispersion (range, standard deviation, variance).

16.2.1.1 Measures of Central Tendency

Measures of central tendency describe the central or average value of a dataset. The common measures of central tendency are as follows:

- *Arithmetic Mean*: The arithmetic mean is calculated by summing all the values in the dataset and then dividing it by the total number of observations. It represents the arithmetic average of the dataset. Let us consider a dataset containing the heights of ten different plant varieties. The height of plants in cm are 21.6, 30.6, 31.6, 36, 29.6, 23.6, 30, 29.3, 29, and 20. The mean height is 28.13, which represents the average height of all plant varieties.
- *Median*: The median is the middle value of a dataset when the values are arranged in ascending order. It divides the dataset into two equal parts, with half of the values lying below it and half lying above it. For the above plant height data, the median height is 29.45, which represents the middle value of the height distribution.
- *Mode*: The mode is the value that appears most frequently in the dataset. In case of discrete frequency, the mode is the value corresponding to the maximum frequency. Let us consider a dataset containing the days to 50% flowering of 20 different plant varieties (91, 81, 91, 81, 96, 102, 96, 96, 75, 105, 81, 91, 91, 105, 91, 81, 81, 96, 91, and 91). Then, the mode of the variable days to 50% flowering is 91, which occurs the maximum number of times, that is, seven times in the dataset.

16.2.1.2 Measures of Dispersion

Measures of dispersion quantify the spread or variability of data points around the central tendency. Common measures of dispersion are as follows:

- *Range*: The range is the difference between the maximum and minimum values in the dataset. It provides a simple measure of the spread of data but is sensitive to outliers. For the above plant height data, the maximum and minimum values are 36 and 20, respectively. Therefore, the range of the plant height is 16.
- *Standard Deviation*: The standard deviation is the square root of the arithmetic mean of the squares of the deviations of the values from their arithmetic mean. It represents the average distance of data points from the mean and is often used to describe the spread of data. For the above plant height data, the standard deviation is 4.66.
- *Variance*: The variance is the square of the standard deviation. The variance measures the average squared deviation of each data point from the mean. It provides a measure of how much the data points deviate from the mean. For the above plant height data, the variance is 21.71.

In the aforementioned plant height example, the variance and standard deviation quantify how much the heights of different plant varieties deviate from the mean height, providing insights into the variability of plant heights.

Graphical representations, such as histograms and boxplots, can be used to visually display the distribution of data and highlight important characteristics of the dataset. A histogram is a graphical representation that displays the frequency distribution of continuous data. It consists of bars that represent the frequency of data points within predefined intervals or bins. Histograms help visualize the shape, central tendency, and variability of the data distribution. A boxplot provides a visual summary of the distribution of data through five key statistics: minimum, maximum, median, and first and third quartiles. It displays the central tendency, spread, and skewness of the dataset, making it useful for identifying outliers and comparing distributions. By creating a histogram of plant heights, we can visually inspect the distribution of heights and identify any patterns or outliers present in the dataset (Fig. 16.1).

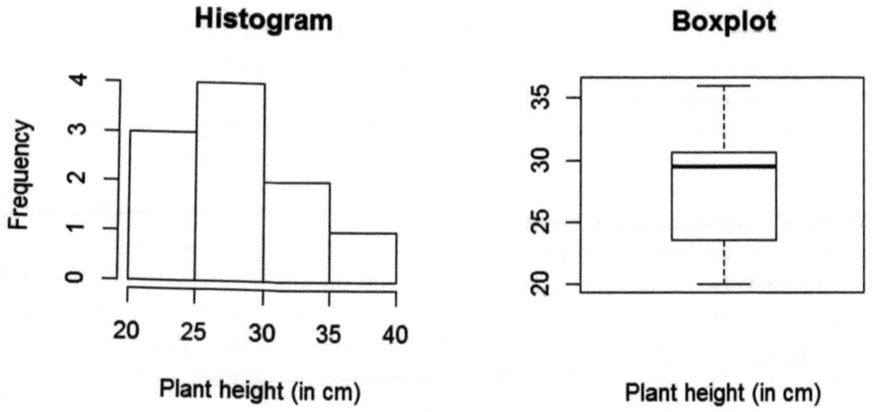

Fig. 16.1 Histogram and boxplot of plant heights

16.2.2 Correlation and Regression

So far, we have discussed the descriptive statistics for a single variable (univariate analysis). Bivariate analysis is a statistical method used to explore the relationship between two variables. It aims to understand how changes in one variable are associated with changes in another variable. In the context of plant genetic resources, bivariate analysis can help researchers understand the relationships between different traits or characteristics of genotypes.

16.2.2.1 Correlation Analysis

Correlation analysis measures the strength and direction of the linear relationship between two quantitative variables. The degree of association is measured by correlation coefficient (Pearson's product-moment correlation coefficient), often denoted by r, which ranges from –1 to 1.

If $r = +1$, there is a complete positive linear relationship between the variables.
If $r = -1$, there is a complete negative linear relationship.
If $r = 0$, there is no linear relationship between the variables.

Example: Let us consider the same dataset containing the heights of ten different plant varieties (21.6, 30.6, 31.6, 36, 29.6, 23.6, 30, 29.3, 29, and 20) and the corresponding yield per plant of these ten plant varieties are 1.1, 1.8, 2.8, 4.3, 1.6, 1.3, 2.1, 2.7, 4, and 0.3. The correlation coefficient between plant height and yield per plant is 0.81, which shows a positive linear relationship between the variables.

16.2.2.2 Simple Linear Regression

Simple linear regression is a statistical method used to model the relationship between two variables by fitting a linear equation to the observed data. In this technique, one variable (the predictor or independent variable) here represented as X is used to predict the value of another variable (the outcome or dependent variable) here represented as Y. It means that changes in X result in proportional changes in Y. Simple linear regression fits a straight line to the data points. This line is called the regression line or the line of best fit.

The regression line is represented by the equation:

$$Y = a + bX$$

Where a is the intercept (the value of Y when X is 0) and b is the slope (the rate of change of Y with respect to changes in X).

The slope (b) of the regression line represents the change in the dependent variable (Y) for a one-unit change in the independent variable (X). It tells us how much Y changes when X changes by one unit. The intercept (a) is the value of Y when X is 0. It gives us the starting point of the regression line. A regression line can be used to make a prediction of Y for a given value of X.

16.2.3 Multivariate Statistical Techniques

Multivariate analysis encompasses a set of statistical methods used to analyze datasets with multiple variables simultaneously. It allows researchers to explore complex relationships among variables and identify patterns that may not be evident through univariate or bivariate analysis alone. The most commonly used multivariate statistical techniques, viz., principal component analysis (PCA) and cluster analysis, have been discussed below.

16.2.3.1 Principal Component Analysis (PCA)

PCA is a widely used multivariate statistical technique for dimensionality reduction and data visualization. It helps in simplifying complex datasets by transforming them into a lower-dimensional space while preserving the essential structure and variation present in the original data. PCA achieves this by identifying the principal components (PCs), which are linear combinations of the original variables.

The following are the steps involved in PCA:

- *Standardization of Data*: Before performing PCA, it is crucial to standardize the data to have a mean of zero and a standard deviation of one across variables. Standardization ensures that variables with larger scales do not dominate the PCA results.
- *Computing the Covariance Matrix*: Once the data is standardized, the covariance matrix is computed. The covariance matrix summarizes the relationships between all pairs of variables in the dataset. It quantifies how much two variables change together.
- *Eigenvalue Decomposition*: Eigenvalue decomposition on the covariance matrix is performed to extract the principal components. The eigenvectors represent the directions of maximum variance in the original data, while the eigenvalues indicate the amount of variance explained by each principal component.
- *Selecting Principal Components*: The eigenvalues are sorted in descending order, indicating the amount of variance explained by each principal component. The first principal component (PC1) captures the most variance in the data, followed by PC2, PC3, and so on. Typically, only a subset of principal components that capture a significant amount of variance in the data is retained.
- *Constructing Principal Components*: Each principal component is a linear combination of the original variables weighted by the corresponding eigenvector. The weights represent the contribution of each variable to the principal component. The principal components are orthogonal to each other, meaning they are uncorrelated.
- *Dimensionality Reduction*: PCA allows for dimensionality reduction by selecting a subset of principal components that capture most of the variability in the data. By retaining only the top principal components, researchers can reduce the dimensionality of the data while preserving most of the important information.

Example: Consider a dataset containing various characteristics of different plant species, such as height, leaf area, flowering time, maturity time, yield, and seed weight. Multivariate analysis techniques like PCA can help in reducing the dimensionality of the data and identifying the most important variables that contribute to the overall variation among plant species.

16.2.3.2 Cluster Analysis

Cluster analysis is a statistical technique used to classify objects or observations into groups, or clusters, based on their similarity. It is a method of unsupervised learning, meaning that it does not require predefined labels or categories for the data. Instead, it identifies inherent patterns and structures within the dataset. The primary goal of cluster analysis is to partition a dataset into groups such that observations within each group are more similar to each other than to those in other groups. The aim is to maximize the intra-cluster similarity and minimize the inter-cluster similarity. Cluster analysis relies on a distance or similarity measure to quantify the resemblance between pairs of observations. Common distance metrics include Euclidean distance and Manhattan distance. Some popular clustering algorithms include hierarchical clustering (agglomerative and divisive) and non-hierarchical clustering (e.g., *K*-means clustering). The process of clustering typically involves initialization, where initial cluster assignments are made, followed by iterative refinement to optimize the clustering criterion. In *K*-means clustering, for example, the algorithm iterates between reassigning data points to the nearest centroid and updating the centroids based on the newly assigned data points. Evaluating the quality of clusters is essential to assess the effectiveness of the clustering algorithm. Validation measures, such as the silhouette coefficient and Davies–Bouldin index, assess the compactness and separation of clusters.

16.2.4 Analysis of Variance

Analysis of variance (ANOVA) is a statistical method widely used in plant breeding experiments to compare the means of three or more groups and determine if they are significantly different from each other. ANOVA allows researchers to assess whether there are any statistically significant differences between the means of multiple groups, such as different plant varieties or treatments.

The procedure involves:

Hypothesis Testing: ANOVA tests the null hypothesis that the means of all groups are equal against the alternative hypothesis that at least one mean is different.

Calculating Variability: ANOVA decomposes the total variability in the data into two components, that is, variation between groups and variation within groups.

F-Statistic: ANOVA calculates the *F*-statistic by comparing the variability between groups with the variability within groups.

Interpretation: If the F-statistic is significant (i.e., if the p-value is less than a predetermined significance level, say 0.05), it indicates that at least one group mean is significantly different from the others.

ANOVA helps researchers determine if there are significant differences in traits (such as yield, plant height, disease resistance, etc.) among different plant varieties or treatments.

16.2.5 Experimental Designs

Experimental design is the systematic planning and organization of experiments to ensure valid and reliable results. The main goals of experimental design are to minimize variability, control potential sources of bias, and maximize the efficiency of experiments. It involves defining treatment groups, randomizing subjects or experimental units into different treatments, replicating each treatment to account for variability, blocking similar units to reduce variability, and including control groups for baseline measurements. Experimental design involves defining treatment groups or conditions that will be compared in the experiment.

Treatments: Various objects of comparison in a comparative experiment are termed as treatments; for example, in field experiments, different fertilizers or different genotypes of crops or different methods of planting are the treatments.

Experiment Unit: The smallest part of the experimental area to which we apply the treatments and on which we take observations on the different variables under study is termed as an experimental unit; for example, in field experiments, the plots of "Land" are experimental unit.

Blocks: When the whole experimental unit (field) is divided into relatively homogenous subgroups or traits that are more uniform among themselves than the entire field, these are known as blocks.

Yield: The measurement of the variable under study on different experimental units, for example plots in field experiments, is termed as yield.

Experimental Error: The error caused by external factors beyond human control is known as experimental error. For example, germination of seed, plant growth, number of pods or ears per plant are the factors beyond human control. All types of variations due to the inherent variability in the experimental unit, lack of standardized experimental methodology, and lack of representatives of the sample to the population under study, all lead to experimental error.

Precision: The reciprocal of the variance of the mean is termed as the precision or the amount of information of a design. Thus, for an experiment replicated r times, the precision is given by

$= 1/\text{var}(\bar{x}) = \frac{r}{\sigma^2}$ Where σ^2 is the error variance per unit.

Efficiency of a Design: Consider the designs D_1 and D_2 with error variances per unit s_1^2 and s_2^2 and replications r_1 and r_2, respectively. The efficiency of a design can be increased by controlling, that is, decreasing s^2, the error variance per unit. This is done by arranging the plots into small homogenous blocks and increasing the number of replication (r).

There are three basic principles of experimental designs, which are given below:

Randomization: Randomization is the process of randomly assigning subjects or experimental units to different treatment groups to reduce bias and ensure that each group is representative of the population.

Replication: Replication involves repeating each treatment multiple times to account for variability and obtain more reliable estimates of treatment effects. The main purpose of replication is to provide an estimate of the experimental error. The greater number of replications reduces the standard error of the treatment mean(s).

Local Control: Local control, also known as error control, is the method to reduce the experimental error, thereby increasing the efficiency of the design. This can be achieved by making the experimental units homogeneous. Blocking is another method that involves grouping the experimental units into several homogeneous groups, allowing variation between the groups. In another way, reducing the experimental error by dividing the relatively heterogeneous experimental area (field) into homogenous blocks (due to physical contiguity as far as field experiments are concerned) is known as local control.

Some of the experimental designs are given below:

Completely Randomized Design: This is the simplest and most commonly used design, where treatments are randomly assigned to experimental units. It is suitable for studies where the experimental unit is relatively homogeneous or uniform and there is no known variation. The principle of "Local control" is not used in this design. Hence this design is suitable for experiments being conducted under controlled conditions, for example in a lab or greenhouse, not under field conditions. For example: germination experiment in lab, pot experiment and quality estimation in lab condition

Randomized Complete Block Design: This design is used when there is known variation in the experimental area, such as differences in soil fertility or sunlight exposure. The experimental units are grouped into blocks, and each treatment is randomly assigned within each block. This helps to control for the variation and increase the precision of the experiment. This is the most common design used in the field experiment.

Latin Square Design: This design is used when two sources of variation need to be controlled, such as row and column effects in a field trial. The treatments are arranged in a square grid, with each treatment appearing once in each row and once in each column.

Factorial Experiment: This experiment is used to study the effects of two or more factors simultaneously. The treatments are combinations of the different levels of the factors, and the experiment allows for the assessment of both main effects and interaction effects.

Augmented Block Design: This design is a modification of the randomized complete block design, where additional experimental units are added to the blocks to allow for the estimation of error variance. This is particularly useful when the number of treatments is large or when the experimental material is heterogeneous.

In plant genetic resources, a mega-characterization experiment requires a huge number of germplasms to be tested (1000–10,000 or even more) where seed quantity is limited and also huge resources and a large experimental area will be required to carry out the experiment. The available quantity of seed is often not sufficient for replicated trials and it is very difficult to maintain the within-block homogeneity.

Augmented (Hoonulaku) designs were introduced by Federer (1956) to fill a need arising in screening new strains of sugarcane at the Experimental Station of Hawaiin Sugarcane Planters Association on the basis of agronomic characters other than yield. Here genotypes or test entries are unreplicated but checks are randomized and replicated in each block.

Split-Plot Design: This design is used when there are two types of treatments, one applied to the whole plot and the other applied to subplots within the whole plot. This is often used in plant breeding experiments where the whole plot treatment is a variety and the subplot treatments are different fertilizers or irrigation regimes.

Repeated Measures Design: This design is used when the same experimental units are measured multiple times over time. This is useful for studying plant growth or development over time.

The choice of experimental design should be carefully considered based on the specific research question, the nature of the plant material, and the available resources. Proper experimental design ensures that the data collected is reliable and can be used to draw valid conclusions about the effects of treatments or the characteristics of different plant varieties. By employing proper experimental design, plant breeders can minimize bias, control sources of variability, and maximize the efficiency of their experiments, ultimately making informed decisions about which plant lines to select for further breeding and which traits to prioritize for improvement.

16.3 Basics of Artificial Intelligence

Artificial intelligence (AI) pertains to the advancement of computer systems capable of executing various tasks, typically requiring human intelligence, such as learning, reasoning, problem-solving, perception, speech recognition, and language understanding. The goal of AI is to create machines that can simulate human intelligence to some extent. It encompasses machine learning (ML) and deep learning (DL) techniques (Fig. 16.2).

16.3.1 Machine Learning

Machine learning stands as a fundamental component within the broader domain of artificial intelligence, focusing specifically on enabling computers to learn and improve their performance on tasks without explicit programming. This means that instead of instructing a computer on how to perform a task step by step, we provide it with data and algorithms that allow it to learn patterns and relationships within that data to make predictions or decisions.

ML algorithms are designed to process large datasets, identifying patterns, extracting insights, and making predictions based on the information contained within the data. These algorithms come in various forms, each suited to different types of problems and data structures.

Following are the steps involved in development of machine learning models:

- *Data Collection*: Data collection involves gathering relevant datasets that contain information about the problem you want to solve. The data may come from various sources such as databases, application programming interfaces (APIs), sensors, or manual data entry. It is crucial to ensure that the collected data is accurate, comprehensive, and representative of the problem domain.
- *Data Preprocessing*: Data preprocessing is the step where the collected data is cleaned, transformed, and organized to make it suitable for machine learning algorithms. This process involves handling missing values, removing outliers, scaling features, and encoding categorical variables. Data preprocessing helps

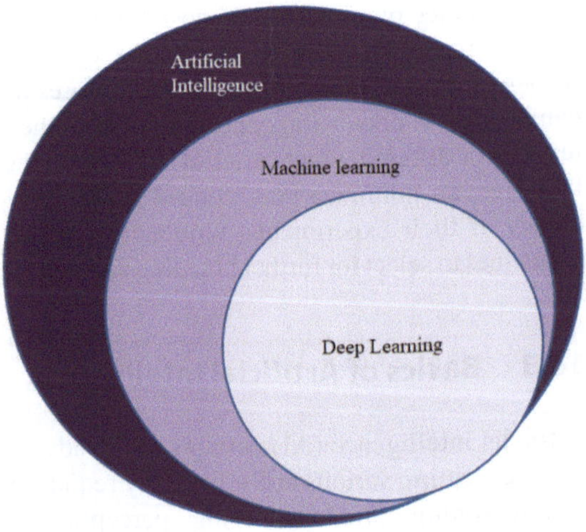

Fig. 16.2 Relationship between artificial intelligence (AI), machine learning (ML), and deep learning (DL)

improve the quality of the data and enhances the performance of machine learning models.

- *Feature Extraction and Selection*: Feature extraction involves transforming raw data into a set of meaningful features that can be used as input for machine learning algorithms. Feature selection is the process of choosing the most relevant features from the dataset while discarding irrelevant or redundant ones. Effective feature extraction and selection can reduce dimensionality, improve model efficiency, and enhance predictive accuracy.
- *Machine Learning Algorithms*: Once the data is preprocessed and features are selected, the next step is to choose a suitable machine learning algorithm. There are various types of machine learning algorithms, including supervised learning, unsupervised learning, semi-supervised learning, and reinforcement learning. The choice of algorithm depends on the nature of the problem, the type of data, and the desired outcome.
- *Performance Evaluation*: Performance evaluation is the final step where the trained machine learning model is tested on unseen data to assess its accuracy and effectiveness. Common evaluation metrics include accuracy, precision, recall or sensitivity, F1-score, and Area under the Receiver Operating Characteristics (ROC) curve (AuROC). Performance evaluation helps determine how well the model generalizes to new data and whether it meets the desired objectives.
- Accuracy is defined as the proportion of correct predictions (both true positive [TP] and true negative [TN]) out of the total number of predictions. It provides an overall measure of how well the model performs across all classes.
- Precision is defined as the proportion of true positive predictions out of all positive predictions made by the model. It indicates the model's ability to correctly identify positive instances without misclassifying negatives.
- Recall or sensitivity is defined as the proportion of true positive predictions out of all actual positive instances in the data. It measures the model's ability to identify all relevant instances, especially important in scenarios where missing a positive instance is costly. F1-score is defined as the harmonic mean of precision and recall, providing a balance between the two metrics. It considers both precision and recall, offering a single metric that balances between false positives and false negatives.
- AuROC is a graphical representation of the trade-off between true positive rate (sensitivity) and false positive rate (1-specificity) across different threshold values. AuROC provides a measure of the model's ability to discriminate between positive and negative instances across various threshold settings, with higher values indicating better discrimination.
- The mathematical expressions of the aforementioned metrics are given below:

  - $\text{Accuracy} = \dfrac{TP + TN}{TP + TN + FP + FN}$

  - $\text{Precision} = \dfrac{TP}{TP + FP}$

- Recall (Sensitivity) = $\frac{TP}{TP + FN}$
- F1-Score = $2\frac{\text{Precision} \times \text{Recall}}{\text{Presicion} + \text{Recall}}$
- Specificity = $\frac{TN}{TN + FP}$

- In the given model, TP represents the true positive, which is the number of correct positive predictions. TN represents the true negative, which is the number of correct negative predictions. FP represents the false positive, which is the number of incorrect positive predictions. FN represents the false negative, which is the number of incorrect negative predictions.

One of the main differences in ML is between supervised and unsupervised learning. In supervised learning, the algorithm is trained using labeled data, where each example has a corresponding label or outcome. The algorithm learns to make predictions or classifications based on the input data and the labels provided. On the other hand, unsupervised learning involves exploring data without explicit guidance or labels, to uncover hidden patterns or structures within the data itself.

Supervised learning tasks include both classification and regression, where models are trained on labeled data to make predictions on unseen data. However, they differ in the type of output variable they predict: categorical for classification and continuous for regression. In classification, the goal is to predict the category or class label of a new instance based on its features. It involves assigning input data into predefined categories or classes—for example, classifying images of plant leaves or seeds into different species for biodiversity conservation and management. In regression, the objective is to predict a continuous target variable based on one or more input features. It involves estimating the relationship between variables and making predictions on new data points—for instance, predicting crop yields based on environmental factors such as temperature, precipitation, soil type, and historical yield data.

In addition to supervised and unsupervised learning, there are other approaches to ML, including semi-supervised learning and reinforcement learning. Semi-supervised learning utilizes a combination of labeled and unlabeled data to train models, while reinforcement learning involves learning optimal decision-making strategies through interaction with an environment, and receiving feedback in the form of rewards or penalties.

Following are some of the popular ML techniques:

1. **Support Vector Machines (SVM):** SVM is a powerful supervised learning algorithm used for classification and regression tasks. It finds the optimal hyperplane that best separates data points belonging to different classes in the feature space, maximizing the margin between classes.

2. **k-Nearest Neighbors (KNN):** KNN is a simple and intuitive algorithm used for classification and regression tasks. It predicts the class or value of a new data point by considering the majority class or averaging the values of its k-nearest neighbors in the feature space.
3. **Naive Bayes:** Naive Bayes is a probabilistic classifier based on Bayes' theorem with the "naïve" assumption of feature independence. It calculates the probability of each class given the input features and selects the class with the highest probability.
4. **Decision Trees:** Decision trees are versatile supervised learning algorithms used for both classification and regression tasks. They partition the feature space into regions and make predictions by traversing the tree based on the values of input features.
5. **Random Forest**: Random forest is an ensemble learning technique that builds multiple decision trees during training and combines their predictions to make more accurate and robust predictions. It reduces overfitting and improves generalization performance.
6. **Gradient Boosting Machines (GBM):** GBM is an ensemble learning technique that builds a sequence of weak learners, typically decision trees, in a serial manner. It fits each new tree to the residuals of the previous ones, gradually improving prediction accuracy.

16.3.2 Deep Learning

DL is a subset of machine learning that employs artificial neural networks (ANNs) with multiple layers (deep neural networks) to learn representations of data at different levels of abstraction. These networks are inspired by the structure and function of the human brain, with interconnected layers of neurons that process information. Deep learning architectures can automatically discover intricate patterns and features from raw data, without the need for manual feature engineering. The key components of deep learning include input layers, hidden layers, and output layers, with each layer performing non-linear transformations on the input data.

Some of the popular DL techniques are mentioned below:

1. **Artificial Neural Networks (ANNs):** ANNs are the foundation of deep learning and are composed of interconnected nodes organized into layers. Input data is passed through multiple layers of neurons, with each layer extracting increasingly abstract features from the data. ANNs can be used for tasks such as image recognition, natural language processing, and predictive analytics.
2. **Convolutional Neural Networks (CNNs):** CNNs are specialized neural networks designed for processing grid-like data, such as images and videos. They use convolutional layers to automatically and adaptively learn spatial hierarchies of features from input images. It can be used for tasks like image classification, object detection, and image segmentation.

3. **Recurrent Neural Networks (RNNs):** RNNs are designed to process sequential data by maintaining an internal memory state. They have connections that form directed cycles, allowing them to exhibit temporal dynamic behavior. RNNs are well-suited for tasks involving sequential data, such as speech recognition, language translation, and time series prediction.
4. **Long Short-Term Memory Networks (LSTMs):** LSTMs are a special type of RNN designed to address the vanishing gradient problem. They use gated units to regulate the flow of information, enabling them to learn long-term dependencies in sequential data. LSTMs are commonly used in tasks that require modeling sequences with long-term dependencies, such as speech recognition, language modeling, and sentiment analysis.

16.4 Role of Statistics and Artificial Intelligence in PGR

The assessment of PGR has encountered challenges in keeping pace with the evolving complexities of agriculture and the expanding realm of available data. The revolutions in genomics, phenomics, and big data are driven by advancements in technologies such as data storage, computing, sensors, and DNA sequencing. These advancements, coupled with innovative data collection methods like high-throughput field phenotyping and crowdsourcing, contribute to a significant increase in the amount of available data. The shift to "big data" methodologies represents a move away from traditional small-sample statistical approaches to adopting data analysis grounded in ML techniques. Understanding statistics and AI is vital for effectively managing and utilizing PGR in various agricultural and conservation contexts. Some of the applications of statistics and AI are discussed below.

16.4.1 Core Collection

In the field of PGR, the development of core collections serves as a crucial strategy for efficient germplasm management. A core collection represents a subset carefully selected from the broader germplasm collection to encapsulate the maximum genetic diversity of the entire collection. ML technologies have become instrumental in shaping these core collections, offering a sophisticated approach to streamline germplasm utilization and support targeted research initiatives. The methodology involved in creating a core collection through ML typically follows a two-step process. Initially, comprehensive data is gathered from the entire germplasm collection of a crop, encompassing data types like genetic markers, biochemical traits, and phenotypic traits. Subsequently, ML algorithms, including clustering or classification models, are employed to analyze this rich dataset and identify accessions that optimally represent the genetic diversity inherent in the entire germplasm collection.

For instance, Borrayo et al. (2016) implemented the core collection strategy in foxtail millet, utilizing the *K*-means clustering algorithm to group accessions based on genetic and phenotypic similarities. The iterative nature of *K*-means clustering efficiently assigned accessions to clusters, resulting in a core collection that robustly captured the genetic variability of the entire germplasm collection. Similar success stories are evident in chickpea, where Upadhyaya and Ortiz (2001) applied ML techniques to develop a mini-core collection using geographic information and quantitative traits. Many software packages are available for sampling of core collection, for example MSTRAT, PowerCore v2.0, Core Hunter 3, and GenoCore.

16.4.2 Crop Diversity

The application of ML/DL techniques significantly enhances the management and understanding of crop diversity. This involves a systematic methodology that starts with the collection of diverse datasets, comprising information about plant traits, environmental conditions, and genetic diversity data. Trait data may include details on yield, disease resistance, and morphological characteristics, while environmental conditions cover factors like temperature, soil composition, and precipitation. Genetic diversity data incorporates molecular markers, genomic sequences, and other genetic information. The collected datasets then undergo a comprehensive integration process, addressing challenges such as missing values and normalizing data to ensure consistency. This step is important for creating a dataset that can be effectively utilized by ML algorithms. For example, the study by Poland et al. (2012) showcased the efficacy of ML approaches in the analysis of genetic diversity in wheat varieties.

16.4.3 Genomic Selection

Genomic selection (GS) is a modern breeding approach that uses genomic information to predict the performance of plants or animals for specific traits of interest. In GS, the entire genetic makeup, or genome, of individuals within a population is analyzed. This involves identifying and studying specific regions of DNA that are associated with important traits like yield, disease resistance, or nutritional content. Using sophisticated statistical and AI techniques, prediction models are developed that correlate the genetic information with the observed performance of individuals for target traits. These models learn from known relationships between genotype (genetic makeup) and phenotype (observable traits) to predict the performance of new individuals based solely on their genomic data. GS relies on collecting genetic data from a diverse population of plants, achieved through DNA sampling and advanced sequencing techniques. The data are analyzed to identify genetic markers associated with desirable traits. Subsequently, prediction models are trained using this genetic information and trait performance data, enabling the prediction of trait values for new individuals based solely on their genetic profiles. This facilitates the selection of individuals with desired genetic traits for further breeding efforts, streamlining the breeding process by identifying superior candidates early on. As a result, genomic selection expedites the development of improved crop varieties by reducing the time and resources needed compared to traditional breeding methods.

16.4.4 Genome-Wide Association Studies (GWAS)

Genome-wide association studies are powerful tools used in genetics to identify genetic variants associated with specific traits or diseases. In GWAS, genetic data from a large number of individuals are analyzed to identify genetic variations, known as single nucleotide polymorphisms (SNPs), that are correlated with a particular trait or phenotype. Simultaneously, individuals are phenotyped for traits of interest, such as height or disease status. Statistical analysis is employed to correlate genotypic data with trait measurements, correcting for factors like population structure. GWAS identifies genetic loci associated with traits, often containing candidate genes influencing trait expression. ML techniques contribute to GWAS through their ability to handle large-scale and high-dimensional genomic datasets. Traditional GWAS approaches might face challenges in handling the complexity and scale of data, and ML algorithms provide a powerful solution to analyze such datasets. The methodology typically involves two main steps: training the ML model, and identifying significant associations. During the training phase, ML algorithms are exposed to a dataset containing genomic information and phenotypic traits. ML models are trained to recognize patterns and associations between genetic markers and traits. The trained models can then predict the potential associations in new, unseen datasets. ML algorithms, ranging from regression models to advanced ensemble methods, are then employed to decipher patterns and relationships within this data. These algorithms can handle the complexity of multifactorial traits, identifying genetic markers associated with specific traits more efficiently than traditional methods.

16.4.5 Decision Support Systems (DSS)

The ML model can serve as decision support for breeders, predicting the success of crossbreeding and identifying trait combinations likely to result in desired outcomes. Decision support tools driven by ML have been applied in crop breeding programs. Some of the examples of ML-based DSS include Nutrient Expert and Weather-Rice-Nutrient Integrated Decision Support System (WeRise). The Nutrient Expert decision support tools for maize and wheat were officially introduced for public use in India on June 20, 2013, during a joint meeting organized by the International Plant Nutrition Institute (IPNI) and International Maize and Wheat Improvement Center (CIMMYT) at the National Agricultural Science Centre Complex in New Delhi. Designed to enhance yields and profits for Indian farmers, the Nutrient Expert tools are user-friendly, computer-based systems that account for spatial and temporal variability. They offer precise nutrient recommendations tailored to the wheat and maize systems of India's smallholder farmers. Developed through dialogues and consultations with local stakeholders, the tools address challenges such as resource constraints, limited access to soil testing, and the absence of tillage-specific nutrient management strategies in India. WeRise is a comprehensive platform designed to provide decision support in rainfed rice production. This system integrates information on weather patterns, rice farming practices, and nutrient management to assist farmers in making informed decisions. By analyzing real-time weather data, WeRise

Table 16.1 Bioinformatics tools for PGR management

Tool	Use in PGR
Genome Databases (NCBI GenBank, Ensembl Plants)	Comprehensive repositories of genetic information for different plant species
Sequence Alignment Tools (BLAST, ClustalW)	Compare DNA or protein sequences from different plant varieties or species
Variant Calling Software (Genome Analysis Toolkit, FreeBayes)	Identify genetic variants such as SNPs, insertions, deletions, and structural variations within plant genomes
Genetic Mapping Software (JoinMap, MapQTL)	Assist in the construction of genetic maps for marker-assisted breeding and Quantitative Trait Locus (QTL) analysis
Phylogenetic Analysis Tools (Molecular Evolutionary Genetics Analysis, PhyML)	Reconstruct evolutionary relationships among plant species based on genetic data
Population Genetics Software (STRUCTURE, Arlequin)	Analyze genetic diversity, population structure, and gene flow within plant populations
Genome Annotation Tools (Augustus, GeneMark)	Predict and annotate gene structures, regulatory elements, and functional elements within plant genomes
Trait Analysis by aSSociation, Evolution, and Linkage (TASSEL)	Perform association studies, marker-trait association analysis, and genetic linkage analysis in plant breeding

offers insights into optimal planting times, irrigation scheduling, and fertilizer application, optimizing crop management for rice cultivation.

16.5 Bioinformatics Tools for PGR Management

Bioinformatics, the interdisciplinary field merging biology, computer science, and statistics, facilitates the analysis and interpretation of biological data, including plant genetic resources. Bioinformatics tools serve as crucial components enabling the integration of statistical methodologies and AI techniques into the management and analysis of plant genetic resources. These tools harness statistical algorithms and AI models to process vast amounts of genomic data, extract meaningful patterns, and derive insights into the genetic makeup of plant species. Some of the bioinformatics tools and their uses are listed in Table 16.1.

16.6 Summary

In this chapter, we explored the critical role of statistics and AI in the context of PGR. Beginning with an introduction to PGR and its importance in agriculture, we explored both traditional and emerging methods employed in plant breeding, emphasizing the pivotal role of statistics and AI in expediting the breeding process. We discussed various statistical techniques such as univariate and bivariate analysis, along with multivariate methods like PCA and cluster analysis, emphasizing their applications in understanding genetic diversity and trait prediction. Additionally, we provided an overview of AI, focusing on ML and DL techniques and their relevance in genotype–phenotype associations and decision support systems for crop breeding. Throughout the chapter, we examined real-world examples and applications, from the development of core collections to the implementation of genomic selection and GWAS. By harnessing the power of statistics and AI, researchers and breeders can unlock the full potential of plant genetic resources, paving the way for sustainable agriculture and food security in the face of global challenges.

Multiple-Choice Questions

1. Which one of the following can be used to quantify the spread or variability of data points around the measures of central tendency?
   (a) Correlation analysis
   (b) Cluster analysis
   (c) Measures of dispersion
   (d) Principal component analysis
2. What is the median of a distribution?
   (a) The arithmetic mean
   (b) The value of the variable that divides it into two equal parts
   (c) The value that appears most frequently
   (d) The difference between the maximum and minimum values
3. What does the range measure in descriptive statistics?
   (a) The arithmetic mean of the dataset
   (b) The central tendency of a dataset
   (c) The spread or variability of data points
   (d) The correlation between two variables
4. What is the range of the Pearson's correlation coefficient (r)?
   (a) $-\infty$ to ∞
   (b) -1 to 1
   (c) 0 to 1
   (d) 0 to ∞
5. Which one of the following is used to model the relationship between two variables by fitting a linear equation to the observed data?
   (a) Simple linear regression
   (b) Cluster analysis
   (c) Correlation analysis
   (d) Analysis of variance
6. Which statistical technique is commonly used to assess the significance of experimental results in plant breeding experiments?
   (a) Cluster analysis
   (b) Principal component analysis
   (c) Analysis of variance
   (d) Simple linear regression

7. What is the primary goal of cluster analysis?
   (a) To determine the strength and direction of the relationship between two variables
   (b) To compare the means of three or more groups
   (c) To classify the observations into groups based on their similarity
   (d) To explore the relationship between two quantitative variables
8. Which statistical technique is used for dimensionality reduction and data visualization in multivariate analysis?
   (a) Correlation analysis
   (b) Principal component analysis
   (c) Cluster analysis
   (d) Analysis of variance
9. What distinguishes supervised learning from unsupervised learning?
   (a) Supervised learning involves labeled data; unsupervised learning does not
   (b) Supervised learning is faster than unsupervised learning
   (c) Unsupervised learning requires human intervention
   (d) Unsupervised learning is more accurate than supervised learning
10. What is the first step involved in the development of machine learning models?
    (a) Feature extraction
    (b) Data preprocessing
    (c) Machine learning algorithm selection
    (d) Performance evaluation
11. Which machine learning approach utilizes both labeled and unlabeled data?
    (a) Supervised learning
    (b) Unsupervised learning
    (c) Semi-supervised learning
    (d) Reinforcement learning
12. Which evaluation metric provides a balance between precision and recall?
    (a) Accuracy
    (b) F1-score
    (c) Precision
    (d) Recall
13. Which machine learning technique involves building multiple decision trees during training?
    (a) Support Vector Machines
    (b) *k*-Nearest Neighbors
    (c) Decision Trees
    (d) Random Forest
14. Which one of the following is not a deep learning technique?
    (a) Convolutional Neural Networks
    (b) Long Short-Term Memory Networks
    (c) Artificial Neural Networks
    (d) Support Vector Machines

15. Which one of the following can be used to perform association studies and marker-trait association analysis?
    (a) MEGA
    (b) ClustalW
    (c) TASSEL
    (d) Augustus

Answers

1. (c)
2. (b)
3. (c)
4. (b)
5. (a)
6. (c)
7. (c)
8. (b)
9. (a)
10. (b)
11. (c)
12. (b)
13. (d)
14. (d)
15. (c)

Suggested Reading

Borrayo E, Machida-Hirano R, Takeya M, Kawase M, Watanabe K (2016) Principal components analysis-K-means transposon element based foxtail millet core collection selection method. BMC Genet 17(1):1–10

Cochran WG, Cox GM (1992) Experimental designs, 2nd edn. John Wiley & Sons Inc. ISBN-13: 978-0471545675

Federer WT (1956) Augmented (or hoonuiaku) designs.

Gomez KA, Gomez AA (1984) Statistical procedures for agricultural research, 2nd edn. John Wiley and Sons, New York. 680 p

Johnson RA, Wichern DW (2015) Applied multivariate statistical analysis, 6th edn. Pearson Education India. ISBN-13: 978-9332549555

Nigam AK, Gupta VK (1979) Handbook on analysis of agricultural experiments, 1st edn. I.A.S.R.I. Publication, New Delhi, pp 16–20

Poland J, Endelman J, Dawson J, Rutkoski J, Wu S, Manes Y et al (2012) Genomic selection in wheat breeding using genotyping-by-sequencing. Plant Genome 5(3)

Snedecor GW, Cochran WG (1989) Statistical methods, 8th edn. Iowa State University Press, Ames. ISBN: 978-0813815619

Upadhyaya HD, Ortiz R (2001) A mini core subset for capturing diversity and promoting utilization of chickpea genetic resources in crop improvement. Theor Appl Genet 102:1292–1298

PGR Informatics for Efficient PGR Management

17

Sunil Archak

Learning objectives

- PGR Informatics and its significance in PGR management
- Purpose of PGR Informatics and clientele
- Status of PGR Informatics at global and national level

Keywords

Plant genetic resources · Germplasm · PGR · Informatics · Database · Documentation · Digitization

17.1 Introduction

It is needless to emphasize the dependence of varietal development programmes on the availability of plant genetic resources (PGRs). Therefore, the conservation of PGR is non-negotiable for sustaining food production in a world increasingly challenged by changing climate and emerging crop pests and diseases. Plant breeders depend upon the ex situ collections as sources of valuable traits they can incorporate into new varieties. Managing PGR collection requires a set of highly specialized and multi-disciplinary skills. PGR activities requiring disparate expertise depend heavily upon data associated with germplasm accessions and in turn generate huge amount of data during experimentation. PGR utilization depends upon access to information. The lack of operational PGR databases and an interactive portal leads

S. Archak (✉)
Germplasm Exchange and Policy Unit, ICAR-National Bureau of Plant Genetic Resources, New Delhi, India
e-mail: sunil.archak-icar@nic.in

K. Tripathi et al. (eds.), *Textbook of Plant Genetic Resources*,
https://doi.org/10.1007/978-981-96-5024-8_17

to redundant investments in PGR characterization and evaluation and notably inefficient utilization of valuable genetic resources.

17.2 PGR Documentation

PGR documentation is recording, storing, managing and using the PGR data. Each genebank accession is attached with a range of data points including passport data, seed health, conservation and regeneration data, characterization data, trait data, utilization data, etc. Data sets that are essential for using a genebank accession include passport data, availability/accessibility data, trait data, images, subsets like core, etc. On the other hand, data that are essential for managing the collections include inventory data, seed health data, indent-supply records, etc. One should remember that, in the course of looking at a germplasm accession for the first time until its eventual utilization, a sequence of perception stages exists (Box 17.1).

Box 17.1 Informatics Pipeline

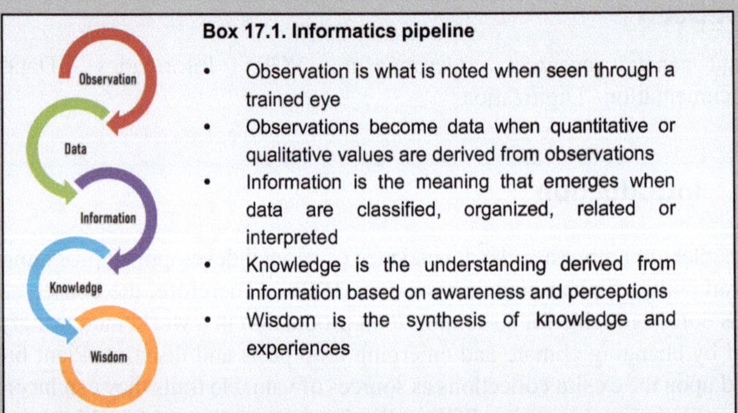

- Observation is what is noted when seen through a trained eye
- Observations become data when quantitative or qualitative values are derived from observations
- Information is the meaning that emerges when data are classified, organized, related or interpreted
- Knowledge is the understanding derived from information based on awareness and perceptions
- Wisdom is the synthesis of knowledge and experiences

Maintaining and updating hard-copy registers is an impossible task for genebank managers. Nearly all genebanks around the world have digitized the PGR information. The way this is done essentially depends on the size of the collection and its composition, expanse of genebank operations, availability of expertise and the extent of integration in the stakeholder accessibility (Box 17.2).

Box 17.2 Digitization, Digitalization and Digital Transformation in PGR

Computerization has ushered in overwhelming changes to the management of PGR information. While all three processes leverage advancements in technology, they essentially differ in aim, scope and outcome.

Whereas digitization focuses on PGR data conversion from hard copy to softcopy (e.g. registers to excel sheets), digitalization aims to improve PGR conservation and utilization operations with digital solutions (e.g. deploying PGR databases and web applications to implement PGR services like issuing import permit, allotment of national identity numbers, germplasm registration, etc.). Digital transformation targets to leverage the former two for a total overhaul of PGR management.

Digitization: Registers to excel sheet

Digitalization: Excel sheet to databases and development of applications

Digital transformation: Building an online portal of PGR management linking processes and partners with services and stakeholders

Smaller collections of fewer crops allow for simple documentation systems. Historically, genebanks around the world started working with MS Excel spreadsheets. However, data processing, data integrity and data security become challenging for larger collections and a greater number of operational events. It is at this stage that the creation of databases became necessary. Genebanks explored the use of specialized database management software, such as SQL, MS Access or Oracle. It must be noted that MySQL or MSSQL can only offer general data management functionalities. Therefore, genebanks needed to develop an application that facilitates the storage of specific information and the desired operations. Such a system called a database management system (DBMS) was developed and operated in CG genebanks and many national genebanks.

Using a DBMS invariably requires data standards for all data sets. Genebanks in the early 1990s used passport and characterization descriptors to manage accessions. By the turn of the millennium, Food and Agriculture Organization of the United Nations (FAO) and International Plant Genetic Resource Institute (IPGRI) standardized the *multi-crop passport descriptors* and *minimum descriptors for characterization*. By 2010, there were profound changes in the way PGR data were managed (Fig. 17.1) and consequently, PGR documentation was transformed into a more sophisticated field named PGR Informatics.

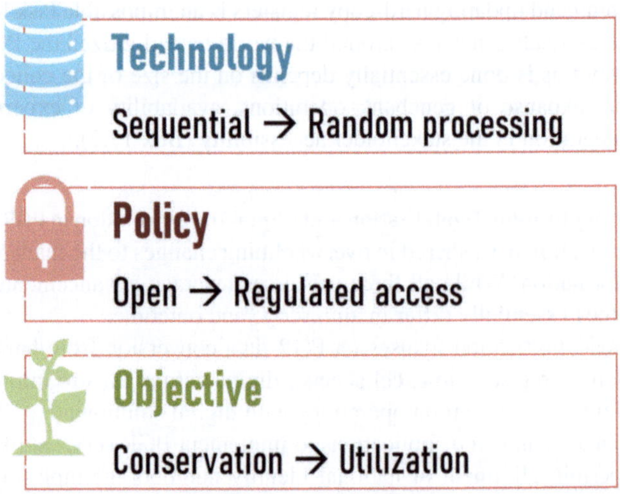

Fig. 17.1 Fundamental changes that affected PGR Informatics

17.3 What Is PGR Informatics?

PGR Informatics is the *management* (creation, storage, retrieval and presentation) and *analyses* (discovery, exploration and extraction) of *diverse* information (facts, figures, statistics, knowledge and news) related to PGR management. PGR Informatics goes beyond mere PGR documentation because it is now user-centric with the inclusion of analytical steps and improved algorithms.

Evolution of diverse germplasm accessions and subsequently their selection and conservation occur at sites far away from centres that are involved in genebanking, research and database management. PGR Informatics has the power to connect farmers and local communities with researchers and ex situ genebanks. Therefore, the fundamental objective of PGR Informatics is that 'an organized digital information system provides fair and just opportunity for all to access'. PGR Informatics provides a decisive edge to PGR conservation and utilization.

PGR Informatics is built upon sound scientific grounds, which go beyond computer expertise to encompass genetic, taxonomic, geo-informatic, bioinformatic and genomic linkages. PGR Informatics revolves around a user-centric approach that includes:

1. Growth – through richer content and metadata, especially field genebanks, horticultural genetic resources and farmers' efforts
2. Scalability – through established architecture and decentralized services

3. Capacity – through development of appropriate tools, products and services
4. Visibility – through the development of a powerful and user-friendly data portal, nodes, thematic portals and rich Internet applications

Research and development in PGR analytics accelerate cognitive decisions that facilitate PGR utilization. PGR analytics, at varying levels of analysis and complexity, includes:

1. Descriptive – reporting and querying of data to identify problems and solutions
2. Predictive – modelling, forecasting and simulating outcomes based on the data
3. Prescriptive – recommend the best course of action based on the data

The above-mentioned developments are not possible without laying down data standards as well as benchmarks for algorithms and modules. These in turn cannot be developed without creating a sufficient knowledge body based on basic research in PGR Informatics.

17.3.1 PGR Informatics in the Spotlight

In recent times, PGR Informatics has come into the limelight because of the following reasons:

- Increased awareness about PGR among the general public about the responsibilities of researchers, the role of policy makers and most importantly rights of farmers and local communities.
- Several international agreements coming into force to which India is a signatory including the Convention on Biological Diversity, the Global Plan of Action on PGR and the International Treaty on Plant Genetic Resources for Food and Agriculture. As a deliberate act, point number 17 in all these texts relates to the exchange of PGR information and therefore building an informatics system (Fig. 17.2).
- Digitization has taken off globally across fields and this has resulted in the availability of information in text, images, maps, videos, etc. that are directly or indirectly related to PGR.
- Newer technologies and hardware have made it easy and universal to record and archive such diverse types of information. Algorithms are available that scan the digital world to assemble and link the relevant information.
- Power of the computers in terms of hardware capacity is ever-increasing and the cost of a decent computer has become progressively affordable. The expansion of the Internet has been transformative in facilitating access and retrieval of PGR information.

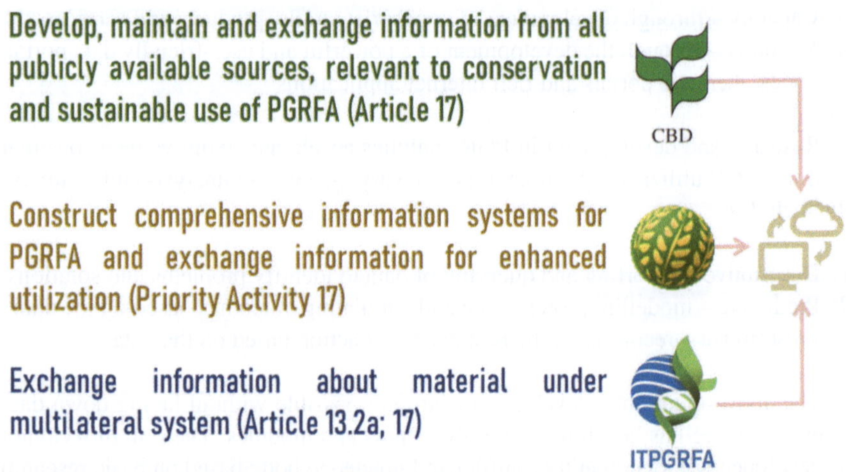

Fig. 17.2 PGR Informatics in various legal texts

17.3.2 Need for PGR Informatics

Following critical gaps necessitate investment in PGR Informatics:

1. Germplasm information management suffers not only because of inadequate digitization, collation and curation of data but also because neither the information is standardized nor the compatible software and data models implemented.
2. The absence of all-encompassing scientific questions to guide PGR Informatics is resulting in developments that have no connection to genuine insights and forward progress.
3. Researching the links between diverse dimensions of plant genetic diversity – conservation status, taxonomy, distributional biology, ecology, interactions, genomics and phylogenetics – will enable a transition from data management to informatics science.
4. Lack of emphasis on PGR analytics to facilitate cognitive decisions.

Typically, data availability and technology have driven many of the ideas and concepts in the PGR Informatics field. The desirable state of the field would see ideas and concepts driving the development of new technology and new data resources.

17.3.3 Basics of PGR Informatics

Considering the significance of PGR data, developing a clear scientific basis and standards for PGR information management becomes paramount. At the outset, it is important to understand the purpose (Why should we manage?), content (What

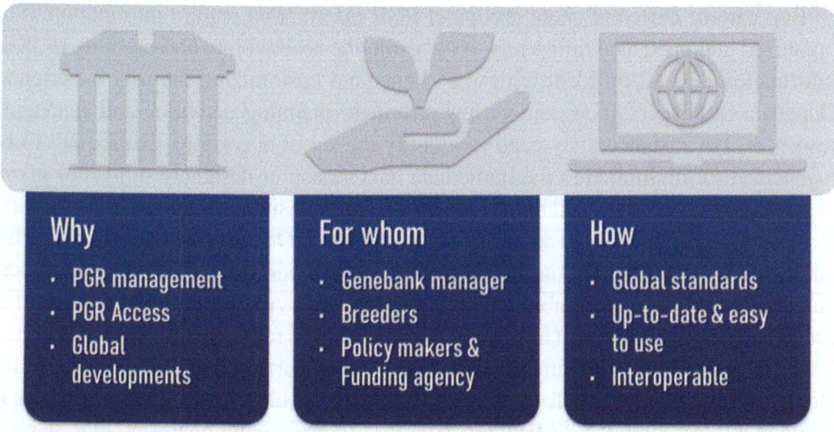

Fig. 17.3 Basics of PGR Informatics

PGR information?), process (How should be the PGR information?) and the clientele (For whom should we manage?) of PGR Informatics (Fig. 17.3).

Why: PGR management involves people with diverse expertise collectively contributing to augmentation, conservation, trait discovery and utilization. These experts require accurate and relevant information to prioritize, plan, execute and review the activities. Thus, PGR Informatics supports efficient PGR management. Further, end-users of germplasm (breeders and plant scientists) depend exclusively on accession information and trait information for shortlisting the accessions to carry out research work. Thereby, PGR Informatics supports enhanced utilization of ex situ conserved germplasm. India, as the signatory to various global agreements and treaties (Fig. 17.2) that are legally binding, has to comply with the requirements of developing a robust information system on PGR conserved ex situ by the state. PGR Informatics facilitates the nation's effective participation in the global developments related to PGR.

What: Acquisition of germplasm accessions either by exploration and collecting or by means of import from foreign sources will be accompanied by passport data. When such accessions undergo quarantine testing, the process generates seed health data. Every accession also has data on the mode of conservation (seed genebank, in vitro and cryo, field genebank, etc.). Trait-level data are generated when the accessions are characterized (based on pre-decided descriptors and states) and evaluated for specific traits such as quality, agronomic utility, resistance to biotic and abiotic stress factors, etc. carried out in the field or lab (biochemical and genomic).

How: PGR information needs to be up-to-date, accurate, reliable, systematically stored and easily accessible in order to ensure that it is available for users and extensively used. It is also important to ensure global standards, wherever available, are followed to facilitate inter-operability. On the contrary, information not possessing the above qualities not only becomes useless but also can be counterproductive.

For whom: Different client groups of PGR information require information to be organized and presented differently. As a result, the accessibility and usability of PGR information depend on identifying the clients and customizing the user experience. Genebank managers focus on setting priorities, planning activities and managing resources. For instance, they need to identify the set of accessions that require to be regenerated or multiplied in the immediate next season or designate subsets of germplasm based on evaluation data or prepare for germplasm acquisition based on gaps. Genebank managers require access to every accession. On the other hand, breeders are interested in only selecting accessions with superior performance along with procedures to obtain them preferably online indenting tools. An entirely different group of users include policy makers (from genebank directors to research managers to secretaries in the ministries) and funding agencies (e.g. Department of Biotechnology and Department of Science and Technology of India or Consultative Group on International Agricultural Research (CGIAR)/Crop Trust). They are usually interested in looking at information of a higher level, particularly the trends that are provided by dashboards with quantitative information. Such information facilitates decision-making during the allocation of funds or for political approval of programmes.

17.3.4 Standards of PGR Informatics

As part of the global information ecosystem, PGR information system must follow minimum technical standards. Every information system is expected to be FAIR (**F**indable, **A**ccessible, **I**nteroperable and **R**eusable). Since PGR information is associated with germplasm accessions, a PGR information system is only a custodian of the information and not the owner. Therefore, every PGR information system must follow CARE principles (**C**ollective benefit, **A**uthority to control, **R**esponsibility and **E**thics).

17.3.5 Components of PGR Information System

Successful PGR information management is characterized by quality, efficiency, sharing capacity, security and compliance (Fig. 17.4). The system functions as an organized combination of people, hardware, software, data, procedures and networks that collects, transforms and disseminates information.

- People: Information Technology (IT) specialists (such as a Database Administrator or Network Engineer), data entry operators as well as end-users (students, breeders, etc.)
- Hardware: all the physical aspects of an information system, ranging from peripherals to computer parts and servers
- Software: system software, application software and utility software
- Data: all the databases
- Networks: communication media and network support
- Procedure: architecture, data flow and access policy

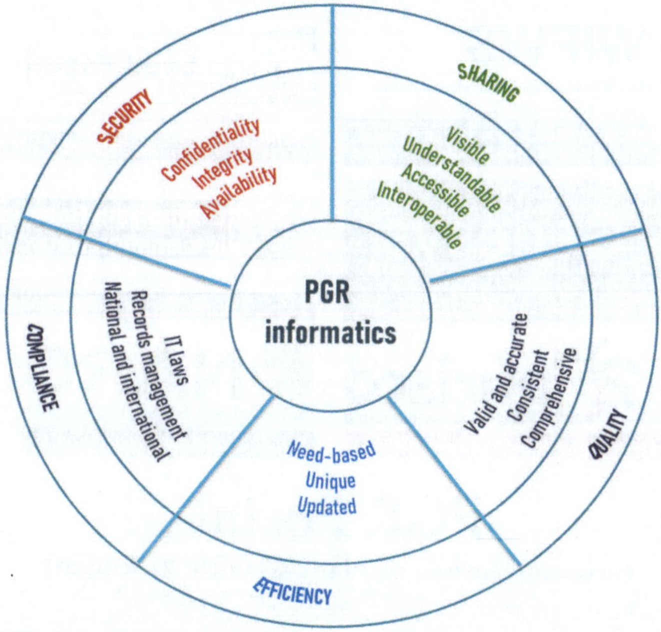

Fig. 17.4 Five characteristics of a PGR Information system

17.4 Global Status of PGR Information Systems

The genebank information systems are limited to 11 CGIAR genebanks, and 7 major national genebanks are implemented at multiple levels (Fig. 17.5):

1. Genebank level (e.g. CGN-PGR, Netherlands; IPK, Germany; CRF, Spain; VIR, Russia; EMBRAPA, Brazil; Kew's, UK; NIAS, Japan).
2. National level (e.g. AusPGRIS, Australia; PGRC, Canada; NORDGEN; ARS-GRIN, USA).
3. Regional level (e.g. EURISCO, Europe; EAPGREN, Eastern African Nations; REMERFI, Latin America; SPGRC, South African Countries; GRENEWECA, West and Central Africa).
4. Global level (System-wide Information Network for Genetic Resources [SINGER] and GENESYS; Global Biodiversity Information Facility [GBIF]). USDA genebank developed an information management system called GRIN Global in 2008 (collaborative efforts of GCDT, Bioversity, USDA-ARS at $2.5 million). In 2010, GeneSys, a web portal for all the CG genebanks as well as the EURISCO information system, was launched.

Unfortunately, at the other end of the spectrum, almost all the gene-rich countries lack informatics facilities. India (Indian Council of Agricultural Research–National Bureau of Plant Genetic Resources [ICAR-NBPGR]) has databases and portals in

Fig. 17.5 Major PGR web portals

place, which cater to requirements of documentation and sharing of genebank information.

17.4.1 Status of Research in Tools and Algorithms

A number of tools and algorithms have been developed in the field of informatics applicable in distinct areas like banking, clinical studies, engineering, social sciences and forensics. Some examples in biology include tools like Darwin Core Archive and communication protocols namely DiGIR, BioCASE, TAPIR–TDWG, etc. However, customizing available algorithms for PGR Informatics continues to remain a challenge.

17.5 PGR Informatics in India

ICAR-NBPGR is the nodal organization for PGR management in India. The Bureau headquarters in New Delhi houses the second largest genebank in the world. Naturally, ICAR-NBPGR is responsible for PGR documentation and also for developing a PGR information system. To cater to internal requirements of PGR management, to enhance the utilization of PGR by breeders and to keep up with the spirit of international instruments of information sharing, ICAR-NBPGR has developed the necessary infrastructure, databases and applications of PGR information management (for timeline, see Fig. 17.6).

PGR documentation work began as early as in 1997 with the computerized assignment of national identities of the germplasm (IC number). By 2003, all the passport data and genebank data were computerized as Excel sheets. Between 2006 and 2010, databases of passport and genebank inventories were developed in MS-SQL and using the dot NET platform, web applications were developed for use by NBPGR researchers primarily for data entry and report generation. After concluding physical verification of genebank inventory (label, quantity and location) in 2011, PGR Portal, an open access single window system for accessing PGR information was launched on 6 December 2012 (Fig. 17.7). This was a watershed moment that led to the birth of PGR Informatics. ICAR sanctioned a project on PGR Informatics in 2014, one of its kind globally, in the form of a National Fellowship. This project, from 2014 to 2024, builds standards, algorithms and processes to streamline PGR documentation in India. The project outcome includes various web apps and mobile apps for routine PGR management operations. A complete list is available with the URL at http://pgrinformatics.nbpgr.ernet.in. PGR students and researchers will be benefited by looking at various PGR Informatics applications. At present, ICAR-NBPGR is maintaining various databases (Table 17.1) and applications (Table 17.2). Databases are managed only internally but accession through various *online open access* applications. The applications are widely used by PGR researchers indicated by over 40K page views annually.

17.6 Challenges and Future Prospects in PGR Informatics

PGR Informatics suffers globally from a lack of funding support, a multiplicity of applications, lack of inter-operability and standards for diverse data sets. In India, additionally, information managers face challenges of delivering in multiple languages, data quality, harmonization and sustainability (Fig. 17.8).

Traditionally, PGR Informatics has dealt with data and information associated with germplasm accessions in order to facilitate their conservation and use. In the past decade, technological advancements in genomics have led to the explosion of sequence data derived from PGR. The genomic data has added an edge to trait-specific evaluation of germplasm. It has also created a new opportunity to create information systems that combine field data, biochemical data and genomic data to make intelligent choices. Further, the advent of genome editing technology allows researchers to make use of sequence information alone in crop improvement. Such use of digital sequence information (DSI) *in lieu* of germplasm in physical form for crop improvement is termed as dematerialized use of PGR. Globally, DSI has been highlighted as a concern in the access and benefit sharing of PGR. Future efforts in PGR Informatics need to internalize all such technological and regulatory developments to remain relevant and contemporary.

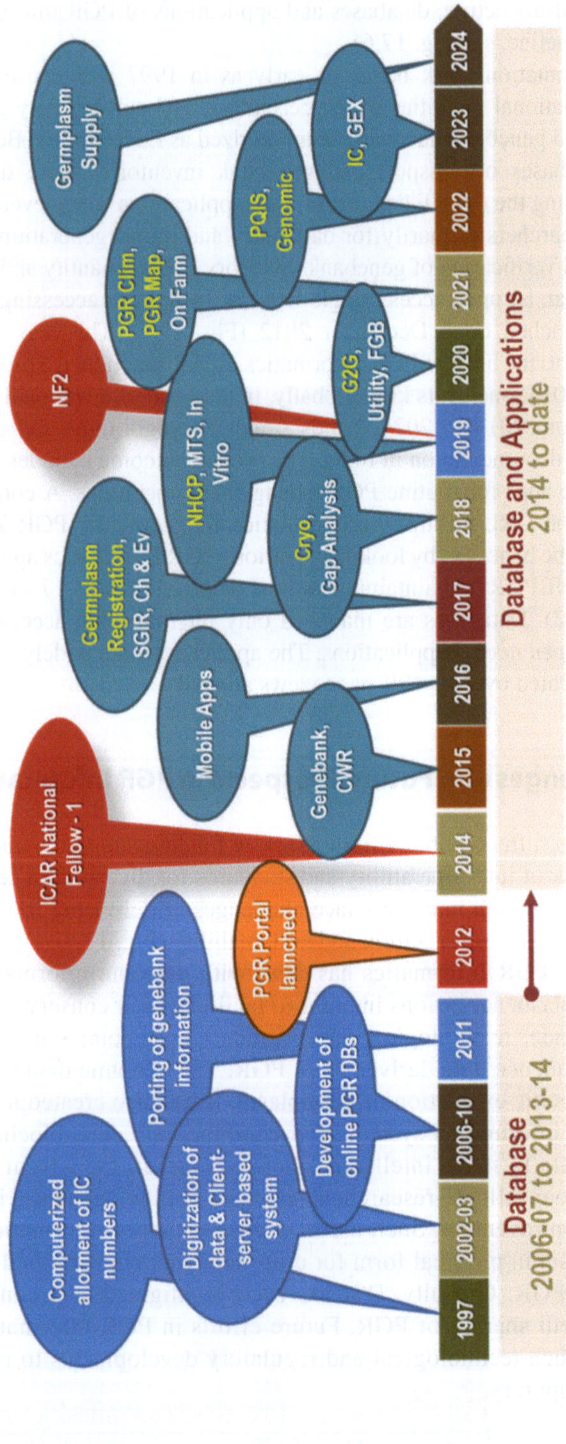

Fig. 17.6 Timeline of PGR Informatics development in India

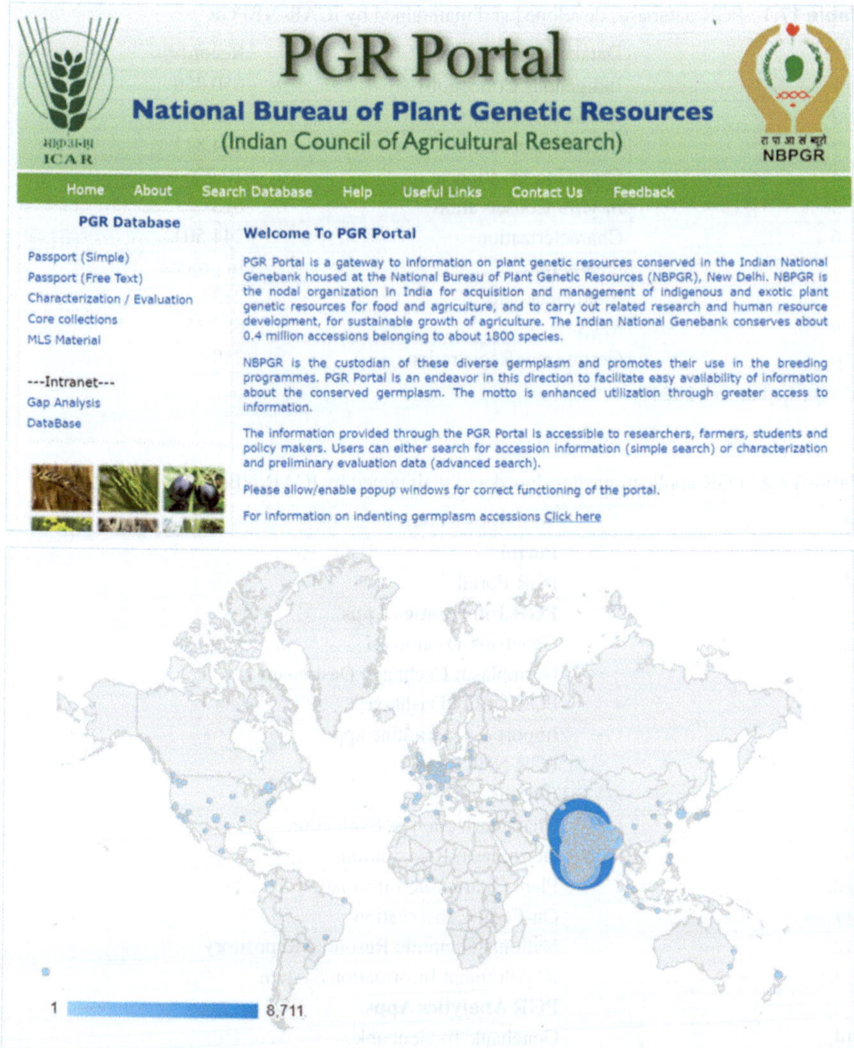

Fig. 17.7 PGR Portal (top) and its access Google analytics (bottom). On an average, PGR Portal has ~200 views per day

Table 17.1 PGR databases developed and maintained by ICAR-NBPGR

#	Database name	Records[a]
1	Indigenous Collections	650,859
2	Exotic Collections	1,200,802
3	Seed Conservation	424,841
4	Cryo Conservation	10,926
5	*In Vitro* Conservation	613
6	Characterization	244,504
7	Quarantine	In process
8	Field Genebank	2133
9	MTS	68,735
10	Germplasm Registration	2198

[a]As on 14 January 2024

Table 17.2 PGR applications developed and maintained by ICAR-NBPGR

#	Web Apps name
	Portal
1.	PGR Portal
	PGR Informatics Apps
2.	Genebank Dashboard
3.	Germplasm Exchange Dashboard
4.	PGR Utility Dashboard
5.	Import & Quarantine app
6.	PGR NHCP
7.	IP PGR
8.	Characterization & Evaluation
9.	Germplasm Registration
10.	Plant Quarantine Information System
11.	On-Farm Conservation
12.	National Genomic Resource Repository
13.	IC Allotment Information System
	PGR Analytics Apps
14.	Genebank To Genbank
15.	PGR Clim
16.	PGR Map
17.	CWR
18.	Sesame Genomic Information Resource
19.	Gap Analysis
	Website
20.	PGR Informatics

Fig. 17.8 Challenges in PGR Informatics

17.7 Summary

PGR Informatics is the management and analysis of diverse information related to PGR management. PGR Informatics aims to facilitate 'an organized digital information system that provides fair and just opportunity for all to access'. Increased awareness about PGR, availability of digitized information, advancements in computer technologies, Internet and hardware, and obligations to the international agreements on the exchange of PGR information have propelled PGR Informatics into prominence. PGR Informatics is expected to enhance the efficiency of PGR management as well as increase PGR utilization. Every PGR information system must follow FAIR and CARE principles. Indian efforts in PGR Informatics have delivered excellent web-based applications and mobile apps. However, information managers in India face challenges of delivering in multiple languages in addition to common problems of data quality, harmonization and sustainability. PGR Informatics must remain relevant by responding to changes in science and technology as well as regulatory framework.

Questions (True/False)

1. In PGR research, exploration generates information, evaluation generates data and genomic study generates insights. True/False
2. Ideally, all PGR data must be made available in English. True/False
3. In the domain of PGR information, equal, fair and just mean the same. True/False

4. In PGR Informatics, building computational and Internet infrastructure is more important than developing human resources. True/False
5. PGR Map is a unique app developed by ICAR-NBPGR for accessing quarantine information. True/False
6. In PGR Informatics, FAIR and CARE principles exist to achieve compliance with global PGRFA agreements. True/False

Answers: All False

Suggested Reading

Agrawal RC, Archak S, Tyagi RK (2012) An overview of biodiversity informatics with special reference to plant genetic resources. Comput Electron Agric 84:92–99

Konzak CF, Dietz SM (1969) Documentation for the conservation, management, and use of plant genetic resources. Econ Bot 23:299–308. https://doi.org/10.1007/BF02860675

Weise S, Oppermann M, Maggioni M, van Hintum T, Knüpffer H (2017) EURISCO: The European search catalogue for plant genetic resources. Nucleic Acids Res 45(D1):D1003–D1008. https://doi.org/10.1093/nar/gkw755

PGR Informatics Related Links

ABSCH – Access and Benefit-sharing Clearing-house Platform of the Nagoya Protocol. https://absch.cbd.int/

Benefit-sharing Fund project coordinated by Indonesia "Multi-country construction of a test platform for the development and allocation of globally unique identifiers for rice germplasm: (W3B-PR-29). http://www.fao.org/plant-treaty/areas-of-work/benefit-sharing-fund/projects-funded/bsf-details/en/c/359491/?iso3=IDN

CAPFITOGEN. http://www.fao.org/plant-treaty/initiatives/capfitogen/en/

CGIAR Open Access and Open Data. http://www.cgiar.org/resources/open-access/

CGIAR Research Program and Platform. http://www.cgiar.org/consortium-news/cgiar-research-program-platform-full-proposals-submitted-for-review/

Cyverse. http://www.cyverse.org/

DataCite. https://www.datacite.org/

DivSeek. www.divseek.org

ECPGR – European Cooperative Programme for Plant Genetic Resources. http://www.ecpgr.cgiar.org/

Eurisco. http://eurisco.ipk-gatersleben.de

GBIF – Global Biodiversity Information Facility. http://www.gbif.org/

Genesys Portal. https://www.genesys-pgr.org/

Global Open Data for Agriculture and Nutrition. http://www.godan.info

GRIN-Global. http://www.ars-grin.gov/npgs/gringlobal/

IBP – Integrated Breeding Platform. https://www.integratedbreeding.net/

IRIS-International Rice Information System. http://irri.org/tools-and-databases/international-rice-information-system

The DOI® System. http://www.doi.org/

WIEWS – World Information and Early Warning System on Plant Genetic Resources for Food and Agriculture. http://www.fao.org/wiews/en/

Description About Five Scientists

Nikolai Ivanovich Vavilov

Born in Moscow on November 25, 1887, Nikolai Vavilov was a merchant's son. During the years 1913–1914, he investigated plant immunology under William Bateson, one of the pioneers of Genetics, at the John Innes Horticultural Institution in London and the University of Cambridge. Following his employment as a teaching staff member in 1917, Bateson adopted the theories of G.J. Mendel, who postulated that certain features are passed down unmodified from generation to generation. In 1917, he was named Deputy Head of the Bureau for Applied Botany. He spent the rest of his career at the Bureau, where he established hundreds of research institutes and experiment stations, employing more than 20,000 people, and turned it into the greatest Crop Research Institute in the world. Vavilov received the title of Professor at Saratov University in 1918. Following his several trips, Vavilov brought samples of 50,000 different wild plant species and 31,000 specimens of wheat to his institute in Russia. He hypothesized that the center of origin for a cultivated plant would be located in the area where the plant's wild relatives exhibited the highest degree of adaptability. After being elected as the head of the Department of Applied Botany 2 years later, he and his colleagues quickly relocated to Petrograd, (now St. Petersburg) and renamed the Institute of Applied Botany and New Crops in 1924, the primary national organization tasked with gathering data on plant diversity worldwide. Afterward, Vavilov postulated eight global centers of plant origin. He was especially interested in the whereabouts of ancient agricultural civilizations. The Institute of Experimental Agronomy was led by Vavilov from 1922 to 1929. It was renamed the V.I. Lenin All-Union Academy of Agriculture in 1930, and he served as its first president from 1930 to 1935. He served as the Director of Institute of Genetics from 1930 till 1940. Vavilov served as president of the Russian Geographic Society from 1931 until 1940. Later, he was charged with rightist conspiracy, spying for England, and numerous other offenses and sentenced to die. Although his sentence was reduced, he passed away from starvation in jail in 1943. The reason for this is that Vavilov, who is renowned for saying, "We shall go into the pyre, we shall burn, but we shall not retreat from our convictions," endorsed the mendelian approach.

K. Tripathi et al. (eds.), *Textbook of Plant Genetic Resources*,
https://doi.org/10.1007/978-981-96-5024-8

Harbhajan Singh

Harbhajan Singh, the great legend of "Plant Genetic Resources" was born on February 6, 1914, in Pusa, Bihar. He was renowned as "The Indian Vavilov." He completed his Master's Degree at the Agra University of Agriculture. His passion for Economic Botany led him to become the Head of the Division of Plant Introduction at IARI. Singh made an invaluable contribution to the Green Revolution in India by bringing in a vast amount of wheat and rice germplasm, as well as a variety of new species like sunflower, soybean, and low-chilling and temperate fruits like apples, peaches, West Indian cherry, and Chinese gooseberry, as well as a broad range of ornamentals from abroad. Singh has made enormous contributions to the introduction of plants, vegetables, and horticultural crops that are important to the country. The introduction of various crop germplasm, such as peas, tomatoes, cauliflower, bhindi, turnip, carrot, French bean, guar, bottle-gourd, luffa, cowpeas, onion, sweet potato, cucumber, watermelon, and garden pea, were added in the collection. His research on okra helped to create cultivars resistant to jassids and the yellow mosaic virus including popular cultivar Pusa Sawani. Additionally, he created oat cultivars that are ideal for use as green fodder and in the breakfast food sector. He was one of the first plant breeders in this country to understand the importance of growing soybeans, and his work also included horticultural crops such as Chinese gooseberry, jujube, and low-chill variants of temperate fruits like peaches and apples. He also made significant contributions to the production of a large variety of attractive plants, including pompon (small-flowered) Chrysanthemums. The Indian President bestowed upon him the Padma Shri Award in 1971 in recognition of his noteworthy achievements. In the same year, he also received an honorary Doctor of Science (Honoris Causa) Degree from Punjab Agricultural University.

Charles Darwin

The man famed for his theory of Evolution by Natural Selection, Charles Darwin, was born in Shrewsbury, Shropshire, England, on February 12, 1809. He attended Edinburgh University to study medicine in 1825 and later pursued theology at Cambridge, where his interest in natural history grew. His professor of botany, John Stevens Henslowe, took note of his passion for collecting plants, insects, and geological objects. He secured a spot for his gifted pupil on the HMS Beagle surveying mission. Darwin made the decision to leave his comfortable circumstances at the age of 22, against his father's misgivings. From 1831 to 1836, Charles Darwin's journey lasted 5 years. Upon his return, Darwin carried with him observations from his travels to Tenerife, the Cape Verde Islands, Brazil, and most notably the Galapagos Islands. Unfortunately, Darwin had caught a tropical sickness on the voyage that would leave him semi-invalid for the rest of his life. After publishing multiple papers based on the voyage's results by 1846, Darwin was elected secretary of the Geological Society. In 1859, Darwin published "On the Origin of Species" which proposed that all species evolve from common ancestors through natural selection. This challenged traditional views on creation and sparked significant scientific and religious debate. Darwin continued to expand on his theories throughout his life, publishing additional works on topics such as the descent of humans and the role of sexual selection in evolution. His research extended beyond biology, influencing fields like anthropology, psychology, and ecology. On April 19, 1882, he passed away at Down House, London, due to "angina pectoris," a heart condition known as "coronary thrombosis."

Jack Rodney Harlan

On June 7, 1917, Jack Rodney Harlan was born in Washington, D.C. He received his Bachelor's Degree from George Washington University in Washington, D.C., in 1938, and his Doctorate in Genetics from the University of California in 1942. He was the first person to complete a doctorate under the supervision of G. Ledyard Stebbins. He was well-known for his contributions to the evolution of plants and his archaeological excavations. Harry V. Harlan, with whom he travelled for the archaeological excavations, had an influence on Harlan. He was employed in Honduras as a research assistant for Tela Rail Road Company. His official career began in 1942 when he was employed by the United States Department of Agriculture (USDA) in Woodward, Oklahoma, where he oversaw the Southern Great Plains Regional Grass Breeding program and the Oklahoma Forage Crop and Rangeland Improvement program. In the United States, Jack Harlan conducted research and brought plants from Australia, Asia, Africa, Central America, and South America. He oversaw a botanical exploration journey to Turkey, Syria, Lebanon, and Iraq in 1948 that was funded by the USDA. Harlan is thought to have gathered over 12,000 accessions for the US from 45 different nations. In addition to forage legumes and grasses, large-seeded legumes, forest trees, ornamentals, and fruits, he also gathered wheat, barley, and maize. Certain accessions have been widely deployed due to their distinct genetic characteristics or as a source of disease resistance. Jack Harlan won numerous accolades and honors over his career such as the Frank N. Meyer Memorial Medal (1971), the International Service in Agronomy Award (1976), the John Simon Guggenheim Memorial Fellowship (1959), the American Grassland Council Merit Award (1962) and the Crop Science Award (1971). The Society for Economic Botany awarded him the Distinguished Economic Botanist Award in 1985. In addition to the American Academy of Arts and Sciences (1975), he was a fellow of the American Society of Agronomy (1962), the American Association for the Advancement of Science (1956), and the Crop Science Society of America (1985). Harlan died in Urbana on October 11, 1982.

Otto Frankel

On November 4, 1900, Otto Herzberg-Frankel was born in Vienna. He was employed as a plant breeder on a private estate close to Bratislava, Czechoslovakia, between 1925 and 1927. After that, he went to Palestine to help set up programs for breeding plants and animals. After that, he relocated to Cambridge, England, where he worked in the Plant Breeding Institute of the university in a temporary capacity. Following the merger of the DSIR's agronomy division and the Wheat Research Institute's breeding department in 1949, he was named Associate Director of the newly formed division and then Director the following year. He received a Doctor of

Science Degree from the University of New Zealand in 1951. By dividing the division into research specialty groups rather than agricultural and pasture programs, Otto greatly improved the Commonwealth Scientific and Industrial Research Organization's (CSIRO) standards. He was assigned to an executive position at CSIRO in 1962.

Frankel's advocacy for the conservation of the genetic resources of wild relatives of crop and pasture plants made him important on a global scale. He participated actively in both the Consultative Group on International Agricultural Research (CGIAR) and the FAO's International Biological Program. He was instrumental in the creation of the CGIAR's International Board for Plant Genetic Resources in 1974. He persisted in promoting the importance and responsible handling of genetic resources even after he retired, especially while working with CSIRO scientists. Frankel, a distinguished member of the Australian scientific community, was elected as a fellow of the Australian Academy of Science (1954), the Royal Society of New Zealand (1948), and the Royal Society of London (1953). In 1988, he was also admitted as a foreign associate of the National Academy of Sciences of the United States. Otto Herzberg-Frankel took his last breadth on November 21, 1998, in Canberra.

Glossary

Abiotic It is related to physical and chemical factors of the environment such as water, temperature and soil.

Accession number A unique identifier that is assigned by the curator when an accession is entered into a genebank. This identifier should never be assigned to another accession.

Accession or entry A sample of a plant, line or population maintained in a germplasm bank or breeding programme for conservation and use.

Acclimatization The adjustment of a species or population to a new environment that is a new habitat, after several generations.

Active collection A group of germplasm samples or accessions stored for the short to medium term and maintained for study, distribution or use.

Addition line When a line has one pair of chromosomes from another variety or species in addition to the normal somatic chromosome complement of the species.

Agriculture The science of cultivating the soil to raise crops and/or rearing animals.

Agrobacterium A genus of soil bacteria that introduces genes into certain plants through their plasmids.

Agroecological zone A geographical area with similar characteristics in terms of climate, landform, soils and/or land cover and having a specific range of potentials and constraints for land use.

A-line The male-sterile parent line in a cross is made to produce a hybrid seed.

Allele Each of two forms of a gene present in the same place (locus) in a pair of homologous chromosomes or one of the alternative states of the same gene.

Acknowledgment
Glossary are compiled and adapted from below given sources:

1. Nagur, T., Oswalt DL and Sing F. 1992. A Glossary for Crop Improvement. Skill Development Series number 1. The International Crops Research Institute for the Semi-Arid Tropics (ICRISAT)
2. Pineda B. and Meja M. Glossary Compiled by CIAT, BI and CTA for Multi-Institutional Distance Learning Course on the Ex-situ Conservation of Plant Genetic Resources
3. Singh B.D. Plant Breeding: Principles and Methods. Kalyani Publishers

K. Tripathi et al. (eds.), *Textbook of Plant Genetic Resources*,
https://doi.org/10.1007/978-981-96-5024-8

Allelomorph: The characteristic specified by an allele.

Alloenzymes Isoenzymes whose variants are coded by different alleles at the same locus. Alloenzymes are also allelic and, as a result, are ideal for studying population.

Allogamy Cross-fertilization in plants. Allogamous plants are those plants that preferentially cross-pollinate.

Alloploidy A condition whereby an organism has more than two sets of chromosomes in its cells, with each set coming from a different species.

Altitudinal transect An imaginary line that is traced from a point in a mountain range to sea level. It is used to study altitudinal variations among plant communities.

Amino acids Organic molecules that contain amino and carboxyl groups. They are protein monomers. An enormous variety of proteins exist because of the large number of combinations and lengths.

Amphiploid An individual that originates from hybridization between species and possesses the total chromosome complement of the parental species. It is usually produced by the duplication of the chromosome number of the F1 plant hybrid.

Aneuploid An individual with a chromosome number that is not an exact multiple of the haploid chromosome complement.

Angiosperms or Angiospermae A taxon of plants, whose principal characteristic is that they present true flowers. Angiosperms are divided into monocotyledons and dicotyledons.

Anther In flowers that part of the stamen which contains pollen.

Anthesis In flowers, the dehiscence of anthers, when pollen is dispersed.

Apetalous flower A flower with no petals.

Apomixis A phenomenon whereby asexual reproduction occurs instead of normal reproduction through reductional division and fertilization. It is common in some plant species where embryos do not result from meiosis and fertilization but from certain asexual processes. Seedlings produced in such a way are called apomictic. Those plants that reproduce only through apomictic embryos are known as obligate apomictic, and those whose embryos are either sexual or apomictic are designated as facultative apomictics.

Arboretum (pl. arboreta) A garden where trees and shrubs are cultivated for study and display.

Autogamy In plants, self-fertilization. Autogamous plants are preferentially self-pollinated.

Autonym An automatically established name in which a generic name or specific epithet is repeated as the final epithet in the name of a subdivision of a genus or of an infraspecific taxon that includes the type of the adopted, legitimate name of the genus or species, respectively; the final epithet of an autonym is not followed by an author citation.

Autopolyploid or autoploid An organism whose somatic cells carry more than two sets of chromosomes, with all sets coming from parents of the same species.

Backcross breeding a cross of a hybrid with one of its parents or with a genetically equivalent organism.

Base collection The widest and most complete collection of germplasm accessions. It is stored over long periods for conservation purposes. It is used only to fill gaps in the active collection.

Basic seed Produced from seed developed by the plant breeder and so managed that the original genetic identity and purity of a given variety is faithfully conserved. The production of basic seed is carefully supervised or used by the representatives of an agricultural experiment station. Basic seed is the starting point for obtaining certified seed, either directly or through registered seed.

Basionym The legitimate, previously published name on which a new combination or name at a new rank is based. The basionym provides the final epithet, name or stem of the new combination or name at the new rank.

Biochemical markers Various isoenzymes that catalyse the substrate itself or other enzymes and which are used to evaluate the enzymatic heterogeneity of plants, that is, the genetic variability between individuals at the level of enzymes and proteins.

Biodiversity or biological diversity The set of all plant and animal species in a given region, including their genetic materials and the ecosystems of which they are part.

Biological containers Containers designed as protection mechanisms in the use of organisms in genetic engineering applications. Their purpose is to minimize the 'ability' of organisms used to survive, persist and self-replicate. The process is also known as 'genetic weakening' and leads to 'engineeringly' diminished organisms.

Biology The science that deals with the study of living things and vital phenomena in all their aspects.

Biometry The science that deals with the application of statistical methods to biological problems.

Biomolecules Formerly called immediate principles, these are the basic architectural elements of living things. Inorganic biomolecules are, above all, water, mineral salts and gases such as oxygen and carbon dioxide. The four groups of organic compounds, exclusive to living things, are carbohydrates, lipids, proteins and nucleic acids.

Biosystematics The study of classification of biological units of organisms on the basis of relationships, variability and interrelationships.

Biotechnology That set of technological applications that use biological systems and live organisms or their derivatives to create or modify products or processes for particular uses.

Biotic In agriculture, biotic factor or agents are frequently associated with three important groups that affect crop yield: pests (including nematodes), diseases and weeds.

Biotype A population in which all individuals have an identical genotype.

B-line The fertile counterpart of the A-line. The B-line does not have fertility-restoring genes and is used as the male parent to maintain the A-line.

Callus The initial tissue formed by the cellular division of explants. It is usually uniform, not having been differentiated into organized tissues.

Carbohydrates Organic biomolecules formed by polyalcohols with an aldehyde or ketone group.

Cell The structural and functional unit of plants and animals that typically consists of a mass of cytoplasm that encloses a nucleus (except in prokaryotes) and is bound by a differentially permeable membrane.

Certified seed The progeny of basic seed, which is produced and used in such a way that it maintains a satisfactory level of purity and genetic identity. It has been approved and certified by an official agency for certification.

Characterization The measurement or evaluation of the presence, absence or degree of specificity of traits whose expression is little modified by the environment.

Chimaera or mosaic An individual that presents two or more genetically different cell lines as a consequence of an anomaly in the first mitosis of the zygote.

Chromatid One of two filamentous structures that form in the duplication of a chromosome to form other chromosomes.

Chromosome number The usual constant number of chromosomes in a somatic cell that is characteristic of a particular species.

Chromosome An elongated intracellular organelle found in the nucleus and consisting of DNA associated with proteins. It constitutes a linear series of functional units known as genes, which conserve their individuality from one cell generation to the next. The chromosome number is typically constant in any given species.

Cleistogamy Pollination and fertilization within a closed flower.

Clone A group of cells or organisms with identical genetic constitutions and backgrounds and which are derived through binary division or asexual reproduction.

Cloned gene A gene copied from an initial gene. It is inserted into a molecular vector through *in-vitro* using recombination techniques.

Co-evolution The joint evolution of two or more organisms that are inter-related either positively or negatively.

Collection The action of gathering together, assembling, or grouping similar things into one place, usually with a particular focus. A plant germplasm collection therefore brings together the germplasm, whether as seeds, propagules or other genetic material, of particular plants. The focus may be conservational, agricultural, research and educational, environmental, historic, aesthetic or economical.

Complete flower A flower that has all four essential organs (sepals, petals, stamens and pistils).

Conifers Ancient plants of the Gymnosperm order whose principal characteristics are that they do not present true flowers but wooden structures, and their leaves are usually needle-like or aciculate.

Consanguinity The pairing of closely related organisms; in plants, this is usually achieved through self-pollination.

Conservation The conservation of plant genetic resources refers to the maintenance of population in their natural habitat (*in situ* conservation) or to samples of these population in germplasm banks (*ex situ* conservation). Conservation presumes that the materials are useful or potentially useful and seeks to maintain and manage them for both current and future benefits.

Core collection A collection that groups, into a minimum number of accessions, the greatest variability existing in a base collection.

Correct name The name that must be adopted in accordance with the rules for a taxon with a particular circumscription, position and rank.

Correlation The mutual relationship between two things in such a way that an increase or reduction in one is usually associated with an increase or reduction in the other. Linear correlation is determined by the coefficient of correlation, the value of which may vary from −1 to +1.

Cotyledons One or a pair of primary leaves of the embryo within the seed and, commonly, the first leaves to emerge in germination.

Crossing over The exchange of segments between the chromatids of two homologous chromosomes during meiosis.

Cross-pollination The transfer of pollen from the anther of one plant to the stigma of a flower of another genotypically distinct plant.

Crown In horticulture, that part of a plant's stem that is located in the soil or below its surface, from which new shoots originate.

Cryopreservation The conservation of materials at very low temperatures (−196°C), usually in containers with liquid nitrogen.

Cultivated species or domesticated species A species whose evolution has been influenced by humans seeking to meet their own needs.

Curator-in-charge That person, who conserves and administers plant genetic resources.

Cytogenetic map The configuration of coloured bands of chromosomes observed under the optic microscope after staining.

Cytology The science that deals with the structure, function and life cycle of cells.

Cytoplasm The protoplasm of a cell, except the nucleus.

Cytoplasmic inheritance Heredity that depends on hereditary units of the cytoplasm.

Darwinism The theory of evolution using natural selection was put forward by Charles Darwin in his book *Origin of Species* published in 1859 and by Alfred Russel Wallace. The theory was based on their observations of the genetic variability that exists within any given species.

Dehiscence The rupture or opening of a fruiting structure or anther.

Descriptors The quantitative or qualitative characteristics, whether morphological, agronomic or ecogeographic, that permit the identification of a plant at different taxonomic levels.

Detasselling The elimination of immature tassels or ears. This practice is followed in the seed production of hybrid maize.

Determinate An inflorescence in which the terminal flowers open first, thus impeding the extension of the floral axis.

Dichogamy In plants, the maturing of male and female organs at different times, thereby ensuring cross-fertilization.

Dicotyledons Those plants or plant species that have two cotyledons or the first pair of leaves forming in the embryo within the seed.

Dihybrid The result of a cross between parents that differ in two specified genes.

Dioecious A plant species that has male and female flowers on different individuals (see also Monoecious).

Diploid An organism that has two sets (genomes) of chromosomes, that is, a chromosome number of 2n, as in a zygote. The somatic tissue is normally diploid, in contrast to the gametes, which are haploid.

Disease The alteration or deviation from the normal physiological state in one or more parts of the plant for generally known causes, and which manifests according to characteristic symptoms and signs, and whose development is more or less foreseeable (see also Pathogen).

DNA cloning or gene cloning A technique in genetic engineering that consists of isolating and multiplying a given gene or fragment of recombinant DNA by incorporating it into a host cell (usually a bacterium or yeast) and then isolating copies of DNA thus obtained (see also Genetic engineering; Molecular cloning).

DNA fingerprint The pattern of DNA fragments obtained in restriction analysis of certain highly variable repeated DNA sequences within the genome. Their number and arrangement are virtually unique to each individual and can be used to identify that individual.

DNA library, DNA bank or genomic library A collection of DNA fragments amplified in cloning vectors. The cloned fragments may come from genomic (chromosomal) DNA or from complementary DNA (cDNA).

DNA or deoxyribonucleic acid Nucleic acid formed by nucleotides in which the sugar is deoxyribose and the nitrogenous bases are adenine, thymine, cytosine and guanine. Except in retroviruses, which have RNA, DNA codifies the information for cell reproduction and operation and for the replication of the DNA molecule itself. It represents the security copy or deposit of primary genetic information that, in eukaryotic cells, is confined to the nucleus.

DNA sequence The order of sequence of the nitrogenous bases of the nucleotides that constitute DNA and which codes for all genetic information. When it is a codifier (exon), it defines the order of the amino acids that form the corresponding protein.

Documentation With reference to plant genetic resources, the procedure by which information (data) on germplasm is identified, acquired, classified, stored, handled and disseminated.

Dominant gene A gene that manifests itself exclusively in a hybrid, that is, to the exclusion of its counterpart (recessive) allele.

Dominant A trait that expresses in phenotype of a hybrid to the exclusion of the counterpart (recessive) trait.

Donor parent Donor parent of a trait in a crossing programme.

Dormancy The state in which a seed, bud or reproductive structure of a plant is found at rest, inactive, quiescent, or dormant but which can initiate activity when the necessary conditions for activation occur.

Duplicate genes Two or more pair of genes that produce identical effects, whether together or separately.

Duplicate Germplasm sample that was mistakenly introduced into a collection as a different accession but which is genetically identical to others already in the collection.

Ear See Spike.

Ecology The science that studies living things at their different levels of organization and their interrelationship among themselves and with the environment.

Ecosystem A dynamic complex of communities of plants, animals and micro-organisms and their non-living environment; which complex acts as a functional unit.

Ecotype A group of biotypes adapted to a specific environmental niche.

Egg In plants, the female gamete.

Emasculation The elimination of anthers from a flower, either closed or open before pollen is released.

Embryo sac See Megasporangium.

Embryo The rudimentary plant within a seed. The embryo originates from the zygote.

Endangered or at risk Taxa whose habitat has been so drastically reduced as to be considered in immediate danger of disappearance, thereby leading to their extinction; or that are possibly already extinct but have been seen in the wild within the last 50 years (see also Categories of species conservation)

Endemism The condition of being endemic, autochthonous or indigenous. A plant or associated animal that originates in a given country or region and is restricted to that region, that is, aboriginal or native to a given geographical area.

Endosperm or albumen A triploid tissue that comes from the triple fusion of a spermatic nucleus with the two polar nuclei in the megagametophyte. In seeds of certain species, the endosperm persists as storage tissue for food reserves, which are used during the development of both the embryo and seedling during germination.

Entry See Accession.

Environment The sum total of external influences acting on the life, development and survival of an organism or group of organisms.

Enzyme A biological catalyst, normally a protein, that mediates and promotes a chemical process without itself being altered or destroyed. Enzymes are extremely efficient catalysts and specifically linked to particular reactions.

Epiphytotic The unexpected development and usually general distribution of a destructive disease of plants.

Epitype A specimen or illustration selected to serve as an interpretative type when the holotype, lectotype or previously designated neotype, or all original material associated with a validly published name, cannot be identified for the purpose of the precise application of the name to a taxon.

Ethnobotany The study of folklore and history of use, with particular reference to plants.

Evaluation The measurement, observation and analysis of a germplasm collection with a view to detecting its potential use. It generally uses descriptors of quantitative traits that are affected by the environment.

***Ex* situ conservation** *Ex* situ literally means out of the original place; hence, the conservation of plant genetic resources outside the areas where they had developed naturally (i.e. outside their natural habitats).

Exons DNA sequences, specific to genes, and which codify for amino-acid sequences in proteins.

Explant A segment of tissue or an organ obtained from a plant (e.g. leaf, root, anther, shoot, bud, embryo and meristem) and used to initiate an *in vitro* culture.

Exploration Searching for a particular plant or animal species to establish its range of agronomically useful variability and geographic distribution.

Extinct in the wild A category of conservation status, where a species is considered extinct in its natural distribution when it has not been located or sighted in the wild state for the last 50 years (a criterion used by CITES).

F1, F2 Symbols used to designate the 1st filial generation, the 2nd filial generation etc., after a cross.

Farmers' rights Those rights attributed to farmers for their contribution (past, current or future) to the conservation, improvement and availability of plant genetic resources.

Fats See Lipids.

Fertilization The fusion of an ovule and sperm cell (male gamete), forming a zygote.

Filament In flowers, the column of the stamen that sustains the anther.

Fixed line Denotes a plant genotype that is characterized by homozygosity in most loci, ideally a so-called double haploid.

Flora A set of plant species that is found in a given place. It is usually described in terms of a systematic or alphabetical list of all the plant taxa recorded in that place.

Floret A small flower of an inflorescence, as in the case of a grass panicle or compound spike.

Gametes or sexual cells Cells that, when fused, form the zygote. In plants, the gametes are the male sperm cell and the female egg.

Gene cloning See DNA cloning.

Gene expression The protein product resulting from the set of mechanisms that decode the information contained within a gene, processing it through transcription and translation.

Gene flow The exchange of genetic material between population through the dispersion of gametes and zygotes.

Gene interaction The modification of gene action through non-allelic genes.

Gene Pool In a population of randomly mating individuals the genes constitute a pool in the form of gametes that combine randomly to give the next generation.

Gene The physical and functional unit of hereditary material that determines a trait or characteristic of an individual and is transmitted from generation to generation. Its material base is constituted by a part of a chromosome known as locus and which codifies information through DNA sequences. By interacting with other genes, cytoplasm and the environment, it affects or controls the development of a trait, it is receptacle of genetic material that is particular to a given species.

Genetic drift A random fluctuation of genetic frequencies of a population from generation to generation, caused by factors such as natural selection. It is more evident in small isolated population, and may lead to the fixation of an allele and to the extinction of the other.

Genetic engineering, genetic manipulation or recombinant DNA technology The process of forming new combinations of hereditary material by inserting nucleic acid molecules, obtained from outside the cell, into any virus, bacterial plasmid or other vector system outside the cell. Thus, the host organism incorporates the new hereditary material in a way that does not appear natural, but where such molecules are able to reproduce continuously.

Genetic enhancement or pre-breeding The concepts refer to activities aimed at transferring genes, gene combinations and/or genetic variability from unadapted sources into more usable breeding materials that can be used as parents in breeding programmes.

Genetic erosion The loss of genetic diversity, that is, of genetic materials, including individual genes or combinations of genes (genetic complexes), genotypes and species.

Genetic identity The characteristic that should be maintained during conservation. This refers to the maintenance, as a set, of all the alleles of all the accession's genes.

Genetic instability Susceptibility of stored seeds to cumulative genetic changes (with age), resulting in the alteration of the initial genetic structure of the conserved sample.

Genetic integration The insertion of a DNA sequence into another through recombination.

Genetic manipulation See Genetic engineering.

Genetic map A descriptive diagram of the genes in each chromosome.

Genetic material All material, whether of plant, animal, microbial or other origin, that contains functional units of heredity.

Genetic recombination A combination of alleles from different parents that produce a recombinant individual. Such an organism or progeny may result from a crossing event or from an independent reorganization of different chromosomes during meiosis.

Genetic resources That set of population samples, whether from plants, animals or microorganisms, that is acquired to provide useful genetic traits with present or potential value.

Genetic stability The maintenance of a certain degree of genetic balance in each individual of a population.

Genetic variability The degree of genetic variation existing in a population or species, as a consequence of the evolutionary processes to which it has been subjected. It is that set of differences present among individuals of a single species. Genetic variability is the basis on which plant breeders develop new varieties.

Genetic variation The heritable variation, derived from changes in genes, usually because of environmental factors.

Genetic vulnerability The condition where the risk of exposure is high for plants that are susceptible to certain pathogens, pests and environmental stress as a result of genetic uniformity, induced by breeding (see also Genetic uniformity).

Genetically modified organism or GMO Any organism whose genetic material has been modified in a way that would not happen naturally in mating (or multiplication) or in natural recombination. GMOs are classified as high or low risk, according to the nature of the receiving or parental organism, and the characteristics of both vector and insert used in the operation.

Genetics The science that deals with reproduction, inheritance, variation and the set of phenomena and problems related to descendancy, that is, the science that deals with heredity.

Genome The set of all genes of an organism, of all the genetic patrimony stored in the set of its DNA or chromosomes. Also, represents as corresponding to the haploid number of chromosomes of a given species.

Genotype The genetic composition of an organism, that is, the total sum of its genes, both dominant and recessive.

Genotypic ratio The proportion of different genotypes of a given progeny (see also Phenotypic ratio).

Germ cell One of two cells found in the pollen grain and which divides by mitosis into sperm cells. This division may occur before or after pollination.

Germination In plants, the resumption of the embryo's growth under favourable conditions after the seed has matured and dispersed, and the emergence of the young root and shoot from the seed. Germination is taken as completed when photosynthesis begins and the plant no longer relies on the food stored in the seed.

Germplasm bank or genebank An entity constituted to conserve genetic resources. For plants, it is the most practical method of safeguarding genetic material, storing samples of landraces, breeding products, varieties not in use, and wild species.

Germplasm The base material of heredity, that is, the structure that carries the total sum of hereditary characteristics of a species. The word 'germplasm' supposes that the structure can give rise to a new generation, transmitting its genetic characteristics. In another words, the sum total of genetic material of plant species.

Glumes Bracts or leaves found on the outside of each spikelet in a grass inflorescence or spike.

Graft A procedure by which two parts of living plant tissue are brought together so that they continue living and later behave as one plant.

Grassland That type of vegetation or plant formation is dominated by grasses and herbaceous plants. Various forms are found such as the Sahel; savannah; steppe; and veldt.

Gymnosperms Plants of ancient origins that present the characteristic of not possessing true flowers. In their place are reproductive structures known as cones.

Habitat (a) A particular place in the environment occupied by organisms or communities of organisms and with which they interact. The habitat is described about those interactions. (b) A place where a plant or animal grows.

Haploid The cell or individual that has one set of chromosomes.

Herbarium Collection of preserved classified plant specimen.

Hereditary disease A disease that has, as its cause, the alteration of genetic material and which is transmitted from generation to generation.

Heredity or inheritance The transmission of genetically based characteristics from parents to progeny or from generation to generation.

Heritability That part of the variation observed in a progeny is due to heredity.

Heterosis or hybrid vigour The increase in vigour, growth, size, yield or functional activity of a hybrid progeny in terms of its parents, resulting from the crossing of genetically different organisms.

Heterotrophs See Primary consumers; Secondary consumers.

Heterotypic synonym (taxonomic synonym) A name based on a type different from that of another name referring to the same taxon.

Heterozygous An organism that has different alleles in the locus corresponding to homologous chromosomes. An organism may be heterozygous for one or more genes.

Hexaploid An organism that has six sets of chromosomes, that is, with a chromosome number of 6n.

Historical varieties Varieties or cultivars, including locally adapted cultivars, which were once registered on formal (official) variety lists but have since been de-listed and made redundant. The epithet 'historical' is not fixed to any certain period, but is subject to each country's perceptions. Often denoted 'heirloom varieties'.

Holotype A holotype of a name of a species or infraspecific taxon is the one specimen or illustration used by the author or designated by the author as the nomenclatural type.

Homologous chromosomes or homologues Chromosomes that pair up during the first division in meiosis. Each member of the pair comes from a different parent and has a sequence corresponding to the locus of genes.

Homologue exchange The exchange of segments between chromatids of two homologous chromosomes during meiosis.

Homonym A name spelled exactly like another name published for a taxon of the same rank based on a different type.

Homotypic synonym (nomenclatural synonym) A name based on the same type as that of another name.

Homozygous An organism or homozygote that has similar genes in the corresponding loci of homologous chromosomes. An organism may be homozygous for one, several or all genes.

Hormone A chemical substance of specialized action that acts as a messenger to those cells that respond to its stimulus, thereby controlling tissues and organs in any part of the organism. The difference between animal and plant hormones is that animal hormones are created in particular organs and regulate almost all organic functions.

Host An animal or plant that harbours or nourishes another organism (e.g. parasite). Or in genetic engineering, that organism, whether microbial, animal or plant, whose metabolism is used to reproduce a virus, plasmid or other form of DNA foreign to that organism and which incorporates elements of recombinant DNA.

Hybrid vigour See Heterosis.

Hybrid The first generation of offspring of a cross between two individuals that differ in one or more genes. Or The progeny of a cross between species of the same genus or distinct genera.

Hybridization A method for creating new varieties, using crosses to obtain genetic recombination.

Illegitimate name A validly published name that is not in accordance with specified rules of ICBN/ICN, principally those on superfluity and homonymy.

Imperfect flower A flower that does not have stamens or pistils (see also Perfect Flower).

***In situ* conservation** *In situ* means 'in the original place'. The conservation of plant genetic resources in the areas where they had developed naturally and, in the case of cultivated species or varieties, in the surroundings of the area where they had acquired their distinctive properties.

***In vitro* (adj.)** literally, in the glass. Said of anything studied and manipulated in laboratory test tubes, that is, outside the live organism.

Incompatibility In plant reproduction, the failure of achieving fertilization and seed formation through self-pollination is usually because of sluggish growth of the pollen tube in the style tissue.

Incomplete dominance The production of an effect by two different alleles. This effect is intermediate between those produced by the same alleles under homozygous conditions (see also Partial dominance).

Incomplete flower A flower that lacks one or more of the four essential organs (sepals, petals, stamens and pistils) (see also Complete flower).

Incorporation, or base-broadening The large-scale development of locally adapted population from unimproved germplasm stocks, through a long-term, population-oriented approach. It is a less commonly used approach to genetic enhancement.

Independent association A random association of two or more pairs of segregating genes in gametes.

Indeterminate (adj.) Said of an inflorescence in which the terminal flower is the last to open. The flowers are formed in axillary buds and the floral axis may elongate indefinitely using a terminal bud. An example is the raceme (see also Determinate).

Indeterminate threat A category of conservation status, describing taxa that are known to be either endangered, vulnerable or rare, but not which one (see also Categories of species conservation).

Indigenous See Endemism.

Infection The invasion of a living being by a pathogen, thereby triggering disease.

Inflorescence A group of flowers growing on a floral axis, and having a characteristic arrangement and form of development. Inflorescence arrangements may be determinate or indeterminate (see also Panicle, Spike).

Inherit To receive from predecessors. In organisms, the chromosomes and genes are transmitted or inherited from one generation to the next.

Inheritance see Heredity.

Inoculate To place an inoculum where it will produce an infectious disease.

Inoculum The spores, bacteria or mycelium fragments of pathogens that can infect plants or soil.

Introgression Transfer of the desired genetic traits into the elite gene pool of adapted material. This is the most common approach to genetic enhancement and is used, for example, in transferring major gene-disease resistances. Introgression can be achieved by repeated crosses (backcrossing) or by using biotechnological techniques.

Introns DNA sequences that do not code for genes and whose function is unknown.

Isoenzymes Multiple molecular forms of an enzyme that occur within an organism. They have the same catalytic function (catalysing the same substrate) but possess different kinetic properties (e.g. reaction speed) (see also Alloenzymes).

Isonym The same name based on the same type, published independently at different times perhaps by different authors. Only the earliest isonym has nomenclatural status.

Isotype An isotype is any duplicate of the holotype; it is always a specimen.

Karyotype The number and morphology of a particular chromosome complement of an individual or a group of related individuals.

Kilobase or kb the unit used to measure the length of a DNA fragment, itself made up of a series of bases. 1 kb = 1000 bases.

Landraces These are crop populations in balance with their environment and remain relatively stable over a long period of time. Dynamic population(s) of a cultivated plant that has historical origin, distinct identity, genetically diverse, locally adapted and associated with traditional farming systems.

Latency (adj. latent) See Dormancy.

Lectotype A specimen or illustration designated from the original material as the nomenclatural type if no holotype was indicated at the time of publication, or if the holotype is missing, or if a type is found to belong to more than one taxon.

Legitimate name A validly published name that is in accordance with the rules of ICBN/ICN.

Lemma In a grass spikelet, the lower bract of two protects the floret (see also Palea).

Line A group of individuals that descend from a common ancestor. Members of such a group are usually more closely related to each other than those of a variety.

Lineage In evolution, a sequence of species, each of which is considered to have evolved from its predecessor.

Linkage group A group of genes distributed linearly in a chromosome.

Linkage map A diagram of a chromosome, indicating the position of genes.

Linkage The relationship that exists between two or more genes that tend to be inherited together because they are located on the same chromosome. This determines that combinations of these genes, like those of the parents in the gametes, are more frequent than their recombination.

Lipids or fats A group of chemically very diverse organic biomolecules with the common characteristics of insolubility in water, solubility in polar organic dissolvent, and little density.

Liposomes Artificially constructed spherical vesicles made up of two or more layers of lipids. Liposomes are used as gene vectors.

Locus (pl. loci) A position on a chromosome where the gene controlling a given trait is located. Or in genetics, the point on a chromosome is occupied by a gene.

Lodicule One of two structures, similar to scales, at the base of an ovary of a grass flower.

Longevity The length of life. In seeds, it refers to the time that these remain alive. Longevity depends on the species and the seeds' storage conditions.

M1, M2 etc Symbols used to designate the 1st generation, 2nd generation etc., following exposure to mutagenic agents (ionizing radiations, chemical mutagens etc.).

Maintainer line Line used for maintaining a cytoplasmic male-sterile line. It possesses the same nuclear genotype as the male sterile (see also B line).

Male sterility In flowering plants, a condition whereby pollen is not produced or is sterile, or that part of the male organ that produces it does not function.

Marker gene That gene whose function and location are known and which expresses certain characteristics or very notable phenotypic differences that permit the analysis of its heredity, establish its presence in the genome and detect recombination events.

Mass selection That system of plant improvement in which the seed of individual plants is selected based on phenotype and then mixed and used to produce the next generation.

Medium-/Long-term conservation Denotes the time perspective about germplasm conservation (c.f. FAO Genebank Standards, 2013).

Megagametophyte In plants, the female gametophyte. Typically, a female gamete of seven cells with eight nuclei. It originates from the megaspore.

Megasporangium or embryo sac The structure in which the megaspores are produced and the megagametophyte later develops. It eventually forms the nucleus.

Megaspore stem cell A diploid cell of the ovary, which gives rise, through meiosis, to four haploid megaspores.

Megaspore One of four haploid spores that originate from meiotic divisions of the stem cell, in plants, also the diploid megaspore is found in the ovule and undergoes three successive meiotic divisions to give rise to the megagametophyte. It is formed in the megasporangium from a stem cell when it undergoes meiosis.

Meiosis The two successive nuclear divisions of a cell. In the first (or reduction) nuclear division, the diploid chromosome number is reduced to a haploid number. The second nuclear division is mitotic (see also Mitosis)

Meristem A region of rapid cellular division (mitosis), Undifferentiated tissue from which cells tend to form differentiated and specialized tissues. Meristems are found in growing areas such as buds and apexes.

Messenger RNA See mRNA.

Microbe See Micro-organism.

Micro-injection A technique, carried out under the microscope, that introduces a gene in solution into a cell, using a micropipette.

Micro-organism or microbe A microscopic organism, usually a bacterium, alga, fungus or protozoan.

Microsatellites, simple sequence repeats or SSRs Short DNA sequences made up of 1 to 6 nucleotides that repeat themselves consecutively 10 or more times. These simple DNA sequences are highly variable and can be studied, using a fast and relatively simple methodology.

Microspore stem cell or pollen stem cell A diploid cell of the anther, which gives rise, through meiosis, to four haploid microspores.

Microspore In plants, one of four haploid spores originates from the meiotic division of the microspore stem cell in the anther and gives rise to a pollen grain.

Mitosis The nuclear division of a cell, whereby chromosomes divide longitudinally, forming two daughter nuclei, each of which has a chromosome complex like that of the original nucleus (see also Meiosis)

Molecular biology (a) This part of biology deals with biological phenomena at the molecular level. (b) In a restricted sense, it includes the interpretation of these phenomena based on the participation of proteins and nucleic acids.

Molecular cloning A technique of genetic engineering that consists of inserting a segment of foreign DNA of a given length into a vector that replicates itself in a specific host.

Molecular markers Gene markers that are used to directly evaluate the genome (DNA), or part thereof without their being affected by the environment, thus conferring greater accuracy.

Monocotyledons Modern plants, whose principal characteristic is to develop only one embryonic leaf (cotyledon) on germinating (see also Dicotyledons).

Monoecious (adj.) A plant species that has both staminate and pistillate flowers on the same individual (see also Dioecious)

Monosomic (adj.) Refers to a chromosome that lacks its homologous partner. Or An individual with one chromosome less than the somatic chromosome number (2n-1).

Morphometry The study of anatomical measures.

Multiple alleles A series of alleles or alternative forms of a gene. A normal heterozygous diploid would have only two genes of an allelic series. Multiple alleles originate from repeated mutations of a gene, in which each mutant produces different effects.

Multiple genes Two or more pairs of independent genes that produce complementary or cumulative effects on a single trait of the phenotype.

Mutation A sudden variation or alteration in an organism, which is then said to be a mutant, especially when such alteration is heritable generation after generations. It may involve changes in genes (genic mutation) or chromosomes (chromosomal mutation). A genic mutation consists of a change in one allele or another of a gene.

Naked DNA DNA that is deprived of its proteinic or lipidic coat.

Native race A population of usually heterozygous plants that were commonly developed in traditional agricultural systems through direct selection by farmers and which, characteristically, are adapted to local conditions.

Native species A species, which is a part of the original flora of the area.

Natural selection The elimination of random alleles, without intervention from humans. The process Darwin called the 'struggle for survival', whereby those organisms least adapted to their environment tend to die and the better adapted to survive. According to Darwinism, natural selection acts on a varied population, causing its evolution. Natural selection appears as the inevitable result of three basic facts of life: overpopulation, variability and heredity.

Naturalized (adj.) Said of a plant that is not native to a country or region but lives there, surviving as if it were indigenous.

Neotropics A biogeographical region, which spans several habitat types, but has strong biogeographic affinities, particularly at taxonomic levels higher than the species level. It extends from southern Mexico, through Central America and the West Indies, and includes the South American continent (see also Afrotropic)

Neotype A specimen or illustration selected to serve as nomenclatural type if no original material is extant or as long as it is missing.

Nomen nudum (nom. nud.) A designation of a new taxon published without a description or diagnosis or reference to a description or diagnosis.

Nomenclature type A nomenclatural type (typus) is that element to which the name of a taxon is permanently attached, whether as the correct name or as a synonym. The nomenclatural type is not necessarily the most typical or representative element of a taxon.

Nucellus The megasporangium, after which it eventually forms the inner layer of the ovule wall.

Nucleic acids Biomolecules formed by nucleotide macropolymers or polynucleotides. Present in all cells, they constitute the basic material of heredity that is transmitted from one generation to another. Two types exist: deoxyribonucleic acid (DNA) and ribonucleic acid (RNA).

Nucleoside A combination of a pentose sugar with a purine or pyrimidine nitrogenous base.

Nucleotide A monomer of nucleic acids, made up of a combination of a nitrogenous base (purine or pyrimidine), sugar (ribose or deoxyribose) and a phosphate group. It is the product of the hydrolysis of nucleic acids through the action of nucleases.

Obsolete variety Those plant varieties that are no longer cultivated commercially but may be kept in collections for use in improvement programmes.

Offshoot or tiller A characteristic type of lateral shoot or branch that develops from the base of the principal stem in certain plants. The term 'tiller' is applied to several lateral shoots that emerge from the crowns of monocotyledons such as grasses.

Operator gene The gene that stimulates the structural gene into functioning. Its activities may be modified by the regulatory gene.

Operator A special segment of DNA, adjacent to the promoter, that is part of the controlling region for operon transcription. The operator interacts with the repressor protein, thus regulating the synchronized transcription of the corresponding operon.

Operon A set of genes, comprising an operator gene and the structural genes that it controls.

Organism A biological entity able to reproduce itself or transfer genetic material. Microbiological entities are included within this concept, whether or not they are cellular. Almost all organisms are formed of cells, which may then be grouped into organs, and these into systems, each of which carries out particular functions.

Orthodox seed Seed that can be dried to low levels of moisture content and stored at low temperatures over long periods without losing viability (see also Recalcitrant seed).

Out of danger A category of conservation status, describing a species or other taxon that had been included in a higher category on the continuum to extinction, but is now considered to be in a relatively safe state of conservation due to the adoption of effective conservation measures or elimination of previously existing threats (see also Categories of species conservation).

Outbreeding Mating between individuals less closely related.

Outcross Cross-pollination, usually by natural means, with a plant different in genetic constitution

Ovary In flowers, the swollen base of the pistil, in which seeds are formed.

Ovule In plants, the structure that contains the female gamete or megagametophyte becomes seed after fertilization.

Palea In a grass spikelet, the upper of two bracts protect the floret (see also Lemma).

Panicle An open and branched inflorescence, with flowers possessing stalks or pedicels.

Parasite (adj. parasitic) A living being that, for all or part of its life, derives its food from another living being (the host). The host is usually harmed to some extent by the association.

Paratype Any specimen cited in the protologue that is neither the holotype nor an isotype nor one of the syntypes if in the protologue two or more specimens were simultaneously designated as types.

Parthenocarpy In plants, the production of fruits without fertilization and normally without seeds.

Parthenogenesis Egg cell of an embryo sac derived from megaspore develops into an embryo without fertilization.

Partial dominance The lack of complete dominance. Or the production of an intermediate hybrid among reproducing types (see also Incomplete dominance).

Patent The exclusive right granted to the ownership of an invention as a social counterpart to the innovation.

Pathogen (adj. pathogenic) The producer or causal agent of a disease.

Pathogenicity The capacity of an organism to cause or incite disease.

PCR or polymerase chain reaction A technique for analysing the genome by an unlimited amplification of minuscule but particular parts of DNA. It is a revolutionary method of exponential amplification of DNA that uses the intervention of a heat-stable enzyme, the Taq polymerase.

Pentaploid An organism that has five sets of chromosomes, that is, with a chromosome number of 5n (see also Ploidy level).

Peptide A polymer or amino acid chain.

Perennial A plant that lives for three years or more. They may be woody such as shrubs and trees or herbaceous. Herbaceous perennials may be evergreen; deciduous (i.e. the aerial organs are annual but the underground organs such as rhizomes and bulbs are persistent); or monocarpic, that is, living for many years until flowering and fruiting, after which they die, for example, Agave spp (see also Annual; Biennial).

Perfect flower A flower that has stamens and pistils (see also Imperfect flower).

Persistent (adj.) said of a plant organ that remains inserted or does not fall at maturity once it has fulfilled its physiological function.

Pests In agriculture, organisms such as insects, nematodes and other plants that attack crops and livestock (see also Disease; Pathogen; Weeds).

Phenotype The final appearance of an individual that results from the interaction of its genotype with a given environment.

Phenotypic ratio The proportion of different phenotypes of a given progeny (see also Genotypic ratio).

Phylogeny The evolutionary history of groups of organisms.

Physiognomy That aspect of a plant community or species that is subject to visual appraisal. It depends on the set of special structures and characteristic forms of its biological constituents.

Physiological race Those pathogens of the same species and variety that are similar structurally, but differ in their physiological and pathological characteristics and, especially, in their ability to parasitize different varieties of a given host.

Phytosanitary quality or plant health quality The set of characteristics that plant germplasm should have concerning the presence or absence of pathogens transmissible in planting materials and/or micro-organisms that cause deterioration during conservation.

Pistil In flowers, the female organ where the seed originates. It comprises the ovary, style and stigma.

Pistillate flower A flower that bears pistils but has no stamens (see also Staminate flower).

Plant community A more or less complex group of plants that occupy a certain area, regardless of the character, composition and structure that the plants present.

Plant Genetic Resources for Food and Agriculture (PGRFA) Any genetic material of plant origin of actual or potential value for food and agriculture (definition from the International Treaty on PGRFA, 2001).

Plant genetic resources Genetic material of plants, including modern cultivars, landraces and wild relatives of crop plants as potential use and resource for present and future generations of people.

Plasmagene The cytoplasmic unit of heredity.

Plasmid A circular fragment of double-stranded DNA that contains some genes and is found within certain bacteria. It acts and replicates independently of bacterial DNA and may pass from one bacterium to another. As with proviruses, they do not produce diseases but induce small mutations in cells. They are used as vectors in genetic engineering.

Ploidy level The complete number of complements or basic sets of chromosomes that a cell or organism has. The living unit may be haploid, diploid, triploid, tetraploid, pentaploid or hexaploid if it possesses 1, 2, 3, 4, 5 or 6 basic sets of chromosomes, respectively. A polyploid has more than two sets of chromosomes, and an aneuploid does not have an exact set.

Point mutation A type of mutation that causes the replacement of a single-base nucleotide with another nucleotide. Or includes insertions or deletions of a single base pair.

Polar nuclei The two central nuclei found within the megasporangium and which join with the second sperm cell in triple fusion. In certain seeds, the product of this triple fusion gives rise to the endosperm.

Pollen tube A tube that, under favourable circumstances, develops from the pollen grain after being placed on the stigma of a flowering plant. It grows down the style to the ovary and eventually to an ovule. The sperm cell is carried to its destination at the tip of the pollen tube.

Pollen A fine powder produced by anthers and male cones of seed plants, composed of pollen grains. Each grain encloses a developing male gamete, itself having originated from a microspore.

Pollination The transfer of pollen from the anther to the stigma in flowering plants or from the male to the female cone in gymnosperms (see also Cross-pollination; Self-pollination).

Polycross An isolated group of plants or clones distributed so that random cross-pollination can occur.

Polymer A chemical compound formed by the combination of repeated structural units (monomers) or linear chains of the same molecule.

Polyploid An organism with more than two sets of chromosomes in its cells (see also Ploidy level).

Population dynamics The pattern of any process, or the interrelationship of factors, which affects growth or changes within the population, including birth and death rates, immigration/emigration etc.

Population structure Refers to how a population is built up, through sub-populations

Population A group of individuals of a species living in a given area. It is the most significant level of organization of a species and is also of evolutionary and conservational significance.

Primary consumers or heterotrophs These organisms take advantage of the chemical energy stored in the organic matter of primary producers. This level is composed of herbivores.

Primary gene pool The gene pool of similar or closely related species with which inter-crossing is easy, yielding fully fertile offspring. The most easily accessed genetic variability. The primary gene pool corresponds to the concept of biological species.

Primary producers or autotrophs These take advantage of the energy from light, using photosynthesis. They can synthesize organic matter from inorganic matter. This level corresponds to that of green plants (see also Primary consumers; Secondary consumers).

Primary production This represents the increase in biomass of the primary or photosynthetic producer organisms. Gross primary production refers to the biomass synthesized through the photosynthetic activity of primary producers.

Production That process which increases biomass per unit of time. It may be measured in mg cm^{-3} day^{-1}, kg ha^{-1} $year^{-1}$ or kcal ha^{-1} $year^{-1}$. It expresses the idea of biomass available per unit of time for use by the next trophic level without endangering the ecosystem's stability. Production = Biomass/Time.

Productivity This is the relationship between production and biomass. In algae, for example, which reproduce daily, that is, they double their mass every 24 h, productivity is 100%. In contrast, the average productivity in land plants does not reach 0.3%. For example, an almond tree forms almonds only once a year. Productivity = Production/Biomass (see also Biomass and ecosystem productivity).

Prokaryotes Organisms whose cells possess a single chromosome and no membrane to isolate it from the cytoplasm. This means that such an organism lacks a true cellular nucleus. The most representative examples are blue-green algae and bacteria.

Promoter A region of DNA that is involved in and necessary for the initiation of transcription. It includes the RNA polymerase binding site (the starting point of transcription) and various other sites at which gene regulatory proteins may bind.

Propagule Any structure that serves to vegetatively propagate or multiply a plant, for example, cuttings, tubers, differentiated tissues and cells.

Prophylaxis or preventive treatment In phytosanitary procedures, a measure or set of measures is taken to prevent the occurrence of disease. This may include the use of protectants, which are usually chemical agents, to prevent a given disease or disease among plants.

Protandry The condition of hermaphrodite plants where male gametophytes mature before female gametophytes (see also Dichogamy; Protogyny).

Proteins Biomolecules formed by amino-acid macro polymers or macro polypeptides. They function as enzymes, hormones and contractile structures that endow organisms with their characteristic size, metabolic potential, colour and physical capacities.

Protocol A document of standardization that establishes the rationale, objectives, design, methodology and foreseen analysis of results, and the conditions under which such activities are to be carried out and developed.

Protogyny In hermaphrodite plants, the condition where female gametes mature before male gametes (see also Dichogamy; Protandry).

Protoplast A cell that is isolated and deprived of its cell wall.

Pteridophyta A taxonomic class of plants of ancient origins. The plants are principally characterized by the absence of true roots (they have rhizomes) and reproduce by spores.

Pure line A genetically pure line where all its members are homozygous, having originated from the self-fertilization of a simple homozygous individual.

Qualitative trait or discontinuous variation A trait whose observed variation is discontinuous, or which presents several states. It is usually controlled by one or a few genes, with little or no influence from the environment (e.g. yellow flower versus white flower).

Quantitative trait A trait that is determined by several independent genes that have cumulative effects and it shows continuous variation with strong influence of environment.

Quarantine A procedure of legal character that consists of confining or isolating plants or other materials introduced from other countries. They are then subjected to inspection to detect plant health problems that could threaten the agriculture of the country in which they are entering.

Rare (adj.) A category of conservation status given to taxa whose world populations are small, but are not currently at risk of extinction or vulnerable, even though they are subject to a certain degree of risk. These taxa are normally located in restricted geographical areas or habitats, or have extremely low density over a more-or-less broad distribution.

rDNA or recombinant DNA A DNA molecule formed by recombining DNA fragments from different origins. The protein that codes is a recombinant protein.

Recalcitrant seed Seed that cannot be dehydrated nor conserved at low temperatures without suffering damage. It can be stored for only a few days or weeks under special treatment. Species that have recalcitrant seeds or do not produce seeds are usually conserved in field germplasm banks. In these areas, collections of live plants are kept, that is, the germplasm is conserved as a permanent live collection (see also Orthodox seed).

Recessive gene That gene that needs a double 'dose' to be expressed. It does not express itself in the presence of a counterpart or dominant allele (see also Dominant gene).

Recombination The formation of new gene combinations as a result of cross-fertilization between individuals that differ in their genotype.

Recurrent parent or donor parent In plant improvement, in a back cross, that parent with which the hybrid material is again crossed.

Recurrent selection That system of genetic improvement designed to increase the frequency of genes favourable for yield or other characteristics through repeated selection cycles.

Regeneration or rejuvenation Within the context of germplasm banks, the cultivation of a sample of an accession (e.g. seed, clone, in vitro plant or other propagule) to produce fresh, viable and sufficient samples of plants from which sexual or asexual seeds with similar genetic constitution can be harvested, and which permits the preservation, in a better state, of the seed or propagule when stored.

Registered seed Progeny of basic seed or certified seed that is produced and used in such a way that it satisfactorily maintains its identity and genetic purity. It is approved and certified by an official certification agency.

Regressive form Concerning crops, a species related to the cultivated form, growing in the wild, but not used in agriculture. It usually shows characteristics of both the cultivated species and its wild relatives.

Regulatory gene An ancient gene whose mutations can influence evolution. It also modifies the action of the operator gene.

Relicts (adj. relict) In the sense of relics, those plants that had been dominant in other times, but which are now scarce. By extension, a country's original vegetation remains or persists.

Replication (a) That process by which a DNA or RNA molecule originates from another identical one. (b) Generally, the duplication of nucleic acid.

Replicon A nucleic acid structure that can replicate. Replicons include chromosomes of eukaryote cells, prokaryotic nuclear DNA, plasmids and viral nucleic acids.

Representative sample A sample that contains at least 95% of alleles (genetic variability) of the sampled population.

Repressor gene That gene which represses the operator gene.

Resilience The capacity of the ecosystem to fluctuate between given limits and thereby restore itself to its original state after disturbance. Such capacity operates within certain limits, beyond which the system is not able to return to the condition of pre-disturbance and, hence, is degraded towards pioneering successional states. The limits of resilience differ for different ecosystems, as does the speed of recovery (see also Stability).

Resistance That characteristic of a host plant that enables it to prevent or delay the development of a pathogen or other harmful factor (see also Susceptibility).

Restriction enzymes Enzymes that bacteria synthesize in defence against the invasion of foreign DNA from, for example, bacteriophages, thereby degrading that DNA while remaining protected through particular methylations. A restriction enzyme always divides the DNA in the same site, specific loci or targeted sequences. Their scissor-like behaviour opened the doors to genetic engineering.

Rhizome An underground stem that is usually horizontal and elongated. It differs from a root by the presence of nodes and internodes, sometimes scale-like leaves and shoots in the nodes themselves.

Ribosomes Small cellular organelles found in all living beings where protein synthesis is carried out.

RNA or ribonucleic acid Nucleic acid formed by nucleotides in which the sugar is ribose, and the nitrogen bases are adenine, uracil, cytosine and guanine. They act as intermediaries and complement the genetic instructions coded in the DNA. Several different types of RNA exist, related to protein synthesis; these are messenger RNA (mRNA, ribosomal RNA (rRNA), transfer RNA (tRNA), and heterogeneous nuclear RNA (hnRNA). RNA is normally the product of the transcription of the DNA template, although, in retroviruses, RNA acts as the template and DNA is the copy.

Ruderal (adj.) Said of environments and plant and animal species that are linked to human activities, either directly or indirectly.

Sample A set of random observations taken from a population.

Scarification Any process to rupture, scratch or mechanically alter seed coats to make them permeable to water or gases.

Secondary gene pool The gene pool of related species with which inter-crossing and gene transfer is possible, yielding hybrids and offspring that sometimes are fertile.

Secondary production This represents the speed of storing energy at the levels of consumers and decomposers. It is therefore the increase in biomass per unit of time and space at these levels.

Seed tubers or seed pieces Small tubers or cut pieces of tubers used for planting new-season crops. They are stored during winter and later transplanted for seed production. Seed tubers are obtained from plants that have not been thinned. Seed pieces are obtained by cutting up tubers so that each piece carries an eye or set of eyes that, on planting, will grow into a new plant, as in a potato.

Seed A mature ovule with its embryo and all the normal seed coats. In some plants, the seed also has an endosperm.

Segregation The separation of homologous chromosomes (and, consequently, their genes) of the different parents during meiosis.

Selection Any natural or artificial process that permits an increase in the proportion of certain genotypes or groups of genotypes in successive generations.

Selective sweep A selective sweep is a process through which a new beneficial mutation that increases its frequency and becomes fixed in the population leads to the reduction or elimination of genetic variation among nucleotide sequences that are near the mutation. In the selective sweep, the positive selection causes the new mutation to reach fixation so quickly that linked alleles can 'hitchhike' and also become fixed.

Self-pollination In plants, the process whereby pollen is transferred from the anther to the stigma of the same flower, another flower of the same plant or clone (see also Cross-pollination).

Self-sterility or self-incompatibility In plants, the incapacity, usually physiological, to incur fertilization and produce seeds after self-pollination.

Sexual reproduction Reproduction based on germ cells and the fusion of gametes (see also Vegetative propagation).

Shoot In vascular plants, that part is derived from the plumule (being the stem) and usually leaves. A sprouted part, branch or offshoot of a plant (see also Crown; Sucker).

Somaclonal variation The variation observed in somatic cells, which divide mitotically in tissue culture. Depending on the species, this variation may be genetic, phenotypic or in habitat. Many of these modifications are transferred to progenies of regenerated plants.

Somatic (adj.) Refers to diploid cells, normally with one set of chromosomes from the male parent and another set from the female parent.

Somatic embryos Those embryos that originate from the fusion of somatic cells, that is, not from gametes.

Speciation The formation of one or more new species from one already existing. It occurs when an isolated population develops certain distinctive characteristics as a result of natural selection and loses the possibility of reproducing with the rest of the population, even when no geographical or physical reasons are apparent to prevent it.

Species A group of actually or potentially interbreeding natural population which normally are reproductively isolated from each other such groups which show common characteristics.

Sperm cell In plants, part of the male gametophyte. The germ cell in the pollen grain undergoes mitosis to produce two sperm cells, one of which will fuse with the egg in the ovule to form the zygote. The other sperm cell will fuse with the two polar cells in the ovule to form the endosperm that later feeds the embryo (see also Egg).

Spike An inflorescence with a more-or-less elongated axis, throughout which the flowers are almost sessile also known as ear or tassel.

Spikelet A unit of a spike, that is, of an inflorescence typical of grasses. It is formed by glumes, the rachis and florets.

SSO The symbol used to designate the original self-fertilized plant. S1, S2 etc. symbols are used to designate the first self-fertilization (progeny of plant S0), second self-fertilization (progeny of plant S1) etc.

Stability The ability of ecological systems to persist over time despite external disturbances, whether of natural or anthropic origin (see also Resilience).

Stamen In plants, the male reproductive organ of the flower comprises the anther and filament.

Staminate flower A flower that bears stamens but has no pistils (see also Pistillate flower).

Static conservation The type of conservation that stops the natural processes of evolution and co-evolution of genetic resources, hence conserving them in isolation and outside their natural habitats. The term is applied specifically to ex-situ conservation.

Stem The main axis of a vascular plant. It bears buds, leaves or scale leaves, and reproductive structures such as flowers. It is usually borne above ground and has a characteristic arrangement of vascular tissue (see also Bulb; Corm; Crown; Rhizome; Stolon; Tuber).

Sterility The impossibility of completing fertilization and acquiring seed because of defective pollen or ovules, other aberrations, or unusual or seasonal plant activity.

Stigma In flowering plants, that part of the pistil that receives pollen.

Stolon or runner (a) A specialized creeping stem, capable of forming roots and shoots at its nodes. (b) Specifically, those prostrated or scattered stems that develop in leaf axils in the crown of the mother plant, grow horizontally over the ground and form new plants from its nodes.

Strain In microbiology, a set of viruses, bacteria or fungi that have the same gene pool.

Structural gene That gene which regulates the formation of an enzyme or other protein required for a cell's structure or metabolism (see also Evolution).

Style In flowering plants, the column that connects the ovary to the stigma.

Subculture The aseptic transfer of part of a plant in a collection to a fresh medium for renewal and strengthening.

Sucker A shoot that originates in a plant but from under the soil surface. This term is more precisely used to designate a shoot that originates from an adventitious bud on a root. Or loosely, the term also refers to shoots originating close to the crown, even though they arise from stem tissue.

Superfluous name A name that, when published, was applied to a taxon that, as circumscribed by its author, definitely included the type of a name that ought to have been adopted, or of which the epithet ought to have been adopted, under the rules.

Susceptibility The extent to which a plant, vegetation complex, or ecological community would suffer if it were exposed to a pathogen or other harmful factor (regardless of whether it receives such exposure). A host plant is said to be susceptible if it cannot prevent or retard the harmful effect of a pathogen or other noxious factor (see also Resistance).

Sustainable development Sustainable development is development that serves the needs of the present generations without undermining the needs of future generations' (Brundtland Report 1986). 'It is a process of sustained and equitable improvement in the quality of life, based on the conservation and protection of the environment in such a way that it does not compromise the expectations of future generations' (Law 19,300, Colombian Congress).

Sustainable use The use of components of biodiversity in such a way and at such a rate that it does not cause long-term reduction in biodiversity, thereby maintaining the possibilities of meeting the needs and aspirations of current and future generations.

Synonym One of two or more names that apply to the same taxon.

Synthetic variety Advanced generations of seed mixtures of free pollination from a group of lines, clones or self-fertilized lines, or of hybrids among them.

Syntype Any specimen cited in the protologue when there is no holotype or any of two or more specimens simultaneously designated in the protologue as types.

System A coherent set of interacting elements that can be isolated from the rest of the universe according to appropriate criteria.

Taxon (pl. taxa) A taxonomic group (e.g. species, genus or family) at any level within a classification system of living beings that are thought to have certain similarities and a given degree of evolutionary relatedness. Taxa are typically organized hierarchically from the largest categories (e.g. kingdom, division or class) where the members are less related to the smallest (e.g. species, subspecies or variety), where the members are closely related.

Taxonomic character Any observable feature or trait, which can be used for the purpose of comparison and classification of the particular organism.

Tertiary gene pool The gene pool of distantly related species, or taxa, with which gene transfer is very difficult and which requires specialized techniques to produce hybrids. The resulting offspring is normally sterile and often lethal.

Test cross That process whereby a hybrid is crossed with one of its parents, or with a genetically equivalent recessive homozygote. The test is used to prove homozygosis or linkage.

Tetraploid An organism that has four sets of chromosomes, that is, a chromosome number of 4n (see also Ploidy level).

Thermotherapy In plants, a treatment to disinfect plant material, using heat. For cassava, for example, the treatment consists of taking selected stem cuttings to the greenhouse or growth room and subjecting them to 3 weeks of temperatures at 40°C during the day and 35°C during the night and a day length of 12 h. To increase effectiveness, the treatment is combined with *in vitro* tissue culture.

Three-way cross A cross resulting from crossing a single cross with a third parent (inbred line, strain or cultivar).

Tiller See Offshoot.

Tissue culture Cultivation of plant cells and tissue in vitro on an artificial media.

Top cross A cross between an inbred line and an open-pollinated variety

Totipotent (adj.) Capable of anything. Said of cells that can give rise to cells of any kind.

Toxins (a) Substances, usually albuminoids, made by living beings, especially microbes, that act as poisons, even in tiny quantities. (b) Proteins are responsible for particular functions in certain bacteria and which are poisonous to certain other organisms.

Transcription The biosynthesis of an RNA molecule by polymerizing complementary nucleotides to a DNA sequence.

Transformation (a) In bacteria, one of the natural processes, together with conjugation and transduction, for transferring genetic material from one bacterium to another, involving the direct integration of DNA. (b) Experimentally, it consists of introducing a DNA fragment into a bacterium to stimulate genetic recombination. (c) By extension (and loosely), the term sometimes designates an identical process that affects eukaryotic cells (yeasts, animal and plant cells).

Transgene A gene has been introduced from another species such that it can usually be transmitted to that organism's offspring.

Transgenesis The artificial introduction of new genetic material into the genome of a plant through genetic engineering such that this new material can be inherited by progeny. This technique permits associations of genes that do not exist in nature, as they have been made to jump barriers between species and even higher taxa such as kingdoms.

Transgenic plant A plant carrying a transgene. That plant whose genome has been altered by *in vitro* manipulation.

Translation An exchange of information contained in the sequence of the four nucleotides of mRNA due to the arrangement of the 20 amino acids in the structure of polypeptide chains.

Translocation The structural modification of chromosomes by which a chromosomal segment changes its relative position within the chromosome itself (intrachromosomal translocation) or between chromosomes (Inter-chromosomal translocation) (see also Transposition).

Transposition The change of position of given pairs of bases in the DNA sequence. The translocation of a chromosomal segment to another position within the same chromosome (see also Translocation).

Transposon A mobile genetic element with a defined DNA sequence that can be transferred to new positions in the cell's chromosome without losing the copy in its original position. Moreover, it behaves as a true intracellular parasite. Transposable elements of eukaryotes are grouped into two categories, according to their mechanism of transposition: class 1 (retrotransposons), which jump to the genome through an intermediate step, that is, through RNA and with the intervention of the enzyme known as inverse transcriptase; and class 2, which transpose directly from one chromosomal site to another using a different enzyme (transposase).

Trihybrid The result of one cross between parents that differ in three specific genes.

Triploid (a) An organism which has three sets of chromosomes, that is, a chromosome number of 3n. (b) Having three sets of chromosomes (see also Ploidy level).

Tuber A modified stem structure that develops underground as a consequence of swelling in the subapical part of a stolon and the subsequent accumulation of reserve materials.

Type specimen Type specimen of name of taxa must be preserved permanently and may not be living organisms or cultures. However, cultures of algae and fungi, if preserved in a metabolically inactive state (e.g. by lyophilization or deep-freezing to remain alive in that inactive state), are acceptable as types.

Under-utilized plants Neglected but seemingly useful plants, wild or domesticated, that have economic potential.

Variety or cultivar Among cultivated plant species, that plant which differs by one or more traits. When it reproduces by seed or asexually, these traits are conserved. Agronomically, a group of similar plants that, by structure and performance, may differ from other varieties within the same species (see also Obsolete variety; Synthetic variety).

Vector A carrier, that is, that which transfers an agent from one host to another. It involves a DNA molecule (bacterial plasmid, artificial microsome of yeast, or bacterium) or a defective virus.

Vegetation That spatial structure or mode of organization of the set of plant species found in a given place. It is usually described by examining stratification and coverage, alluding to the species present and the dominant forms of life.

Vegetative propagation, clonal multiplication, asexual reproduction or agamic reproduction A type of reproduction that does not involve the formation and fusion of gametes, leading to the constitution of homogeneous clones (see also Apomixis; Sexual reproduction).

Veldt or veld Open temperate grasslands found in southern Africa (see also Grassland).

Vernalization That treatment of plants with heat or cold to modify successive stages leading to maturity. For some species, vernalization can be achieved by exposing germinating seeds at temperatures slightly above the freezing point.

Viability In seeds, the capacity to germinate when they possess all they need to do so. The fact that a seed is alive does not guarantee that it will germinate, even under optimal conditions, as phenomena such as dormancy or inactive states can occur.

Virion The extracellular form of the virus, that is, before it has entered the host cell to replicate.

Viroids Causal agents of certain plant diseases. Their name derives from their similarity to viruses, from which they are differentiated by a lack of a capsid.

Virulence The relative capacity of a pathogen to incite disease.

Virus An infectious cell-free entity that, even though it can survive extracellularly as a virion, is an obligate parasite because it can replicate only within specific live cells, generating no energy or metabolic activity.

Voluntary release of GMOs The deliberate introduction of a GMO or combination of GMOs into the environment without measures of containment such as physical barriers or a combination of these with chemical or biological barriers having been adopted to limit the GMOs' contact with the human population and the environment.

Vulnerable A category of conservation status describing taxa that are believed to become at risk of extinction. Vulnerable taxa are those whose populations have been reduced to such critical levels or whose habitats have been so drastically reduced that they are at imminent risk of extinction.

Weeds Any undesirable or troublesome plant, especially one that grows profusely where it is not wanted.

Wide cross Cross between two species of the same genus or of different genera.

Wild species Those groups of organisms that are regularly found in nature and have not been domesticated.

Wild type In genetics, a species or organism that carries the normal form of a gene or genes, as opposed to a mutant.

Wild-type gene The allele commonly found in nature or arbitrarily designated as 'normal'.

Working collection or breeder's collection That collection of germplasm accessions used for crop research and improvement (see also Active collection; Base collection; Core collection).

World collection A collection of germplasm of a particular species from different geographic locations, used as source materials in plant breeding.

X1, X2, X3 etc 1st, 2nd, 3rd generation after irradiation (X0) obtained through selfing or clonal multiplications.

Xenia The immediate effect of pollen on the characteristics of endosperm.

Xerophyte A plant structurally adapted for life and growth with a limited water supply, especially using mechanisms that limit transpiration or that provide for the storage of water.

Xylem The vascular tissue that conducts water and mineral salts and provides mechanical support in vascular plants.

Yield plateau A temporary stable state in yield reached in the course of increased production.

Zein Seed protein of maize classified as prolamin. It is low in tryptophane and lysine.

Zygomorphic Flowers that are symmetrical on one plane only (e.g. papilionaceous flowers).

Zygote The cell that results from the fusion of gametes.

Correction to: Textbook of Plant Genetic Resources

Kuldeep Tripathi, Veena Gupta, Gopalareddy Krishnappa, Jyoti Kumari, and Gyanendra Pratap Singh

Correction to:
Chapters 9 and 14 in: K. Tripathi et al. (eds.), *Textbook of Plant Genetic Resources*, https://doi.org/10.1007/978-981-96-5024-8

The original version of this book was inadvertently published with the below error:

In chapter 9, the term Confidence Ratio was misspelled in the initially published version of this chapter and has been corrected to Coincidence Rate.

Chapter 14 originally published with the author name Pooja Rathode has now been corrected and updated as Pooja Rathore.

The updated versions of these chapters can be found at
https://doi.org/10.1007/978-981-96-5024-8_9
https://doi.org/10.1007/978-981-96-5024-8_14

K. Tripathi et al. (eds.), *Textbook of Plant Genetic Resources*,
https://doi.org/10.1007/978-981-96-5024-8_18

MIX
Papier aus verantwortungsvollen Quellen
Paper from responsible sources
FSC® C105338

If you have any concerns about our products,
you can contact us on
ProductSafety@springernature.com

In case Publisher is established outside the EU,
the EU authorized representative is:
Springer Nature Customer Service Center GmbH
Europaplatz 3, 69115 Heidelberg, Germany

Printed by Libri Plureos GmbH
in Hamburg, Germany